MW01327446

H. Asama · T. Fukuda · T. Arai · I. Endo (Eds.)

Distributed Autonomous Robotic Systems

With 370 Figures

Springer-Verlag
Tokyo Berlin Heidelberg
New York London Paris
Hong Kong Barcelona Budapest

Hajime Asama
Research Scientist, The Institute of Physical and Chemical Research (RIKEN), 2-1 Hirosawa, Wako, Saitama, 351-01 Japan

Toshio Fukuda
Professor, Department of Mechano-Informatics and Systems, Nagoya University, Furo, Chikusa-ku, Nagoya, 464-01 Japan

Tamio Arai
Professor, Department of Precision Machinery Engineering, University of Tokyo, 7-3-1 Hongo, Bunkyo-ku, Tokyo, 113 Japan

Isao Endo
Chief Researcher, The Institute of Physical and Chemical Research (RIKEN), 2-1 Hirosawa, Wako, Saitama, 351-01 Japan

ISBN 4-431-70147-8 Springer-Verlag Tokyo Berlin Heidelberg New York
ISBN 3-540-70147-8 Springer-Verlag Berlin Heidelberg New York Tokyo
ISBN 0-387-70147-8 Springer-Verlag New York Berlin Heidelberg Tokyo

Printed on acid-free paper

Printing and binding: Permanent Typesetting & Printing Co., Ltd., Hong Kong

Preface

Recently, new robotic systems with higher robustness and intelligence have been demanded for various applications such as flexible manufacturing processes and maintenance in nuclear plants, where flexibility and fault tolerance of the system are required. Distributed Autonomous Robotic Systems (DARS) is emerging as a new strategy to develop such robotic systems. The conventional strategy is directed at a single sophisticated robot, whereas this new strategy aims to develop robust and intelligent robotic systems through the coordination and self-organization of multiple robotic agents. These trends are being accelerated by the background of rapid progress of computer and network (communications) technology.

DARS has become important as a field of robotic research, and the number of researchers interested in this area has increased rapidly in the last few years. DARS is related to various new topics such as decentralized autonomous systems, multi-agent systems, self-organization, distributed artificial intelligence, and artificial life. Therefore, DARS is quite interdisciplinary, and has been furthered not only by robotics scientists but also by manufacturing engineers, computer scientists, and biologists. The broad nature of DARS is also underlined by Prof. H. Yoshikawa (President of the University of Tokyo) in his presentation on the international project of Intelligent Manufacturing Systems when he explains the DARS technologies that will be applied.

This book serves as an introduction to DARS and shows how DARS can be implemented. First, the fundamental technologies of DARS are explained. Second, the crucial issues of coordination and organization in DARS are discussed from the viewpoint of both synthesis and analysis. Last, the specific problems encountered in the application of DARS are described.

All the papers included in this book were presented at the second International Symposium on Distributed Autonomous Robotic Systems (DARS '94), which was held in Wako-shi, Saitama, Japan, in July 1994. This is the first book to be published on DARS. International symposia on DARS are held every two years. The first International Symposium on Distributed Autonomous Robotic Systems (DARS '92) was also held in Wako-shi in September, 1992. The papers presented in DARS '92 are included in the proceedings which was published by The Institute of Physical and Chemical Research (RIKEN).

H. Asama

Table of Contents

Chapter 7: Multi-Robot Behavior

Chapter 8: Coordinated Control

Chapter 1
Introduction

Trends of Distributed Autonomous Robotic Systems

HAJIME ASAMA

The Inst. of Physical and Chemical Research (RIKEN), Wako, 351-01 Japan

1 Introduction

Though flexible and robust robotic systems have been required in various fields, sufficient functionality has not yet been realized with currently available technologies in spite of all the efforts to develop intelligent robots. Instead of single-sophisticated-robot-oriented researches, distributed autonomous robotic systems (DARS) have recently attracted the attention of many researchers as a new approach for flexible and robust systems. In this chapter, the trends of DARS are described introducing their background and motivations. Then, the subjects on DARS are mentioned, and approaches and assumptions in DARS researches are classified.

2 What is DARS?

Distributed Autonomous Robotic Systems are the systems which consist of multiple autonomous robotic agents into which required function is distributed. In order to to achieve given missions, the agents work cooperatively to operate and/or process tasks.

The most important feature of DARS is that each system is a distributed system composed of multiple agents. Depending on what the agent is and how the system is distributed, there are various types of DARS. Function of processing or control, management, and even decision making may or may not be distributed into multiple agents. On the other hand, the information and knowledge which are necessary for agents' action or processing may also be distributed into multiple agents.

Figure 1 (a) represents the concept of DARS. Regarding the following autonomous units as agents, we can assume various types of systems:

(1) Agent = Robot (Equipment)
(2) Agent = Module (Cell)
(3) Agent = Actor (Object)
(4) Agent = Computer (CPU)
(5) Agent = Process (Task)
(6) Agent = Controller (Device)
(7) Agent = Sensor

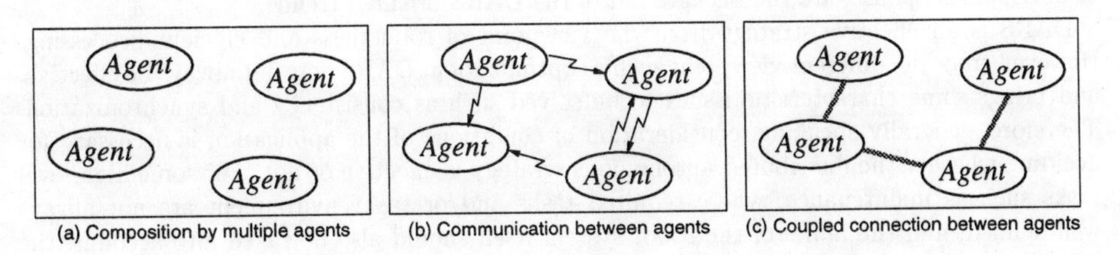

Figure 1: Concept of DARS

Moreover, by combining some of these types, more diverse and complex systems can be designed. In the decentralized autonomous concept, various entities, which have no subjective abilities, might be regarded as agents.

Another important feature of DARS is that each agent does not act independently, but interactively and cooperatively with some consistency towards the common target. For the cooperative action, agents may communicate each other as shown in Fig. 1 (b), or connect with each other by coupling as shown in Fig. 1 (c). One of the most important technologies for DARS is to achieve global behavior with consistency with only limited information exchange between agents, where not only physical inter-agent relation, but also logical relation should be considered. In addition, dynamic change of the system states is also a dominant factor for the behavior of DARS.

3 Trends of Distributed Systems

As a background of distributed autonomous trends, it is first noted that the performance of computers has been improved rapidly as well as the network capability, providing us with convenient distributed environment.

On the other hand, higher flexibility and robustness are required for manufacturing systems, which is motivated by large-variety small-lot production, improvement with new technology transfer, and high fault tolerance and maintainability. To cope with these requirements, some new concepts of manufacturing system, such as the bionic (holonic) manufacturing system and the self-organizing manufacturing system, are proposed. such a concept is also conducted in the international project of IMS (Intelligent Manufacturing System). In addition, all the practical applications such as multi-purpose pipeless batch plants using mobile reactors[1], autonomous decentralized control systems for factory automation, engine assembly lines using a large number of automatic guided vehicles, nuclear fuel reprocessing facilities with modularized replaceable racks[2], are the examples of the distributed autonomous trends. The common feature of these trends are:

(1) Mobilazation for dynamic change
(2) Descretization of continuous system
(3) Modularization of complex systems
(4) Localization of processing

These trends also affected on robotic systems. The intelligence of robots is originally defined as the ability of initiating action according to the situation, which denotes the ability for dynamic task requirements, in changing environment, and even in faulty conditions. Nevertheless, the functionality for the ability has not sufficiently been realized, and the technological limitation of functional enhancement of a robot has become apparent. Thus, instead of the conventional, sophisticated-robot-oriented strategy, a new strategy for cooperative multi-agent robot systems in a distributed manner has become attractive. Especially, for operation of robots in a narrow space, downsizing is an essential issue, for which the distribution and modularization is an effective strategy. Such requirements for robots as efficient processing, maintainability and extensibility of the robot system itself, utilization of limited resources, flexibility taking account of development phase, are the background of the DARS oriented trends.

DARS is an effective strategy from the viewpoint of robustness and efficient processing. However, from the different viewpoint such as optimization, DARS may not always be effective, and brings some characteristic issues to be solved such as consistency and synchronization. Therefore, generally speaking, consideration of conditions of the application is necessary for deciding whether the distributed autonomous strategy is effective or not. For some high-level tasks such as maintenance, where required tasks and/or task environment are not always well defined, and the faults of the robot system itself should also be taken into account, the distributed autonomous or multi-agent strategy is inevitably adopted.

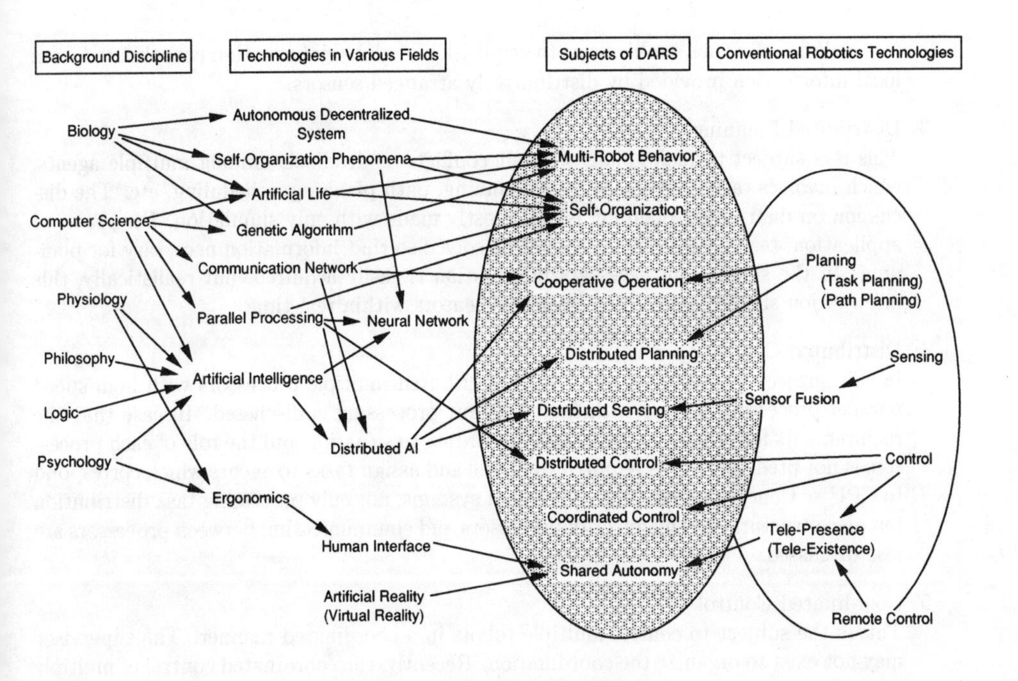

Figure 2: Relation of technologies

4 Subjects of DARS

The subjects of DARS include many divergences, with variety of background disciplines, which is the reason why DARS is an interdisciplinary research field, and researchers with different backgrounds are interested in this field. Figure 2 illustrates the relation between subjects of DARS and technologies related to them. The subjects are summarized as described below.

1. Multi-Robot Behavior
   In this subject, main concern is how multiple robots form a group (it is also called a team or population), and behave with some consistency. This is often discussed after the model of or in contrast with natural lives which form group or society such as school of fish, flock of birds, society of ants or bees, etc. Therefore, most of the approaches are based on analogy of such lives, and genetic algorithms may offer us powerful means to deal with the problems. The discussion is rather analytical. For example, the typical discussion is which kind of consistent behaviors is obtained, or how the behaviors are achieved, by incorporating simple and local mechanisms in each individual (agent). Here, the consistent behaviors mean basic formation of individuals such as line-up, gathering, avoidance, etc. The discussion on multi-robot behaviors is mostly made with only simulation, though a large number of individuals are assumed.

2. Distributed Sensing
   This is a subject to sense, detect, or recognize a target object by distributed multiple agents which work cooperatively. In comparison with sensor fusion where fusion of multiple sensors possibly with different functions, distributed sensing involves cooperative problem solving issues related to sensing, assuming autonomy of each agents. This sub-

ject includes also such problems as to acquisition of global information consolidated from local information provided by distributively arranged sensors.

3. Distributed Planning
This is a subject to make plans without conflicts or deadlocks among multiple agents, which involves task assignment, task planning, path planning, scheduling, etc. The discussion on distributed planning is also mostly made with only simulation. For practical application, technical problems should be solved so that information necessary for planning. In the simulation, necessary information is easily acquired, but realistically, this information should be detected by actual sensors within real-time.

4. Distributed Control
In this subject, control of a robot with distributed multiple processors with high speed parallel processing and/or with fault tolerant processing is discussed. In case that the requirements for processing tasks arise according to situation and the role of each processors is not predefined, it is required to divide and assign tasks to agents which correspond to CPU's. Concerning modular-type robot systems, not only processing task distribution but also structural arrangement of processors and communication between processors are also discussed.

5. Coordinated Control
This is the subject to control multiple robots in a coordinated manner. The supervisor may not exist to organize the coordination. Recently, the coordinated control of multiple arms has been discussed actively, which has common aspects with parallel manipulators. In this case, since the task to be achieved is tacitly presumed, the mechanism for coordinated control is considered incorporated in each robots a priori. Not only multi-robot control for manipulating a common object, but also consistent motion control for avoiding conflicts between robots (i. e. collisions avoidance), are included in the coordinated control in a wider sense. To achieve a common task by a group of multiple agents with the coordinated control is sometimes called collaboration.

6. Cooperative Operation
In this subject, various general problems for operating multiple robots cooperatively are discussed, such as conflict resolution, collaborative team organization, role assignment, resource sharing, etc. Though these problems are strongly related to the distributed artificial intelligence, the realistic issues such as control, sensing, and communication characterize this subject of DARS. In order to solve the problems, communication between robots is essential. Cooperative motion without communication have also been discussed to some extent. However, from the viewpoint of realizing realistic applications, it is considered to involve many technological problems, especially on sensing, to be solved.

7. Self-Organization
This subject aims at realization of self-organizing phenomena by mechanical systems, which has been discussed in science. As represented by cellular robotics[3], it is discussed to distribute the function required to the total system into modules (sometimes called cells) and reconfigure the system autonomously and dynamically according to the situation. As for mechanical structures, it is a metamorphic system where the modules can connect with each other by coupling mechanisms in a building block manner. As for information processing function for self-organization, collective intelligence (intelligence by collecting individual intelligence) and swarm intelligence (intelligence by collecting distributed non-intelligence) are investigated. Not only physical organization to connect

each other to structure a robot but also logical organization to determine the roles of modules so as to play harmoniously are considered as issues of self-organization.

8. Shared Autonomy
The shared autonomy is a concept of cooperative system between machines and humans. Not only machines but also humans are regarded as agents in this subject. The interface between machines and humans plays very important part, so that this subject is strongly related to human interface, tele-presense (tele-existence), and virtual reality.

5 Classification of Approaches and Assumptions

Discussion on DARS is developed with diverse approaches, where various assumptions may be introduced. The following is the classification of the approaches and assumptions from some viewpoints:

1. Top-down approach vs bottom-up approach
In the top-down approach, required tasks are divided and distributed into multiple agents. In contrast, in the bottom-up approach, the available agents are collected and organized.

2. Analytic approach vs synthetic approach
In the analytic approach, it is investigated what can be achieved if each agent is equipped with some mechanisms with some relation between agents. Conversely, in the synthetic approach, in order to achieve the requirements, it is investigated what kind of function is required for each agent, and which kind of relation is required between agents.

3. Homogeneous system vs heterogeneous system
In research studies on autonomous decentralized systems, the homogeneous system is often assumed, where every agent has the same attributes and performance. In contrasr, in the case of functional distribution, the heterogeneous system is assumed, where each agent has individual characteristics.

4. With communication vs without communication
Depending on communication availability, the conditions for the agent differ in assumption. In the strategy to utilize communication, there are two cases, one is to make use of specific communication utility such as radio and/or optical communication, and another is to utilize sensors such as vision for flag communication or odor sensor for pheromone communication. If the communication is assumed to be used, it means intensional information exchange is enabled. On the other hand, if the communication is not assumed to be used, each agent should be provided with a mechanism to infer the cooperative behavior based on sensed information.

5. With centralized agent vs without centralized agent
The decentralized systems have no centralized agent. However, in some distributed systems, a centralized agent may be assumed to be presented. In some cases, the centralized agent is predetermined and fixed, where the system is mostly hierarchically structured. However, in other cases, the centralized agent is determined according to the situation. For collaborative tasks, such a centralized agent is called a leader, a master, a manager or a coordinator. The centralized manager is not always an active supervisor to give instructions, but just a passive agent to manage ordered information as seen in a blackboard model.

or ability of each agent depends on the assumption. For example, low information processing ability of each agent is assumed for swarm intelligence, but high intelligence is assumed for dynamically reconfigurable robot systems.

7. Tight relation vs loose relation
For distributed systems, the localization is essential, which means the relation between agents is loose. Generally speaking, if each agent becomes more autonomous, the relation between agents may become loose. However, in some subjects such as coordinated control, distributed control, and distributed sensing, the tight dependence on the other agents must be assumed, which requires frequent communication between agents. If synchronization between agents is necessary, the relation between them must be tight.

6 Summary

DARS are introduced with analysis of research studies on DARS so far. DARS have not a long history. The more precise discussion on general issues of DARS such as how to reconcile the local autonomy and global consistency and coordination is expected as well as development on fundamental technologies for practical applications. Though behavior-based approach[4] which attracts the attention of people in the field of artificial life and robotics is viewed as a new strategy to realize reactive intelligence of agents in a low level, further synthetic discussion is considered indispensable for high level intelligence (i.e. task processing with organization and coordination). Technology on down-sizing will become important for DARS, because the merits of DARS will be amplified with this technology. On the contrary, researchers in the field of micro machines are interested in group behaviors of multiple micro robots. In addition, another problem is to establish criteria to evaluate the efficiency of DARS quantitatively, which is necessary to decide explicitly whether the distributed autonomous strategy is effective or not. More active interdisciplinary discussion is expected to advance DARS technology in future.

References

[1] K. Tanaka, M. Hirayama, "Multi-Production Batch Plant Oriented Control System by KAYAKU M-POCS-K1," *Chem. Economy Eng. Review*, **17**, 4(1985)187. (1989)283.

[2] T. Koizumi, et. al., "Remote Maintenance Test of Two-Arm Bilateral Servo-Manipulator System," *in Proc. 37th Conf. on Remote Systems Tech.*, (1989)129.

[3] T. Fukuda and T. Ueyama, "Cellular Robotics and Micro Robotic Systems," World Scientific, (1994).

[4] R. Brooks, "Intelligence without Reason," *in Proc. IJCAI-91*, (1991)569.

Manufacturing in the Future
-Some Topics of IMS-

HIROYUKI YOSHIKAWA

President, The Univ. of Tokyo, Tokyo 113, Japan

The feasibility study phase for the Intelligent Manufacturing Systems Program (IMS) has just been completed, and the program is now being reviewed for ratification by the countries and regions that are involved. It is expected that the full program will be ready to start by the beginning of next year, that is, 1995. There are still some uncertain points, however, regarding the understanding of the purpose of the program. It is necessary, therefore, to elucidate the purpose of the program and the possible impact that it may have on manufacturing technology.

The title of this Symposium is "Distributed Autonomous Robotics Systems." This type of system will become a very important technology in the future, and will play a central role in the improvement in manufacturing technology. This paper does not deal with the technology itself, but with the framework in which new technology will be utilized. I have some doubts about wide application of new high technologies such as distributed autonomous robotics systems. Of course the technology is very useful, but there still remains uncertainty about whether the technology will be utilized within an adequate social, economic, or technological framework. A suitable framework in which newly developed technologies can be applied should be constructed in the future.

Let us start with discussing a few inventions from the Industrial Revolution, almost 200 years ago. People say that the steam engine was invented, but it must be stressed that it was not only the steam engine itself that was invented; some kind of "social engine" was also invented at the same time. Technology itself is very important, but it is also very important to develop some sort of social environment that enables us to utilize that technology efficiently. The social engine that was invented concurrently with the steam engine could be considered the popularization of the technological knowledge that was used in the steam engine, and of course the invention of the modern corporation. Before the Industrial Revolution, society did not have such a highly structured system of corporation. After the Industrial Revolution, a new corporation system was established. People could be hired; they received money, and could purchase goods as consumers. This in turn created an impetus for manufacturers to produce goods, and an entire market cycle involving the flow of goods and money was set into motion. So, this type of social mechanism helped to enrich society itself. Of course, the steam engine itself was very important, but the social engine was vital for the production of wealth, employment, division of manufacturing among corporations and companies, and free market competition. All of these things started, for the most part, at the time of the Industrial Revolution.

At this moment, however, we are facing a very serious question. Is this social engine that was developed during the Industrial Revolution still operating perfectly? Can we safely say that this device will continue to be adequate to help construct a fully rich globe? Unfortunately, we must say that these social and technological engines are somewhat obsolete, and are not entirely suitable to develop the kind of rich society that emerged in the Industrial Revolution. There are many reasons for this, but one of the most important is certainly the uneven distribution of wealth. The social engine developed during the time of the Industrial Revolution is working very well to improve the social living standards in some countries, but it cannot cover all the world. There are many countries which have not been able to improve their living standards; some of them are actually starving. So at best, some people can say that this social engine is "on the way" to improving the living standards of the whole world, but unfortunately, because of new global environmental problems, etc., many people have started saying that the present manufacturing system cannot be maintained forever, and that we must convert our direction in developing countries towards a new kind of "sustainable development." When we think of the direction of the future of manufacturing technology, no one has reached any sort of conclusion on this point, so it needs to be discussed in further detail. Today's topic is connected with this point.

If we try to amend the social engine, of course it is also necessary to amend the technological engine, and there are many methodologies and policies designed to cope with these tasks. One of those methodologies is the "shift wealth paradigm." That is, there are rich countries on the earth, but on the other hand there are poor countries there, and it is necessary for human beings, for moral and many other kinds of reasons, to try to resolve this uneven distribution of wealth. This may take the form of official development aid or something like that. Of course this is absolutely necessary, and we should continue working in this direction, but it must be said that there are limitations to this policy in that it involves the transfer of only wealth from the A countries to the B countries. This is useful, but it cannot solve everything. So, as an alternative to this paradigm, a "shift knowledge paradigm" may be proposed. The concentration of wealth in certain countries may be due in part to the concentration of manufacturing knowledge. If a company or country monopolizes some manufacturing knowledge, then that company or country can get a lot of money, which in turn can increase the wealth in that region, company, or corporation. So by shifting manufacturing knowledge, we are in effect shifting wealth, and in the long run this might be a more effective method of balancing the distribution of wealth among countries than the shift of money itself. The shift of manufacturing knowledge will not directly detract from the wealth of the country that is donating it, and from this point of view, we can say that the final result will be something like the total increase of wealth.

It is easy to refer to this diffusion of manufacturing knowledge in words, but of course there are many obstacles and impediments to this process. One obstacle, for example, is the view that manufacturing knowledge is the cutting edge of a company's competitiveness; it should be kept secret, and not be diffused outside of the company. I think this is legitimate. According to our free market system, companies are allowed to protect their own knowledge, although they can sell their products anywhere. Keeping valuable manufacturing knowledge and universal sale of the fruits of that manufacturing knowledge is, in fact, one of the fundamental principles of the free market system. This is one of the biggest impediments that manufacturing knowledge must overcome.

However, if we forget about protecting the secrecy of knowledge -- for example, if a certain corporation tries to diffuse its manufacturing knowledge to other countries and companies-- there are still many impediments. Some people say that the total transfer of a manufacturing technology from country A to country B cannot be accomplished in as short a time as 10 years. One of the main reasons for this is the less systematized nature of manufacturing technology. Other factors are differences between the countries in ways of life, ways of thinking, cultural and traditional barriers, political differences, inconsistent standards, language barriers, etc. So manufacturing knowledge is very difficult to diffuse even when great efforts are being made, and we must do everything that we can to overcome each of these obstacles.

The first obstacle is the secrecy of the free market system. Within a company, the technology developed by the research and development department can be very easily transferred to the business department, and this in turn is very effective in improving the product and increasing the competitiveness of the company. From this stage, the product moves out into the market, and it becomes evident that the effort initially made by the research and development department towards developing the product will be usefully applied or used to develop or improve the competitiveness in the market. This kind of flow is currently operating without any difficulty in many companies, but it is very difficult to exchange this kind of information between these companies, and this of course is the result of market competition. If we could establish some kind of intellectual property rights institution, if there were some institution to buy the good knowledge and sell it to others, then this would create a new market besides the product market -- a so-called "knowledge market." With the creation of this type of market, knowledge would become a commodity; knowledge could be bought and sold, and the diffusion of knowledge would be accelerated. This is another form of social engine, and there are of course many difficulties in controlling the flow and making the distribution of knowledge more even. How to evaluate and control the price or value of manufacturing knowledge is a very difficult problem. Despite these problems, however, the creation of this type of institution is a dream that we should pursue in the future to promote and accelerate the diffusion of manufacturing knowledge among companies. Of course, this is just one example of a social engine, and all social engines have to be introduced by someone at some time. I will not touch upon this point too deeply at this time, however, because it is closely related to politics, and I am not so politically oriented.

The next point is very important for us researchers and engineers, and it is that the lack of systemization of manufacturing knowledge will impede the dissemination of manufacturing know-how from some points to other points. How can we systematize knowledge? Let me show you a very short history of knowledge. In the past there was a system of knowledge that dealt with the motion of stars, astrology. Astrology has a long history of many theories and discussions about the motion of stars. In the case of astrology, however, the knowledge on the subject was highly monopolized by astrologers, and could not be utilized by common people. This is a typical example of the monopolization of knowledge. Throughout history, particularly in medieval times, many kinds of so-called "scientific" knowledge have been monopolized by certain people, and its usefulness has therefore been limited. Alchemy was also monopolized by rich men. I do not know if alchemy was really useful to human beings, but many people thought that the results of alchemy were very useful to make human life longer. This very useful knowledge could never be distributed to others, however. Because of the less systematized nature of the so-called sciences of astrology and alchemy, knowledge about them could not have been effectively disseminated, even if they had been openly distributed by their practitioners; even if books had been written, they would have been very difficult for common people to understand, and from that point of view the systemization of these bodies of knowledge was too limited.

After several hundred years, however, astrology and alchemy both evolved into the newer sciences of chemistry, physics and astronomy, etc. Once these fields of knowledge had evolved, they became far more accessible to common people. Chemistry was a highly systematized science, and the science of physics was very easy to understand. The dynamics of Newton have been very rationally and clearly systematized, with no uncertainty in scientific terms, and many people can easily learn to grasp its fundamental principles without spending money. A textbook on the subject can be purchased very cheaply at this time. Trying to monopolize this knowledge at this point would be an extremely expensive venture. If you look at university curricula, you can learn not only about the basic sciences of chemistry and astronomy, you can also obtain a great deal of knowledge that can be usefully applied to manufacturing activity through the engineering departments. Mechanical engineering, electrical engineering, etc., are all very systematized; textbooks on them can be obtained very easily; and they can be learned with very little money. This kind of knowledge is public these days. If you look at manufacturing know-how, however, we do not have any good textbooks at this moment. It is absolutely necessary to exactly duplicate manufacturing experience in order to establish some sort of knowledge system for a company. Of course, as earlier, some companies try to disseminate this manufacturing know-how to other companies. Information can be disseminated within a single company much faster, but if you try to disseminate it from company to company, the process takes a long time. So, from that point of view, manufacturing technology is still somewhat monopolized. As pointed out earlier, there are two reasons for this monopolization. The first is the intentional reason. At the same, if a company tries to disseminate manufacturing knowledge, there are other difficulties, and this is the second reason, which is what we are concerned about.

We must think about the process of knowledge development. There are some types of basic research concerned with manufacturing technology, be it material science, mechanical mechanisms, etc. It is possible for researchers to invent new ideas without introducing knowledge from other relevant fields. So, if you are trying to develop some good ideas in physics, you can only think about physics; thinking about other fields such as psychology, etc., is not necessary. That is the reason why we have such kinds of domains; and within the domains we can conduct relevant research activities.

By applying many kinds of knowledge which were developed by conducting basic research, we can develop some kind of "pre-competitive research" that might be applied for structural applications, designing and implementing new equipment, etc., and might be useful for enhancing the competitiveness of many corporations. Many kinds of knowledge must be collected and integrated to make them useful for certain purposes. Just by using the results of this pre-competitive research, the competition, or the competitive research will be conducted within a firm, so of course the use of monopolized research is very useful in this case. This is a very simple linear model of the development of knowledge from basic research to pre-competitive and competitive research. At the same time it is necessary to think about post-competitive research. If you think about manufacturing technology, you can see that the main parts of manufacturing technology are now being created within firms. There are many researchers engaged in the field of research and development of technologies in universities and other institutions, but a huge amount of knowledge concerned with manufacturing technology is now being created within firms. However, as mentioned earlier, these firms are monopolizing the information; they are not disseminating it. If a firm goes bankrupt, then the knowledge is dissipated into thin air, and that is a real loss from a social point of view. It is necessary for us to structure this kind of manufacturing knowledge into another form. One way to do this would be putting it into a textbook, for example. If we could systematize the manufacturing knowledge into a textbook the knowledge would still remain when the company disappears; it would remain in society as a public asset. This type of activity may be called "post-competitive" research, and unfortunately, we are not seeing a lot of it. This type of post-competitive research would be very useful to systematize manufacturing knowledge which is created within firms and to put it on the level of basic research, and a textbook of this kind could play a role in encouraging basic research itself.

So, the circle of basic research -- precompetitive research -- competition -- post-competitive research -- basic research will become very important. Unfortunately, the arc from basic research to competition is very strong but the arc from competition to basic research is very weak because of the social engine under which we are living now.

A very important question is, what is manufacturing knowledge? Before we go into manufacturing knowledge, however, we must first think about post-competitive technology. We have many experiences that follow a basic pattern, and one good illustration of this pattern can be provided by fluid dynamics. Fluid dynamics is, of course, very closely related with the water wheel, and if you look at the history of the water wheel, you can see that human beings have possessed its technology for over 2,000 years, i.e., knowledge about the utilization of water-flow energy for mechanical purposes. Unfortunately, just before the Industrial Revolution many people started to ask why the efficiency of the water wheel was so low, and many researchers were encouraged to study the mechanisms of how to extract energy from flowing water to mechanical systems. So through this competition, many firms accumulated and maintained a great deal of information on the subject of fluid dynamics at this time, but unfortunately, their knowledge was not systematized. At the same time, many basic researchers also pursued the field, and their knowledge was systematized. The systematic textbooks that they put together made it possible to educate engineers and designers about the field of fluid dynamics. This was useful not only for education purposes, however; it was also very useful in developing the technology of fluid dynamics itself. That is why there is this sort of circle leading up to this post-competitive technology.

So now, the very important question, what is manufacturing technology? Let us begin with thinking of a fried egg. In order to cook a good fried egg, you must maintain some know-how. You must have the knowledge to perform the basic steps, i.e., putting the oil in the frying pan, heating the oil, cracking the egg open and putting it in the pan, adjusting the flame, and finally, turning off the flame. These steps, of course, are very simple, but the final result is not so simple. When I fry an egg, I try to make a good fried egg. I try to keep it somewhat clean; I try to keep the yolk intact and surrounded on all sides by the white, and if possible, I try to make the cooked egg come out as round as possible. If I make a mistake in calculating the height from which I release the egg into the pan, and drop it from too high, then the yolk will be destroyed, and the final product will look more like a scrambled egg than a fried egg. If I drop the egg into the pan from too low a height, the eye will not open, and I will have something like a sleeping eye. So, to perform a task like this it is necessary to know many kinds of things, but this kind of know-how is not included in this systematized knowledge. If you look at the cooking process of a fried egg, there are many technological theories at work. There are the theories of heat transfer, heat conduction, protein solidification, theology, surface tension, dynamics, and geometry. All of these theories play a vital role in the frying of an egg, but if you know everything there is to know about each one of them, this still will not help you to fry a good egg. There is still some sort of missing link. The same also applies to the manufacture of a motor car, for example. Of course it is absolutely necessary to be thoroughly well versed in all of the scientific theories at work in the manufacturing process, but again, that alone is not enough to successfully design and manufacture a high-quality, popular motor car. The systematization of manufacturing knowledge, of course, is indispensable.

Again, let us go back 200 years ago. There are some pictures in the Encyclopedia drawn by Mr. Diderot at the end of the 18th century. They show the manufacturing process of glass. There are about 10 or 20 figures that show exactly how to obtain the final product, something like plate glass. If we look at these figures, then we can understand what the properties of glass are. For example, if you look at this figure of the man blowing the glass, you can see by the shape of his face that he is blowing as strongly as possible, so you can assume that high pressure is required to perform this task. After 200 years, the necessary pressure to blow this kind of bulb was calculated precisely by applying a scientific equation about the viscosity of glass. We have developed the knowledge of the properties of glass very well -- from the heat point of view, the mechanical point of view, the flow point of view, etc. -- but at the same time, there should be some other kind of knowledge that covers the sequence of the making of glass. Put it into the furnace, take it out, blow it; this kind of thing has not yet been established as systematized knowledge. Surprisingly, it is found that there is no good methodology on how to describe the sequence of operations. The missing link for manufacturing technology is something like this. The sequence of action cannot be described in terms of differential equations. There is usually some kind of record or some kind of manual that we can read when we need to, although it is often very difficult to do so. The education included in the sequence of operation is very difficult.

The IMS has now been implemented to go into this kind of process of know-how. Of course there should be many other kinds of research within the IMS program, but I would like to stress that this point is also very important. The IMS is something like an inter-corporate collaboration that is necessary for the integration of knowledge, avoidance of duplicate investment, and the avoidance of meaningless innovation. Now we are confronted with very heavy investment amounts due to the increasingly sophisticated manufacturing facilities. Our automated robotics systems are very expensive, and if they are secretly carried out by many companies, then maybe only one or two of them will be successful. So, from that point of view, duplicate investment will be a very big burden on society itself. Today the investment amounts today alone make the social engine that I previously mentioned obsolete. So, this kind of inter-corporate collaboration will be increased in the future to avoid this problem.

There will also be university-industry collaboration and industry-university collaboration. For industry-university collaboration, which will comprise a kind of pre-competitive research, the intention will originate in the industries, and the material will be provided by the universities. Conversely, university-industry collaboration will be useful for post-competitive research. The motivation for this kind of collaboration will mainly come from the universities, which will want to produce good textbooks, etc., but the material will always be found within the firms. So, there will be two types of collaboration that involve universities and industries.

Of course, international collaboration will be very important. As I have already said, it will become necessary to write a good textbook on manufacturing technology which can be used by any country. If a textbook is written on the subject now in Japanese, the book might be useful in coming generations, but only in Japan. In order to establish a good textbook which can be applied in any country, international collaboration will be essential. So, as you can see, there are many types of collaboration which will be implemented by the IMS program.

I have based my reason for proposing the implementation of this type of IMS program on the Japanese experience of the 1970s. Some people may recall the historical program originating in 1970, which was called MUM, or method of unmanned manufacturing. A feasibility study was conducted from 1972 to 1975, and a national project study was conducted by MITI from 1977 to 1985. The study involved 30 companies, a great many professors from universities, and a total budget of $0.3 billion. The final result produced was a prototype for an unmanned factory, the new concept of "metamorphic manufacturing systems," totally computerized control, and intelligent elements. All of these developments were very useful, but unfortunately they were only national, and they were not well documented. The people who were involved in the program enjoyed it very much, but it is not easy for us to find the documentation. This illustrates that we still are lacking in post-competitive research, and need to pursue it further in the future.

The feasibility study for the IMS program was started in February 1992, and was completed in December of last year, 1993. The purpose of the feasibility study was the identification of necessary conditions for international cooperation in manufacturing. When this proposal was put forward, it elicited many strange reactions. Many people said that we could never conduct collaborative research in the manufacturing field, because manufacturing technology is vital for the competitive strength of companies, and must be kept secret. This is true, and we should never try to control the mechanism of competition. At the same time, however, we should try to decrease the cost of competition. We are now confronting very excessive costs to maintain our competition. In order to win, many companies have to pay a great deal of money to invent excessive and sometimes useless options for new products.

Finally, I would like to give you some examples. I have been saying that the manufacturing industry should change its structure from the open-loop to the closed-loop. A traditional open-pass industry takes material, perhaps a natural resource from the earth, converts it into useful manufacturing material, components, and then the product is assembled and it is utilized by users. No one knows about the waste after utilization. But now people say we must change our concept, converting the developing industry into a sustainable industry, which raises the question, what is a sustainable industry? Very simply, sustainability should be considered the circulation of a material without even a drop of waste. All of the material should be reused and we have already been implementing the new activity of recycling of material in our society. Still, however, this inverse pass is very weak from many points of view. From an economic point of view, we cannot enjoy good profits from this kind of industry, and more importantly, we have very limited technology and no good technological system about constructing this kind of inverse industry. If the restoration of our social engine and the restoration of our technological engine are conducted concurrently, however, then it is possible to establish this kind of thing. Of course it is a kind of ideal state, and it will take a long time to realize, but we should consider it a target to pursue. We are now trying to establish this kind of clean manufacturing system in the IMS, and we have just begun the evaluation of the burden on our environment which is produced by one artificial product. Unfortunately, such endeavors are often not supported economically within a free market system, so it is necessary for us to deliberately analyze the problem and develop a new concept which will harmonize the technological and social engines in the future.

Chapter 2

Distributed System Design

Hierarchical Control Architecture with Learning and Adaptation Ability for Cellular Robotic System

TSUYOSHI UEYAMA[1], TOSHIO FUKUDA[2], ATSUSHI SAKAI[2], and FUMIHITO ARAI[2]

[1]Nippondenso Co., Ltd., Aichi, 470-01 Japan
[2]Dept. of Mechano-Informatics and Systems, Nagoya Univ., Nagoya, 464-01 Japan

Abstract

This paper deals with a hierarchical control architecture of mobile robots for Cellular Robotic System (CEBOT). The CEBOT is a distributed autonomous robotic system composed of a number of robotic units called "cells." Since the CEBOT provides variable structure, flexibility and extendibility are required in the control system for the CEBOT. To design the mobile robots for the CEBOT, we have adopted a parallel processing control system, which makes it easy to add new rules according to the change of organization of the system. This paper proposes a method to integrate multiple processes for decision making of the behavior of the mobile robots. We define two relation matrices that denote the relationship between the processes: a priority relation matrix and an interest relation matrix. These matrices are used to adjust the output of the processes and optimize the behavior of mobile robots. That is, the conflicts of inter-processes are solved by coordination between the priority relation matrix and the interest relation matrix. The definition of the priority relation matrix and the interest relation matrix make it possible to realize a learning and adaptation ability for the behavior. To obtain the most suitable priority relation matrix, especially, this paper introduces a learning method for the mobile robots. Several simulation results present the effectiveness of the proposed matrices and the proposed learning method.

Keywords: Cellular Robotic System, Mobile Robots, Hierarchical Control Architecture, Priority Relation Matrix, Interest Relation Matrix, Learning and Adaptation

1. Introduction

Research on decentralized autonomous robotic systems has become active and attractive, because of the advantages of the robotic system. The advantages come from the configuration of the robotic system. Recently, many robotic manipulators have been used in manufacturing systems. According to the extension of the robotic application fields, the dimensions and the complexity of the robotic systems have been increased. The large dimension and complexity of the systems make it difficult to control the whole system optimally or feasibly. As an approach method for the problem, the decentralized autonomous robotic systems have been proposed to control the systems distributively, not individually. The decentralization and autonomy of the robotic systems make it possible to reduce the dimension and the complexity of each system, and it also improves the reliability of the systems. The decentralization aims at the improvement of the ability of robotic groups instead of the improvement of the ability of robotic unit. That is, in the decentralized autonomous robotic systems, several functions are decomposed into robotic units, and any required function or performance ability is organized by the combination of the robotic units.

According to this advantage, we have proposed Cellular Robotic System (CEBOT) as one of the self-organizing robotic systems[1, 2]. The self-organizing robotic systems consist of a number of autonomous robotic units, that is, the self-organizing robotic systems have the advantage and characteristics of decentralization and autonomy of robotic systems. The concept of the CEBOT has been derived from the organization of biological organization[3]. The CEBOT consists of a large number of robotic units called cells comparing with biological cells. The cell has a simple function, such as rotation, bending, extension, handling, and so on. The CEBOT can be organized dynamically depending on its given tasks or its environment.

This paper deals with a hierarchical control architecture of mobile robots for the CEBOT. Especially, the proposed architecture is designed for multiple mobile robots or a dynamically reconfigurable robotic system. The proposed architecture has been influenced by the conceptual organization of the subsumption architecture from the viewpoint of the build-up ability. Since the CEBOT is constructed by many autonomous robotic units called "cell," we must provide build-up and reconfiguration ability to the design of the control architecture for the CEBOT. Figure 1 represents the analogy between creature and the CEBOT. The hierarchical configuration of the CEBOT is closely related to a proposed architecture. A hierarchical control architecture with learning and adaptation ability is proposed. This paper denotes that multiple mobile robots can behave with learning and adaptation to their environment or given tasks. To provide the adaptation and learning ability to the multiple mobile robots, we organize the control architecture by multiple processes or multiple agents, and introduce a priority relation matrix and an interested relation matrix. Each process has a simple function to control a mobile robot. By using the proposed matrices, we can see that the coordination of these processes or agents makes it possible to drive the mobile robots optimally. These matrices are used to define the relation between processes that have different functions individually. The adaptation and learning ability is provided by changing the values of these matrices depending the task execution results. By using the matrix representation between the processes, we can extend the system configuration and build up the control processes or control layer on the previous designed control layers. This advantage is designed for the reconfigurability and self-evolution of the CEBOT. We represent several simulation results where the adaptation and learning performance is carried out effectively.

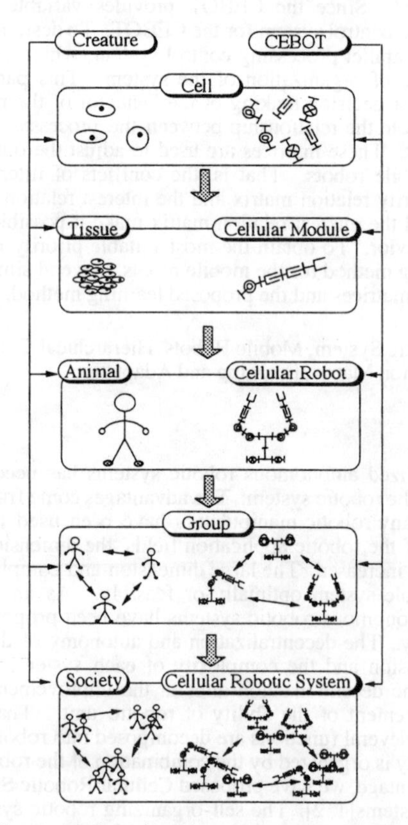

Fig. 1 Analogy between Creature and CEBOT

2. Decision Making of Behavior

Autonomous robots must determine their behavior by themselves according to their situation. Approach methods of decision making of the behavior, especially mobile robots, are divided into two categories, such as traditional approach and behavior-based approach. In the traditional approach, information flows from sensors as data to actuators as energy through planning or inference engine. The traditional control architecture has a closed feedback loop in a system, that is, the architecture can be designed under the consideration of a whole system. On the other hand, the behavior-based architecture, such as the subsumption architecture[4], decomposes a control architecture into several layers that correspond to elemental behavior or tasks. In the behavior based approach, the control architecture is composed by several processes or agents that are classified by task achieving behaviors. That is, the behavior-based architecture can be regarded as a hierarchical control architecture that provides a parallel performance ability in each layer. The similar formation is shown in the hierarchical decentralized parallel architecture proposed by Albus[5]. To control multiple robots cooperatively, Parker presented a motivational behavior as an adaptive action selection for agent teams, where the subsumption architecture was adopted[6]. As an evolutional behavior based on the behavior-based architecture, Maes and Brooks indicated a learning algorithm for six legs robots, in which binary and dual feedback was treated[7]. Arkin proposed the AuRA (Autonomous Robot Architecture) and implemented it to a mobile robot[8].

The proposed architecture in this paper adopts the behavior-based approach, because of the flexibility and reconfigurability of the control architecture. Especially, the proposed control architecture is designed under the consideration of the configuration of the CEBOT. Since the configuration of the CEBOT allows the addition of functional modules/robots flexibly, the control architecture of the CEBOT must be designed to make it possible to cope with additional functions. As shown in Fig.2, the relation between input and output of a system, which includes cells, individuals and groups, should be designed according to the configuration of the system. That is, the configuration of the control architecture for the groups must be designed as the extension of the control architecture in the cells. The proposed control architecture consists of parallel distributed processes corresponding to task achieving behaviors. This approach makes it easy to design each process individually. Yet, it is necessary to integrate or fuse the output of each process optimally. That is, each process indicates a behavioral vector, therefore, the conflict of the vector must be arranged to emerge optimal behavior. To arrange the conflict, priority relationship between the behavior vectors must be considered. This paper deals with the behavior emergence of mobile robots. A conceptual diagram of the control system is shown in Fig. 3, where input devices encourage the processes to generate output vectors. The output vectors are summed to emerge the behavior of the robot, that is, an optimal behavior depends on the synthesis strategy. Especially, the proposed architecture is designed for a control system of the CEBOT Mark V, where each process is carried out asynchronously. The flow of the processes is shown in Fig. 4. The asynchronous progress of the processes makes it possible to add a new process.

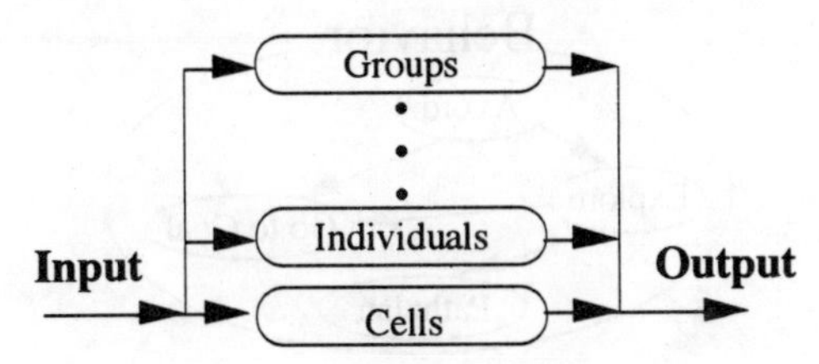

Fig. 2 From Cell to Group

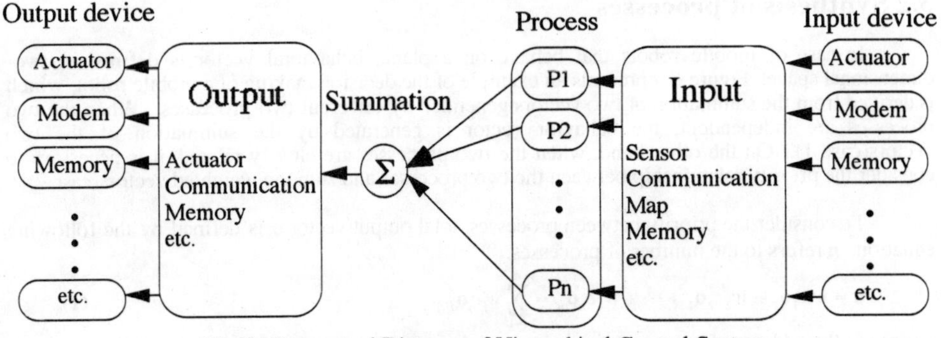

Fig. 3 Conceptual Diagram of Hierarchical Control System

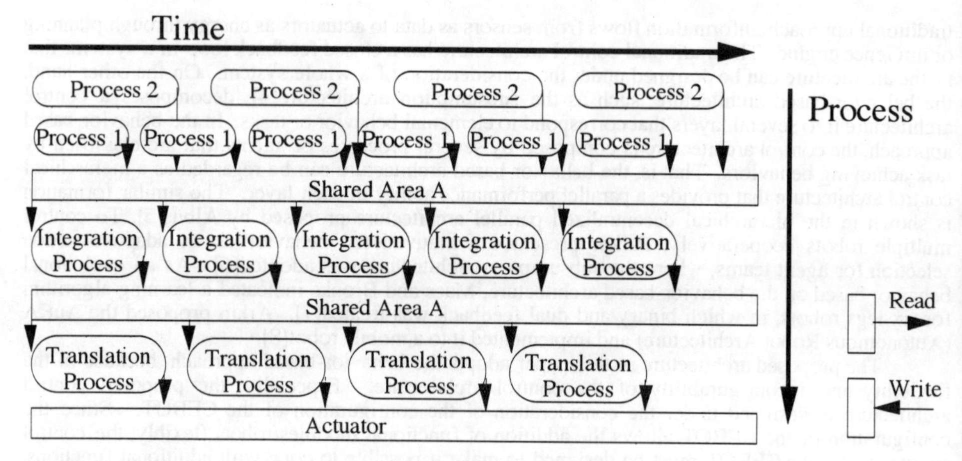

Fig. 4 Flow of Processes

3. Priority Relationship

3.1 Priority relationship between processes

The following discussion deals with the processes that generate behavior vectors for a mobile robot. Figure 5 represents an example of the priority relationship between behavioral processes. That is, in order to add a behavioral process, it is necessary to adjust the priority relationship between the behavioral processes. In order to define the priority relationship, the following section proposes a priority relation matrix and an interest relation matrix. The definition of the priority relation matrix and the interest relation matrix makes it possible to realize a learning and adaptation ability for the behavior. These proposed matrices define the relationship between the processes, especially, a learning method is described to acquire an optimal priority matrix that makes it possible to generate an optimal behavior in multiple robots.

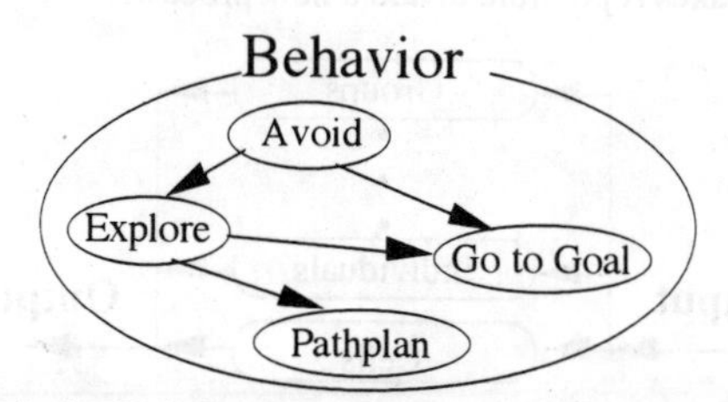

Fig. 5 Priority Relationship between Behavioral Processes

3.2 Synthesis of processes

In case of mobile robots that behave on a plane, behavioral vector is defined in two-dimensional space. Figure 6 represents an example of the decision making of a mobile robot, which is derived from the summation of two vectors generated by different two processes. When the two processes are independent, the behavior vector is generated by the summation of the two vectors(case 1). On the other hand, when the two processes are closely related, it is necessary to consider the priority relationship between the two processes and sum the weighted vectors(case 2).

To consider the priority between processes, total output vector $\mathbf{o}$ is defined by the following equation. n refers to the number of processes.

$$\mathbf{o} = w_1 c_1 \mathbf{o}_1 + w_2 c_2 \mathbf{o}_2 + \cdots + w_n c_n \mathbf{o}_n = \sum_{i=1}^{n} w_i c_i \mathbf{o}_i \qquad (1)$$

where i is the address of the process. w_i is a weight coefficient, and c_i is defined as follows:

$$c_i = 1 - \max_i \{p_{ij} \cdot i_{ij}\} \tag{2}$$

p_{ij} is an element of a priority relation matrix **P**, and i_{ij} is an element of a interest relation matrix **I**.

$$\mathbf{P} = [p_{ij}], \quad 0 \le p_{ij} \le 1 \tag{3}$$

$$\mathbf{I} = [i_{ij}], \quad 0 \le i_{ij} \le 1 \tag{4}$$

By adjusting the elements of these matrices, the robotic behavior can be optimized. This paper presents a learning method of the priority relation matrix. Figure 7 represents an example of the relation between the proposed matrices and the output of two processes.

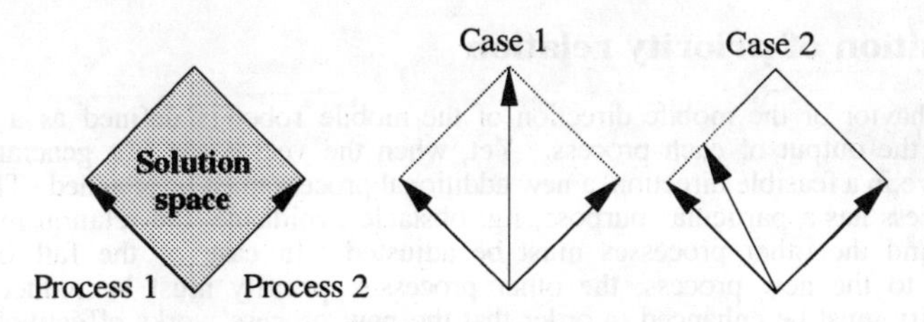

Fig. 6 Decision Making of a Mobile Robot

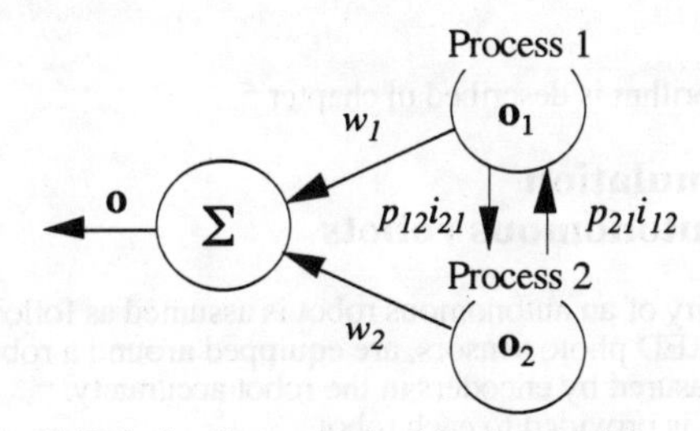

Fig. 7 Relation between the Matrices and Processes

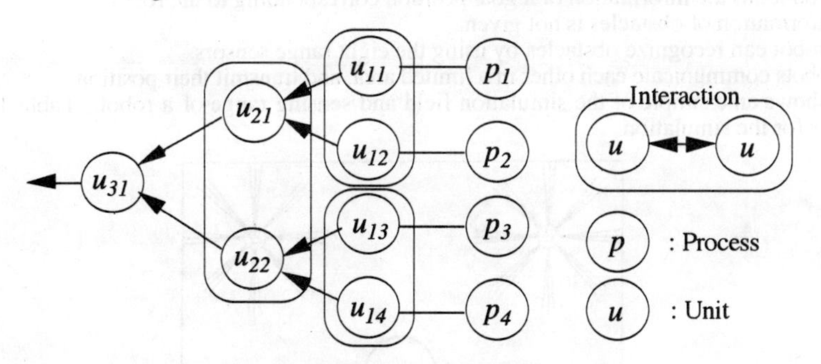

Fig. 8 Three-Layer Control System

3.3 Extension to multiple layers

According to the increase of processes, the relationship between processes will be complex. Therefore, it is necessary to synthesize the processes hierarchically. Figure 8 represents a three-layer control system, where a unit refers to the synthesis of the output of units or process in a lower layer. A unit in the bottom layer corresponds to a process. An output vector of the unit *mk* ${}_m\mathbf{o}_k$ is defined as follows:

$${}_m\mathbf{o}_k = {}_mW_1\,{}_mC_1\,{}_m\mathbf{o}_1 + {}_mW_2\,{}_mC_2\,{}_m\mathbf{o}_2 + \cdots + {}_mW_{n_k}\,{}_mC_{n_k}\,{}_m\mathbf{o}_{n_k} = \sum_{i=1}^{n_k} {}_mW_i\,{}_mC_i\,{}_m\mathbf{o}_i \tag{5}$$

where *m* refers to the layer of the unit, and *k* is the unit number. n_k presents the number of units that relate to the unit *mk*. The total output vector of the layer *m* $\mathbf{O}m$ is presented as follows:

$$\mathbf{O}_m = ({}_m\mathbf{o}_1, {}_m\mathbf{o}_2, \cdots, {}_m\mathbf{o}_{Nm}) \tag{6}$$

where Nm refers to the number of units in the layer m. The relation between the total output vector of the layer m and layer $(m-1)$ $\mathbf{O}m$ is given by,

$$\mathbf{O}_m = \mathbf{O}_{m-1}\mathbf{C}_m\mathbf{W}_m \tag{7}$$

Therefore, the total output of a system $\mathbf{o}$ is defined as follows:

$$\mathbf{o} = \mathbf{O}_0\prod_{k=1}^{M}\{\mathbf{C}_k\mathbf{W}_k\} \tag{8}$$

where M refers to the number of the layers.

3.3 Acquisition of priority relation

The behavior or the mobile direction of the mobile robot is defined as a vector by the summation of the output of each process. Yet, when the vector can not generate an adequate solution to move in a feasible direction, a new additional process must be attached. That is, the new additional process has a particular purpose, i.g. obstacle avoidance, the relationship between the new process and the other processes must be adjusted. In case of the fail of the purpose corresponding to the new process, the other processes' priority must be reduced or the new process's priority must be enhanced in order that the new process works effectively. In case of Eqs. (9) and (10), the role of the process j is enhanced.

$$p_{jk} \rightarrow p_{jk} - \Delta p_{jk} \tag{9}$$

$$p_{kj} \rightarrow p_{kj} + \Delta p_{kj} \tag{10}$$

The detail of the learning algorithm is described in chapter 5.

4. Assumptions in Simulation

4.1 Assumptions of autonomous robots

The performance ability of an autonomous robot is assumed as follows:
(1) Eight range sensors, e.g. LED photo sensors, are equipped around a robot at regular intervals.
(2) Moving path length is measured by encoders in the robot accurately.
(3) A communication module is provided to each robot.
(4) Each robot has the information of a goal position corresponding to the robot.
(5) The information of obstacles is not given.
(6) Each robot can recognize obstacles by using the eight range sensors.
(7) The robots communicate each other in a limited area, and transmit their position.
Figure 9 shows an example of the simulation field and sensing range of a robot. Table 1 indicates parameters for the simulation.

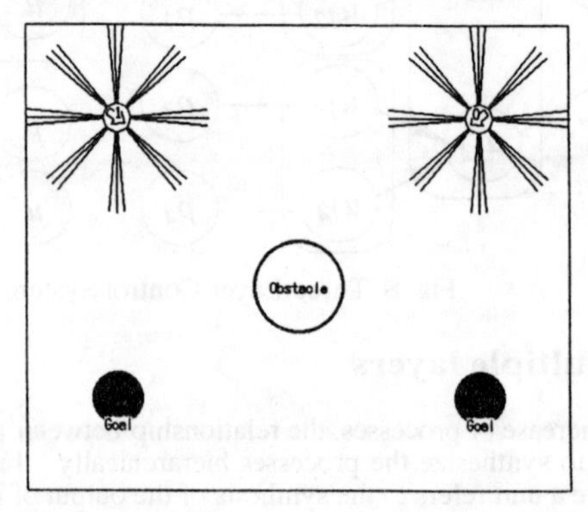

Fig. 9 Simulation Field and Sensing Range of a Robot

In order to reach the goal individually, we have provide three processes, as follows:
(1) Goal direction
(2) Obstacle avoidance
(3) Turn right to avoid collision between robots
The followings describe how to define the behavior for each process.

Process 1: The process generates a vector to achieve the goal. The vector is defined as follows:

$$\begin{pmatrix} x_1 \\ y_1 \end{pmatrix} = \frac{1}{(x_g - x_r)^2 + (y_g - y_r)^2} \begin{pmatrix} x_g - x_r \\ y_g - y_r \end{pmatrix} \tag{11}$$

where, $(x_g, y_g)^T$ is the position of the goal. $(x_r, y_r)^T$ is the coordinates of the robot.

Process 2: This process is used to avoid obstacles, which are recognized by eight range sensors. In this simulation, a weight vector is adopted to consider the forward direction of the robot. Eq. (12) presents a vector of the process 2.

$$\begin{pmatrix} x_2 \\ y_2 \end{pmatrix} = \sum_{k=1}^{8} \left\{ -\frac{w_k R^2}{r_k^2} \vec{s}_k \right\} \tag{12}$$

where, $w_k = \{1.0, 1.0, 1.0, 0.5, 0.5, 0.5, 1.0, 1.0\}$,

r_k: measuring value of range sensor k,

s_k: direction vector of range sensor k,

R: radius of the robot.

Process 3: This process is added to avoid the collision between robots. As a function of this process, a vector that is in order to turn right itself is generated when another robot exists in the moving direction. The definition of the vector is given by,

$$\begin{pmatrix} x_3 \\ y_3 \end{pmatrix} = \begin{pmatrix} x_d \\ y_d \end{pmatrix} \begin{pmatrix} \cos\left(-\frac{\pi}{4}\right) & -\sin\left(-\frac{\pi}{4}\right) \\ \sin\left(-\frac{\pi}{4}\right) & \cos\left(-\frac{\pi}{4}\right) \end{pmatrix} \tag{13}$$

where, $(x_d, y_d)^T$ refers to a moving direction of the robot.

Figure 10 represents the motion of the processes.

Table 1 Parameters for simulation

Field dimension	600 x 500
Number of Robots	2
Number of Goals	2
Diameter of Robot	30
Max. Moving Speed of Robot	3/Step
Radius of Sensing Range	50
Radius of Communication Area	100

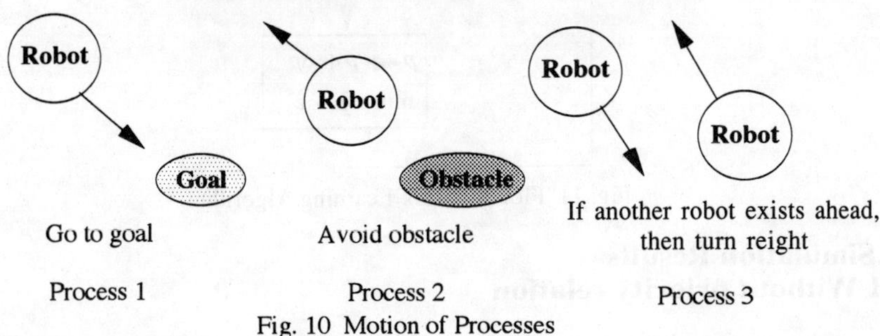

Fig. 10 Motion of Processes

4.2 Definition of interest relation matrix

Elements of an interest relation matrix are defined, which are based on cosine of the vectors for the processes. This definition is lead from the relation between the direction of the vectors and the interest or intention of the process. That is, the direction of the vectors is similar to the interest of the process.

$$i_{jk} = \frac{1 - \dfrac{\mathbf{o}_j \cdot \mathbf{o}_k}{|\mathbf{o}_j||\mathbf{o}_k|}}{2} \tag{14}$$

where, $\mathbf{o}_j$ refers to the output value of the process j.

5 Learning of Priority Relation

This chapter denotes how to apply a learning and adaptation of the priority relation matrix. This simulation focuses on the relation between the process for obstacle avoidance and the other processes. Consequently, the priority relation matrix is defined as follows:

$$\mathbf{P} = \begin{pmatrix} 1 & p_{12} & 0 \\ 0 & 1 & 0 \\ 0 & p_{32} & 1 \end{pmatrix} \tag{15}$$

Here, it is assumed that the parameter p_{12} is equal to the p_{21}, and is refereed as p. The learning and adaptation make the parameter p optimize to arrive at a goal first. The learning strategy is as follows: When the collision between a robot and another robot or an object, the priority of the process (obstacle avoidance) is enhanced. On the other hand, when the robot can get to the goal without collision, the priority of the process is decreased. Figure 11 shows the flow chart of this algorithm. The described algorithm is applied in section 6.2, where two cases are carried out.

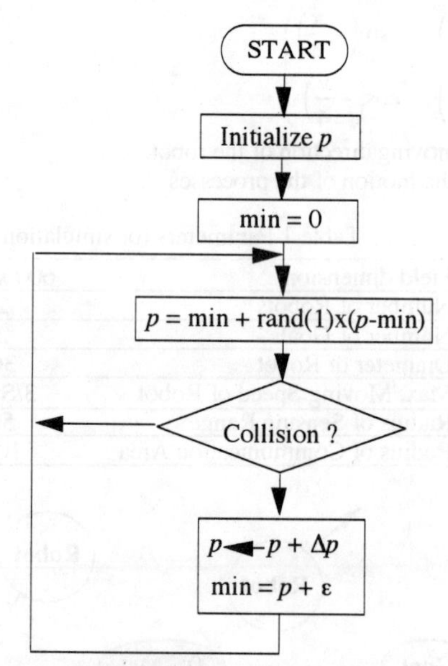

Fig. 11 Flow Chart of Learning Algorithm

6. Simulation Results

6.1 Without priority relation

This section shows a simulation result to compare the following results, where the simulation is carried out without the consideration of priority relation. Figure 12 represents the simulation result, where a robot collides with an object. Figure 13 shows the vectors that are generated by three processes. From the result, the vector of the process 2 (obstacle avoidance) is canceled by the integration of the vectors of the process 1(goal achievement) and the process 3 (turn right). That is, in order to avoid the collision, it is necessary to consider the priority among these processes, e.g., the priority of the process 2 must be increased.

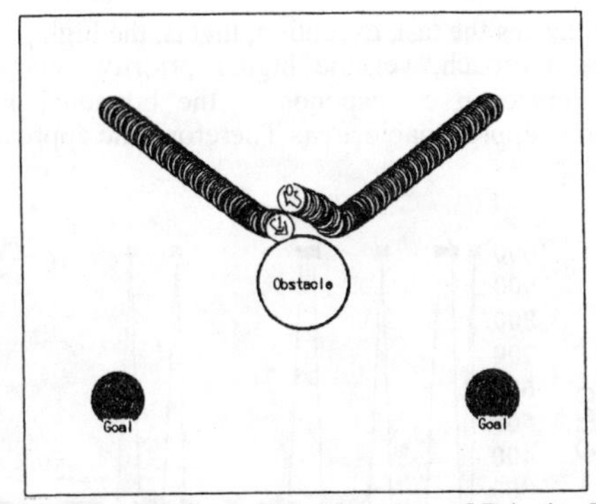

Fig. 12 Simulation Result without Consideration of Priority Relation

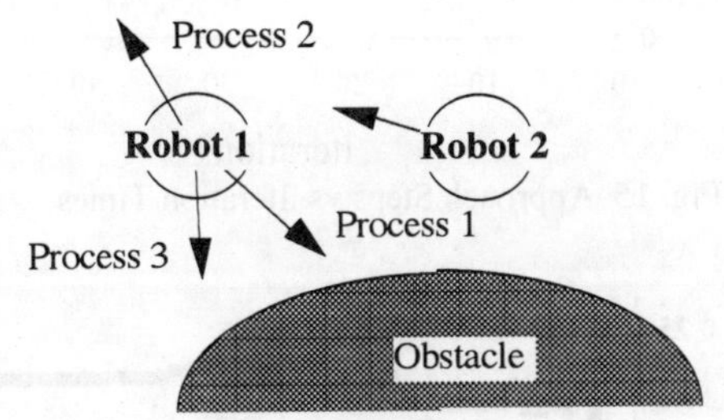

Fig. 13 Vectors of Three Processes

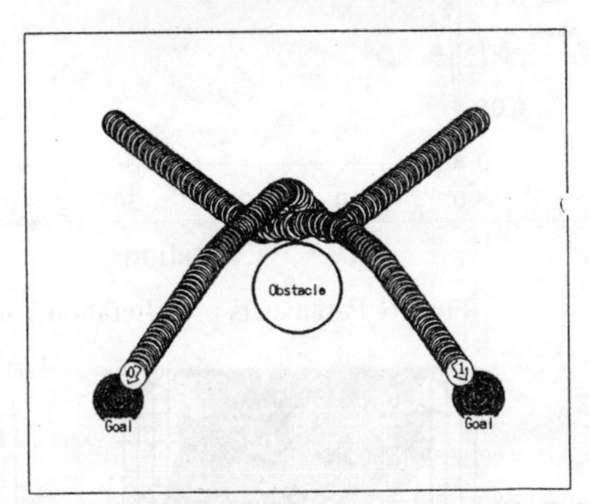

Fig. 14 Simulation Result with Consideration of Priority Relation

6.2 With priority relation and learning ability

(a) In case of a fixed obstacle

Simulation result is shown in Fig. 14, in which two mobile robots move to the goals separately with a fixed obstacle. By comparing it with the result described in previous section, it can be seen that the purpose of this task is carried out successfully. Figure 15 presents the relation between the iteration of the task and the step to reach the goal, where the case that a robot collides is refereed as 999 steps. In addition, Fig. 16 shows the relation between the iteration times and the parameter p. From both Figs. 15 and 16, it can be seen that the convergence of the parameter p and the step is closely related, where the parameter p converges on 0.21.

Figure 17 shows the relation between the parameters $p12$ and $p32$, and the approach steps. The approach steps are categorized into several areas. In the figure, white areas represent a fail case, where the collision between the robots or the obstacle is occurred. Increasing the values of parameters, the step is also enlarged. From this result, we can see that the priority of the process 1 (obstacle avoidance) influences the task execution, that is, the high priority of the process 1 makes it possible to success the approach, yet the higher priority enlarges the approach step. The convergence of the parameter p corresponds to the boundary between the limitation of an approachable area and an unapproachable area. Therefore, the approach step is reduced according to the learning process.

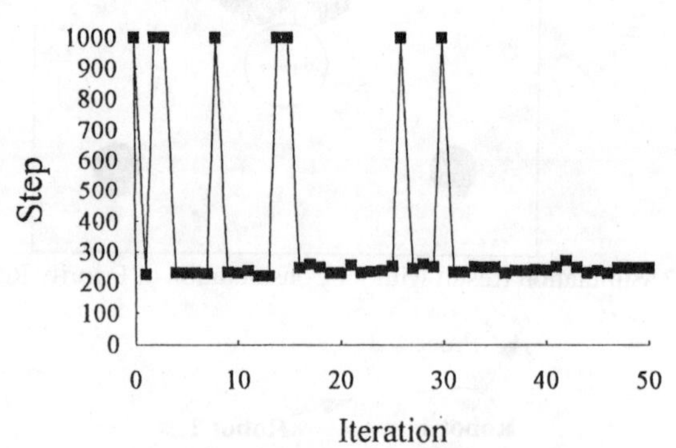

Fig. 15 Approach Steps vs. Iteration Times

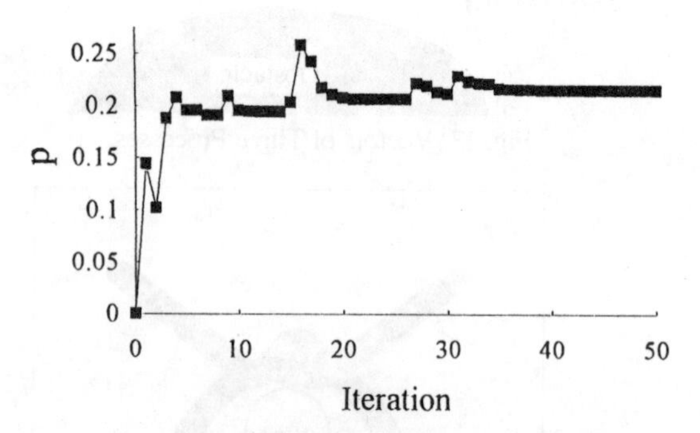

Fig. 16 Parameters p vs. Iteration Times

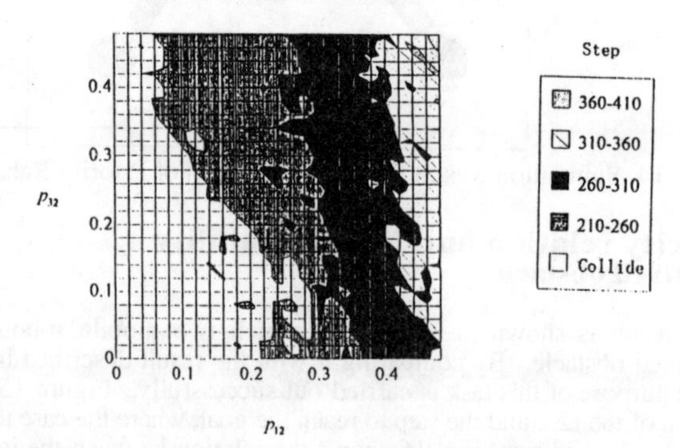

Fig. 17 Approahable Step vs. Combination of Parameters

(b) In case of a moving obstacle

As the second case, this simulation indicates the learning to more complex environment, where an obstacle moves in the field. Therefore, the two robots must avoid the moving obstacle without collision and reach to the goal. As the learning data to the above mentioned parameters, six patterns of a moving obstacle are provided. The six patterns are shown in Fig. 18. The provided learning and adaptation patterns are randomly simulated, because of the generalization of the parameters. Figure 19 presents the relation between the iteration times and the parameter *p*. In this case, the parameter p converges on larger value of the parameter *p* in case of a fixed obstacle. That is, in this case, the process of obstacle avoidance is more important than it in the case of a fixed obstacle. Figure 20 represents a trajectory of the robots and a moving obstacle. Figure 21 indicates the relation between the parameters *p12* and *p32*, and mean approach steps, where the seven patterns are adopted as learning patterns, the six patterns shown in Fig. 20 and a fixed obstacle. By comparison with the results shown in Fig. 17, the case of a moving obstacle decreases the approachable area apparently. In the same way, in the case of a moving obstacle, the learning parameter *p* converges on the boundary between the approachable area and the unapproachable area.

7. Conclusions

As a method to integrate multiple processes for decision making of the behavior of the mobile robots, this paper proposed the priority matrix and the interest relation matrix. These matrices were used to adjust the output of the processes and optimize the behavior of mobile robots. The effectiveness of the proposed matrices and the learning algorithm was indicated in the simulation.

References

[1] T.Fukuda and S.Nakagawa, A Dynamically Reconfigurable Robotic System (Concept of a system and optimal configurations), IECON'87, 588/595, 1987
[2] T.Fukuda, T.Ueyama, Y.Kawauchi, and F.Arai, Concept of Cellular Robotic System (CEBOT) and Basic Strategies for its Realization, Computers Elect. Engng Vol.18, No.1, 11/39, Pergamon Press, 1992
[3] T.Fukuda, and T.Ueyama, Self-Evolutionary Robotic System - Sociobiology and Social Robotics -, Journal of Robotics and Mechatronics Vol. 4, No. 2, 96/103, 1992

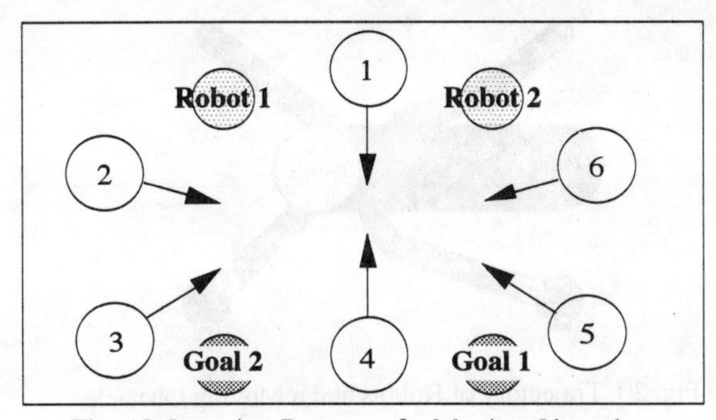

Fig. 18 Learning Patterns of a Moving Obstacle

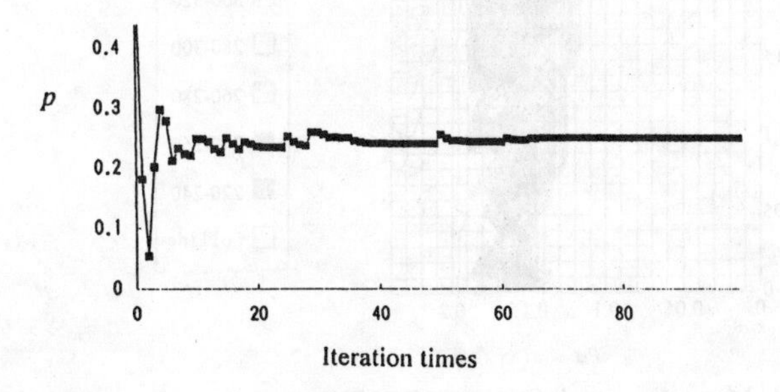

Fig. 19 Parameters p vs. Iteration Times

[4] R.A.Brooks, "A Robust Layered Control System For A Mobile Robot," IEEE Journal of Robotics and Automation, Vol.RA-2, No.1, 14/23, 1986
[5] J.S.Albus, Outline for a theory of Intelligence, IEEE Transactins on Systems, Man, and Cybernetics, Vol.21, No.3, 473/509, 1991
[6] L.E.Parker, Adaptive Action Selection for Cooperative Agent Teams, From Animal to Animat II, The MIT Press, 442/450, 1993
[7] P.Maes and R.A.Brooks, Learning to Coordinate Behaviors, Proc. AAAI-90, 1990
[8] T.R.Collins, R.C.Arkin, and A.W.Henshaw, Integration of Reactive Navigation with a Flexible Parallel Hardware Architecture, Proc. IEEE ICRA, 271/276, 1993

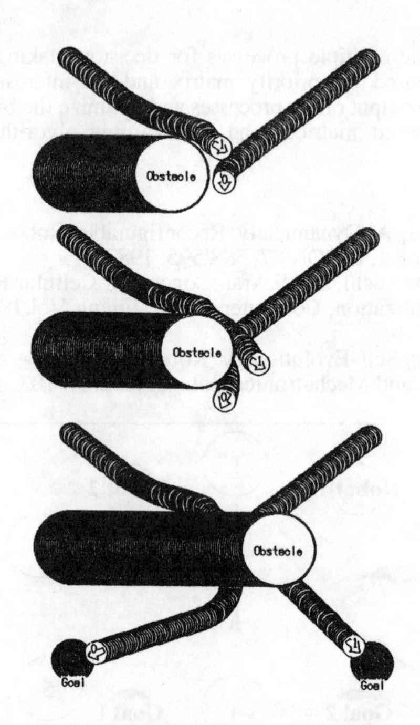

Fig. 20 Trajectory of Robots and a Moving Obstacle

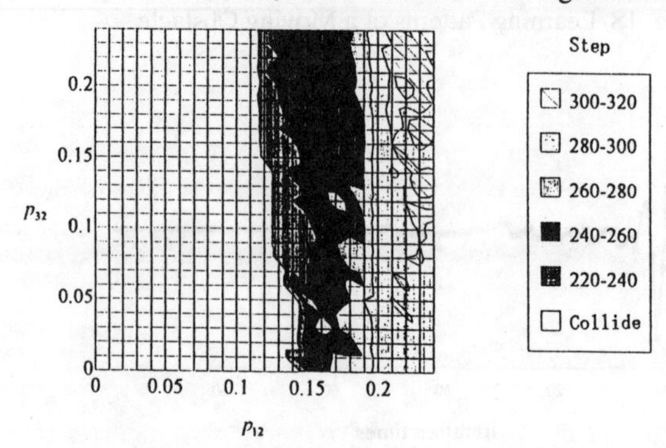

Fig. 21 Mean Approach Steps vs. Combination of Parameters p

Optimization of the Distributed Systems by Autonomous Cooperation
- Distributed Maximum Principle -

KOICHI SHIBAO and YUJI NAKA

Research Laboratory of Resource Utilization, Tokyo Inst. of Tech., Yokohama, 227 Japan

Abstract

This paper proposes a methodology for optimizing the whole distributed systems by their autonomous activities by means of the cooperative work among them. Although the proposed method is developed for the design of a large-scale complex system such as chemical process plants, its wide applicability is expected to realize the optimum collaborative control of the distributed autonomous robotic systems. The method intorduces market mechanism to provide optimizing prices by which individual subsystems make decisions to maximize their profits autonomously so as to achieve the optimization of the total system. The optimizing prices are obtained through the cooperation among subsystems. The completely decentraized and distributed framework of the method provides the concurrent, robust solution for a large scale problem as well as distributed systems.

Keywords:
Optimization of Distributed Systems, Decentralized Optimization, Distributed Maximum Principle, Optimizing Price

1.Introduction

Nowadays, the capacity and the complexity of the industrial production facilities are increasing and market competitiveness forces us to keep optimal operating conditions. Also, we must take into account the robustness of the production technologies, minimization of the communication load in the real time system, and reduction of the computation load to obtain optimal solution for the complicated and gigantic systems.

Most of the optimization problems we practically face is too complicated to be solved all at once, because we must deal with many distributed systems which composes a total system and are inter-related with each other. It was the worst case to try to solve a whole complicated problem all at once that the centralized planning of the national economy previously employed by the Soviet Union. Considering this result, it proved gravely important to reduce the cost and labors for the information processing and communication for planning and optimization .

Compared with the centralized planning, the price mechanism might be regarded as a better paradigm for the whole economic system by decomposing it to the distributed systems such as suppliers and/or buyers. According to the market price, suppliers and buyers autonomously make decisions so as to maximize their profit, which reduces the cost for information processing and communication.

However, the partial optimization based only on the market mechanism can not assure the total optimization of the whole system. The distributed maximum principle(DMP) outlined in this paper provides a mathematical base for optimizing price which leads to the total optimization. Based on such optimizing prices of input and output, individual subsystems can autonomously and concurrently make decisions to maximize their respective profits. By offering such price information, the subsystems make a negotiation with each other over the price and amount of input and output products, so as to find the optimizing prices (Detailed proof of the DMP will be reported at a later date.)

2. Distributed Maximum Principle

2.1 Problem Statement

The subsystems in the complicated systems are interrelated with each other as shown in Fig. 1. For the simplification of the discussion, a transferred variable from one system to another is conceptually devided into two variables: an input variable and an output variable as shown in Fig. 2.
The definition of variables used below is as follows: $X=\{x^i_j\}$,$Y=\{y^k_{j'}\}$ denotes the input and output variable, respectively. S denotes a subsystem, superscript i and k denotes the idendtification number of subsytems, and subscript j of variables denotes jth variable of the subsystem denoted by the superscript.

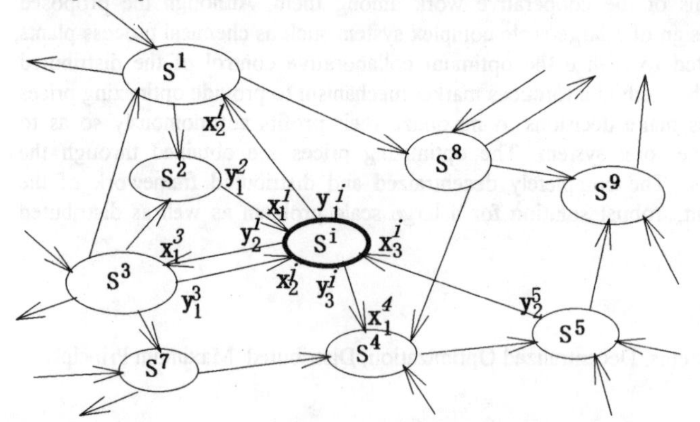

Figure 1 Interrelation Among Distributed Subsystems

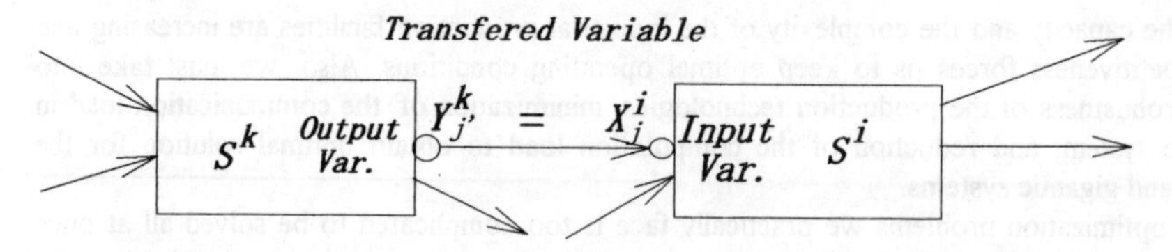

Figure 2 Relation of Input and Output Variables

The distributed optimization problem can be formulated by employing separable objective functions and constraint functions for each subsystem i as shown follows: The summation of each objective function of subsystem corresponds to the objective function of the whole system. The separable problem has been investigated because of its easiness and simplicity of the caluculation of the objective function in each subsystem. For transferred variables between subsystems, corresponding variables should share the same values, $y^k_{j'}=x^i_j$ as shown in Fig.2.

Problem Statement

$$\max_{u_i} imize\ H_0 = \sum_{i=1}^{N} f^i(x^i, y^i, u^i) \tag{2.1}$$

subject to:

$$g^i(x^i, y^i, u^i) \le 0 \quad for\ i = 1, N \tag{2.2}$$

$$h^i(x^i, y^i, u^i) = 0 \quad for\ i = 1, N \tag{2.3}$$

$$x^i{}_j = y^k{}_l \quad i \ne k, \quad for\ \ i = 1, N \tag{2.4}$$

for all corresponding input and output variables

where

$u^i{}_j$ is the j th internal variable in subsystem i.

For each subsystem, only its internal constraints and boundary conditions of the transferred variables are taken into consideration. The constraints among several subsystems are not explicitly handled, empirically it may be preferable the closely inter-related constraint problems to be solved at once as a single problem. Furthermore, by employing the virtual subsystem with any form of constraints, however, arbitrary constraints among subsystems can be handled in the proposed formulation. Sometimes it was more preferable than the formulation using ordinary decomposition principle imposing artificial constraints among subsystems. For example, in place of imposing upper limitation to the available amount of utilities, the proposed method models the virtual subsystems after utility facilities so as to yield more profound information on utility facility design and operation.

In order to develop the methodology, the total system is separated into two subsystem groups : the target subsystem, and all the other subsystems surrounding the target system.

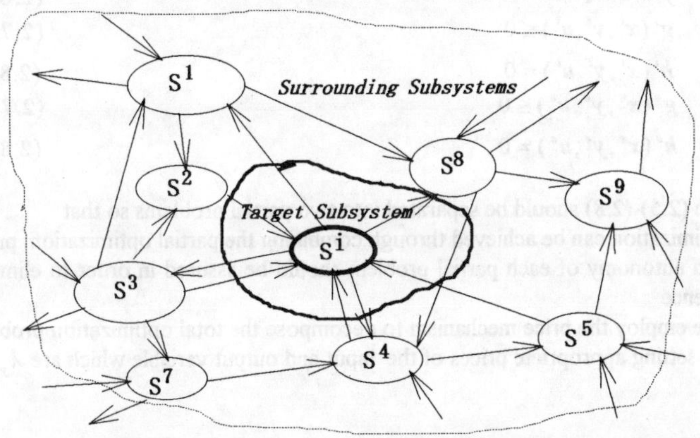

Figure 3 Separation of Target Subsystem and Surrounding Subsystems

Surrounding subsystems could be regarded as a single system around the target system, a surrounding subsystem.

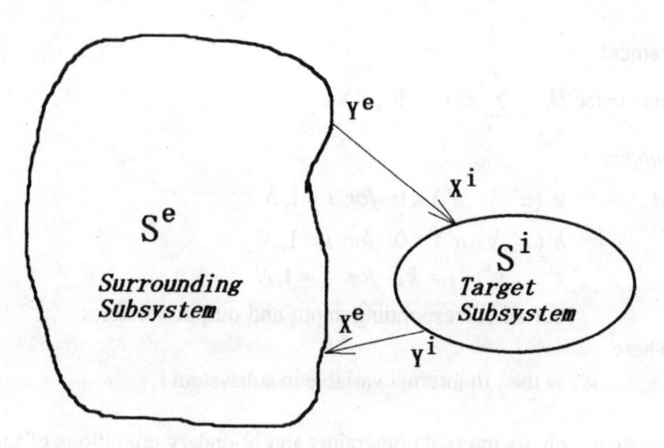

Figure 4 Variable Transfer between Target Subyestem & Surrounding Subsystem

2.2 Formulation of Distributed Autonomous Problems

Let us consider the optimization problem of the total system (2.1)-(2.4) as described below:

$$\max imize\,\{f^i(x^i,y^i,u^i)+f^e(x^e,y^e,u^e)\} \quad (2.5)$$

subject to

$$x^i{}_j = y^e{}_l \quad (2.6a)$$

$$y^i{}_j = x^e{}_k \ . \quad (2.6b)$$

$$g^i(x^i,y^i,u^i)\le 0 \quad (2.7a)$$

$$h^i(x^i,y^i,u^i)=0 \quad (2.8a)$$

$$g^e(x^e,y^e,u^e)\le 0 \quad (2.7b)$$

$$h^e(x^e,y^e,u^e)=0 \quad (2.8b)$$

The original problem (2.5)-(2.8) should be separated into two partial problems so that

1)Total optimization can be achieved through combining the partial optimization problems.

2)Maximum autonomy of each partial problem should be assured in order to eliminate the unnecessary interference.

For the purposes, we employ the price mechanism to decompose the total optimization problem into two subproblems by setting appropriete prices of the input and output variable which are λ_y and λ_y, respectively.

Partial Problem A

$$\max_{x^i,y^i,u^i} imize\,\{f^i(x^i,y^i,u^i)+\lambda^e{}_x y^i(x^i,y^i,u^i)-\lambda^e{}_y x^i(x^i,y^i,u^i)\} \quad (2.9a)$$

suject to

$$g^i(x^i,y^i,u^i)\le 0 \quad (2.7a)$$

$$h^i(x^i,y^i,u^i)=0 \quad (2.8a)$$

Partial Problem B

$$\max_{x^e,y^e,u^e} imize\,\{f^e(x^e,y^e,u^e)+\lambda^i{}_x y^e(x^e,y^e,u^e)-\lambda^i{}_y x^e(x^e,y^e,u^e)\} \quad (2.9b)$$

subject to

$$g^e(x^e, y^e, u^e) \le 0 \tag{2.7b}$$
$$h^e(x^e, y^e, u^e) = 0 \tag{2.8b}$$

Both partial problems are symmetrical, the methodology developed for the target system can be applied to individual subsystems successively.

2.3 Necessary Conditions for Optimality

The first-order necessary conditions for the optimality of the whole distributed systems, the distributed maximum principle(DMP), are discussed in brief. We assume the objective and constraint functions of the subsystem are first order differenciable with respect to the input and output variables. Then, each subsystem sholud satisfy the follwoing conditions to optimize the total system.

1) Constraints of Subsystem

The constraints of the subsystem i are:

$$g^i(x^i, y^i, u^i) \le 0 \tag{2.7a}$$
$$h^i(x^i, y^i, u^i) = 0 \tag{2.8a}$$

We represent the feasible region of variables (x^i, y^i, u^i) satisfying above-mentioned constraints of the subsystem i as $(x^i, y^i, u^i) \in G^i$

2)Equality of Transferred Variables

The transferred variables, i.e., an output variable from one subsystem and a corresponding input to another subsystem, should share the same value.

$$x^i{}_j = y^e{}_l \tag{2.6a}$$
$$y^i{}_j = x^e{}_k \tag{2.6b}$$

3)Partial Optimality Conditions:

If appropriete prices of the inputs $\lambda^e{}_y$ and outputs $\lambda^e{}_x$ are given, the optimum point$(\hat{X}^i, \hat{Y}^i, \hat{U}^i)$ sare obtained from the follwoing conditions.

$$(\hat{X}^i, \hat{Y}^i, \hat{U}^i) = \left\{ (X^i, Y^i, U^i) \middle| \max_{(x^i, y^i, u^i) \in G^i} imize\,(f^i + \lambda^e{}_x Y^i - \lambda^e{}_y X^i) \right\}. \tag{2.10}$$

We represent the partiall optimal region of variables (x^i, y^i, u^i) satisfying the partial optimality above-mentioned under the given prices and partial constraint condițions of the subsystem i as :

$$(x^i, y^i, u^i) \in O^i \ (\subset G^i)$$

4) Optimizing Price Conditions

The partial optimization could result in the total optimization only if the prices of the input and output variables introduced rightly indicate the changes of the added value of the objective of the surrounding system caused by the change of the amount of input and output variables. As the change of variables is to be always varied satisfying constraint conditions and maximizing the objective, such change should satisfy the partial optimality condistions.

a.Marginal price $\lambda^i{}_{x_j}$ of the input variable $x^i{}_j$ of the subsystem i

$$\lambda^i{}_{x_j} = \{\frac{d}{dx^i{}_j}(f^i + \lambda^e{}_x Y^i - \lambda^e{}_y{}' X^i{}')\big|(x^i, y^i, u^i) \in O^i\} \qquad (2.11a)$$

b.Marginal cost $\lambda^i{}_{y_j}$ of the output varible $y^i{}_j$ of the subsystem i

$$\lambda^i{}_{y_j} = \{-\frac{d}{dy^i{}_j}(f^i + \lambda^e{}_x{}' Y^i{}' - \lambda^e{}_y X^i)\big|(x^i, y^i, u^i) \in O^i\} \qquad (2.11b)$$

where,

$\lambda^e{}_x$:Marginal price vector of the input variables of the external subsystem corresponding to the output variables from subystem i (y^i).

$\lambda^e{}_y$:Marginal cost vector of the output variables of the external subsystems corresponding to the input variables to the subsystem i.(x^i).

$X^i{}'$:Input variables to subsystem i except $x^i{}_j$

$Y^i{}'$:Output variables from subsystem i except $y^i{}_j$

$\lambda^e{}_x{}'$Marginal price vector of the input variables of the external subsytems corresponding to $X^i{}'$

$\lambda^e{}_y{}'$:Marginal cost vector of the output variables of the external subsystems corresponding to $Y^i{}'$

5) Total Optimality Conditions

Let us assume the problem is totally optimized and the marginal price of the input variable in the target system is higher than the marginal cost of the corresponding output variable of the external system,$\lambda_x{}^i > \lambda_y{}^e$. If such variable is not bounded by the constraints, we can increase the objective value of the total system by increasing the input variable of the target system $\delta y^e (= \delta x)$ or the corresponding output variable of the surrounding system as follwos: Total increase of the objective is $\delta o = \{\delta f^i + \delta f^e\} = (\lambda_x{}^i - \lambda_y{}^e)\delta x$, where δx are the change of value of the total objective and the input variable respectively.

In the case of $\lambda_x{}^e \geq \lambda_y{}^i$, similarly we can increase the objective value of the total system by increasing the value of the input variable of the surrounding system or the corresponding output variable of. the target system.

As this contaradicts the the assumption mentioned above, we obatain the total optimality conditions of the marginal price and cost of a transferred variable in a subsystem and in its surrounding systems as follows:

(1) if a transferred variable is bounded by the upper limit, $\lambda_x \geq \lambda_y$ $\quad (2.12a)$

(2) if a transferred variable is bounded by the lower limit, $\lambda_x \leq \lambda_y$ $\quad (2.12b)$

(3) if a transferred variable is not bounded by the limits, $\lambda_x = \lambda_y$ $\quad (2.12c)$

where,

λ_x:Marginal price of an input variable in one subsystem which is either the subsystem i or the external subsystem.

λ_y:Marginal cost of the output variable in another subsystem corresponding to the input variable of the subsystem above.

3. Numerical Example

We illustrate the principle of the above-mentioned method by solving the following example.

[Example 1]

The plant A produces intermediate raw material(Y1=X2) which is supplied to the down stream plant B. Using the intermediate raw material, Plant B produces a final product(Y2) which is sold to the market as shown in figure 5. The superscript 1 and 2 of variables correspond to plant A and plant B respectively.

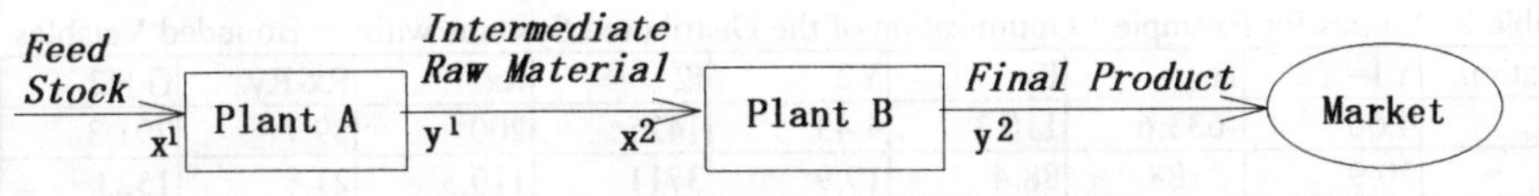

Figure 5 Configuration of Subsystems

The price(P) of the final product varies according to the sale volume(Y2). The relationship between P and Y2 is called 'price-elasticity'. The Y2 should be optimized to maximize the profit of the total systems composed of plant A and plant B. The prblem is formulated as follows.

Problem Statement

$$\underset{y1=x2}{\max imize}\{f^1+f^2\}$$

Plant A

$$f^1=-c_1x^1-c_2(y^1)^{0.7}$$

subject to

$$x^1=a_1y^1$$

PlantB

$$f^2=py^2-c_3y^2-c_4(y^2)^{0.7}$$

subject to

$$x^2=a_2y^2$$

$$p=p_0(y^2)^{-0.3}$$

The values of the constants are shown in table 1

Table1 Values of the Constants

Constants	c1	c2	c3	c4	a1	a2	Po
Values	16.6666	136.617	50	102.463	3.0	1.05	683.086

Accordingly, with the proposed method the total optimization problem is separated into the following partial(local) optimization problems for individual plants to optimize autonomously.

for Plant A maximize $\{H^1=R_xy^1-c_1x^1-c_2(y^1)^{0.7}\}$ s.t $x^1=a_1y^1$

for Plant B maximize $\{H^2=py^2-c_3y^2-c_4(y^2)^{0.7}-R_yx^2\}$ s.t $x^2=a_2y^2, p=p_0(y^2)^{-0.3}$

Optimizing prices of the intermdeiate raw material ,i.e, Ry and Rx, are evaluated as:

$$Ry=\frac{d}{dy^1}\{c_1x^1+c_2(y^1)^{0.7}\big|x^1=a_1y^1\}$$

$$Rx=\frac{d}{dx^2}\{py^2-c_3y^2-c_4(y^2)^{0.7}\big|\,x^2=a_2y^2, p=p_0(y^2)^{-0.3}\}$$

Figure 6 shows the optimization process in line with the object-oriented paradigm which describes the computation module as the plant object.

Case1

Table 2 shows the result of optimization process without bounded variables under the initial assumption of the marginal cost of the intermediate raw material(Ry=Ry=200). Plant B optimize its size(sale volume:Y2=4.43) by taking into account the price-elasticity. Then plant B calculates the respective amount(X2=4.66) and the marginal price(Rx=200) of intermediate raw mateiral and convey them to the plant A. As X2 is not bounded, the calculated marginal price Rx is same as the previous Ry.

Table 2 Values for Example -Optimization of the Distributed Systems without Bounded Vaiables

iteration	Y1=X2	f1	Ry	Y2	f2	Rx	Rx-Ry	f1+f2
1	4.66	-633.6	110.3	4.43	1425	200	89.7	791.3
2	20.9	-2188	88.4	19.9	3711	110.3	21.8	1523
3	34.3	-3333	83.1	32.6	5027	88.4	5.3	1693
8	40.9	-3878	81.4	38.9	5588	81.4	0.0045	1710

The plant A designs itself based on the received Y1(=X2=4.66), Rx(=200), and calculates the maginal cost (Ry=110.3), evaluates the convergence of the solution by comparing previous and present values of the marginal prices(Ry),Y1 values. The optimality of the total systems is made assure by comparing the margianl cost(Ry) and the marginal price(Rx) received. If the plants A and B are found to be converged and totally optimized, the computation finishes. Otherwise, Ry is transferred to the plant B which repeats the computation process.

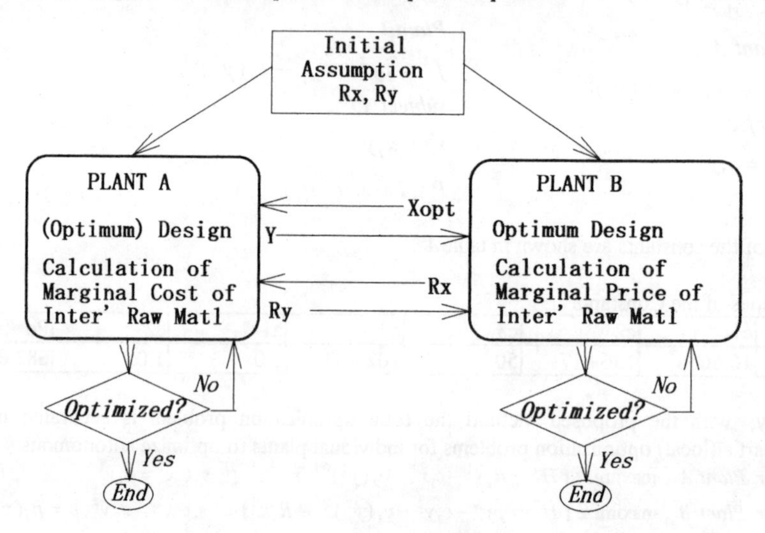

Figure 6 The Problem Solution Process

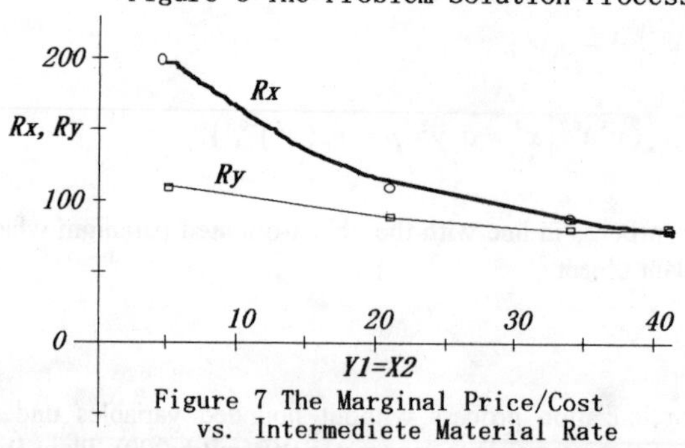

Figure 7 The Marginal Price/Cost vs. Intermediete Material Rate

The plant B continues to optimize itself and calculates X2 and Rx and evaluates its convergence and optimality in the same manner witch the plant A takes. If all plants involved in the problem are found to be converged and to be totally optimized, the computation ends. If X or Y is not bounded, Rx and Ry successively takes the same value due to the total optimality condition(2).

Case 2

The next example shows the case there are bounded variables ,20=<Y2=<30, in Table 3.

Based on the initial assumption of Ry=200, plant B calculates its size(sale volume:Y2=4.43) by optimization. However, Y2 is less than lower limit 20, so Y2 is set at 20. Then plant B calculates the respective amount(X2=21) and the marginal price(Rx=110.) of the intermediate raw mateiral and sends them to plant A. The plant A designs itself based on the receieved Y1(=X2=21), Rx(=110.), and calculates the maginal cost (Ry=88.4). As the marginal price Rx is higher than the marginal cost Ry, optimum Y2 is larger than the lower limit. So the plant A sends Ry to the plant B to recursively optimize the whole system as well as plant B. In 2nd and 3rd iteration, in spite of the convergence of Y1(=X2=84) still Rx(91.9) is higher than Ry(84.). Accordingly Y1 is found to be bounded by the upper limit. When X2 or Y1 is directly or indirectly bounded, Ry differs from Rx.

Table3 Values for Example -Optimization of the Distributed Systems with Bounded Variables

iteration	Y1=X2	f1	Ry	Y2	f2	Rx	Rx-Ry	f1+f2
1	21	-2200	88.4	20	3727	110.0	21.6	1526
2	31.5	-3104	84.0	30	4779	91.9	7.9	1675
3	31.5	-3104	84.0	30	4779	91.9	7.9	1675

4. Discussions

4.1 Characteristics of the Distributed Maximum Principle

1) Comparison with the Discrete Maximum Principle

The multistage decision making process may be regarded as the decision making of different subsystems because the multistage decision optimization is similar to the distributed system optimization. Let us we compare the distributed maximum principle with the discrete maximum princple in order to clarify the features of the proposed DMP.
The discrete maximum principle [2][3][4] was developed so as to apply Pontryagin's maximum principle[1] to the optimization of multistage decision process. The typical formulation of the problems is as follows:

$$\max_u imize \sum_{j=1}^{N} x_0{}^j = \max_u imize \sum_{j=1}^{N} T_0^j(x^{j-1}, u^j)$$

subject to

$$x_i{}^j = T_i^j(x^{j-1}, u^j) \quad for\ i = 0,..s \quad j = 1, N$$

where
$x_i{}^j$:state variables
u^j:decision variables
$T_i^j(x^{j-1}, u^j)$:transformation operator

The problem can be solved by using Hamiltonian functions and auxiliary(adjoint) variables as follows:

$$H^j = \sum_{i=0}^{s} \lambda_i{}^j x_i{}^j$$

$$\lambda_i^{j-1} = \frac{\partial H^j}{\partial x_i^{j-1}}$$

$$u^j = \{u^j \mid \max_{u^j} imize\ H^j\}$$

where
H^j:hamiltonian functions

$\lambda_i{}^j$:vectors of auxiliary(adjoint) variables

With the discrete maximum principle, a state variable(taransfered variable) has only one auxiliary variable to express the influence of the successive decision stage with respect to the change of the input state variable ,i.e., the marginal price of the input variable. Respective flow directions of the material and the price information(auxiliary variable) always takes one way as shown in Fig 8.

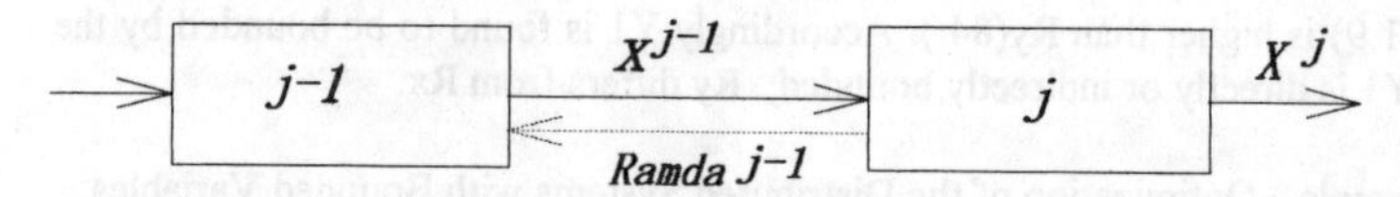

Figure 8 Flow of Transfered Variable and Its Auxiliary Variable in Discrete Maximum Principle

On the contrary, the DMP provides a transferred variable(a state variable) with two auxiliary variables, i.e., the marginal price,Rx , and the marginal cost,Ry, as shown in Fig 9.

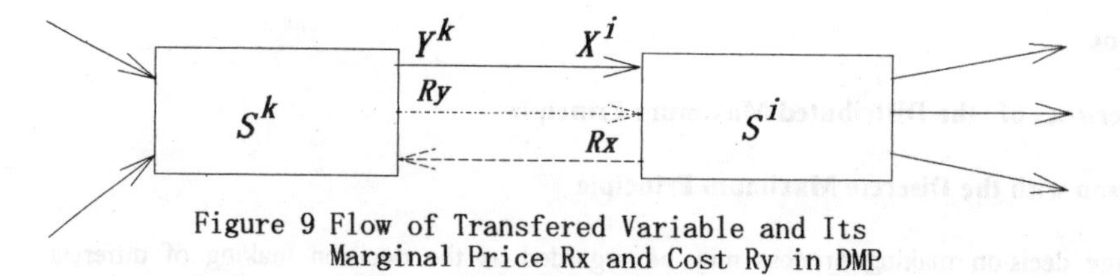

Figure 9 Flow of Transfered Variable and Its Marginal Price Rx and Cost Ry in DMP

The optimization process of DMP can be represented as the cooperative task of multi-agents : a supplier agent proposes the minimum unit cost(marginal cost) of his output to a buyer agent who will decide the amount of the transferred variable(intermediate resource) to be purchased so as to maximize its profit. It is noteworthy in the MDP framework that the buyer agent also can proposes the maximum unit price(marginal price) of the product to the supplier agent which will decide the amount to sell so as to maximize its profit. During such negotiation among multi-agents, an agent can locally determine whether it takes optimum or not through comparing the marginal cost and cost to investigate the total optimal contitions. If every agent satisfys its total optimal conditions, the whole system is optimized.
As the discrete maximum principle provides only one anxiliary variable(unit price or unit cost), each agent can not localy assure its total optimality. The discrete maximum principle deal with only the case of $\lambda_x = \lambda_y$ of the total optimal conditions stated in equation (2.12c)

2) Characteristics of the DMP

The DMP employing two types of the auxiliary (adjoint) variables, has several advantageous features mentioned below and might be useful for the distributed autonomous robotic systems.

1) Completely Decentralized and Distributed Optimization Framework.
Historically, large-scale problems have been solved by hierachical decomposition methods employing the supervisory system and subsystems (Dantzig and Wolfe, etc. [5][6][7][8]). Instead, the DMP addressed with mathematical foundations to optimize total systems in completely decentralaized formulation, which does not require any control mechanism such as master problem, planning bureaucratic center, supervisory organization, etc. However, if such hierarchical architecture is found to be more efficient to solve some types of problems, the DMP solves the problems by formulating an arbitural hierachical subsystem structure .

2) Reduction of Computaion and Communication Load.
In general, the partial optimization is not so difficult to be calculated, but the total optimization of large-scale systems, in a practical sense, is rather impossible to be solved with sufficient precision and accuracy. It is noteworthy that the proposed method makes it possible to achieve the total optimization by adding a small amount of task, i.e, evaluation of the marginal cost and price to the partial optimization task. individual subsystems requires a very small amount of information regarding the price and the quantity of input and output variables to be communicated with each other among subsystems.

3) Robust, Concurrent and Parallel Optimization
Every subsystem is able to concurrently solve its partial problem in parallel. In addition to the total optimality conditions, the marginal price and cost empirically tend to take nearly the similar values with respect to the given systems as shown in table 2 of case 1. Accordingly, each subsystem can expect to achieve total optimization by considering only interaction among adjacent subsystems. The localizability of the total optimization discussed above makes it possible to apply concurrent and parallel optimizations which requires robustness. Especially, in consideration of the practical problem being faced with ever changing surrounding conditions, the characteristics of the proposed method mentioned above is advantageous.

4) Harmonization
The proposed method does not intend to compete with or supersede the existing optimizing method . All types of optimization methods such as LP, SQP, ILP,..etc., even a rule of thumb, manual studies conducted on desks can be employed for solving partial problems in subsystems. However, additional function to evaluate the marginal price and cost of the input and output variables is needed. As the amount of information to be communicated with is small, the integration of the different type of tasks is possible without so much difficulty.

4.2 Application to the Dsitributed Autonomous Robotic Systems

1) Autonomous, Robust and Parallel Behaviors
The completely decentralized and distributed architecture might be strongly required for the distributed autonomous robotic applications. Due to the localizability of the optimization, the DMP can be applicable to the distributed robotic systems in which each robot is expected to behave independently and autonomously so as to maximize the shared objectives. The DMP does not seem suitable for the robotics whose process sequence is linked strictly with the other robotic systems. Autonomous and rapid behavior of the possible distributed robots realized by applying DMP might be advantageous, for example, to maintain their safety as well as stable posture at the dangerous situation.

2) Self-Organization
Self-organization of the cellular robotics is one of the most interesting field. Fukuda et. al [10] attempted to optimize the allocation of the local knowledge to the distributed subsystems so as to equalize the communication frequency of each subsystem. In general, most of self-organizing problems can be formulated employing the concept of "optimality" in place of the adhoc criteria in terms of the multi-agents framework. Although the DMP does not directly generate the process sequence for the robotic systems, it provides quantiative information to autonomously achieve the shared goal for the distributed autonomous robotic systems. By using such qantitative information, the distributed robotic systems are expected to be optimally self- organaized.

3) Possibility of Process Sequence Generation
In general the DMP is mathematically applicable mainly to the parametric maximization. As we do

not need any restriction on the type of the internal variables, individual subsystem is able to optimize itself systematically as well as parametically. Therefore, there is a possibility that the flexibility of the of the DMP enables it to generate process sequence, or to apply to system optimization(system synthesis) in some fields, although we cannot identify in which fields the proposed method may be applicable at present.

5. Conclusions

This paper has presented a completely decenterized methodology for the optimization of the whole distributed system. To optimize the whole system, the separated subsystems should cooperate each other to find the appropite price of the input/output products. The DMP proposed such optimizing price of the input and output products of subsystems, i.e, the marginal price and cost . The use of two kinds of such adjoint variables makes it possible to localize the evalutation of the total optimality of the target system to a comparatively large extent. This leads to the possibilty of concurrent and parallel optimizations with robustness.

Due to the indirect control based on the price mechanism, especially the localizability of the DMP assures maximum autonomy of the subsystems. Therefore, if the proposed method is used for the control of distributed autonomous robotic systems, the following characteristics are expected: Maximum robustness of the individual subsystems, minimum requirements for the communication among member systems, the simplicity of the modeling and its maintenance for the subsystems.

Acknowledgement

We wish to express our thanks to Dr.H.Asama of RIKEN, who kindly encouraged and helped us to prepare this paper.

References

[1],L.S.,V.G. Boltyanskii,R.V.Gamkrelidze,and E.F. Mishchenko, The Mathematical Theory of Optimal Processes, English translation by K.N.Trinogoff, Interscience, New York, 1962
[2]Liang-Tseng Fan,Chiu-Sen Wang, The Discrete Maximum Principle, John Wiley & Sons, Inc., 1964
[3]Chang,S.S.L., Digitized Maximum Principle,Proc.,IRE,Dec 1960,pp.2030-2031.
[4]Katz,S.,A Discrete Version of Pontryagin's Maximum Principle, J.Electronics and Control, 13, 179, 1962.
[5]G.B.Dantzig, Linear Programming and Extensions,Princeton University Press, 1963
[6]G.B.Dantzig, P.Wolf, Decomposition Principle for Linear Programs, Opns. Res. Vol.8, No.1, 1960
[7]L.S.Ladson,Optimization Theory for Large Systems, MacMillan, New York, 1970
[8]Mesarovic,MD., D.Macko, Y.Takahara, Theory of Hierachical, Multilevel Systems, Academic Press, New York, 1970
[9] Kuhn,H.W.,A.W.Tucker, Nonlinear Programming, Proc.2nd Berkley Symposium on Mthematical Statistics and Probability,J.Neyman(ed),University of California Press,1959.
[10] T.Fukuda,Y.Kawauchi,H.Asama, Analysis and Evaluation of Cellular Robotics(CEBOT),IEEE Int.Workshop on Intelligent Robotics and Systems,1990

Fault-Tolerance and Error Recovery in an Autonomous Robot with Distributed Controlled Components

TIM C. LUETH and THOMAS LAENGLE

Inst. for Real-Time Computer Systems and Robotics • IPR, Univ. of Karlsruhe,
D-76128 Karlsruhe, Germany

Abstract

Most of the existing autonomous robot systems have a centralized hierarchical control architecture. In such robot systems, all planning, execution control, and monitoring tasks are performed by a single control unit on a defined level. In case of an error that occurs during the execution, this central control unit has the complete knowledge about the past executed actions and is able to reason on the error situation. Besides the centralized control architectures, distributed and decentralized control architectures have been developed to overcome some problems with the centralized systems. Because of the missing overall control, error recovery is more difficult than in centralized systems. This paper presents concepts to obtain fault-tolerance behaviour and error recovery in a distributed controlled robot system. As an example for such a robot system, the Karlsruhe Autonomous Mobile Robot KAMRO is considered that is being developed at IPR. Many experiments were performed with the former centralized control architecture. Our intention is to achieve the same and better results with the distributed control architecture KAMARA.

Keywords: Distributed controlled robot systems, autonomous robots, fault-tolerance, error recovery, reliability

1 Introduction

In contrast to hierarchical and centralized controlled robot systems [1], which have advantages at the initial system design process, decentralized and distributed controlled robot systems reveal their main advantages when it is necessary to enhance the system, to integrate components, and to maintain the system. For this reason, most of the following properties are well-known in the area of computer architectures:

- *Modularity:* By having a predefined framework for information exchange processes, it is possible to develop and test single parts of the system independently (e.g. the drive and navigation module for a mobile platform).
- *Decentralized knowledge bases:* Each subsystem of the overall system is allowed to use a local knowledge base, that stores relevant information for the subsystem, and is capable exchanging this information with other subsystems if required (e.g., for mobile manipulators, information about obstacles on the platform's path is not important for the manipulator itself).
- *Fault-tolerance and redundancy:* If system-inherent redundancy exists, this redundancy should be usable without any error model, in case of a broken down subsystem or another error situation (e.g., if one of several vehicles in a transportation system is damaged, nothing other should happen than slower task termination).

- *Integrebility:* Without any change in the control architecture, cooperation of subsystems is possible and all synergetic effects can be used (e.g., any kinematic chain in a multi-manipulator system or support of a manipulator by an existing mobile platform).
- *Extendibility:* New system components can be added to the original system and any improvements (such as reduced task completion time) come about without any change in the system architecture. New components inform other components about their existence and capabilities (e.g., extension of a sensor system).

Up to now, different robot systems like the CEBOT [2] have been designed to achieve this properties. On the other hand, the control of existing robot systems has been improved by using these concepts in a special communication network ACTRESS [3] among several robot system. An other idea is to improve the behaviour of individual manipulators by this methods [4].
Although the overall behaviour of distributed or decentralized controlled systems is assumed as more powerful as the behaviour of a centralized control system, some aspect like error recovery seem to be more difficult. For a serious estimation of the capabilities of distributed control architectures concerning this topic, a comparison with a centralized system can be helpful.

2 The KAMRO as Testbed

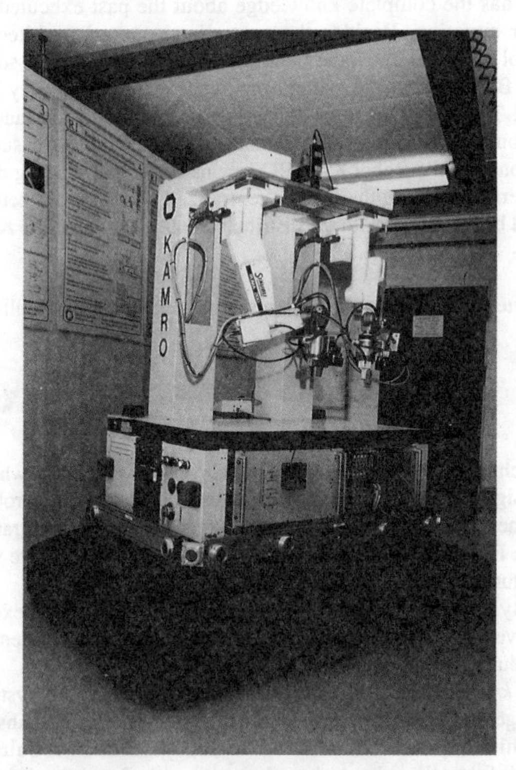

Fig. 1: The KAMRO System

The Karlsruhe Autonomous Mobile Robot KAMRO is being developed since 1985. In this project an autonomous mobile two-arm robot is being built, the task of which is to take over in an industrial manufacturing setting in areas of transportation, fabrication, maintenance, and process control. The robot consists of a mobile platform with an omnidirectional drive (3 DoF) on which there are mounted two PUMA-type manipulators (6 DoF) in a hanging configuration (Fig. 1).

Further mounted on this platform are several different sensor systems for CCD-cameras, ultrasonic range sensors, and force-torque sensors. The robot is controlled by a hierarchical and function oriented planning and control architecture and is capable of executing assembly tasks. In order to do this task it is necessary to generate an assembly precedence graph by means of an off line cell control system [5, 6] describing the sequence of an assembly. The assembly precedence graph is then translated into a condition event Petri net (CEP) [7] and transmitted to the KAMRO-Robot. After that the robot is ready to approach a workstation containing the assembly parts. Having determined the assembly part position by means of the overhead vision system, the plan execution system generates then by interpreting the Petri net (implicit task specification) a sequence of elementary robot operations (EEO) which in turn is being translated into reactive manipulator movements by the robot control system [8].
Experiments in 1990 have shown [9] that the process of interpreting an off line generated assembly precedence graph and of its situation dependent execution is principally suited to solve the task of robot assembly. However, the analyses of these experiments have also shown that it became absolutely necessary to improve the reliability of the task related performance of the robot and to expand on its manipulative capabilities. The different improvements of the control architecture to achieve this goal are described in [10]. On the other hand, a new distributed control architecture is in preparation for this robot system.

3 Control Architectures

The centralised control architecture FATE [8] consists of a blackboard planning level that generates situation-dependent manipulator-specific elementary operations. The real-time robot control system RT-RCS executes the elementary operations. The real-time controller is able to control the manipulators independently or in a closed kinematic chain. Therefore, the overall system can be described as a centralized execution architecture with distributed action selection (Fig. 2a).

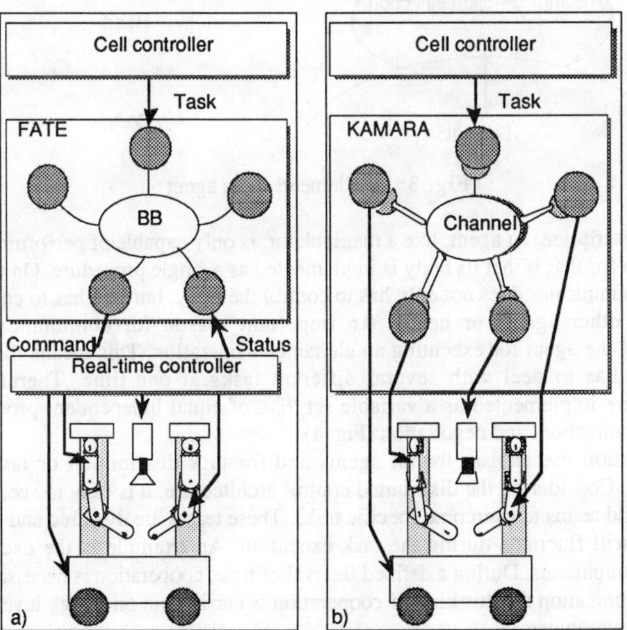

Fig. 2: a) Centralized control architecture FATE and b) distributed control architecture KAMARA

For the independent movement of the manipulators and for the kinematic chain, two different kinds of elementary operations are necessary. It was already shown that this control architecture is principally suitable for solving the problem of autonomous assembly by robots. On the other hand, many difficulties have arisen by the extending of the system with miniature hand-eye cameras or through the integration of a mobile platform and manipulators for mobile manipulation. Similarly, the automatic execution (replacing) of a damaged manipulator's task through a cooperation of the platform and the functioning manipulator is only realizable by completely redesigning the centralized control architecture. These difficulties are avoidable by implementing a new architecture, that does not only have one executive agent (as FATE uses RT-RCS), but has multiple agents like the image processing system, the two manipulators, and the mobile platform (Fig. 2b). These agents have to communicate and negotiate which each other to collect the missing information, that is required for autonomous assembly, and for performing the desired assembly. The new control architecture, KAMARA (KAMRO's Multi-Agent Robot Architecture), for distributed intelligent robot systems and their components allows easier control in many directions and also easier component integration. Different types of cooperation for coupled agents, like closed kinematic chains or camera-manipulator coupling, are also considered in this architecture.
An agent consists of three parts (Fig. 3): *communicator*, *head* (for planning and action selection), and *body* (for action execution). The communicator connects the head to other agents on the same communication level or higher. The head is responsible for the action selection process for the body (centralized approach), organizes the communication among the body's agents actively or passively (distributed approach), or is only a frame to find a principle of order for the decentralized approach. The body itself consists of one or more executive components, which can be similarly considered as agents.

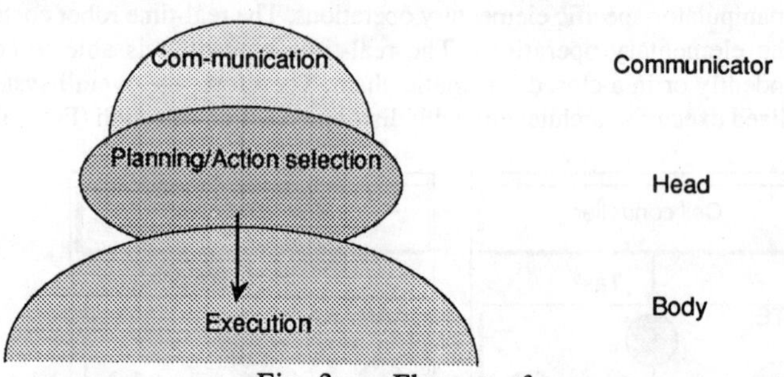

Fig. 3: Elements of an agent

In our system description, an agent, like a manipulator, is only capable of performing one task at a time. The reason for that is that its body is implemented as a single procedure. On the other side, a head with a communicator does not only has to control the body, but also has to communicate and negotiate with other agents or heads. An important reason for communication is to the determination of the agent for executing an elementary operation. This means the head (and the communicator) has to deal with several different tasks at one time. Therefore, head and communicator are implemented as a variable set $\mathcal{H}$, $\mathcal{C}$ of equal independent processes H, C for planning, communication, and negotiation (Fig. 4)
The communication mechanism for all agents and for task distribution or task allocation is Blackboard-like. Considering the distributed control architecture, it is easy to see, that the agents often have to build teams to overcome specific tasks. These teams are dynamic and the number and kind of agents will fluctuate during the task execution. An example is the exchange of parts between both manipulators. During a defined interval of time, cooperation is necessary to reach this goal. The communication for this kind of cooperation is carried out on a high level of abstraction by the agent's communicator.

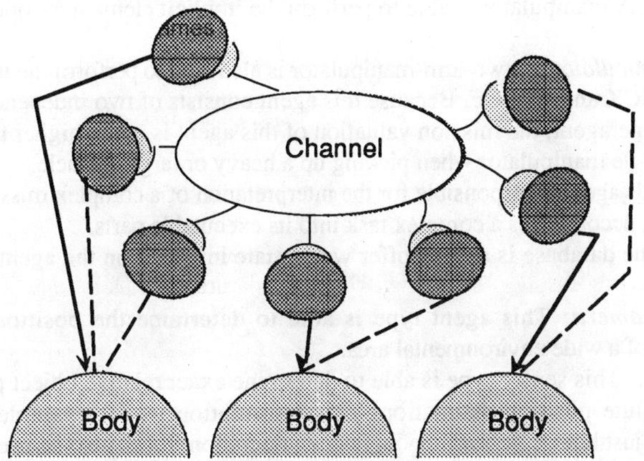

Fig. 4: Head and communicator can be several processes

A different situation arises if two manipulators grasp a large or heavy part and close a kinematic chain. In this case, depending on the desired control concept for the kinematic chain, a decentralized architecture (simple reflexive behavior), a distributed architecture (master slave tasks), or a centralized architecture is required. In some cases, for example complex two-arm manipulation tasks, a centralized robot controller is better than any other approaches at the moment. This is the reason for an extension of the distributed control concept by the introduction of special executive agents. These special agents *SA* have, like all other agents, a head $\mathcal{H}$ and a communicator C. The body is allowed to allocate bodies of other agents, if available, and control them by special communication channels with high transfer rates. During this time the „normal" agents have no access to their bodies, used by „special agents". Because special agents change the structure of the control architecture while they are active, they should be used only if no other type of cooperation is suitable.

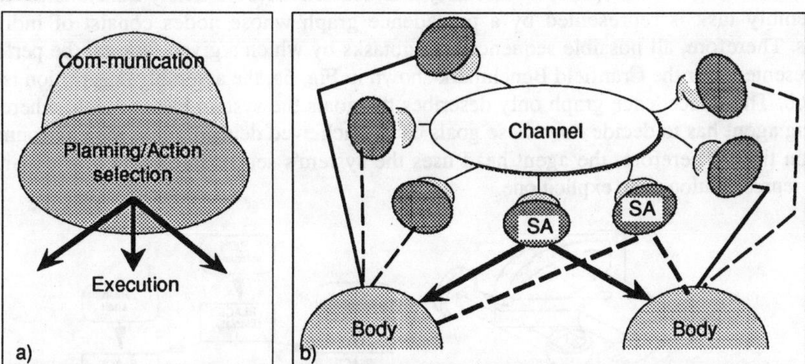

Fig. 5: a, b) Special agent: centralized planning for other agents' bodies

4 Agents and Tasks

In the KAMARA system, there exist several agents that work together to perform the desired task. Consequently, a communication protocol between the agents is required. This language consists of operations that address an agent to perform a task and could be used by other agents to involve other agents in the solution process. In KAMARA, the following system components can perform the following operations:

- *Manipulator:* A manipulator is able to perform the implicit elementary operations PICK and PLACE.
- *Two-arm-manipulator:* A two-arm-manipulator is also able to perform the implicit elementary operations PICK and PLACE. Because this agent consists of two independent actuators that make up a superagent, the mission valuation of this agent is much higher than the calculated value of a single manipulator when picking up a heavy or large obstacle.
- *Manager:* This agent is responsible for the interpretation of a complex mission the system has to perform. It decomposes a complex task into its executable parts.
- *Database:* The database is able to offer world state information the agents need to perform their tasks.
- *Overhead camera:* This agent type is able to determine the position of obstacles by examination of a wide environmental area.
- *Hand camera:* This sensor type is able to determine exact relative object positions based on inexact absolute position estimation. This information is, for example, necessary for a manipulator, just before performing a grasping operation. It can also be used to extract object positions like the overhead camera.
- *State controller:* This agent is responsible for the blackboard structure and the state of the other system components. One important task is to control the time a mission waits on the blackboard for execution. If the time stamp increases above a given threshold, the state controller can search the protocols to determin whether a system component has recently performed a similar mission. If so, this component may be damaged, is overloaded with tasks or is blocked. Thus, the mission could be given to the cell controller, so another system independent of KAMARA's control can perform the task. The state controller also controls system component evaluation: if a system component estimates its ability to perform a mission as very high, and execution of the mission often fails, the state controller is able to inform the corresponding agent, and this way reduces its evaluation coefficient to zero.

The communication and negotiation concept between agents to perform a mission will now be demonstrated using the example of an assembly task, the Cranfield Assembly Benchmark.
An assembly task is represented by a precedence graph whose nodes consist of individual subtasks. Therefore, all possible sequences of subtasks by which a given task can be performed are represented. For the Cranfield Benchmark shown in Fig. 6a, the assembly description is given in Fig. 6b. This precedence graph only describes the goals the system has to reach, whereas the executing agent has to decide how these goals can be achieved depending on the environment at execution time. Therefore, the agent head uses the system's sensor information to expand this implicit representation to an explicit one.

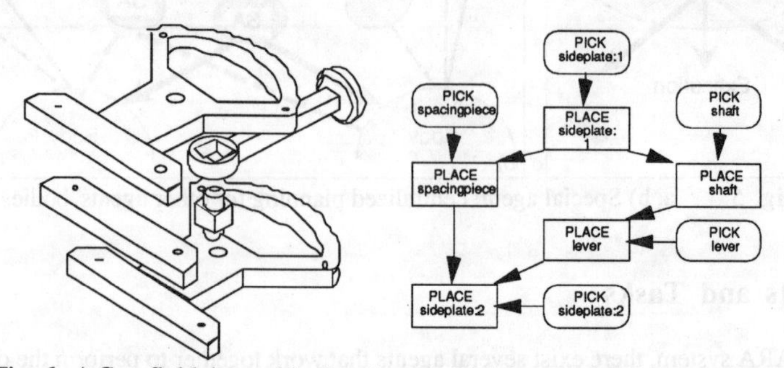

Fig. 6: a) Cranfield Assembly Benchmark, b) Assembly Precedence Graph

5 Error Situations

In the assembly domain, we can distinguish two main classes of errors:
1. Errors that can be avoided by the system's fault tolerance [12], and
2. Errors that occur by inconsistency of the world model and have to be recovered [13].

To be more specific, we have considered three different situations that require fault-tolerant behaviour:

1a. Broken down system components, for example a manipulator or a camera system doesn't work any longer because a computer controlled light source is damaged or a cable is squeezed,

1b. Errors that occur by uncertain information, for example inconsistency of the world model if an assembly part is not at it's predefined position (Fig. 7a), and

1c. Reduced reliability of system components, for example if their accuracy is not sufficient any longer.

By the use of error recovery modules, it should be possible to guarantee correct system behaviour after an error situation is already occurred. To obtain this behaviour, three steps like model-based error detection, error diagnosis and error recovery are required. We distinguish four types of errors:

2a. information error (uncertainties of the world model, for example a slipped obstacle in the gripper, see Fig. 7b),

2b. operational error (for example collision between an obstacle and a manipulator),

2c. constraint error (for example an immovable obstacle that has to be gripped), and

2d. precondition error (not correctly executed preceding operations).

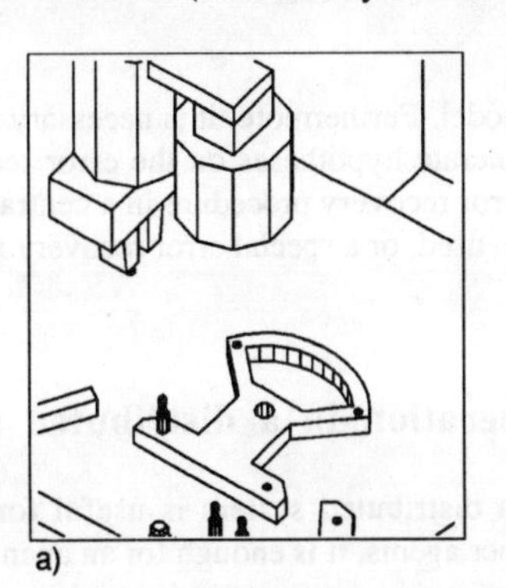

a)

b)

Fig. 7: a) Error that is caused by uncertain position information, b) error that must be recovered

The previously described centralized system FATE is already possible to handle uncertain information (1b), information error (2a) and precondition error (2d) [10].

To handle all above presented error situations with the distributed control system KAMARA, it is necessary to focus on the following problems:

- Detection of reduced reliability of system components
- Detection and explanation of occurred error situations (2a, 2b, 2c, 2d)
- Recovery of the occurred error
- Determination of the preceding operation in a distributed controlled robot system (1c)
- Extension of the communication protocol between the system components
- Tuning of the problem solving validation function of damaged system components
- Execution of inspection tasks before critical assembly operations

5.1 Detection of reduced reliability

During the negotiation process for allocating open tasks, all subsystems can bid for getting the contract. Sometimes, the reliability of a system is not as high as expected or, a system is damaged and it was not possible to notice it by specific sensors. The only information subsystems have is a success information by supervising sensors. Therefore, the subsystems can be forced to reduce their bid evaluation. The second possibility is to use the state controller agent to evaluate agents after task execution and reduce their bid evaluation. Using this method, unreliable agents will get contracts only in situations when no other agents are available.

5.2 Detection and explanation of occurred error situations

In case of an unsuccessful completed operation, it is necessary to determine the reason. In case of a constraint error, no possibility exist to avoid this error in future. Avoiding such errors requires the ability to learn which is not connected to the question of distributed control. Precondition errors are errors that have been occurred in an earlier situation. This type of error has a high probability if it is not possible to find an error reason caused by the last executed operation. In such a case the agent has to ask other agents whether or not they may modified an object position with high probability. Operational error shouldn't occur, because a model is used. So the most important error is the information error. In this case the unsuccessful agent can verify all parameters by submitting inspection tasks.

5.3 Recovery of the occurred error

The recovery algorithm is based on an error cause model. Furthermore, it is necessary to have an global history of all past executed operations to generate hypotheses on the error reason. This means that at least one agent has to coordinate the error recovery procedure in a centralized way. Thus, in a distributed system such activities cannot be used, or a special error recovery agent must support the state controller.

5.4 Determination of the preceding operation in a distributed system

The determination of the preceding operation in a distributed system is useful for different applications. To avoid to store action protocols of other agents, it is enough for an agent to have a history of it's own activities with time stamps. By reasoning on space and time it is possible to detect the desired action.

6 Conclusion

In this paper some problems concerning fault-tolerance and error recovery in distributed controlled autonomous robot were discussed. Some aspects such as the redundancy of agents are better supported by distributed approaches. On the other hand, error recovery is a task, which can be easier solved by a centralized controlled system. The presented concept is more a rough idea what should be investigated with more effort than a closed theory. There is a need for future work. Up to now, the concepts has not been proven with the real robot but only with the robot's simulation system.

7 Acknowledgement

This research work was performed at the Institute for Real-Time Computer Systems and Robotics (IPR), Prof. Dr.-Ing. U. Rembold and Prof. Dr.-Ing. R. Dillmann, Faculty for Computer Science, University of Karlsruhe. The project is part of the nationally based research project on artificial intelligence (SFB 314) funded by the German Research Foundation (DFG).

8 References

[1] Albus, J.S., Barbera,A.J., Nagel,R.N.: Theory and practice of hierarchical control. *Proceedings of IEEE Comp. Soc. Int. Conf.*, 1981, pp. 18-39.

[2] Fukuda, T.; Buss, M.; Hosokai, H.; Kawauchi, Y.: Cell Structured Robotic System CEBOT - Control, Planning and Communication Methods-. *Proceedings of IAS Intelligent Autonomous Systems*, Amsterdam, December, 11-14, 1989, pp. 661-671.

[3] Asama, H.; Ozaki, K.; Ishida, Y.; Habib, M.K., Matsumoto, A.; Endo, I.: Negotiation between Multiple Mobile Robots and An Environment Manager. *Proceedings of 5th ICAR Int. Conf. on Advanced Robotics*, Pisa, Italy, June, 4-6, 1991, pp. 533-538.

[4] Kotoska, S.; Asama, H.; Kaetsu, H.; Ohmori, H.; Endo, I.; Sato, K.; Okada, S.; Nakayama, R.: Development of a functionally adaptive and robust manipulator. *Proceedings of DARS Int. Symp. on Distributed Autonomous Robotic System*, Wako, Japan, Sep. 21-22, 1992, pp. 85-90.

[5] Hörmann, K.; Rembold, U.: A Robot Action Planner for Automatic Parts Assembly. *Proceedings of IEEE/RSJ Int'l Workshop on Intelligent Robots and Systems,* Tokyo, Japan, 1988, pp. 311-317.

[6] Frommherz, B.; Werling, G.: Generating Robot Action Plans by Means of an Heuristic Search. *Proceedings of IEEE Int. Conf. on Robotics and Automation*, Cincinnati, Ohio, May 13-18, 1990, pp. 884-889.

[7] Hörmann, A.; Meier, W.; Schloen, J.: A Control Architecture for an Advanced Fault-Tolerant Robot System. *Proceedings of Intelligent Autonomous Systems*, Amsterdam, December, 11-14, 1989, pp. 576-585.

[8] Cheng, X.; Kappey, D.; Schloen, J.: Elements of an Advanced Robot Control System for Assembly Tasks.*Proceedings of Int'l Conf. on Advanced Robotics*, Pisa, Italy, June, 4-6, 1991, pp. 411-416.

[9] Hörmann, A.; Rembold, U.: Development of an Advanced Robot for Autonomous Assembly. *Proceedings of IEEE Int. Conf. on Robotics and Automation*, Sacramento, California, April 1991, pp. 2452-2457.

[10] Lueth, T.C.; Rembold, U.: Extensive Manipulation Capabilities and Reliable Behavior at Autononomous Robot Assembly. *Proceedings of IEEE Int. Conf. on Robotics and Automation*, San Diego, CA, May 8-13, 1994, pp. 3495-3500.

[11] Lueth, T.C.; Laengle, Th.: Task Description, Decomposition, and Allocation in a Distributed Autonomous Multi-Agent Robot System. *Proceedings of IROS IEEE/RSJ Int. Conf. on Intelligent Robots and Systems*, Munich, Germany, Sep. 12-16, 1994, pp. to appear.

[12] Trevelyan, J.P, Nelson, M.: Adaptive robot control incorporating automatic error recovery. *Proceedings of IInternational Conference on Advanced Robotics,* 1987

[13] Srinivas, S: Error recovery in robot systems. PhD Theses, California Institute of Technology, 1977

A Human Interface for Interacting with and Monitoring the Multi-Agent Robotic System

TSUYOSHI SUZUKI[1], KAZUTAKA YOKOTA[2], HAJIME ASAMA[3], YOSHIKI ISHIDA[4], AKIHIRO MATSUMOTO[5], HAYATO KAETSU[3], and ISAO ENDO[3]

[1]Graduate School of Science and Eng., Saitama Univ., Saitama, 338 Japan
[2]Dept. of Mechanical Systems Eng., Utsunomiya Univ., Tochigi, 321 Japan
[3]The Inst. of Physical and Chemical Reseach (RIKEN), Wako, 351-01 Japan
[4]Computer Center, Kyushu Univ., Fukuoka, 820 Japan
[5]Dept. of Mechanical Eng., Toyo Univ., Saitama, 350 Japan

Abstract

We are developing an autonomous and decentralized robot system ACTRESS. There are two classes of humans who deal with ACTRESS. One is the developer who develops robotic agents of the system, and the other is the operator who runs the system for specific tasks. We have developed a human interface system for the multi-agent robotic system to be used in both development and run-time environments.

The paper discusses communication between operators and agents which is essential to realize the human interface system which lets the operator monitor the system and give or get some information. In ACTRESS, an agent cooperates with other agents using wired and wireless communication systems. Likewise, the human operator communicates with the agents using the same communication systems. He also uses communication to monitor the system. However, in ACTRESS, communication bandwidth is limited. Therefore, the communication of monitoring must not degrade the performance of the communication in the system. We discuss efficient communication to monitor the behavior of agents. And we modeled and evaluated the amount of communication and information for various communication types. It is found that eavesdropping message exchange among agents will reduce necessary communication for the human operator and ease communication traffic. The human operator can make efficient monitoring with the agents using both explicit communication and eavesdropped messages.

Keywords: Decentralized Robotic System, Human Interface, Communication, Monitoring, Mobile Robot Simulation

1. Introduction

In recent years, human-machine interaction has been studied in various research areas. Among these studies, there are few studies on the human interface system for the multi-agent environment [1] [2]. To handle many agents, there are some issues like communication between a human operator and many agents or handling of information in a system to be addressed.

We have been discussing the human interface system which enables communication between the human operator and many agents in the distributed autonomous robotic system. In distributed autonomous system, an operator usually does not have to intervene in the system because he relies on the system's autonomy. However the operator must intervene in the system in some cases as follows:

- when he gives start or stop command of tasks to the system,
- when he assigns tasks to the agents,
- when he is required for some cooperation from agents,
- when he is asked to solve problems which the agents cannot do so themselves.

To realize these communication between the operator and agents, we are developing a human interface system.

In this paper, monitoring methods to gather information of the system using communication is discussed. The human interface system of ACTRESS is introduced. The System Monitor of the human interface system for the ACTRESS is designed. Communication types for the monitoring system are proposed. Monitoring methods in each communication type are discussed. Communication frequency and reliability of information in each method are analyzed. Finally, characteristics of monitoring methods are discussed.

2. Human Interface System of ACTRESS

2.1. Concept of ACTRESS

To execute high level tasks such as maintenance in nuclear power plants or flexible automation in manufacturing plants, ACTRESS has been developed. ACTRESS stands for ACTor-based Robots and Equipments Synthetic System [3]. ACTRESS is the multi-agent robotic system based on the concepts of function distribution and cooperation [4].

In ACTRESS, robots and equipments which have various functions are all regarded as agents. All agents have a wired or wireless communication system. Each agent works autonomously, or cooperates with other agents using the communication system.

Figure 1 shows the prototype system of ACTRESS. The system comprises mobile robots and workstations providing various services.

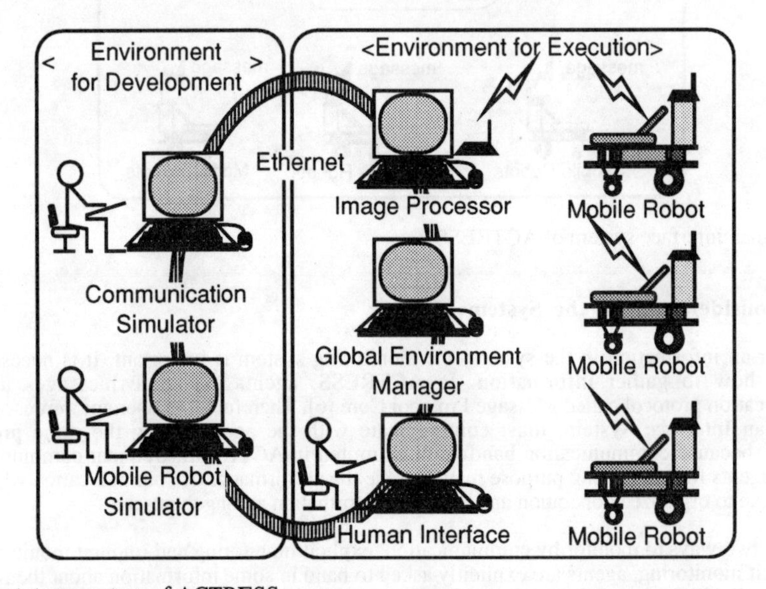

Fig.1 Prototype system of ACTRESS

2.2. Human Interface System of ACTRESS

Though the agents work autonomously in ACTRESS, it requires help from the human operator in several cases. We have designed the architecture of the human interface system in ACTRESS [5]. An operator is regarded as one agent through this interface system. The human interface system consists of three modules (Fig.2).

The ***Presentation Interface*** module accepts input of the operator and shows the condition of system.

The ***Operator Dialog Manager*** module is responsible for coordinating message exchange between the operator and other agents. When the operator explicitly wants to send a command to the agents, he does so by manipulating on-screen mechanisms, such as menus, provided by the Presentation Interface. The Presentation Interface routes his action to the Operator Dialog Manager. The Operator Dialog Manager in turn translates the operator's action into appropriate messages in accordance with ACTRESS's message protocol [6], and dispatches them to the target agents. When the agents needs help of the operator, the Operator Dialog Manager receives its messages and shows them to the operator through Presentation Interface.

The ***System Monitor*** module gathers information in ACTRESS for monitoring purpose. It collects the information from the agents in the system using communication. The gathered information is presented to the operator through the Presentation Interface.

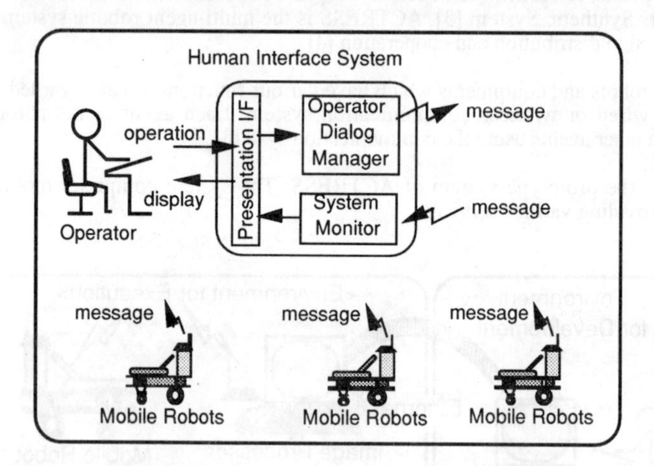

Fig.2 Human interface system of ACTRESS

3. Consideration on the System Monitor

To gather an information in the system, the monitoring system is important. It is necessary to consider how to gather information. In ACTRESS, agents exchange messages using a communication protocol called Message Protocol Core [6]. Therefore, the operator, represented by the Human Interface System, must communicate with the agents using the same protocol. However, because communication bandwidth is limited in ACTRESS, frequent communication with the agents for monitoring purpose may degrade the performance of communication which the agents make to organize cooperation and exchange information among themselves.

There are two ways to monitor by communication: explicit monitoring and implicit monitoring. In the explicit monitoring, agents are explicitly asked to hand in some information about themselves Thus, the reliability of the information is high. However, the explicit monitoring imposes additional load on communication traffic of the system, and may degrade its performance if the System Monitor tries to communicate with the agents too frequently.

In contrast, the agents are not inquired for information in the implicit monitoring; information is gathered by eavesdropping a stream of messages exchanged among the agents. Thus, the agents are not burdened by system monitoring activities and the impact on the performance of communication traffic in the system is expected to be nominal. However, the reliability of the gathered information depends on the number of messages currently exchanged in the system and their contents. It is also not easy to ensure the information is correct and up to date. In order for the System Monitor to retrieve information in an efficient way with nominal impact on performance of the system's communication load, the above two monitoring types and their characteristics need to be examined.

4. Evaluation of system monitoring methods

4.1. Evaluation methods

We evaluated the communication load and the reliability of monitored information for several types of monitoring methods.

We defined the following indices to evaluate the performance of each monitoring methods:
Communication frequency *F*: the number of communication in an unit time between the System Monitor and the agents,
Reliability of information *R*: validity of obtained information.

The issues of gathering information is needed to monitor the system. We discussed what kind of information should be offered to the operator [5]. Table 1 shows the information which the System Monitor collects.

Table 1 Monitored Information

Per agent information	System-wide information
- the ID of the agent - the position of the agent - the current status of the agent - the capability of the agent - the cooperation group, if the agent is currently participating in one	- communication - current status of all tasks - cooperation groups currently in effect - changes in environment - operator actions

In this evaluation, we take the current locations of agents as an example of monitored information. In this case, the reliability R is defined as the proportion of the number of the agents whose locations are known, against the total number of the agent in the environment.

To obtain the frequency F and the reliability R of this information, we modeled the monitoring methods assuming a simple environment (Fig.3). The environment has crisscrossed roads. Each segment has the same length L. Agents are continuously moving at fixed speed v. Agents must move at fixed headings in one segment. And we assume the System Monitor can collect all the messages without a collision of communication.

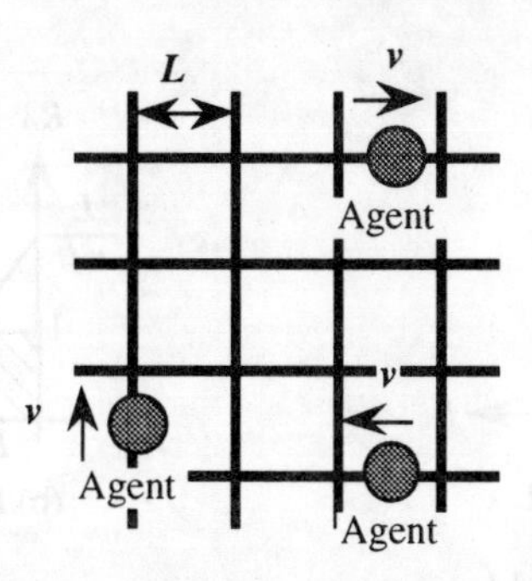

Fig.3 Environment for monitoring

The evaluated monitoring methods and their indices are shown as below:

Type A (explicit monitoring): The System Monitor inquires each agent for its current location and heading at fixed interval time ti.

$$F_A = \frac{N}{ti}$$

$$R_A = \begin{cases} \dfrac{L}{2 \cdot v \cdot ti} & \left(\dfrac{L}{v \cdot ti} \le 1\right) \\ 1 - \dfrac{v \cdot ti}{2 \cdot L} & \left(\dfrac{L}{v \cdot ti} > 1\right) \end{cases}$$

where N, v, and L denote the number of the agents in the environment, the fixed velocity of the agents, and the length of the segment in the environment.

The indices of **Type A** are obtained as follows:

The System Monitor must monitor N agents within the time ti. Then, the number of communication in an unit time is calculated as $F_A = N/ti$.

If the agent has moved the distance l in a segment when the System Monitor communicates with the agent, the System Monitor can correctly assume the location of the agent thereafter until it reaches the end of the segment, as it is moving at the fixed velocity (Fig.4).

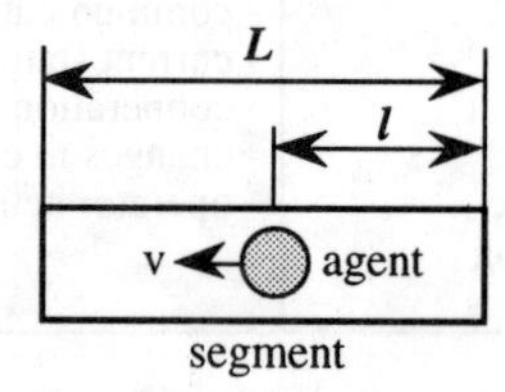

Fig.4 Definition of segment

In other words, for the next period of $(L-l)/v$, the location of the agent is known. The reliability of the information R_A is, thus, defined as $(L-l)/(v*ti)$. Figure 5 shows the relation between R_A and l. Because the velocity of the agent is constant, the probability where the agent exists in one segment is uniform. Therefore, the reliability R_A is average of the shaded portion of Fig.5 (a).

$$R_A = \frac{1}{2} \cdot L \cdot \frac{L}{v \cdot ti} \cdot \frac{1}{L} = \frac{L}{2 \cdot v \cdot ti} \qquad \left(\frac{L}{v \cdot ti} \le 1\right)$$

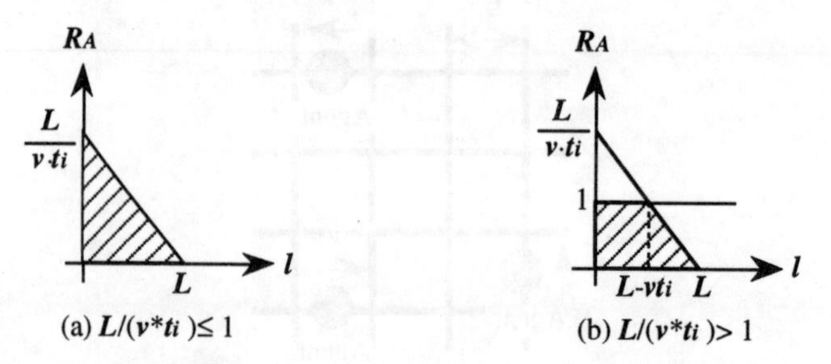

Fig.5 The relation between R_A and l

If $L/(v*ti)$ is over 1, the relation between R_A and l is shown Fig.5 (b). Therefore, the reliability R_A is average of the shaded portion of Fig.5 (b).

$$R_A = \frac{1}{2} \cdot (L + L - v\,ti) \cdot 1 \cdot \frac{1}{L} = 1 - \frac{v \cdot ti}{2 \cdot L} \quad \left(\frac{L}{v\,ti} > 1\right)$$

In this type of communication, it takes 2**T*, ***T*** being the duration of communication from the System Monitor to an agent or vice versa, for the System Monitor to obtain information on an agent, because a message must be sent from the System Monitor and answering message is sent back from the agent. Therefore, in order to monitor ***N*** agents, ***ti*** must satisfy $ti \geq 2 * T * N$, .

Type B (explicit monitoring): The agent reports its current location and heading to the System Monitor at fixed interval time ***ti***.

$$F_B = \frac{N}{ti}$$

$$R_B = \begin{cases} \dfrac{L}{2 \cdot v \cdot ti} & \left(\dfrac{L}{v \cdot ti} \leq 1\right) \\ 1 - \dfrac{v \cdot ti}{2 \cdot L} & \left(\dfrac{L}{v \cdot ti} < 1\right) \end{cases}$$

The indices of **Type B** are obtained in the same way as **Type A**. However, one way communication is performed in this type rather than two way communication as in **Type A**. In order to monitor ***N*** agents, ***ti*** must satisfy $ti \geq T * N$.

Type C (explicit monitoring): The agent reports its current location and heading to the System Monitor when it enters the next segment.

$$F_C = \frac{v}{L} \cdot N$$

$$R_C = 1$$

In this type, the agent enters a new segment every ***L/v***. Therefore, communication frequency F_C is represented as above. The agents move at fixed speed and don't change its headings in a segment. Therefore, it is sufficient for the System Monitor to know when an agent enters a segment in order to infer the location of the agent, provided F_C does not exceed ***1/T***. If F_C is over ***1/T***, there is possible loss of messages and R_C will be smaller than the above.

Type D (explicit monitoring): The agent reports its current location and heading to the System Monitor when it makes a turn at a crossing.

$$F_D = P_D \cdot \frac{v}{L} \cdot N \qquad (P_D\text{: the probability of turning at a crossing})$$

$$R_D = 1$$

The equations of **Type D** is similar to **Type C**. However, F_D includes the probability of turning at a crossing in this type. The System Monitor knows the agent moves straight until next report because the agent always reports to the System Monitor when it makes a turn. Therefore, the System Monitor can correctly estimate the position of agent until the next report comes in. As with **Type C**, if F_D exceeds 1/***T***, the System Monitor does not see some messages, and R_D will be smaller.

Type E (implicit monitoring): The System Monitor eavesdrops messages between agents, provided that agents communicate their current positions and headings to each other when they detect collision.

$$F_E = P_E \cdot \frac{v}{L} \cdot N \qquad (P_E\text{: the probability of collision among } N \text{ agents})$$

$$R_E = N \cdot P_E$$

F_E is similar to the F_D. However, F_E includes the probability of collision. And in this type, the System Monitor has no communication with the agents. For R_E, the reliability of information depends on the communication between agents. In this case, R_E depends on the probability P_E. If F_E is greater than 1/***T***, R_E will be smaller.

In the equations of **Type A**, **B**, and **C**, these values can be calculated analytically. But **Type D** and **E** include probability terms, therefore, cannot be calculated easily. We conducted a computer simulation to obtain those probabilities.

4.2. Simulation

Mobile robotic agents are working in the environment shown in Fig.6. They are continuously moving from one segment to another at fixed speed. The agents don't change its headings in a segment. Each segment is assigned numbers 1 through 12, and corners have numbers 13 through 21. Only one agent occupies a segment. The System Monitor monitors the locations (segment number) and headings of the agents. The start and goal segments of an agent are randomly given.

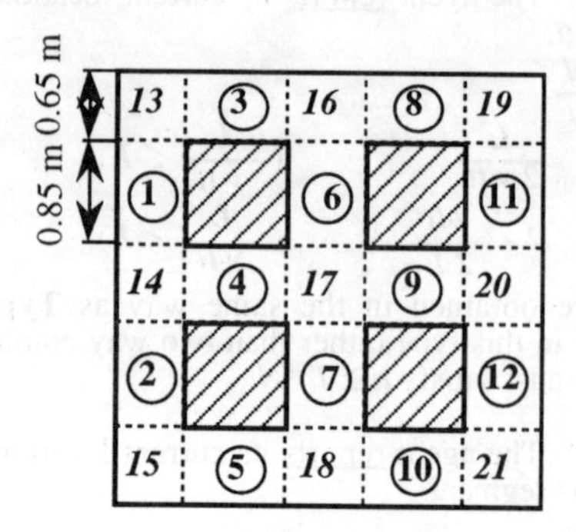

Fig.6 Simulation Environment

We provided two algorithms to select a path between the start and goal segments. The algorithm 1 chooses the shortest path to the goal whereas the algorithm 2 selects a path with the least number of turns. We counted the number of turns exercised by the agents and the number of occurrences of collision between two or more agents. In this simulation, we defined the collision as the situation such that two or more agents enter the same segment or the same crossing at the same time. The conditions of the simulation are shown in Table 2. In this Table, ***T*** is timelag of communication radio wireless modem we are currently using [7]. ***v*** is the average velocity of the mobile robot we have developed.

Table 2 The conditions of simulation

The duration of one-way communication between the System Monitor and an agent (*T*)	0.18 sec
Velocity of an agent (*v*)	0.2 m/sec
The Length of a segment (*L*)	1.5 m

For example, The result of simulation in the case of four agents is shown in Table 3. The steps is the number of segments the agents passed.

Table 3 The result of simulation of 4 agents

		Algorithm 1	Algorithm 2
The number of steps (**steps**)		73,800	83,752
The number of turning (**turns**)		26,510	25,534
The number of occurrences of collision (**c**$_n$)	2 agents (**c**$_2$)	7,795	7,057
	3 agents (**c**$_3$)	654	331
	4 agents (**c**$_4$)	37	5

P_D and P_E were calculated as follows:

$$P_D = \frac{\text{turns}}{\text{steps}}$$

$$P_E = \frac{\sum_{n=2}^{N} n*c_n}{\text{steps}*N} = \frac{2*c_2 + 3*c_3 + 4*c_4}{\text{steps} * 4}$$

(**n** is the number of agents in collision)

Likewise, we calculated indices for other numbers of agents too. And the results are summarized in the next section.

4.3. Evaluation

We calculated communication frequency ***F*** and reliability of information ***R*** of all types. In this calculation, ***N*** changed three through eleven agents, ***L/v*** changed one second through ten seconds. ***L/v*** is the time that the agent is taken in a segment .

In the case of **Type A**, the communication frequency F_A is affected by interval time ***ti*** and ***N***. The reliability R_A is affected by ***ti*** and ***L/v***. The ***ti*** depends on the number of agents. Figure 7 shows the graph of F_A and R_A. As shown, F_A is always less than $1/(2*T)$. In this type, when few agents move slowly, the R_A is high.

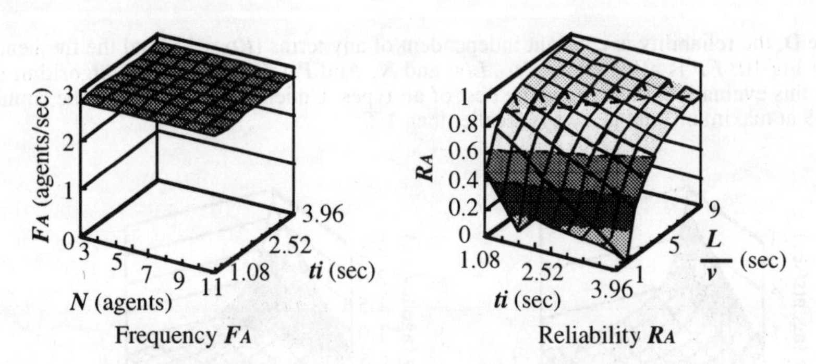

Fig.7 Index/indices of **Type A**

For **Type B**, the graph of F_B and R_B is shown as Fig.8. In this case, We evaluate same things with **Type A**. But, F_B is always less than $1/T$. R_B is also high when the few agents move slowly.

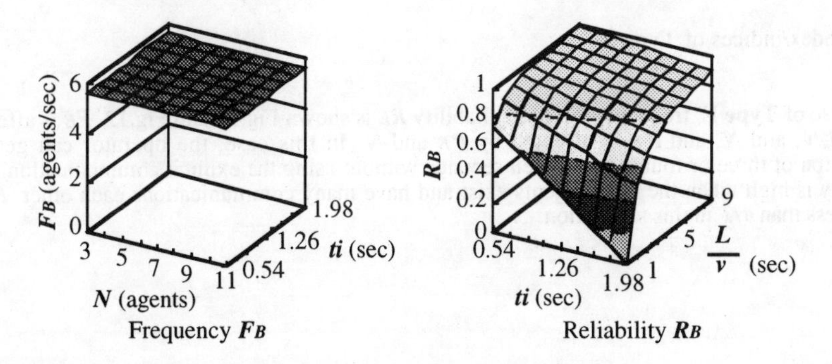

Fig.8 Index/indices of **Type B**

For **Type C**, the reliability is constant ($R_C = 1$). And the graph of frequency F_C is shown Fig.9. F_C is affected by N and L/v. F_C is large when many agents move fast. In this evaluation, communication takes place more frequently than other types. In the part that N is large and L/v is small, F_C is over $1/T$ in this simulation.

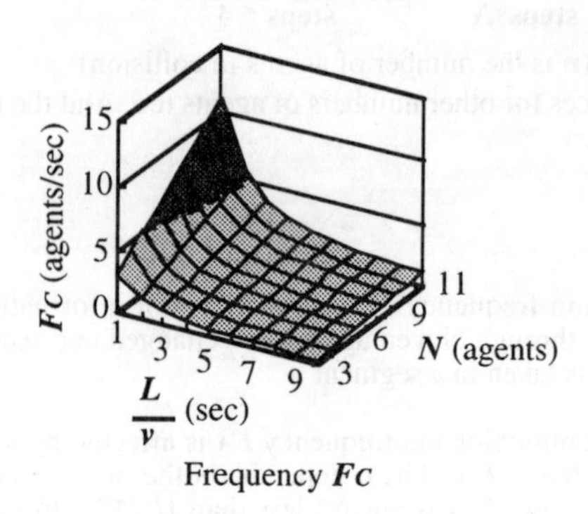

Fig.9 Index/indices of **Type C**

For **Type D**, the reliability is constant independent of any terms ($R_D = 1$). And the frequency F_D is shown Fig.10. F_D is affected by P_D, L/v, and N. And P_D depends on the algorithm of the agent. In this evaluation, this type is the best of all types. Under the conditions of the simulation F_D is 2.59 at maximum, and is always smaller than $1/T$.

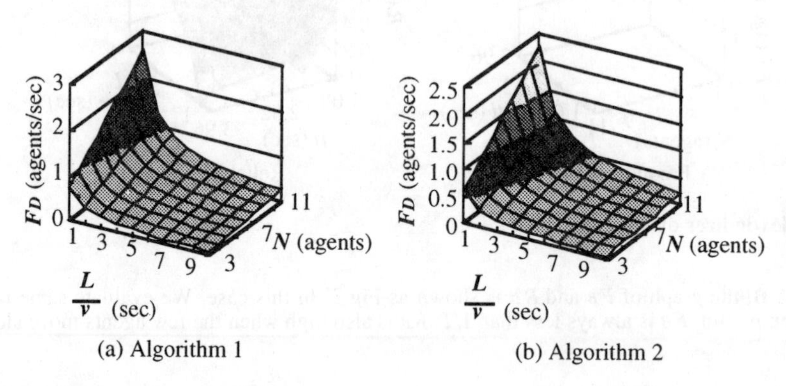

Fig.10 Index/indices of **Type D**

In the case of **Type E**, frequency F_E and reliability R_E is shown Fig.11 and Fig.12. F_E is affected by P_E, L/v, and N, and R_E is affected by P_E and N. In this case, the operator can get the information of three or four agents in ten seconds without using the explicit communication. The reliability is high when the many agents exist and have many communications each other. F_E is always less than $1/T$ in this simulation.

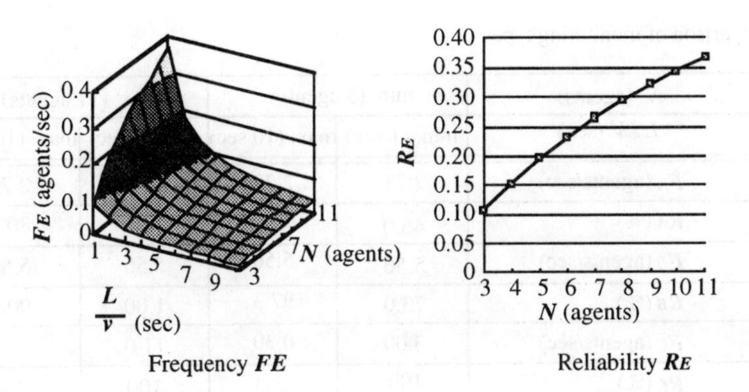

Fig.11 Index/indices of **Type E** (Algorithm 1)

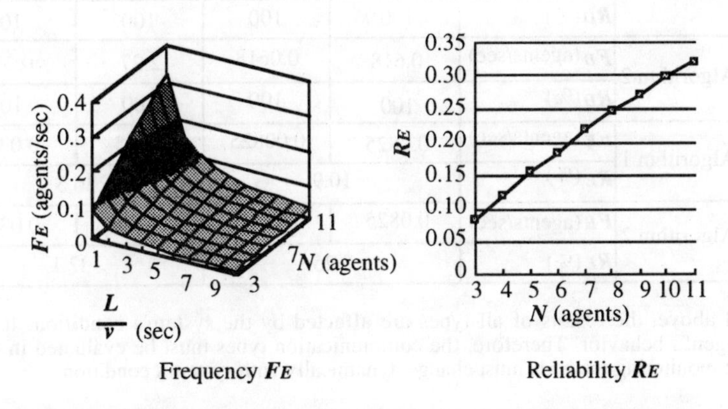

Fig.12 Index/indices of **Type E** (Algorithm 2)

For the **Type C**, **D**, and **E**, the frequency communication is not over $1/T$ in this simulation's condition.

In these result, the indices of five monitoring types are summarized in Table 4.

As shown, monitoring at fixed interval time as in **Type A** and **B** is not efficient because the agent always communicates with the System Monitor for monitoring purpose. Therefore, it is likely to degrade the performance of the communication in the system.

In **Type C**, the reliability is high but the frequency is also high because the agent must communicates more frequently with the System Monitor than **Type A** and **B**.

The **Type D** performed best with the highest reliability and the least frequent communication of all types. But the results of **Type C** and **D** depend on the characteristics of the environment in this evaluation. Therefore, this result changes if the premise of simulation, e. g. the topology of the map and the velocity of the agent, is different.

While the reliability index for the **Type E** is not as good as the others, it is achieved by occasional implicit communication rather than explicit communication in the other types, thus, the communication load of monitoring is considered minimal. This communication is achievable without degrading the performance of the system's communication. However, the cost to interpret the messages must be considered.

Table 4 Comparison of monitoring types

		N (agents)	min. (3 agents)		max. (11 agents)	
		L/V (sec)	min. (1 sec)	max. (10 sec)	min. (1 sec)	max. (10 sec)
Type A		F_A (agents/sec)	2.78	2.78	2.78	2.78
		R_A (%)	46.0	94.6	12.6	80.2
Type B		F_B (agents/sec)	5.56	5.56	5.56	5.56
		R_B (%)	73.0	97.3	1.00	90.1
Type C		F_C (agents/sec)	3.00	0.30	11.0	1.10
		R_C (%)	100	100	100	100
Type D	Algorithm 1	F_D (agents/sec)	0.702	0.0702	2.59	0.259
		R_D (%)	100	100	100	100
	Algorithm 2	F_D (agents/sec)	0.618	0.0618	2.27	0.277
		R_D (%)	100	100	100	100
Type E	Algorithm 1	F_E (agents/sec)	0.0825	0.00825	0.365	0.0365
		R_E (%)	10.9		36.5	
	Algorithm 2	F_E (agents/sec)	0.0825	0.00825	0.321	0.0321
		R_E (%)	8.25		32.1	

As discussed above, the results of all types are affected by the system's condition. It is also affected by agent's behavior. Therefore, the communication types must be evaluated in various cases, and the monitoring methods must change dynamically with system's condition.

5. Conclusion

A monitoring system for human interface was discussed. The system Monitor of the human interface system for the ACTRESS was designed. Communication types for the monitoring system were proposed. Monitoring methods in each communication way were discussed. Communication frequency and reliability of information of each monitoring method were analyzed. Characteristics of monitoring methods were discussed.

We will try various simulation types and evaluate the communication types to design the monitoring system for human interface.

References

[1] Y. Nakauchi, T. Okada, N. Yamasaki, Y. Anzai, Multi-Agent Interface Architecture for Human-Robot Cooperation, *Proceedings of the 1992 IEEE Int. Conf. on Robotics and Automation*, 1992, pp.2786-2788.

[2] T. Fujita, H. Kimura, The Design of Software Development System for Multiple Robots, *Proceedings of the 1993 IEEE/RSJ int. Conf. on Intelligent Robots and Systems*, 1993, pp.1119-1125.

[3] H. Asama, A. Matsumoto, Y. Ishida, Design of an Autonomous and Distributed Robot System: ACTRESS, *Proceedings of 1989 IEEE/RSJ Int. Workshop on Intelligent Robots and Systems*, 1989, pp.283-290.

[4] H. Asama, A. Matsumoto, K. Ozaki, Y. Ishida, Maki. K. Habib, I. Endo, Functional Distribution among Multiple Mobile Robots in An Autonomous and Decentralized Robot System, *Proceedings of 1991 IEEE Int. Conf. on Robotics and Automation*, 1991, pp1921-1926.

[5] K. Yokota, T. Suzuki, H. Asama, A. Matsumoto, I. Endo, A Human Interface System for the Multi-Agent Robotic System, *Proceedings of the 1994 IEEE Int. Conf. on Robotics and Automation,* 1994, pp.1039-1044.

[6] Y. Ishida, H. Asama, K. Ozaki, K. Yokota, A. Matsumoto, I. Endo, A Communication System for a Multi-Agent Robotic System, *Proceedings of the 1993 JSME Int. Conf. on Advanced Mechatronics* , 1993, pp.424-428.

[7] K. Ozaki, H. Asama, Y. Ishida, A. Matsumoto, K. Yokota, H. Kaetsu, I. Endo, Synchronized Motion by Multiple Mobile Robots using Communication, *Proceedings of the 1993 IEEE/RSJ int. Conf. on Intelligent Robots and Systems*, 1993, pp.1164-1169.

Chapter 3

Distributed Sensing

Real Time Robot Tracking System with Two Independent Lasers

ABDELLAH GHENAIM, V. KONCAR, S. BENAMAR, and A. CHOUAR

GEMTEX, Ecole Nationale Supérieure des Arts et Industries Textiles,
59070 Roubaix Cedex 01, France

Abstract

In this paper we propose a control algorithm using two independent lasers and CCD camera for the 2D trajectory tracking. Using this algorithm, the real time trajectory tracking of an articulated robot arm can be realized with some conditions imposed on the robot end effector.

Keywords: Robot, Tracking, Laser, Vision system.

1 Introduction

Industrial robots are mostly programmed to perform repetitive tasks, following certain trajectories. In many cases, laser systems are used to assure the good tracking properties. In the past decades, various tracking systems using lasers and optical sensors have been developed.
T.A.G. Heeren and F.E. Veldpaus [1] have proposed an optical system to measure the end effector position for on line control purposes. Two rotatable lasers are mounted on the working table and a retroreflector pair is mounted on the end effector.
S. Venkatesen and C. Archibald [2] have proposed the real time tracking in five degrees of freedom using two wrist mounted laser range finders. Compact lightweight laser range finder have been developed by M. Rioux and al. [3].
The real-time tracking method is an extension of earlier works done by C.C. Archibald and al. [4] and S. Verkatesen and al [5].
In this paper we propose a 2D trajectory robot tracking system using two independent lasers. The first laser system S1 is used to generate the 2D trajectory, which could be programmed in different forms (mathematical expression, tables,...). The second laser system S2 is mounted on the robot end effector and it tracks the trajectory defined by the system S1. The tracking is realized by the two laser spots superposition.
To achieve this aim (superposition), a fixed CCD camera has been used to locate spot positions.
We also studied the utilization of an optical sensor mounted on the end effector which permits spots superposition detection. The feasibility study has been realized where the laser control signals are modulated. This technic is used to enable more efficient superposition detection of two spots by the optical sensor.
Moreover, some conditions are imposed on robot end effector (angle α between the end effector and the working table, motion speed, ...).

2 Description of real time robot tracking system

The robot tracking system is composed of 6 degrees of freedom articulated robot arm controlled by the CNC. The CNC is connected to the PC computer equipped with a vision system which is used as the feedback loop (Fig. 1).

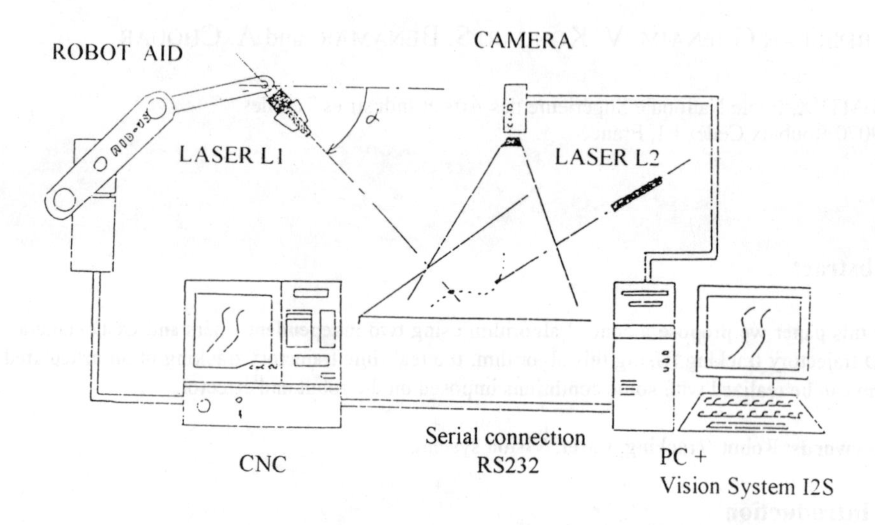

Fig.1. Arrangment of equipement for real time R.T.S.

3 Real time laser spots positions computation

We have used I2S PCSCOPE vision system made up with a black and white CCD camera with 800 points resolution, digital processing running time images MATROX, a monitor and a PC computer which contains a MATROX card IDS 542 (1024*1024 pixels), and manages the different states of digital processing.
This system enables two spots image acquisition, binarisation, computation and finally the spots gravity centres positions location. These positions (spot1 u_1,v_1, spot2 u_2,v_2) are given in pixels and the global system resolution depends on the camera position.
A major use of machine vision is expected to be the detection of object (target) position. Many applications exist in automation, including robotics, in which position detection is required; the case of automated manufacturing is an example. A robot would be required to inspect parts on the assembly line and perform measurements on them for quality control [6].
This paper describes a closed-loop vision system. The system is supposed to detect S1 and S2 spot positions. We needed to achieve a software of image processing in C++ language, in order to obtain spots gravity centres positions, and so to have better accuracy in u_1,v_1 and u_2,v_2 computation. The major problems faced were to interface the camera to the PC computer, and derive the proper equations to do the tracking.
The problem of object segmentation is simplified by making the object to be tracked of enough contrast with the background to be easily distinguished by thresholding. In general, segmentation algorithms have been proposed in the literature. The histogram-directed thresholding remains an effective segmentation method [7]. Several methods for automatic threshold selectors are available (refer to [8], [9] and [10] for details). We used a simple thresholding method to demonstrate the performance of the tracking system; such a method does not necessarily work well in general. Statistical image segmentation techniques may offer advantages in situations such as automatic object extraction from noisy background [11], [12].

Two spots positions are then used to compute the angular setting points for the robot which permit the spots superposition (two spots positions are the same $u_1=u_2$ and $v_1=v_2$).

4 Transformation method

4.1 Basic theory and notations

- Articulation variable j is noted θ_j
- Rigid link is noted C_j

Robot arm is composed of 7 rigid links C_0, C_1,........,C_6 and 6 articulated joints (RRRRRR, 6 degrees of freedom).

The articulation j connects the C_j rigid link to C_{j-1} rigid link.

- Z_j axis is linked to the articulation j,
- X_j axis is defined as shown in Fig.2.

The passage from R_{j-1} reference to R_j reference is given as a function of four parameters (Fig. 2) :

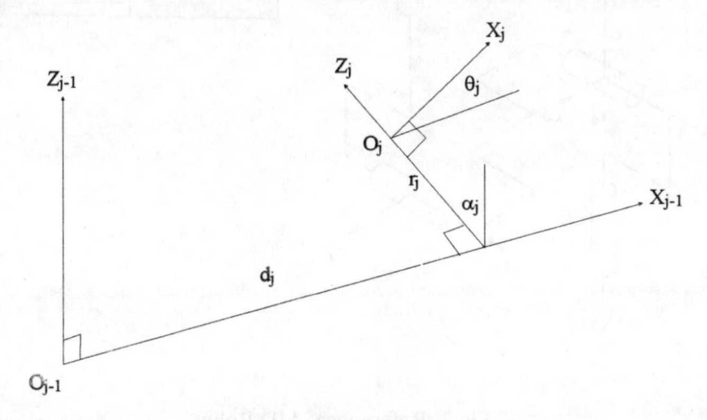

Fig.2. Geometrical parameters

α_j angle between axes Z_{j-1} and Z_j,
d_j distance between Z_{j-1} and Z_j along X_{j-1},
θ_i angle between axes X_{j-1} and X_j,
r_j distance between X_{j-1} and X_j along Z_j.

Transformation matrix defining R_j reference in R_{j-1} reference is given below:

$${}^{j-1}T_j = Rot(x,\alpha j)\ Trans(x,dj)\ Rot(z,\theta j)\ Trans(z,rj)$$, (Trans=translation, Rot=rotation)

Hence ,
$$ {}^{i+1}T_i = \begin{bmatrix} C\theta_j & -S\theta_j & 0 & dj \\ C\alpha_j S\theta_j & C\alpha_j C\theta_j & -S\alpha_j & -rjS\alpha_j \\ S\alpha_j S\theta_j & S\alpha_j C\theta_j & C\alpha_j & rjC\alpha_j \\ 0 & 0 & 0 & 1 \end{bmatrix} \quad (1)$$

4.2 Description of robot arm

The second laser S2 is mounted on the end effector.
The robot references assignment is given in the Fig.3. and robot parameters are given in table 1.

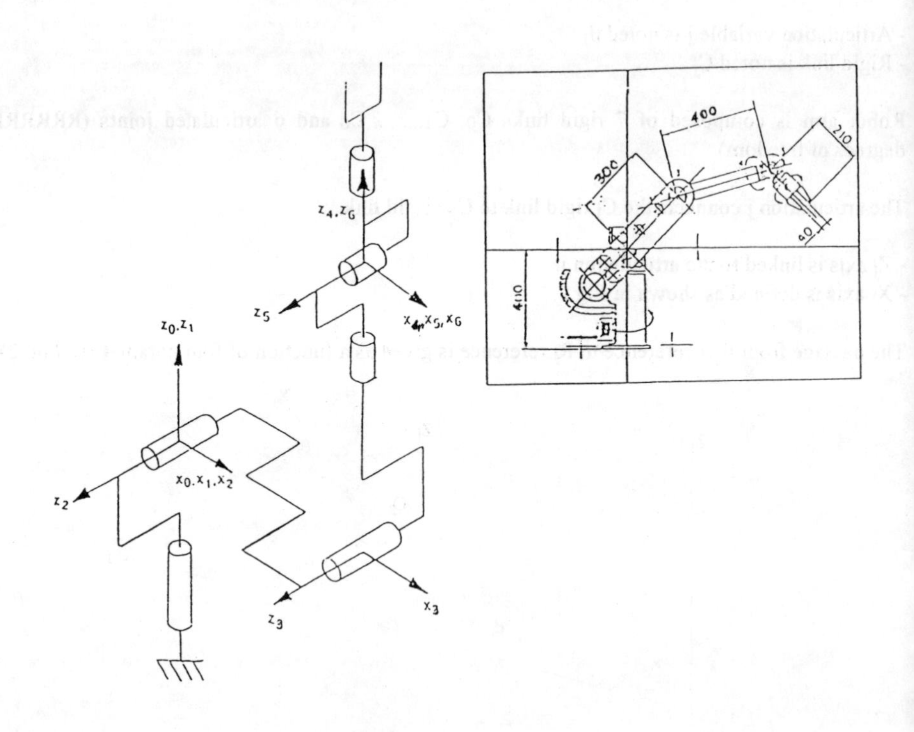

Fig.3. References, AID Robot

J	αj	dj	θj	rj
1	0	0	θ_1	0
2	90	0	θ_2	0
3	0	300	θ_3	0
4	-90	0	θ_4	400
5	90	0	θ_5	0
6	-90	0	θ_6	0

Table 1. Geometrical parameters (Denavit-Hartenberg)

Direct kinematics model X=f(Q):

Direct kinematics model gives relations between the position and orientation of the vector X and the generalized coordinate vector Q, where $Q = (\theta_1, \theta_2, \ldots \theta_6)$.

From the Table 1. and from the matrix (1), we obtain elementary transformation matrixes $^{i-1}T_i$:

$$ {}^{0}T_1=\begin{bmatrix} C1 & -S1 & 0 & 0 \\ S1 & C1 & 0 & 0 \\ 0 & 0 & 1 & 0 \\ 0 & 0 & 0 & 1 \end{bmatrix},\quad {}^{1}T_2=\begin{bmatrix} C2 & -S2 & 0 & 0 \\ 0 & 0 & -1 & 0 \\ S2 & C2 & 0 & 0 \\ 0 & 0 & 0 & 0 \end{bmatrix},\quad {}^{2}T_3=\begin{bmatrix} C3 & -S3 & 0 & 300 \\ S3 & C3 & 0 & 0 \\ 0 & 0 & 1 & 0 \\ 0 & 0 & 0 & 1 \end{bmatrix}, $$

$$ {}^{3}T_4=\begin{bmatrix} C4 & -S4 & 0 & 0 \\ 0 & 0 & 1 & 400 \\ -S4 & -C4 & 0 & 0 \\ 0 & 0 & 0 & 1 \end{bmatrix},\quad {}^{4}T_5=\begin{bmatrix} C5 & -S5 & 0 & 0 \\ 0 & 0 & -1 & 0 \\ S5 & C5 & 0 & 0 \\ 0 & 0 & 0 & 1 \end{bmatrix},\quad {}^{5}T_6=\begin{bmatrix} C5 & -S6 & 0 & 0 \\ 0 & 0 & 1 & 0 \\ -S6 & -C6 & 0 & 0 \\ 0 & 0 & 0 & 1 \end{bmatrix}, $$

Hence

$$ Z=\begin{bmatrix} Sx & nx & ax & Px \\ Sy & ny & ay & Py \\ Sz & nz & az & Pz \\ 0 & 0 & 0 & 1 \end{bmatrix}=\begin{bmatrix} A & P \\ 000 & 1 \end{bmatrix} \tag{2} $$

A is the orientation matrix and P is the position vector of the end effector.

The associeted matrix is computed as follows $Z={}^{0}T_1\,{}^{1}T_2\,{}^{2}T_3\,{}^{3}T_4\,{}^{4}T_5\,{}^{5}T_6$

Sx=C1(C23(C4C5C6-S4S6) - S23S5C6) - S1(S4C5C6+C4S6)
Sy=S1(C23(C4C5C6-S4S6) - S23S5S6) + C1(S4C5C6+C4S6)
Sz=S23(C4C5C6-S4S6) + C23S5C6

nx=C1(-C23(C4C5S6+S4C6)+S23S5S6) + S1(S4C5S6-C4C6)
ny=S1(-C23(C4C5S6+S4C6)+S23S5S6) + C1(S4C5S6-C4C6)
nz=-S23(C4C5S6+S4C6)+C23S5S6

ax=-C1(C23C4S5+S23C5) + S1S4S5
ay=-S1(C23C4S5+S23C5) - C1S4S5
az=-S23C4S5+C23C5

Px=-C1(S23*400-C2*300)
Py=-S1(S23*400-C2*300)
Pz=C23*400+S2*300

Ci=cos(θ_i)
Si=sin(θ_i)
Cij=cos($\theta_i+\theta_j$)

The direct kinematics model has been optimized in order to minimize the computation time of the associated matrix Z.

The optimized direct kinematics model is given bellow:
Tijrs is the element (r,s) of matrix ${}^{i}T_j$, then we obtain instrumental variables:

T4612=-C5S6
T4632=-S5S6
T3612=C4T4612-S4C6

T3613=-C4S5
T3632=-S4T4612-C4C6
T3633=S4S5
T1314=300*C2
T1334=300*S2
T1612=C23T3612-S23T4632
T1613=C23T3613-S23C5
T1614=-S23*400+T1314
T1632=S23T3612+C23T4632
T1633=S23T3613+C23C5
T1634=C23*400+T1334
T0612=C1T1612+S1T3632 ----------------------> nx
T0613=C1T1613+S1T3633 ----------------------> ax
T0614=C1T1614 ----------------------> Px
T0622=S1T1612-C1T3632 ----------------------> ny
T0623=S1T1613-C1T3633 ----------------------> ay
T0624=S1T1614 ----------------------> Py
T0632=T1632 ----------------------> nz
T0633=T1633 ----------------------> az
T0634=T1634 ----------------------> Pz

Total number of multiplication = 30,
Total number of additions = 12.

These equations represent the complete geometrical model. However, 6 more multiplications and 3 additions are necessary to compute the S vector.

Sx=ny*az-nz*ay
Sy=nz*ax-nx*az
Sz=nx*ay-ny*ax

Inverse kinematics model $Q=f^{1}(X)$:

Inverse kinematics model gives relations between the generalized coordinate vector Q and the position and orientation vector X.
In the case of AID robot, used in our application, the inverse kinematics model gives eight possible solutions Q_i, i=1,2, ...8.

a) θ_1 estimation:

θ_1=Atan2(Py,Px)
$\theta_1'=\theta_1+\pi$ (two solutions)

b) θ_2 and θ_3 estimation:

we consider,
B1=Px*cos(θ_1)+Py*sin(θ_1)
X=-600*Pz
Y=-600*B1
$R=70000-Pz^2-B1^2$

$D=X^2+Y^2$
$H=D-Z^2$
$W=(H)^{1/2}$

$C2=(YR+\varepsilon XW)/D$
$S2=(ZR-\varepsilon YW)/D$

$\varepsilon=\pm 1$

$\theta_2=Atan2(S2,C2)$ (two solutions)

$S3=-(Pz*\sin\theta_2+B1*\cos\theta_2-300)/400$
$C3=-(B1*\sin\theta_2-Pz*\cos\theta_2)/400$

$\theta_3=Atan2(S_3,C_3)$

c) θ_4 estimation:

$\theta_4=Atan2[(S1ax-C1ay), -C23(C1ax+S1az)-S23az]$
$\theta_4'=\theta_4+\pi$ (two solutions)

d) θ_5 estimation:

$-S5=C4[C23(C1ax+S1ay)+S23az]-S4(S1ax-C1ay)$
$C5=-S23(C1ax+S1ay)+C23az$

$\theta_5=Atan2(S5,C5)$

e) θ_6 estimation:

$S6=-C4(S1ax-C1sy)-S4[C23(C1sx+S1sy)+S23sz]$
$C6=-C4(S1nx-C1ny)-S4[C23(C1nx+S1ny)+S23nz]$

$\theta_6=Atan2(S6,C6)$

The transformation matrix Zi of the AID robot is defined bellow:

$$Z_i = \begin{bmatrix} S_{x_i} & n_{x_i} & a_{x_i} & P_{x_i} \\ S_{y_i} & n_{y_i} & a_{y_i} & P_{y_i} \\ S_{z_i} & n_{z_i} & a_{z_i} & P_{z_i} \\ 0 & 0 & 0 & 1 \end{bmatrix}$$, the matrix Z_i corresponds to the spot Si (i=1,2) (3)

where the 3x3 orthogonal matrix formed by the [S_i, n_i, a_i] vectors represent the orientation of the end effector and Pi vector represents the end effector position.

Then $X = f(q)$ with $X = [P_x, P_y, P_z, \ldots, S_z]$, $Q = [q_1, q_2, q_3, q_4, q_5, q_6]$ (4)

X is position and orientation vector and
Q is generalized coordinate vector.

4.3 Task coordinate system

The inverse kinematics of the robot arm and the two spots positions (u_1,v_1), (u_2,v_2) given by the vision system are used to obtain the optimal generalized coordinate vector which enables two spots superposition ($u_1=u_2$, $v_1=v_2$).

${}^{2}T_1$ matrix computed by the geometrical transformation method is defined by the robot tracking system coordinate configuration given in the Fig.5.

$${}^{i+1}T_i = Rot(n,-\alpha)\ Trans(a,d)\ Rot(s,\theta)\ Trans(a,(dx-d))\ Rot(n,\alpha);\ \ (Trans=translation,\ Rot=rotation)$$

$${}^{i+1}T_i = \begin{bmatrix} C^2\alpha - S^2\alpha C\theta & S\alpha S\theta & C\alpha S\alpha + S\alpha C\alpha C\theta & -S\alpha S\theta(d-dx) \\ S\theta S\alpha & S\theta & -S\theta C\alpha & -S\theta(d-dx)0 \\ S\alpha C\alpha & C\alpha S\theta & S^2\alpha + C^2\alpha C\theta & S\alpha[C\theta(d-dx)+d] \\ 0 & 0 & 0 & 1 \end{bmatrix} \quad (5)$$

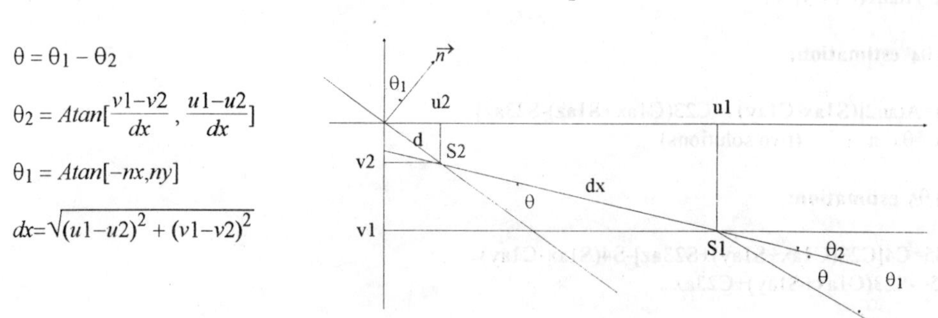

$$\theta = \theta_1 - \theta_2$$

$$\theta_2 = Atan[\frac{v1-v2}{dx}, \frac{u1-u2}{dx}]$$

$$\theta_1 = Atan[-nx, ny]$$

$$dx = \sqrt{(u1-u2)^2 + (v1-v2)^2}$$

Fig.4. Computation of parameters

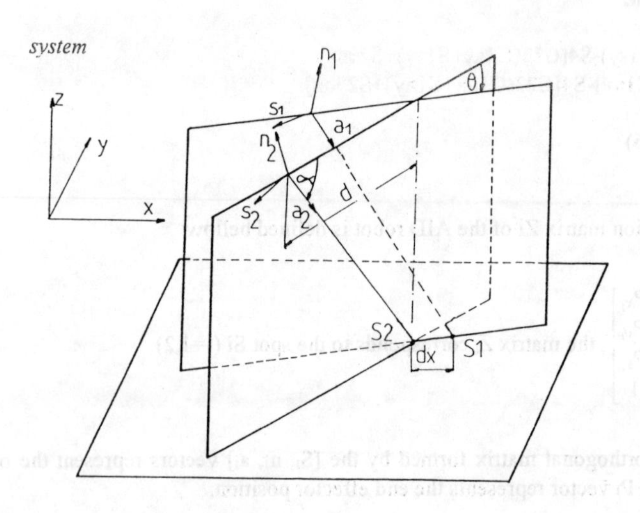

Fig. 5. Robot tracking system coordinate assignment

a) Transformation algorithm

At each sampling period (kT), the data acquisition is done by the vision system and two spots positions (u_1v_1 and u_2v_2) are obtained. Then, the new generalized coordinate vector is computed by the algorithm given in Fig.6.

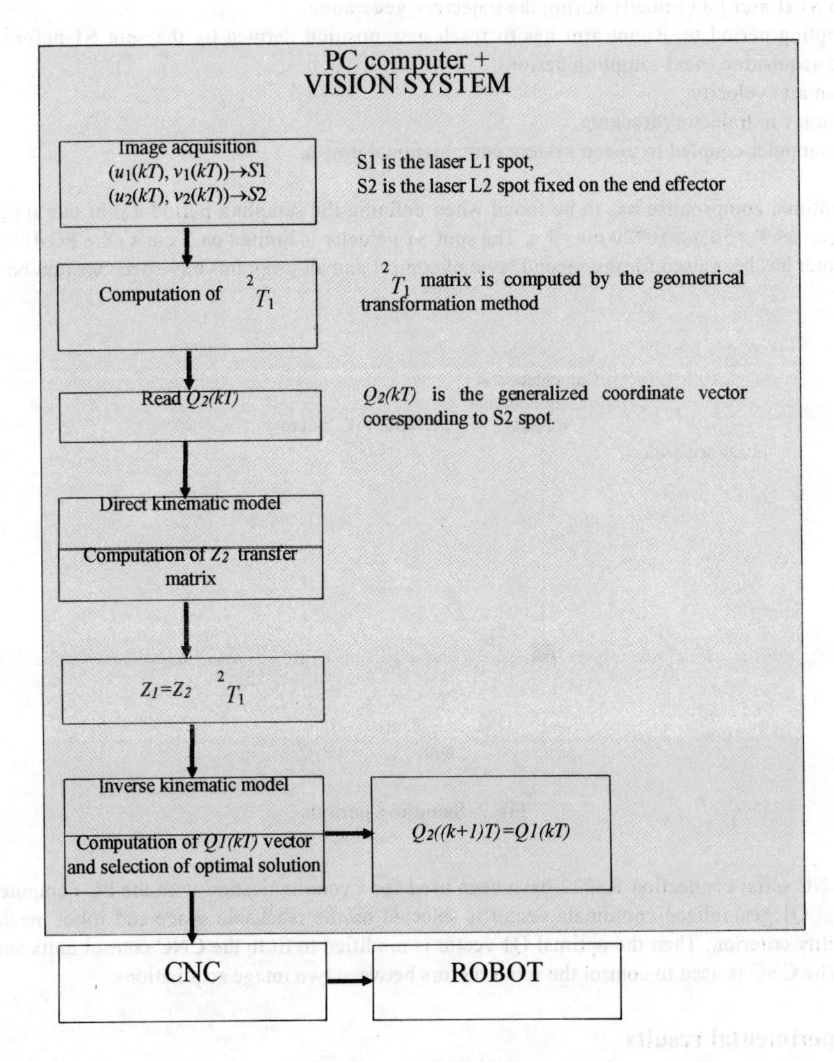

Fig. 6. Design of the tracking algorithm (t=kT)

b) Sampling periods

Real time robot tracking system described in this paper has two levels of control and each level has its own sampling period. The robot arm motors are controlled by the CNC with the sampling period t_e, and CNC is controlled by PC computer, coupled to a vision system and CCD camera, with the sampling period T_e ($T_e > t_e$) Fig. 7.

The CNC sampling period is fixed and it depends on the CNC computational power, on the robot arm mechanical properties and on the DC motors, driving robot arm articulations, technical characteristics. In our application we used t_e=20 ms.

The second level sampling period Te has to be defined in order to respect following criterion:

- Spot S1 (Laser L1) velocity during the trajectory generation,
- Sampling period te. Robot arm has to reach new position defined by the spot S1 before a new image acquisition (next sampling period),
- Robot arm velocity,
- Accuracy in trajectory tracking,
- PC computer coupled to vision system computational power.

The optimal compromise has to be found when defining the sampling period T_e. In our application we have set $T_e=50*t_e=50*20$ ms.=1 s. The spot S1 velocity is limited on 3 cm/s., the PC 486 DX 66 computer has been used for the second level of control and all programs have been written on C++.

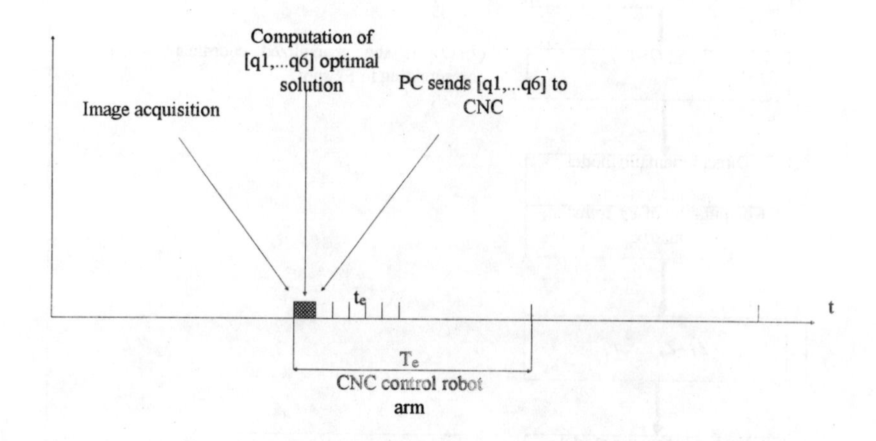

Fig.7. Sampling periods

The CNC serial connection RS232 have been used for a communication with the PC computer. The optimal Q1 generalized coordinate vector is selected on the reachable space and robot mechanical flexibility criterion. Then the optimal Q1 vector is modified to fit to the CNC control units and send to it. The CNC is used to control the robot motors between two image acquisitions.

5 Experimental results

The experimental layout is shown in Fig. 8. The idea was to generate a trajectory with the L1 laser (Spot S1), to detect robot initial position, trajectory starting position and trajectory final position (S2 Spot), by the vision system. The tracking is realized in real time. The velocity of the spot S1 is 3 cm/s.

Experimental results are shown in Fig. 9. We can see two superposition of spots.

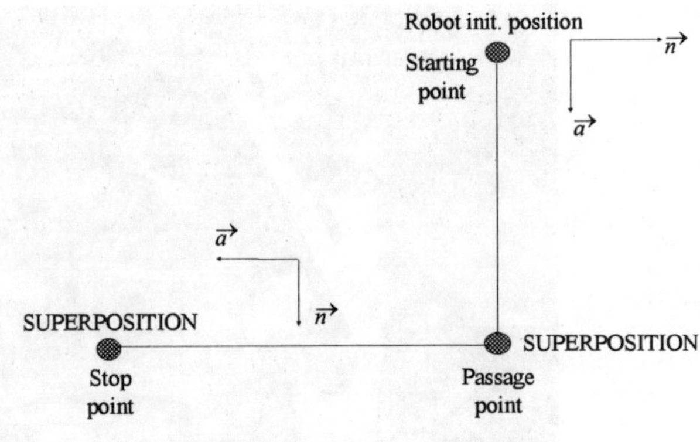

Fig.8. Experimental layout

Fig.9. Trajectory tracking

6 Conclusion

A control algorithm using two independent lasers and CCD camera for the 2D trajectory tracking has been proposed in this paper. This algorithm is a case study of CCD camera and the vision system feedback for the control of both position and orientation of the end effector. Using this algorithm, the trajectory tracking of an articulated robot arm can be realize without tracking operation. No interpolation of given trajectory is necessary. The trajectory is generated by the laser L1 spot on the 3D surface. This tracking is realized by two superposition, the second spot S2 is obtained from the laser L2 fixed on the robot end effector. Two spots superposition tracking principle eliminates the image deformation problems and calibration of the vision system. The laser beam tracking system has been realized by the proposed algorithm. The validity of the control scheme has been confirmed through the experiment.
Figures 10 and 11 are the photographs of the real time robot tracking system with two independent lasers.

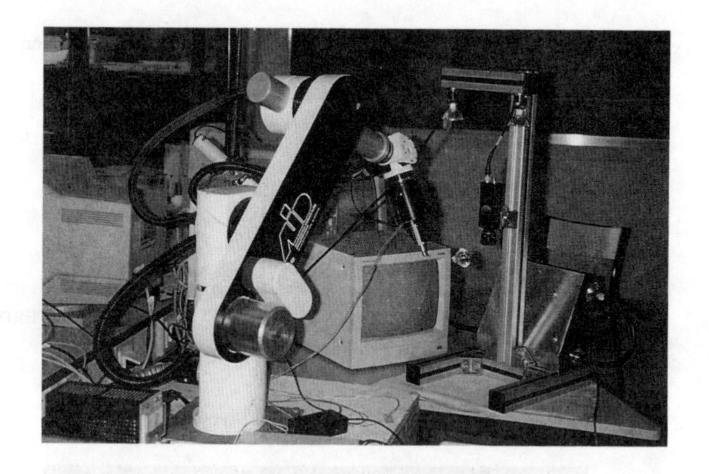

Fig.10. Real time robot tracking system

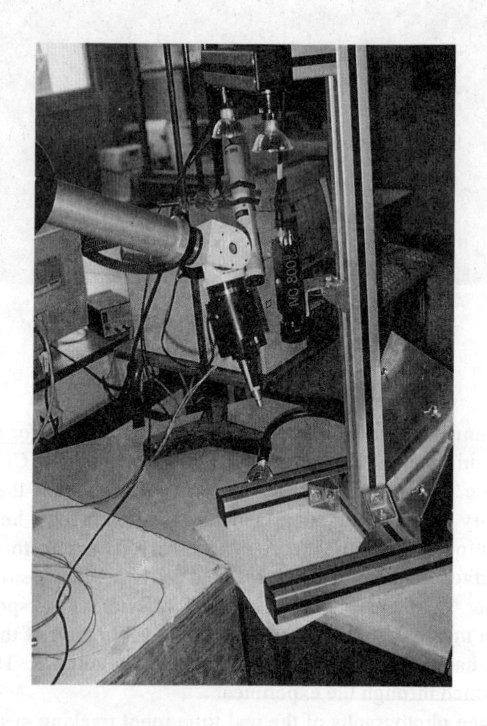

Fig.11. Real time robot tracking system

REFERENCES:

[1] T.A.G. Heeren and F.E. Veldpaus, "An Optical System to Measure the End Effector Position for On-line Control Purposes", The International Journal of Robotics Research, Vol. 11, No. 1, February 1992.

[2] S. Venkatesen and C. Archibald, "Real-time Tracking in Five Degrees of Freedom Using Two Wrist-mounted Laser Range Finders", IEEE Compute. Soc. Press., 1990, Vol.3, pp. 1004-1010; Conference Cincinnati, OH, USA, 13-18 May 1990.

[3] M. Rioux and al., "Design of a Large Depth of View Three-Dimensional Camera for Robot Vision", Optical Engineering, Vol. 26, No. 12, p. 1245, 1987.

[4] C. Archibald and al., "Real-time Feedback Control Using a Laser Range Finder and Harmony", Proc. of the 7th Canadian CAD/CAM and Robotics Conference, pp. 6-56, 1988.

[5] S. Verkatesan and C. Archibald, "Three Degree of Freedom Tracking in Real Time Using a Wrist-mounted Laser Range Finder", Proc. of the 2nd Conference on Intelligent Autonomous Systems, p. 386, 1989.

[6] M. Kabuka, J. Desoto and J. Miranda, "Robot Vision Tracking System", IEEE Transactions On Industrial Electronics, Vol. 35, N° 1, February 1988.

[7] S. Ranade and J.M.S. Prewitt, "A comparison of some segmentation algorithms for cytology", in Proc. 5th Int. Conf. Pattern recognition, pp. 561-564, December 1980.

[8] J.S. Weszka, "A survey of threshold selection techniques", Comput. Graphics and Image Processing 7, pp. 259-269, 1978.

[9] R. Kohler, "A segmentation system based on thresholding", Comput. Graphics and Image Processing 15, 99. 319-338, 1981.

[10] A. Rosenfeld, "Image pattern recognition", Proc. IEEE, Vol. 69, May 1981.

[11] C. Shevington, G. Flochs and B. Schaming, "A statistical approach to image segmentation", in Proc. Pattern Recognition Image Processing, pp. 267-272, August 1981.

[12] C.H. Chen, "On the statistical image segmentation techniques", in Proc. Pattern recognition Image Processing, pp. 262-266, August 1981.

Fusing Image Information on the Basis of the Analytic Hierarchy Process

TOSHIYUKI KOJOH, TADASHI NAGATA, and HONG-BIN ZHA

Faculty of Eng., Kyushu Univ., Fukuoka, 812 Japan

Abstract

We propose a method for recognizing 3-D objects by use of images taken at several fixed viewpoints. In the method, we suppose that each input image possesses an agent which has object models in its model database and can execute feature extraction, reconstruction and model matching processes. Under such a supposition, each agent is intended to implement the analysis on the associated image. In this processing level, however, the method can't recognize objects, and even if it can derive some results, they are possibly unreliable because of the restriction of view directions or occlusions. Therefore, it is necessary to integrate information from other images to derive a correct solution.

We introduce a new mechanism in which the agent being able to provide the largest amount of valuable information about the observed objects will be chosen as the supervisor and play the role of integrating results from other agents. However, it is difficult to formulate rules for evaluating if the chosen agent can produce a good result for the recognition, because various factors should be combined for the evaluation. To resolve the difficulty, we use the analytic hierarchy process (AHP) as the evaluation rules.

Keywords: Image recognition, Sensor fusion, Multi-agent, AHP

1 Introduction

In the research of computer vision, multiple images are frequently used for analyzing environment of intelligent machines, such as robots. The reason for the usefulness of multiple images is that the use of them can get various information such as range data, velocity vectors, to make up the part which a single image can't play. In this paper, we propose a method of fusing or integrating visual information for recognizing environments through cooperative processing on multiple images[1].

In the researches using multiple images, there are studies for reconstruction of general scenes based on the triangulation principle for depth computation[2][3][4]. There are also studies such as generating an object model using a graph or an incremental model from multiple input images[5][6]. These methods differ from our approach basically is that they have no consideration on the cooperative processing. As an approach which is similar to our concept, there are studies such as recognizing 2-D or 3-D object based on multiple knowledge sources or blackboard models[7][8][9]. However, the systems may derive a solution that is globally incorrect, even if sometimes locally reasonable, because they don't have a mechanism for integrating processing results of subsystems.

When a system is composed of some subsystems, there is a centralized or a decentralized approach for the processing in the system. The centralized approach is applied to various systems until now, but it has a critical limit of processing capacity with increasing of system scale. Recently, the decentralized methods have received great attention instead. However, such a system has also some problems such as dead locks among subsystems. In general, it has advantages and disadvantages at the same time and advantages can be exploited to overcome disadvantages if working in a cooperative manner. We expect that utilization of the property can realize a flexible effective recognition system. To put it concretely, for example, each agent analyzes each input image and executes cooperative processing with other ones. The general system executes the decentralized processing at this time. Any agent that obtains effective information in comparison with others can be selected as the supervisor and plays the role of integrating results of the others. If the supervisor breaks down, the general system can continue the processing, because the other subsystems are combined together in a decentralized way. In our research, we expect that a flexible and effective system can be realized by combining subsystems that will behave according to changes of the environment or results at some intermediate stages.

The difficult problem in this concept is how to judge if one agent obtains more effective information than other ones. To resolve this problem, we introduce the analytic hierarchy process (AHP) for the evaluation. We have made some experiments on real images using the proposed method and will discuss the results to show the usefulness of the method.

2 System structure

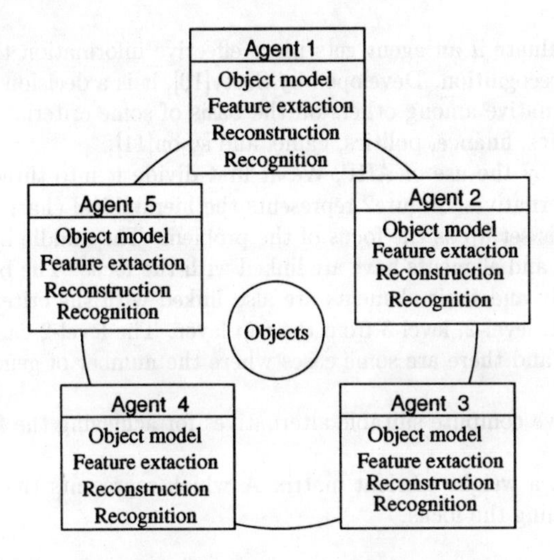

Fig. 1: System structure

The system structure is illustrated in Fig.1. We suppose that the objects to be recognized are polyhedra and some geometric object models exist already in the system. Input images are acquired from viewpoints around the objects to be recognized. In this paper, since five input views are chosen, each viewpoint is at a vertex of a regular pentagon.

Images gotten from CCD cameras are regarded as the agents which have the object models and can execute feature extraction, reconstruction and recognition processes respectively. In the processing, it is difficult to recognize the general environment, because of the restriction of view directions or occlusion. Therefore, fusing or integrating visual information from other agents is necessary. We have proposed a fusion method that communicates each agent with information on the analysis results of other agents, and then updates the agent's state by using the information. If it has more effective information than other agents, any agent can be chosen as a supervisor for further data integration. Because it has some advantages for both centralized and distributed processes, this technique promises to be more flexible and adaptive than a general method. In the method, however, there are still problems how to evaluate the agent's performance to determine the supervisor for integrating the information.

In the next sections, we will present a solution to it by employing the analytic hierarchy process, which is one of the decision methods usually used to determine a best way in some alternatives.

3 A brief description of AHP

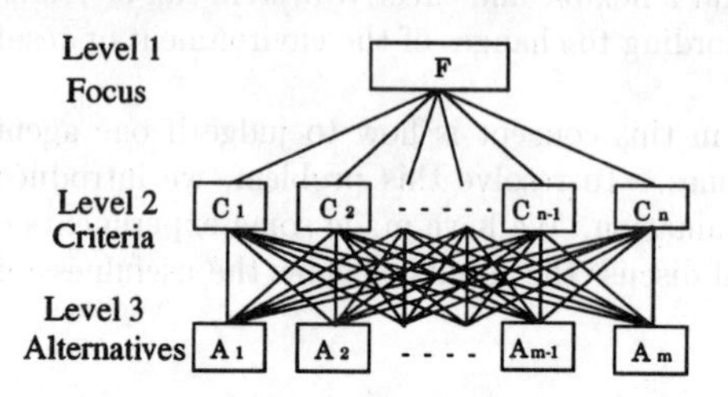

Fig. 2: AHP structure

We use AHP to evaluate if an agent gets more effective information than some others for the purpose of object recognition. Developed by Satty[10], it is a decision method that selects the most suitable alternative among others on the basis of some criteria. It has been applied in the fields of economics, finance, politics, game, and so on[11].

To solve a problem by the use of AHP, We at first divide it into three levels, that is, the focus, criteria, and alternatives. Figure2 represents the hierarchical chart of the division. The top layer in this chart is set up as the focus of the problem. The middle layer lines up criteria to resolve this problem and elements have are linked with the focus. The bottom layer lines up alternatives horizontally and their elements are also linked with the criterion. We call them, respectively, the level-1, level-2, level-3 from the top layer. The level-2 can be subdivided into more detailed criteria, and there are some cases where the number of general layers is four or five.

After the division, we compute suitable alternatives for achieving the focus based on their criteria.

To do so, we derive a weight efficient matrix **A** which represents the importance of each criterion for accomplishing the focus.

$$\mathbf{A} = \begin{bmatrix} a_1 & a_2 & \cdots & a_n \end{bmatrix}^T, \tag{1}$$

where,n is the number of criteria. We also determine a weight efficient matrix **B** that represents how desirable each alternative is against others.

$$\mathbf{B} = \begin{bmatrix} b_{11} & b_{12} & \cdots & b_{1n} \\ b_{21} & b_{22} & \cdots & b_{2n} \\ \vdots & & & \vdots \\ b_{m1} & b_{m2} & \cdots & b_{mn} \end{bmatrix} \tag{2}$$

where, m is the number of alternatives. At last, we calculate an importance matrix **C** of alternatives for accomplishing the focus as

$$\mathbf{C} = \mathbf{BA}. \quad (3)$$

In other words, the elements of **C** represent the utility values of the alternatives.

4 Object models and the reconstruction processing

4.1 Generating model

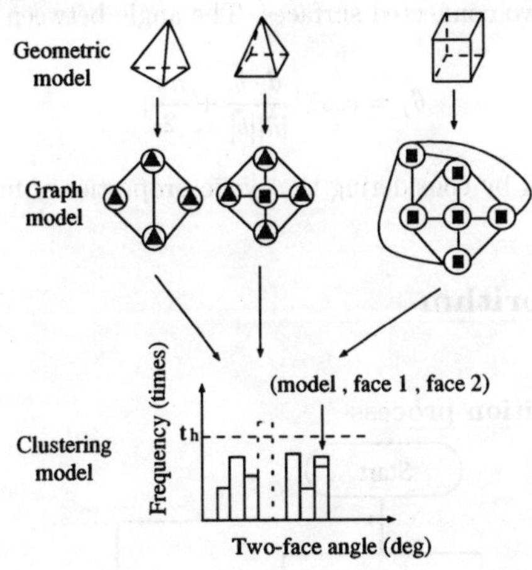

Fig. 3: Model structure

To recognize an object, it is necessary to make use of knowledge about the object. The representation of knowledge depends on recognition processes to be utilized in the recognition system. Because our recognition processes are executed based on two-face angle, we introduce two types of model which represent the model features.

We show in Fig.3 the basic concept for generating object models. A type of models is the graphs model which represent shapes of the object surfaces. Normal vectors of the surfaces are entries in nodes, and adjacent surfaces and the angles between the surface normals are entries in arcs.

Another type is clustering models that are represented in the form of a frequency distribution on the appearance of two-face angles. We consider the recognition agents can realize effective recognition by use of such a model, because they need not to search all graph models by cutting off a large part of the search space. The horizontal axis in the graph on clustering models represents the two-face angle, and the vertical axis represents the times of the angle appearance.

If the frequency is larger than a threshold value t_h, the angle entry should be eliminated because it is not much helpful for limiting search space.

4.2 Feature extraction

Each image processing agent extracts the normal vector for each face of the observed objects and the edge segments to make model matching possible. We use photometric stereo method[12] here to compute the normal vector on each image element by using intensity values and light position parameters when the objects is irradiated from three different positions. As we suppose that the objects to be recognized are polyhedra, the normal vector of an face is derived by clustering same normal vectors at image elements.

Next, the image edges are detected by a filtering process. The edge parameters are computed by using Hough transformation[13], and segmentation of the edges is carried out with some other techniques.

4.3 Reconstruction processing

The goal of reconstruction processing is to make the agents organize the viewed scene as a form similar with the graph models. Two-face angles are calculated as in the following. Extracted edge segments are determined as a ridge line or a contour line. If an edge segment is a ridge line, the faces sharing it is connected. Here, we consider two vectors, $\vec{a}$ and $\vec{b}$, which are normal vectors of the two connected surfaces. The angle between them is computed by

$$\theta_f = cos^{-1}\frac{\vec{a}\cdot\vec{b}}{|\vec{a}||\vec{b}|} + \frac{n\pi}{2}. \tag{4}$$

The variable n is introduced by considering the cyclic properties of function $cos\varphi$.

5 Recognition algorithm

5.1 Flow of the recognition process

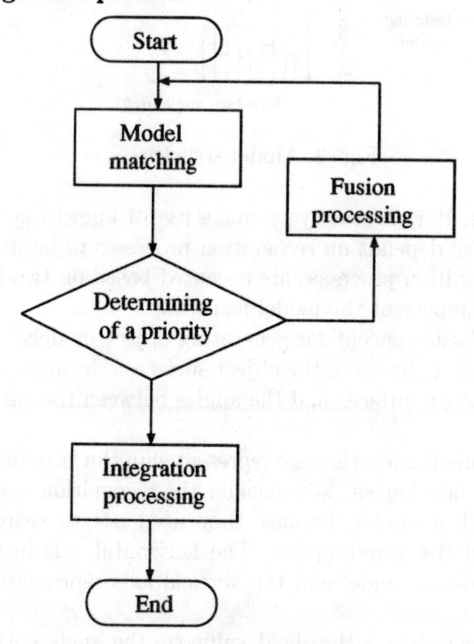

Fig. 4: Flow of recognition process

Objects are recognized by matching between models and features mentioned in Section4. Figure4 shows the flow of the recognition process in this system. At first, each agent executes the model matching and estimates viewed objects. Using the result, it evaluate if the agent obtains more effective information than adjacent ones. This judgement is executed using AHP. If it has gotten effective information, it asserts a priority to them, supervises other ones, and recognizes objects by integrating these analysis results. If there is not enough information for making decision, it obtains information from other agents and iterates the model matching and fusiong processes.

5.2 Model matching

The model matching aims to estimate model objects corresponding to viewed objects by matching the reconstructed scenes with the two types of model.

From the scene described by using the method given in Section4, we generate a frequency distribution with respect to two-face angles the same as the clustering model. Only the features having the smallest value of frequency is matched with the clustering model. Next, the features around the considered one is compared with the graph model.

5.3 Determining a supervisor

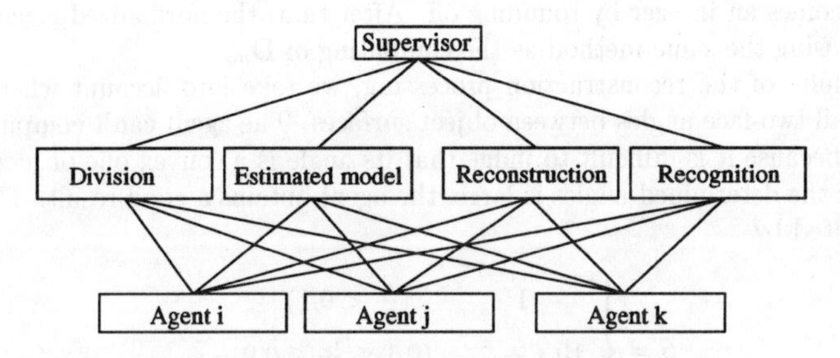

Fig. 5: Determination of a supervisor

We illustrate the structure of our AHP-based method in Fig.5.

The top layer in this figure is set up as the focus of this problem. The focus of our method is to determine a supervisor for general information integration. The middle layer lines up criteria used to evaluate each element of the problem from different aspects. Each criterion has a link with the focus. In the system, the criteria to support the focus are formed on the basis of a division ratio, an estimated model ratio, results of the reconstruction processing, and results of the recognition processing. The bottom layer lines up horizontally the alternatives that can be chosen and they have a link with each criterion. The alternatives in the method are the image agents corresponding to the considered viewpoints. At first, we derive a weight coefficient matrix $\mathbf{A}$ which represents the importance of each criterion for achieving the focus. To derive $\mathbf{A}$, the matrix $\mathbf{V}$ is given as the pairwise comparison matrix using an integers from 1 through 9. The integer implies the importance of each criterion when compared with other ones for the achieving of the focus. $\mathbf{A}$ is derived by computing the normalized eigenvector corresponding to the maximum eigenvalue of the matrix $\mathbf{V}$. In this paper, matrix $\mathbf{A}$ is determined empirically on the properties of general object recognition problems and consists of four column matrices.

Next, we determine a weight coefficient matrix $\mathbf{B}_a$, which represents how each alternative is desirable for each criterion, with a implying the corresponding image agent.

The importance ratio of alternatives concerning each criterion is determined by the results of analysis in each agent. The division ratio implies the number of separated clutter of objects in the scene. Since a separated object is easier to recognize than totally cluttered ones, we consider it is effective information for a recognition task. The importance ratio is computed by

$$D = \begin{cases} 2n-1 & (1 \leq n \leq 5) \\ 9 & (n > 5), \end{cases} \tag{5}$$

where n is the number of clutter and it is counted by scanning contour edges. The computed value D is sent to the adjacent agents. The matrix $\mathbf{D}_{nor}$ similar to $\mathbf{A}$ is derived by using the values. For example, when the division ratios of agent i and j are given as D_i,D_j, respectively. The importance ratio D_{ij} equals to D_i/D_j. The normalized eigenvector $\mathbf{D}_{nor}$ is derived using a pairwise comparison matrix $\mathbf{D}$.

The estimated model ratio implies a ratio of models identified by model matching. If the number of identified model is small, the agent obtains a good results for the subsequent recognition processes. The importance ratio is computed by

$$E = \begin{cases} 1 & (\frac{m}{n} > 9) \\ 10 - \frac{m}{n} & (1 \leq \frac{m}{n} \leq 9) \\ 9 & (\frac{m}{n} < 1), \end{cases} \tag{6}$$

where n is the number of object clutter and m is the number of the identified models. The value of E becomes an integer by rounding off. After that, the normalized eigenvector $\mathbf{E}_{nor}$ is computed by using the same method as the computing of $\mathbf{D}_{nor}$.

For the results of the reconstruction processing, we take into account whether the agent can compute all two-face angles between object surfaces. The agent can't compute all two-face angles easily, because it is difficult to judge that its angle is a convex one or a concave one. If the number of the determined angles is large, the agent obtains a good result. The importance ratio is computed by

$$S = \begin{cases} 1 & (\frac{t_p}{t_n} < 0.1) \\ 10 * \frac{t_p}{t_n} & (0.1 \leq \frac{t_p}{t_n} \leq 0.9) \\ 9 & (\frac{t_p}{t_n} > 0.9), \end{cases} \tag{7}$$

where t_n is the number of the extracted two-face angles and t_p is the number of the computable two-face angles. The value of S becomes an integer by rounding off, and the normalized eigenvector $\mathbf{S}_{nor}$ is computed by using the same method as the computation of $\mathbf{D}_{nor}$.

In the fourth criterion concerning results of the recognition processing, we use the ratio of the faces which have possible matches with a model faces and all of the extracted faces. The importance ratio is computed by

$$G = \begin{cases} 1 & (\frac{f_p}{f_n} < 0.1) \\ 10 * \frac{f_p}{f_n} & (0.1 \leq \frac{f_p}{f_n} \leq 0.9) \\ 9 & (\frac{f_p}{f_n} > 0.9), \end{cases} \tag{8}$$

where f_n is the number of the extracted surfaces and f_p is the number of the recognizable surfaces. The normalized eigenvector $\mathbf{G}_{nor}$ is computed by the same method as mentioned above.

By computing the normalized eigenvectors, the matrix $\mathbf{B}_a$ is represented as

$$\mathbf{B}_a = [\mathbf{D}_{nor}{}^T \mathbf{E}_{nor}{}^T \mathbf{S}_{nor}{}^T \mathbf{G}_{nor}{}^T], \tag{9}$$

which is a 4*3 matrix.

At last, the agent calculates the utility matrix $\mathbf{C}_a$= $[c_{ai}\ c_{aj}\ c_{ak}\ ]^T$ of alternatives for accomplishing the focus by

$$\mathbf{C}_a = \mathbf{B}_a \mathbf{A}, \tag{10}$$

where c_{ai} is the utility value of the agent under consideration, and c_{aj}, c_{ak} are the values of the adjacent ones. When the values are satisfied with

$$c_{ai} - \max(c_{aj}, c_{ak}) > T_{th}, \tag{11}$$

the considered agent asserts a priority over others and becomes a supervisor. T_{th} in (11) is a threshold value.

5.4 Fusion and integration processing

Each agent executes the fusion or the integration processing according to the result of AHP calculation. The agent that is chosen as the supervisior obtains information from other agents and eliminates incorrect models by the model matching. Data transformation among agents is executed by use of homogeneous coodinate transformation. Any agent that is not the supervisor sends its own information to the supervisor and continues their processing again. After the supervisor analyzing the data, it sends the results to other ones. If the solution is correct, the general processing ends. In the case of fusion processing, the relationship among agents is the same. If an agent can not calculate a two-face angle which but can be derived by others, it obtains the information from these agents. Then the agent can take into account the new two-face angle and continue to execute the model matching for the AHP computation.

6 Results of experiments

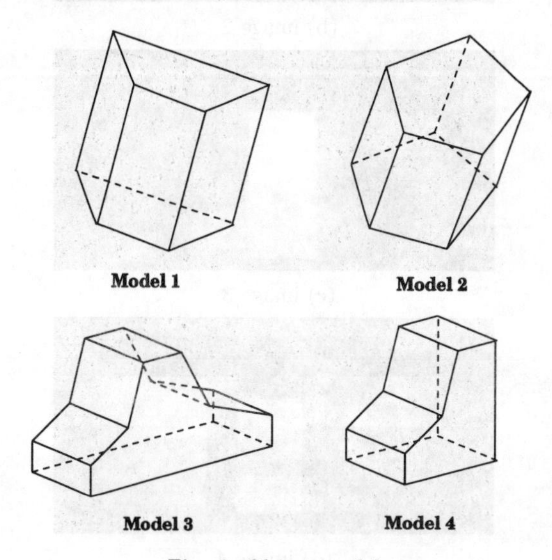

Fig. 6: Object models

We have made an experiment using real images based on the proposed algorithm. Figure6 illustrates the object models used in this experiment and Fig.7 shows each viewpoint image. The following weight coefficient matrix is obtained as

$$\mathbf{A} = (0.15, 0.35, 0.22, 0.28). \tag{12}$$

The threshold value T_{th} is set at 0.2. Table1 shows the utility values after the initial model matching. After the fusion processing is executed, the agent4 becomes the supervisor. The utility values derived at this stage is shown in Table2. Figure8 shows the recognition result.

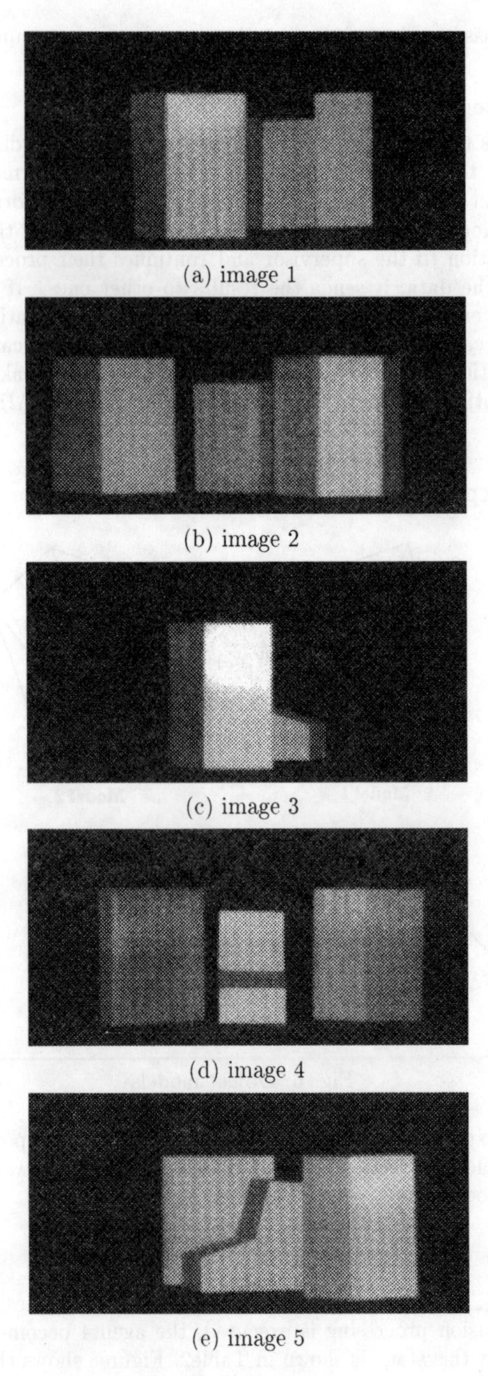

(a) image 1

(b) image 2

(c) image 3

(d) image 4

(e) image 5

Fig. 7: Input images

Table 1: The utility value (No cooperative processing)

C_a	Ag1	Ag2	Ag3	Ag4	Ag5
C_{ai}	0.31	0.38	0.35	0.42	0.32
C_{aj}	0.37	0.33	0.41	0.32	0.40
C_{ak}	0.32	0.29	0.24	0.26	0.28

Table 2: The utility value (Cooperative processing)

C_a	Ag1	Ag2	Ag3	Ag4	Ag5
C_{ai}	0.31	0.42	0.29	0.56	0.26
C_{aj}	0.37	0.38	0.48	0.27	0.56
C_{ak}	0.32	0.20	0.23	0.17	0.18

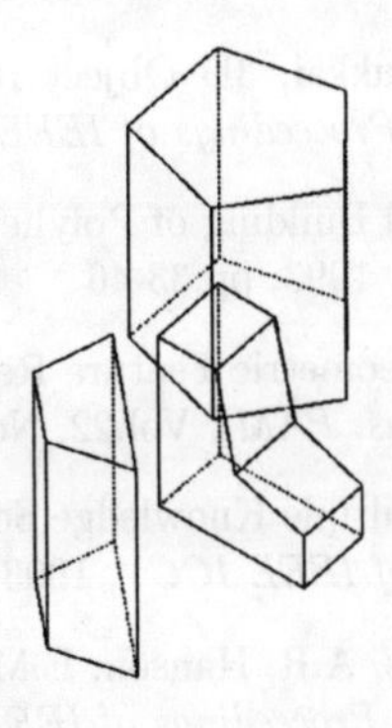

Fig. 8: Recognition result

7 Conclusions

We described a fusion method of visual information for recognizing 3-D objects. The method has been tested by performing experiments on real images. The key contributions are summarized as follows.

- It is shown that it is possible to design a flexible recognition system which can control some agents according to the environment or partial processing results.
- AHP is proven an effective rule for evaluating the agent ability in the information integration. The introduction of it into a recognition process can resolve the complex problems in which various factors should be taken into consideration.

By formulating suitable rules for achieving different purposes, we believe that our approach can be applied to not only 3-D object recognition system, but also to the complicated systems such as multiple robots system.

References

[1] T.Kojoh, T.Nagata, H.B.Zha, Cooperative Recognition using Multi-View Images *Proceedings of Korea Automatic Control Conference ('93 KACC, Int. Session)*, 1993, pp.70-75

[2] M.Ito, A.ishii, Three View Stereo Analysis, *IEEE Trans. PAMI*, Vol.8, No4, 1986, pp.524-533

[3] R.Chiou, K.Hung, J.Guo, C.Chen, T.Fan, J.Lee, Polyhedron Recognition Using Three-view Analysis, *Pattern Recognition*, Vol.25, No.1, 1992, pp.1-16

[4] R.Mohan, D.Weinshall, R.R.Sarukkai, 3D Object Recognition By Indexing Structural Invariants from Multiple Views, *Proceedings of IEEE ICCV*, 1993, pp.264-269

[5] Z.Chen, S.Ho, Incremental Model Building of Polyhedral Invariants from Multiple View, *Pattern Recognition*, Vol.26 No.1, 1993, pp.33-46

[6] Y.C Tang, C.S George Lee, A Geometric Feature Relation Graph Formulation for Consistent Sensor Fusion, *IEEE Trans. PAMI*, Vol.22, No.1, 1992, pp.115-129

[7] B.Besser, S.Estable, B.Ulmer, Multiple Knowledge Sources and Evidential Reasoning for Shape Recognition, *Proceedings of IEEE ICCV*, 1993, pp.624-631

[8] B.A.Draper, J.B.Brolio, T.Collins, A.R. Hanson, E.M Riseman, Image Interpretation by Distributed Cooperative Process, *Proceedings of IEEE CVPR*,1988, pp.129-135

[9] E.W.Kent, M.O.Shneier, T.H.Hong, Building Representations from Fusions of Multiple Views, *Proceedings of IEEE CVPR*, 1998, pp.1634-1639

[10] T.L.Saaty, *The Analytic Hierarchy Process*, McGraw-Hill, 1981

[11] T.L.Saaty, L.G. Vargas, *Prediction Projection and Forecasting*, Kluwer Academic Publishers, 1991

[12] P.J.Woodham, Analyzing Images of Curved Surface, *Artificial Intelligence*, Vol.17, 1981, pp.117-140

[13] R.O.Duda, Use of the Hough Transformation to Detect Line and Curves in Pictures, *Comm.ACM*, Vol.15, No.1, 1972, pp.11-15

Chapter 4
Distributed Planning and Control

Rule Generation and Generalization by Inductive Decision Tree and Reinforcement Learning

KEITAROU NARUSE and YUKINORI KAKAZU

Dept. of Precision Eng., Hokkaido Univ., Sapporo, 060 Japan

Abstract

In this paper, we attempt to construct a planning mechanism composed of distributed agents for autonomous vehicle navigation in an unknown workspace. Each agent decides a direction that the vehicle should move without any communication to the other agents, by only observing the workspace and the other agents. The agent includes the reinforcement learning mechanism for generating rules for the navigation. However, the rules depend on a state observation method. For generalizing the rules, an inductive decision tree is introduced to the agent. In a new workspace, the agent plans a path efficiently by learning a specific rule to the new workspace, and using the generalized rules. Some computational simulations have been carried out for verifying the proposed agent.

Keywords: Multiple agent system, reinforcement learning scheme, inductive decision tree, mobile robots, path planning

1 Introduction

In this paper, we attempt to construct a planning mechanism composed of distributed agents for autonomous vehicle navigation in an unknown workspace. Each agent decides a direction that the vehicle should move without any communication to the other agents, by only observing the workspace and the other agents. For achieving the task, it is needed to acquire a map for the workspace and rules for arriving at a goal point in dynamically changing environments including mobile obstacles and the behavior of the other vehicle.

We employ the reinforcement learning scheme as a rule generation mechanism because it can learn the rules by a simple mechanism without an external teacher. An agent determines a state by observing the workspace, selects an action for the state based on a *policy*, evaluates the selected action and updates the policy. This process is iterated until an adequate policy is obtained. In this paper, Q-Learning, one of the reinforcement learning schemes, is introduced to the agent, because it can learn a sequence of actions to the goal. The policy of Q-Learning is represented as a utility value at a state for an action, called Q-value.

However, the rules generated by the Q-Learning depend on a state observation method, because the policy is represented as a utility value for a state and action. Even if the workspace has been changed slightly, it needs to learn again for the new workspace, although most of the old rules are useful for the new workspace. Therefore, it is expected to extract efficient rules from the old rules and apply them to the new workspace. The rule extraction is regarded as rule generalization in this paper.

For attaining the rule generalization, we introduce an *inductive decision tree* to the agent as rule representation. The decision tree has merits on readability and translatability compared with the one of the reinforcement learning. The decision tree is generated to classify an action for a state and constructed from the state-action pairs acquired by Q-Learning. In the process, it is important to extract rules that are useful for a general case, not ones for a specific case. Then, the rules represented as the tree are fed back to the planning mechanism and it continues learning to acquire to the specific rules for the new workspace.

In the following, the reinforcement learning scheme and the inductive decision tree algorithm are described briefly. Next, efficiency of a tree generation for the planning task is discussed. Then, an agent with the rule generation mechanism is proposed. Finally, the results of some computational simulations are shown for verifying the proposed agent.

2 Reinforcement Learning

In the reinforcement learning, a decision maker selects an action according to a *policy*, evaluates the selected action, and updates the policy based on the evaluation. By iterating the process, the decision maker acquires an adequate policy. In the following, we employ one of the reinforcement learning scheme, Q-Learning [1].

Q-Learning can work in an environment such that each action can not be evaluated immediately. It preserves the policy as an estimation value of an expected reward, termed *Q-value*, for all pairs of a state and action. It selects an action with a high Q-value frequently.

Let $Q(x,a)$ be the Q-value for an action a in a state x. We employ the Boltzmann distribution as an action selection probability among a number of methods for selecting the action. Namely, the action is selected obeying the following probability.

$$\Pr(x|a) = \frac{\exp(Q(x,a)/T)}{\sum_{b \in A} \exp(Q(x,b)/T)}, \tag{1}$$

where T is a temperature parameter and A is a set of actions.

If the state transits to a state y and the decision maker receives a reward r, which is 1 if y is a goal state or 0 otherwise, after performing the selected action, the Q-value is updated by the following equations.

$$Q(x,a) \leftarrow (1-\alpha)Q(x,a) + \alpha(r + \gamma U(y)), \tag{2}$$

$$U(y) = \max_{b \in A} Q(y,b), \tag{3}$$

where α and γ is a learning ratio and discounts ratio (0,1) respectively. $U(y)$ is a maximum Q-value obtained in a state y, and it means utility of that state.

Inductive Decision Tree

ID3 [2] is one of methods for generating an inductive decision tree. It generates a classifier in the form of a tree for given training instances, which is composed of attributes and a class, to identify them correctly.

Let T be a training instance set and $\{C_1, C_2, \cdots, C_k\}$ be classes. If T contains one or more instances, all belonging to a single class C_j, then the decision tree for T is a leaf identifying the class C_j. If T contains no instances, the decision tree must be determined from information other than T. If T contains instances that belong to a mixture of classes, T is partitioned into subsets T_i where T_i contains all the instance in T with an attribute value O_i in a specific attribute. The decision tree for T consists of a decision node identifying the attributes, and a branch for each attribute value. The same procedure is applied recursively to each subset of the training instances, until the subset contains only a single class.

For classifying an un-learned instance accurately, it is expected for the decision tree to be small, because a large tree can include a number of exceptions. Therefore, the order of selecting an attribute is important. In ID3, the attribute is selected the following criterion, called *gain*.

Let $freq(C_i,S)$ stands for the number of instances in any sets of instances S that belongs to class C_i. Consider T has been partitioned in accordance with the n attributes of an attribute X. Then, the gain criterion is given as

$$gain(X) = info(T) - info_X(T), \tag{4}$$

$$info_X(T) = \sum_{i=1}^{n} \frac{|T_i|}{|T|} \times info(T_i), \tag{5}$$

$$info(T) = -\sum_{j=1}^{k} \frac{freq(C_j,S)}{|S|} \times \log_2\left(\frac{freq(C_j,S)}{|S|}\right). \tag{6}$$

Then, ID3 selects an attribute to maximize the gain criterion and partition T into subsets recursively.

3 Proposed Method

3.1 Planning task

In the planning task, we consider multiple vehicles with their own goal in a same workspace. Then, we must generate an action sequence for all the vehicles avoiding any collision and deadlock. The vehicle can move four directions or wait in the same position. The agent can observe only its current position and four neighbors, up, down, left and right. The agent can know where is a goal and where is an obstacle only if it reaches to the point. Figure 1 shows an example of the planning task [3].

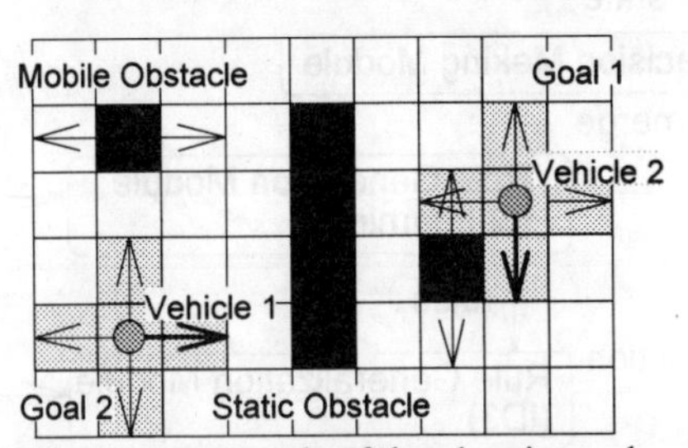

Fig. 1: An example of the planning task.

3.2 State and Action

The proposed agent decides a direction that the vehicle moves as an *action* for a *state* determined by observing the workspace. The agent attempts to acquire rules that represent an adequate action for all the states through an iterative process. They are represented as follows:

$$State = (x, y, Obs(Up), Obs(Down), Obs(Left), Obs(Right)), \tag{7}$$

$$Action = \{GoUp, GoDown, GoLeft, GoRight, Wait\}, \tag{8}$$

$$Obs(x) = \begin{cases} 1, & \text{if there is an obstacle at } x, \\ 0, & \text{otherwise.} \end{cases} \tag{9}$$

In the example shown in Fig. 1, there are $9*6*2*2*2*2 = 864$ states.

3.3 Rule and Strategy

The agent knows nothing about the rule and policy in its initial state. Therefore, it must learn a simple rule such that the vehicle should go right if there are obstacles at the up, down and left of the vehicle. Then, the agent acquires an adequate action for all the states. In this paper, this selection of the action for each state is referred as to a *rule*, and a set of the rules for a given task is referred as to a *strategy*. For an instance of the rule, if the vehicle is at (x, y) and there is obstacle at the left in the neighbor, then the vehicle should go up. The strategy is specific to each task, although the rules can be common for several tasks. In the above example, we must generate the rules for all the states, namely 864 rules as the strategy.

Even if a workspace has been changed slightly, the acquired strategy can not match a new workspace because the acquired strategy depends on the original map. Then, the agent must learn the new workspace spending a long time, although the most of the original rules match to the new workspace. If the rules are represented as generalized manner, they are likely to match the new workspace, compared with the original one. We consider extraction of applicable rules to another task from the acquired rules.

3.4 Proposed Agent

Figure 2 shows a schema of the proposed agent. The agent is composed of five modules. The state observation module observes location of the vehicle and obstacles at its neighbor and decide a state. The decision making module decides an action managing action candidates from the rule generation module(Q-Learning) and the rule generalization module(ID3). The decided action is evaluated by the evaluation module after the action is performed. The rule generation module and the rule generalization module are updated using an evaluation signal.

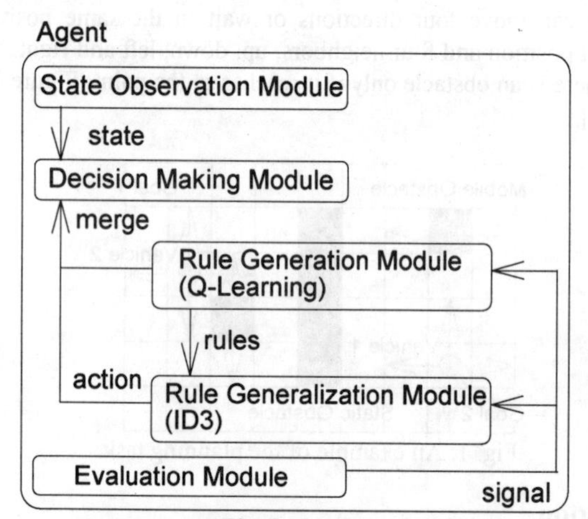

Fig. 2: Agent.

In initial state, the agent has no information on a workspace, a goal point and obstacle's location. The agent searches a workspace without the rule generalization module. The agent acquires a map of the workspace and a policy for arriving at the goal through trails and errors. A reward for Q-learning, r is given as follows:

$$r = \begin{cases} 1, & \text{if } (x(t), y(t)) \text{ is a goal,} \\ -1, & \text{if } (x(t), y(t)) \text{ is an obstacle,} \\ 0, & \text{otherwise.} \end{cases} \tag{10}$$

Once a strategy is obtained by the rule generation module, it is inputted to the rule generalization module as training instances. In the previous example, 864 rules are regard as the training instance set T. T is processed and generalized by the ID3 mechanism. The training instance in T is composed of the states as attributes and the action as a class.

Finally, we propose the decision making module that merges the rules acquired by the rule generation module with the one by the rule generalization module. Let $\Pr_Q(x|a)$ be a probability of selecting an action a at a state x in the Q-learning module, and $\Pr_{ID3}(x|a)$ be the one in the ID3 module, which is 1 for a specific action and the others are 0 because of the acquired rules by ID3 are deterministic. Then, the merged probability of selecting the action a $\Pr_{DM}(x|a)$ is expressed as

$$\Pr_{DM} = p_{merge} \Pr_Q(x|a) + (1 - p_{merge}) \Pr_{ID3}(x|a), \tag{11}$$

where p_{merge} is a parameter for merging two schemes and its interval is [0,1]. If p_{merge} is 0, the agent works as only the Q-Learning module; If p_{merge} is 1, it works as only the ID3 module. By selecting and scheduling p_{merge} adequately, the planning mechanism explores and exploits the rules for the workspace, storing generalized and specific rules.

4 Computational Simulation

For verifying the behavior of the proposed mechanism, we carried out two computational simulations. Simulation 1 was performed for a single vehicle in order to observe a basic behavior of the proposed agent. Then Simulation 2 was carried out for multiple vehicle case.

4.1 Simulation 1

Figure 3 shows workspace configuration. It includes a mobile obstacle and two static obstacles, which are expressed as the black blocks. The mobile obstacle moves a block in a step along a direction represented by the arrow. In this simulation, the agent operates from an initial state without any rules. Ten test phases are carried out after every trial, because the agent performs probabilistically as shown in the previous definition.

Figure 4 represents the result of rules acquired after 300 trials. The white triangle expresses a direction the vehicle should go, and the hatched blocks represent that there is the obstacle at either blocks changing to the other block for every step. Because there are 16 states for neighbor in each location, they are represented by imposing them. Figure 5 shows the changes in average arrival steps. From these figures, we can see that it is obtained the adequate rules and the optimal path through a learning process.

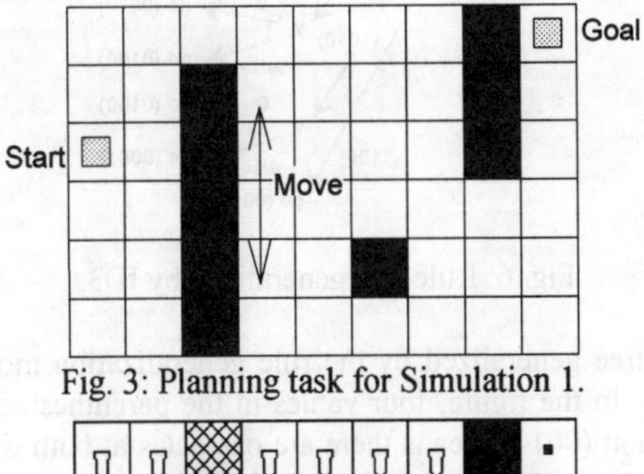

Fig. 3: Planning task for Simulation 1.

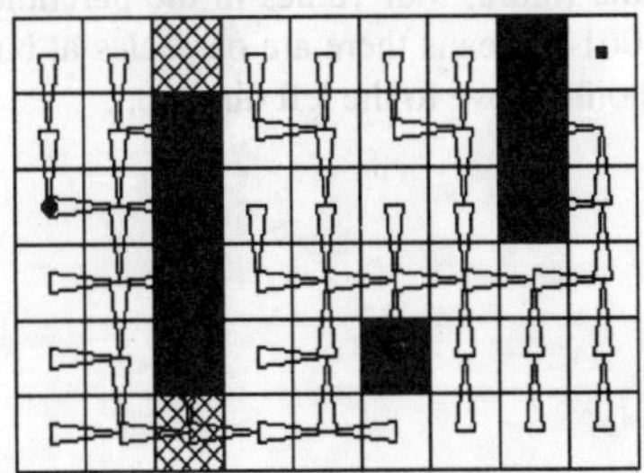

Fig. 4: Graphical representation of rules acquired after 300 trials.

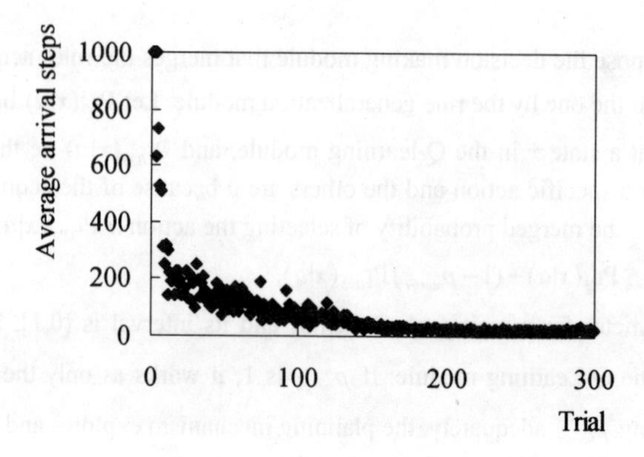

Fig. 5: Changes in average arrival steps.

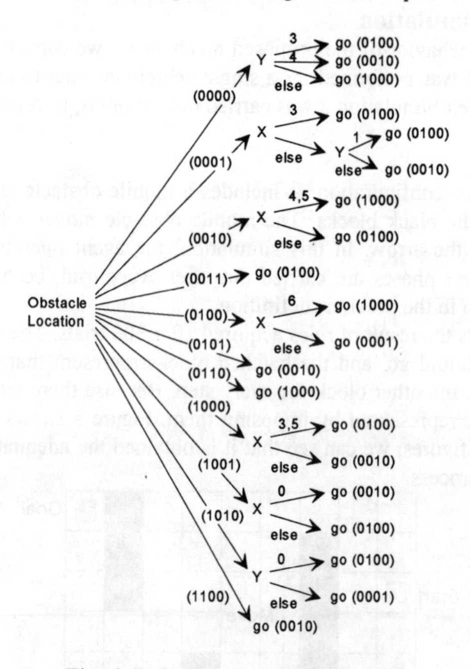

Fig. 6: Rule tree generalized by ID3.

Figure 6 shows a tree generalized by the rule generalization module using ID3 from the rules represented in Fig. 4. In the figure, four values in the parenthesizes depicts a direction. For example, Obstacle Condition (0011) means there are obstacles at both top and left of the vehicle; Go (0100) means the vehicle should move to the left direction.

If Obs(0011) then Go(0100). If Obs(0110) then Go(1000).
If Obs(0101) then Go(0010). If Obs(1100) then Go(0010).
If Obs(0000) If Y=3 then Go(0100)
If Y=4 then Go(0010)
Else Go(1000).
(Down, Right, Up, Left)

Fig. 7: Example of rule expression generalized by ID3.

Figure 7 shows the part of the generalized rules in the form of if-then rule. First four rules represent the one for the simple obstacle avoidance. On the other hand, the other rule is for the case there are no obstacles in the neighbor. The rule means the one for arriving at the goal point, in other words, achieving the global goal. From this figure, we can see that the rules for both general and specific cases are obtained.

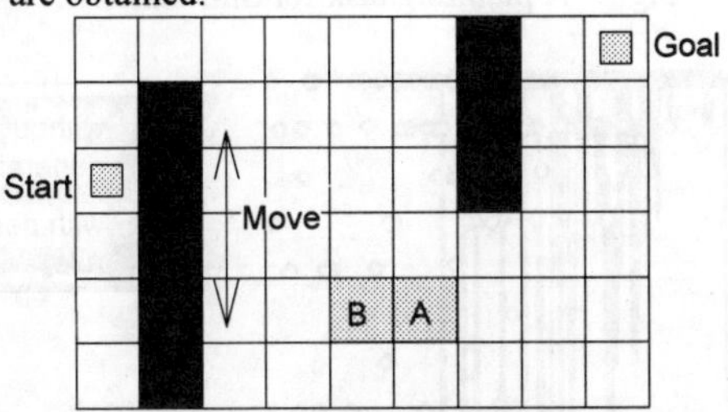

Fig. 8: Modified planning task for verifying the rules acquired in the original one.

Tab. 1: Comparison of the modified task with the original one at first trial.

Workspace	Original (Fig. 3)	Obstacle at A (Fig. 8)	Obstacle at B (Fig. 8)
Goal arrival steps	Failure	16	537
Collision	Yes	No	No
Optimum	No	Yes	No

For verifying the generalized rules, we carried out two simulations using them in the workspace changed slightly shown in Fig. 8. The right long static obstacle is shifted left for a block. The small obstacle at the middle of the workspace is changed into two cases: one is an obstacle is located the point 1 and the other is located at 2. Table 1 shows the result of the original simulation and these ones. For checking whether the acquired rules are applicable for the modified workspace, they are compared by the result of the first trial. In the case A, we can see the shortest path is planned. In this case, the generalized rule works very well. In the case B, although the shortest path can not be planned, there are no collisions, while a path can not be found at the first trial in the original simulation. In this case, we can think that search performance is increased by avoiding any collisions using some of the generalized rules.

4.2 Simulation 2

We carried out a simulation for checking the behavior of the proposed agent in a workspace including two vehicles. Figure 9 shows a planning task for this simulation: There are two vehicles and the start point of one vehicle is the goal point of the other vehicle. In the simulation, the agent for the vehicle 1 includes the generalized rule used in Simulation 1 and the one for the vehicle 2 has no generalized rules. As a comparison result, we also performed the case that the agent for the vehicle 1 has no generalized rules. Figure 10 shows the changes in average arrival steps for the agent with/without the generalized rules. In the case of the agent with the generalized rules, the average goal arrival steps converge more quickly than the one without them. Therefore, the generalized rules acquired in a single vehicle case are efficient to the multiple vehicle case.

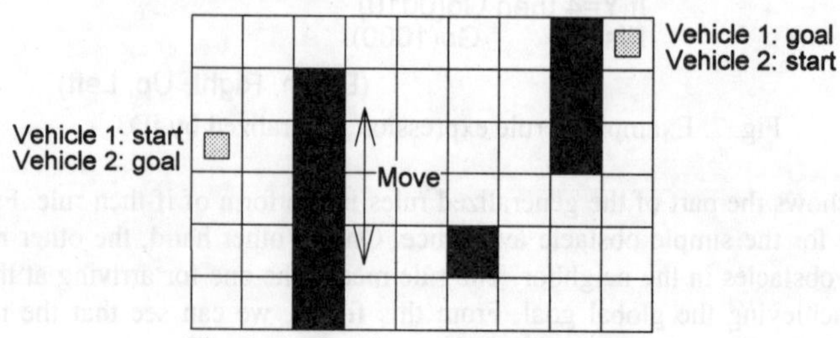

Fig. 9: A planning task for Simulation 2.

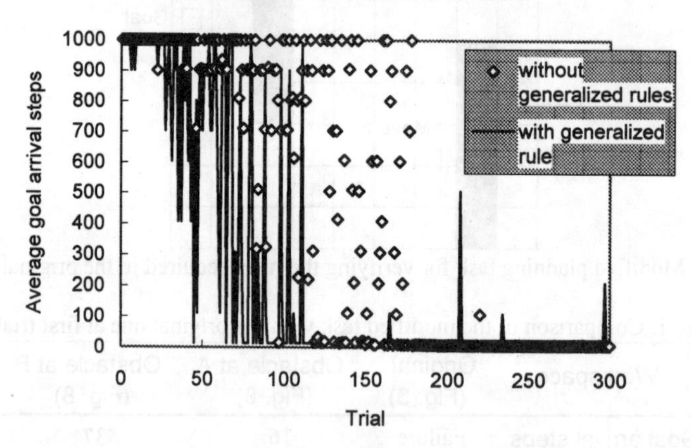

Fig. 10: Changes in average arrival steps.

5 Conclusions

In this paper, for autonomous vehicle navigation task in an unknown workspace, we have proposed a planning mechanism based on simple navigation rules using a reinforcement learning scheme. For generalizing the rules acquired by the reinforcement learning scheme, we have introduced the inductive decision tree to the planning mechanism. Some computational simulations have been carried out for verifying the proposed agent.

References

[1] C. J. Watkins, and P. Dayan, Technical Note Q-Learning, Machine Learning 8, 279-292, 1992.

[2] J. R. Quinlan, Induction of Decision Trees, Machine Learning 1, 81-106.

[3] R. S. Sutton, Integrated Architecture for Learning, Planning, and Reacting Based on Approximating Dynamic Programming, Proceedings of the seventh international conference on Machine Learning, 216-224, 1990.

Fusion Strategy for Time Series Prediction and Knowledge based Reasoning for Intelligent Communication

TOSHIO FUKUDA and KOUSUKE SEKIYAMA

Dept. of Mechano-Informatics and Systems, Nagoya Univ., Nagoya, 464-01 Japan

Abstract

This paper describes a predictive decision making strategy for "Intelligent Communication" in Cellular Robotic Systems (CEBOT). The CEBOT is an autonomous distributed robotic system composed of a number of robotic units called "cell" with limited functions, which configures its optimal structures in terms of hardware and software, according to the task or environment. Intelligent Communication aims at communication reduction through intention reasoning. It regards communication as not only data transmission, but total intelligent activities supposing interaction with other agents. Intelligent Communication synthesizes learning, reasoning and prediction. Additional learning and knowledge acquisition enables more precise inference, and consequently contributes reduction of excessive communication. Since the real world is dynamical, time series analysis will be required and also it will give beneficial clues for intention reasoning to the other robot behavior. However, time series data observed by robot behaviors are nonstationary in general. Because robot behaves based on the planning procedure to execute its tasks, or it might be generated as a result of adaptive motion to the current condition and environment, such as collision avoidance. In order to consider both the trend of state transition and static rules that affect behavior of the robot, we will propose a strategy to synthesize time series prediction and rule / knowledge based reasoning.
Keywords: Cellular Robotic Systems(CEBOT), Intelligent Communication, predictive decision making

1. Introduction

In the multiple robot system, communication plays significant role in achieving cooperative behavior. There are many types and classes of communication, from data transmission to negotiation. However, since agents must cope with many problems which occurs in the group, we deal with rather high level communication, as a part of the decision making process. One of the main role communication is information acquisition to reduce uncertainty in the decision making. Actual environment where robots must work is dynamic and full of the uncertainty. They have to behave under lack of the sufficient information, and therefore, rely on communication to obtain necessary knowledge if possible. However, we should consider that communication load will be explosively high as the number of the agent increases. The CEBOT assumes that a number of robots are engaged in the task[4,5]. Although, Distributed Robot System (DRS) is advantageous in load and knowledge distribution which reduce load to the individual, but on the other hand, agents tend to depend on communication which is noting bur load for the system. Since complete information cannot be obtained in general, if agents continue to communicate with others until they are satisfied, communication delay causes further delay, the system will not work efficiently. Improving efficiency in terms of data transmission will not give the fundamental solution to this problem. We should explore the way to compensate the lack of information when uncertainty remains and communication load is high. "Intelligent Communication" is a strategy to realize communication reduction by intention reasoning. Standing from the view point of that communication should be treated in the context of total intelligent activities which are closely related to interaction with other agents, we attempt to reduce communication load in the whole system through the intention reasoning based on the knowledge and the risk estimate to the decision as a self-regression. Conceptual architecture for the intelligent communication is shown in fig.1. It consists of three main parts.

1) Inner model: Data from sensor inputs first enter the inner model. This model stores necessary information and organizes rules or strategies. Also, it is responsible for communication using the

1) Inner model: Data from sensor inputs first enter the inner model. This model stores necessary information and organizes rules or strategies. Also, it is responsible for communication using the knowledge and rules. Particular important role is to convert correlation of numerical data to symbolic logical relation and abstract rules. The results are reflected to prediction. Realization of these is our future challenging problem.

2) Predictive Decision Making: Actual environment is dynamic. Therefore, intention reasoning should be made considering change of time. Also, only adaptive motion cannot follow rapid change of the world. To be more flexible system, predictive decision making is essential. In particular, unification with knowledge based reasoning makes powerful prediction model. Predictive decision making is a main topic in this paper. So, further details are described later.

3) Risk estimate: As a feedback is effective in the autonomous control, the same mechanism should be applied to the decision making procedure. This means that robot should take into account the result of actions. If sufficient information is not obtained, many candidates of the action can be considered. Our approach is to estimate decision risk. If the result of the decision may cause deteriorating effects over the task execution, the robot must inquire unknown information through observation and communication. On the other hand, if the result is trivial whatever actions may be taken, the robot can neglect communication and take an action. By trading off between decision risk and communication load, we can set an inner criterion as to whether communication is necessary or not [1]. Also, knowing intention of the other cell's behavior pattrens / strategies or some global rules, and restrictions will contribute to reduceing the uncertainty in the process.

In this paper, we rather focus on the fusion process of knowledge and time series data, which is also able to learne and adapt from the actual event.

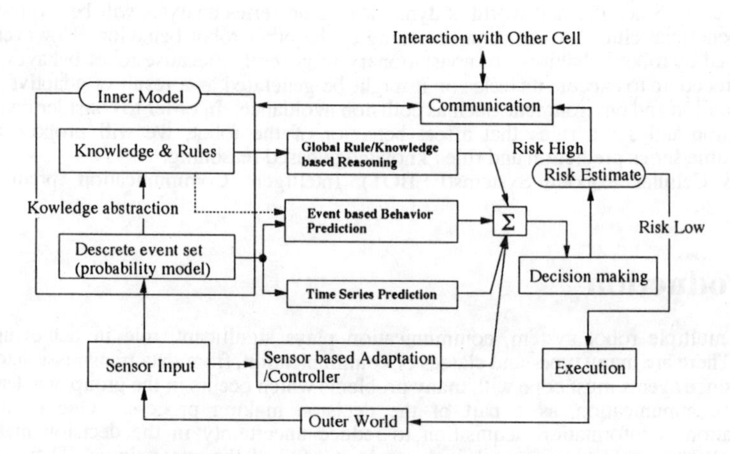

Fig.1 Conceptual Architecture for Intelligent Communication

2. Predictive Decision Making

Most of the risk avoidance or navigation systems rely on the reactive strategies [5]. However, human can avoid many risks by prediction. For example, we comfirm a car is coming or not when we cross the road. We observe not only the distance between object and us, but also change of time series. In conventional decision making strategies, few research works have been found that actively adopt prediction so far. The merits of predictive decision making are following.

1)Since there is a limitation for adaptability in robot reactive motion, predicting next condition is effective in escaping from some risks which cannot be avoided by adaptive motion. Even if it is not, the robot can gain time to cope with the problem.

2)In order to solve a conflict problem, complicated procedures are required. However, predictive decision making will be possible to reduce the cost of these procedures and also communication, by avoiding such problems in advance.

However, in spite of these merits, the following fact hinders to apply prediction to the real use.. That is, behavior of the robot is nonstationary and the model is unknown. For example, if the robot is moving as shown in fig.2, prediction result based on the conventional time series analysis will be dashed line, that is, collision with the wall. However, the robot will perhaps turn right using its collision avoidance algorithm. As is discussed, in considering prediction of the robot behavior, we should notice that state oriented rules, which cannot be observed directly, exist and influence over the behavior. Therefore, predictive decision making should be made based on both the time series prediction and the knowledge based inferences.

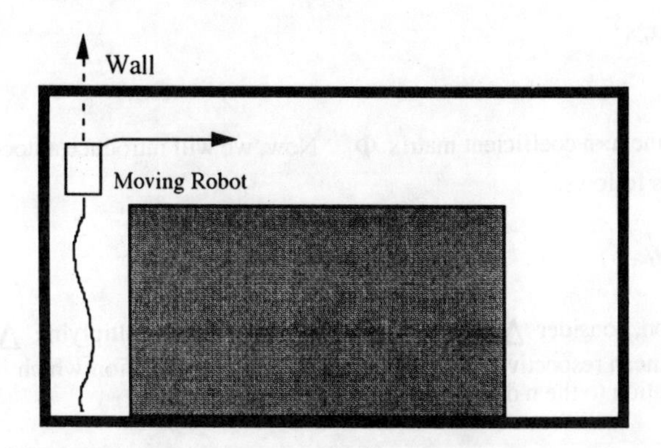

Fig.2 Will a robot collide to the wall?

3. Fusion of Time Series Prediction and Knowledge based Inference

We assume that a robot behavior is determined based on its strategy and planning results, or as a result of adaptation to the environment like a collision avoidance. However, when we try to predict robot behavior, state transition is generally nonstationary. Even if it is seemingly stationary process, it may predict impossible event or state because no rules and environmental constraints are taken into account by conventional statistical time-series prediction model. Therefore, we propose an algorithm to synthesize the time series prediction and knowledge, such as behavior rules or environmental restrictions. Fig.3 shows general operation in this algorithm.

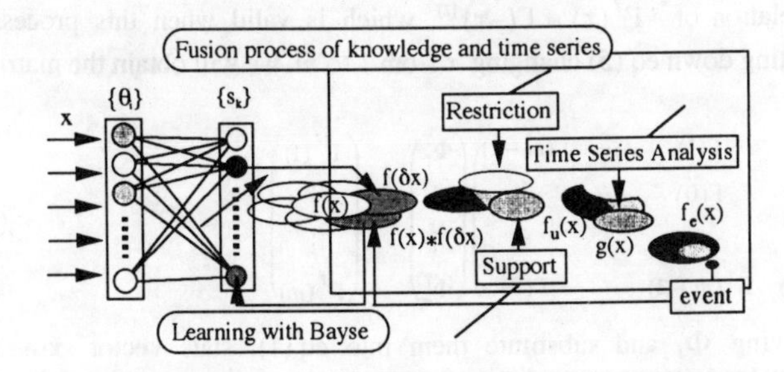

Fig.3 Fusion Process of Knowledge and Time Series Prediction

3.1 Prediction with Time Series Model

Firstly, we will formalize time series prediction model. The process is in general nonstationary, however, many processes result in auto regressive (AR) model by taking difference of neighboring data. This operation cancels trend components which are cause of nonstationary process. So, we can expect to remove the nonstationary component to some degree by adopting vector integrated auto regressive (IAR) model as an approximate model as follows,

$$\Delta^d x_t + \sum_{i=1}^{m} \Phi_i \Delta^d x_{t-i} = e_t \tag{1}$$

Where, $\Delta^d x_t$ is n dimensional vector defined as follows,

$$\begin{cases} \Delta^d x_t = \Delta^{d-1} x_t - \Delta^{d-1} x_{t-1} \\ \Delta^1 x_t = x_t - x_{t-1} \\ \Delta^0 x_t = x_t \end{cases} \tag{2}$$

Our aim is to determine n×n coefficient matrix Φ_i . Now, we will introduce autocovariance matrix of which time lag is τ as follows,

$$\Gamma(\tau) = E\left\{ y_t y_{t-\tau}^T \right\} \tag{3}$$

For simple description, consider $\Delta^d x_t$ as y_t here. Therefore, by multiplying $\Delta^d x_{t-\tau}$ to each term of eq.(2) and taking mean respectively, we will obtain following equation which is actually extension of Yule=Walker equation to the n dimensions,

$$\Gamma(\tau) + \sum_{i=1}^{m} \Phi_i \Gamma(\tau - i) = 0 \tag{4}$$

where, we assumed that $E\left\{ e_t y_{t-\tau}^T \right\} = 0$. Φ_i is i th n×n coefficient matrix. Eq.(8) is also expressed as follows,

$$\Gamma(\tau) + \Phi_1 \Gamma(\tau - 1) + \cdots + \Phi_m \Gamma(\tau - m) = 0,$$
$$\therefore \Phi_1 \Gamma(\tau - 1) + \Phi_2 \Gamma(\tau - 2) + \cdots + \Phi_m \Gamma(\tau - m) = -\Gamma(\tau).$$

By taking transpose, this becomes as follows,

$$\Gamma(1-\tau)\Phi_1^T + \Gamma(2-\tau)\Phi_2^T + \cdots + \Gamma(m-\tau)\Phi_m^T = -\Gamma^T(\tau) \tag{5}$$

where, we used relation of $\Gamma^T(\tau) = \Gamma(-\tau)$ [2], which is valid when this process is regarded as stationary. By writing down eq.(5) changing τ from 1 to m, we will obtain the matrix to solve Φ_i as follows,

$$\begin{pmatrix} \Gamma(0) & \Gamma(1) & \cdots & \Gamma(m-1) \\ \Gamma^T(1) & \Gamma(0) & \cdots & \vdots \\ \vdots & \vdots & \ddots & \vdots \\ \Gamma^T(m-1) & \Gamma^T(m-2) & \cdots & \Gamma(0) \end{pmatrix} \begin{pmatrix} \Phi_1^T \\ \vdots \\ \vdots \\ \Phi_m^T \end{pmatrix} = - \begin{pmatrix} \Gamma^T(1) \\ \vdots \\ \vdots \\ \Gamma^T(m) \end{pmatrix} \tag{6}$$

Therefore, by solving Φ_i and substitute them into eq.(1), state vector $\mathbf{x}_{t+1}$ is solved. The distribution given by the time series prediction is expressed as following normal distribution,

$$g_{t+1}(x) = \frac{1}{(2\pi)^{\frac{n}{2}} |\Gamma(0)|^{\frac{1}{2}}} \exp\left(-\frac{1}{2} \left(x - \overline{x}_{t+1} \right)^T \Gamma(0)^{-1} \left(x - \overline{x}_{t+1} \right) \right). \tag{7}$$

Fig.4 shows an example of prediction to test this model. When an object moves on the parabolic trajectory, this model sufficiently follows the process.

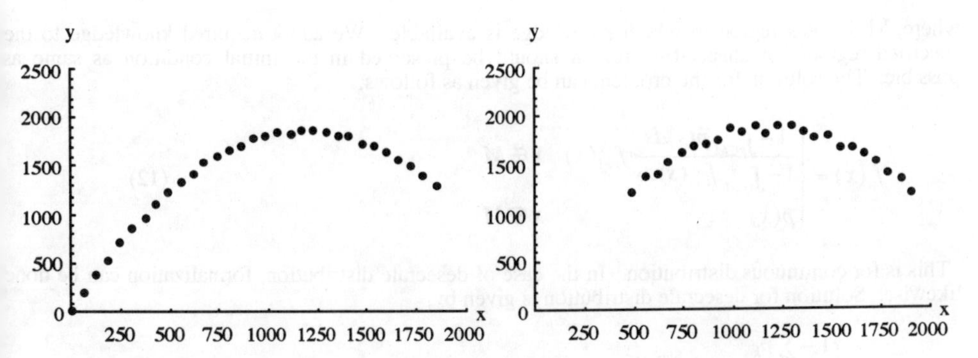

Fig. 4 Example of time series prediction with IAR

3.2 Knowledge based Prediction

In this section, we will formalize knowledge based prediction. We assume that the state will change because of the conditionally dependent robot behavior. Consider that we obtained some information on the state such as geometrical map or rules of thr robot behavior. Acquired information and prediction results are introduced by conditional probability distribution in order to treat their uncertainty quantitatively. We firstly classify acquired information through the sensors or communication into respective conditions, and express it as θ_i. So, subjective probability for the conditionally dependent behavior S_k can be expressed as $P(S_k|\theta_i)$. Suppose that the current state is described by vector **x**, and it will shift to **x**+**δx** by taking action S_k per unit time, as shown in fig.5. **δx** belongs to the following distribution which is defined as total sum of probabilities caused by possible actions,

$$f_{\delta t}(\delta x) = \sum_{k=1}^{n} f_s(\delta x|S_k)P(S_k|\theta_i). \tag{8}$$

Where, $f_s(\delta x|S_k)$ indicates the distribution of small change of the state when action S k is taken. Here, we assume that this is known, but it dose not mean we must know that because it can be learned by the method described later. If the current state **x** belongs to the $f_t(x)$, the next state **x**+**δx** after taking an action will belong to the distribution $f_{t+1}(x)$, which is given by the following convolution, that is ,

$$f_{t+1}(x) = \int_{-\infty}^{\infty} f_t(u) f_{\delta t}(x-u)du \tag{9}$$

Eq.(9) is behavior based state probabilistic distribution.
Now, suppose that rules, constraints, and knowledge etc. are acquired as a subjective probability. We attempt to synthesize these information and behavior based prediction given by eq.(2). This process can be formalized by entropy maximize principle as follows,

Conditions: $$p = \int_{x \in M} \overline{p}(x)dx, \quad \int_{x \notin M} f_u(x)dx = 1-p \tag{10}$$

Target function: $$\int f_u(x) \log \frac{f_u(x)}{f_{t+1}(x)} dx \rightarrow \min \tag{11}$$

where, M denotes region in which knowledge is available. We adopt acquired knowledge to the specified region, but unspecified region should be preserved in the initial condition as same as possible. The solution for the problem can be given as follows,

$$f_u(x) = \begin{cases} \dfrac{1 - \int_{s\in M} \overline{p}(s)ds}{1 - \int_{s\in M} f_{t+1}(s)ds} f_{t+1}(x) & x \notin M \\ \overline{p}(x) & x \in M \end{cases} \tag{12}$$

This is for continuous distribution. In the case of desecrate distribution, formalization can be done likewise. Solution for desecrate distribution is given by,

$$p_i = \begin{cases} \dfrac{1 - \sum_k \overline{p}_k}{1 - \sum_k p_k^0} p_i^0 & (i \neq k) \\ \overline{p}_k & (i = k) \end{cases} \tag{13}$$

where, p_i^0 is initial probability for event i.

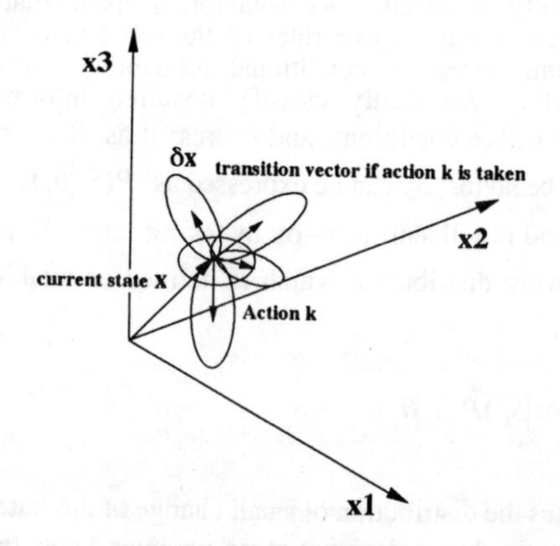

Fig.5 State Description

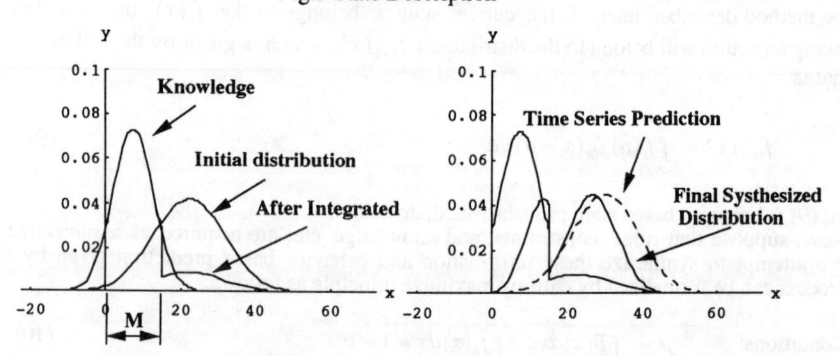

Fig.6 Simple case of knowledg integration

3.3 Synthesized Prediction

Time series prediction and knowledge based state changes are different in its meaning. Former gives quantitative state description and later gives degree of belief for the state condition. However, these can be synthesized by probability distribution that reflects "experience" (statistical analysis) and "belief" (subjective probability). According to eq.(7) and eq.(12) or eq.(13), synthesized distribution of knowledge based prediction and time series prediction is obtained as follows,

$$f_e(x) = \frac{f_u(x) g_{t+1}(x)}{\int f_u(x) g(x) dx}. \tag{14}$$

For pursuing on-line algorithm, calculation of integration is a problem because of its computational cost. However, we can avoid numerical integration by assuming that $f_u(x)$ is expressed as normal distribution. Also, we can avoid integration in eq.(12), by correctly estimating action probability $P(S_k|\theta_i)$ which weights respective distribution. This means eq.(13) is more recommended than eq.(12). If each distribution is assumed to be normal distribution, $f_u(x)$ is linear combination of weighted normal distribution. Suppose, $f_u(x)$ is expressed as,

$$f_u(x) = \sum_{k=1}^{n} f_{t+1,k}(x). \tag{15}$$

So, let us calculate the k th component. Suppose the component is weighted normal distribution give by,

$$f_{t+1,k}(x) = \frac{P(S_k|\theta_i)}{(2\pi)^{\frac{n}{2}} |\Sigma_k|^{\frac{1}{2}}} \exp\left(-\frac{1}{2}(x - \bar{x}_k)^T \Sigma_k^{-1} (x - \bar{x}_k) \right), \tag{16}$$

and also, according to eq.(7), $g_{t+1}(x)$ is given by,

$$g_{t+1}(x) = \frac{1}{(2\pi)^{\frac{n}{2}} |\Gamma(0)|^{\frac{1}{2}}} \exp\left(-\frac{1}{2}(x - \bar{x}_{t+1})^T \Gamma(0)^{-1} (x - \bar{x}_{t+1}) \right). \tag{7}$$

Therefore, the k th component of integrand of eq.(14) becomes

$$\begin{aligned} (fg)_k &= \frac{P(S_k|\theta_i)}{(2\pi)^n |\Sigma_k|^{\frac{1}{2}} |\Gamma(0)|^{\frac{1}{2}}} \exp\left(-\frac{1}{2}\left((x-\bar{x}_k)^T \Sigma_k^{-1} (x-\bar{x}_k) + (x-\bar{x}_{t+1})^T \Gamma(0)^{-1} (x-\bar{x}_{t+1}) \right) \right) \\ &= \frac{|C_k|^{\frac{1}{2}} P(S_k|\theta_i) \exp\left(-\frac{1}{2}\left((\bar{x}_k - m_k)^T \Sigma_k^{-1} (\bar{x}_k - m_k) + (\bar{x}_{t+1} - m_k)^T \Gamma(0)^{-1} (\bar{x}_{t+1} - m_k) \right) \right)}{(2\pi)^{\frac{n}{2}} |\Sigma_k|^{\frac{1}{2}} |\Gamma(0)|^{\frac{1}{2}}} \cdot \frac{\exp\left(-\frac{1}{2}(x - m_k)^T C_k^{-1} (x - m_k) \right)}{(2\pi)^{\frac{n}{2}} |C_k|^{\frac{1}{2}}}. \end{aligned} \tag{17}$$

Where, m_k, C_k is defined as,

$$\begin{aligned} m_k &= (\Sigma_k^{-1} + \Gamma(0)^{-1})^{-1} (\Sigma_k^{-1} \bar{x}_k + \Gamma(0)^{-1} \bar{x}_{t+1}) \\ C_k^{-1} &= \Sigma_k^{-1} + \Gamma(0)^{-1} \end{aligned} \tag{18}$$

which denotes k th meanvector and inverse of covariance matrix of integrated distribution of $f_{t+1,k}(x)$ and $g_{t+1}(x)$ respectively. By integrating eq.(17) and replacing it to A_k, we will obtain following,

$$\int (fg)_k dx = \frac{|C_k|^{\frac{1}{2}} \exp\left(-\frac{1}{2}\left((\bar{x}_k - m_k)^T \Sigma_k^{-1}(\bar{x}_k - m_k) + (\bar{x}_{t+1} - m_k)^T \Gamma(0)^{-1}(\bar{x}_{t+1} - m_k)\right)\right)}{(2\pi)^{\frac{n}{2}} |\Sigma_k|^{\frac{1}{2}} |\Gamma(0)|^{\frac{1}{2}}} P(S_k|\theta_i)$$
$$= A_k \tag{19}$$

Therefore, eq.(14) is reformed as follows,

$$f_e = \sum_{k=1}^{n} \frac{A_k}{\sum_i A_i} \cdot \frac{1}{(2\pi)^{\frac{1}{n}} |C_k|} \exp\left(-\frac{1}{2}(x - m_k)^T C_k (x - m_k)\right). \tag{20}$$

Therefore, final synthesized distribution is also normaldistribution whose mean vector and covariance matrix is defined as,

$$m_F = \sum_{k=1}^{n} \frac{A_k}{\sum_i A_i} m_k, \quad C_F = \sum_{k=1}^{n} \left(\frac{A_k}{\sum_i A_i}\right)^2 C_k \tag{21}$$

4. Inductive Learning of Behavior

It is possible to specify our optimal strategy using the result of eq.(14), but the result will not be reliable enough if we cannot predict other robots' behavior $P(S_k|\theta_i)$. Therefore, we will estimate what actions contribute to the actual state transition by using Bayesian theorem. Since integrand of Eq.(14) is a linear combination with regard to k, the component of $f_{t+1}(x)$ which is caused by action S_k is shown as follows,

$$f_{t+1,k}^{0}(x) = \int_{-\infty}^{\infty} f_t(u) f_s(x - u|S_k) du \tag{22}$$

where,

$$f_{t+1}(x) = \sum_{k=1}^{n} f_{t+1,k}^{0}(x) P(S_k|\theta_i).$$

The k th component corresponds to a prior probability in Bayesian theorem, but it is difficult to specify clearly $f_t(x)$ which the observed data applied most. So, to facilitate inductive learning according to the fitness values, we adopt fuzzy Bayesian method. We solve relative fitness values χ_k for respective $f_{t+1,k}^{0}(x)$ by taking likelihood a_k which is defined as follows,

$$a_k = f_{t+1,k}^{0}(x). \tag{23}$$

$$\chi_k = \frac{a_k}{\sum_i a_i} \tag{24}$$

Therefore, learning equation is obtained as

$$P(S_k|x) = \frac{f_{t+1,k}^{0}(x) \chi_k P(S_k)}{\sum_k f_{t+1}^{0}(x) \chi_k P(S_k)}. \tag{25}$$

This is a statistical fuzzy Bayesian learning equation for the subjective probability of the action which caused observed state changes. By repeating observation and above learning, more precise prediction is expected.

5. Simulation Results

In order to test the performance of this algorithm, we made simulations. This chapter describes comparison between time series model and proposed method in terms of prediction error, and also effect of learning algorithm to the prediction result.

5.1 Estimation of prediction error

We compare prediction performance of IAR model based prediction and proposed method, suppose that a robot is moving along a path surrounded gray block where robot cannot enter, the robot will pass through designated points but some random noise is compounded in the motion. We tried to predict its trajectory. In the simulation, a robot behavior is assumed to be discrete moving direction as shown in Fig.7. Fig.8 (a) shows true trajectory. Fig.8 (b) and (c), shows predicted results of IAR based and knowledge integrated method respectively. In IAR model, we set m=2,n=2,and d=2, in eq.(1). Abrupt changes of moving direction and velocity sometimes cause insufficient results such as a point in the block area in case of time series model only. On the other hand, the proposed method restricts to predict impossible state. Table.1 shows average prediction error. In terms of average prediction error, we can see that the fusion strategy is superior to time series model only.

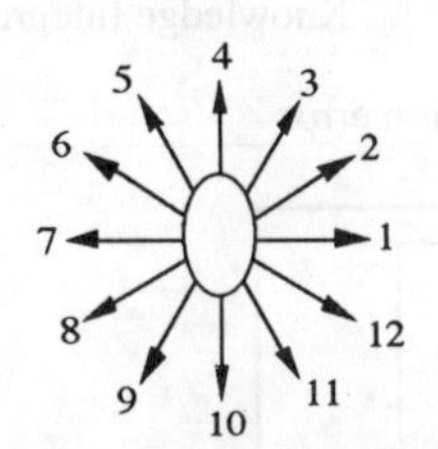

Fig.7 Strategy number

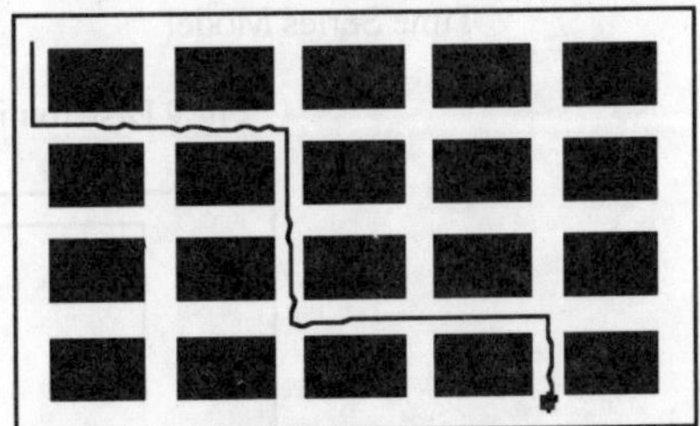

Fig.8 (a) True trajectory

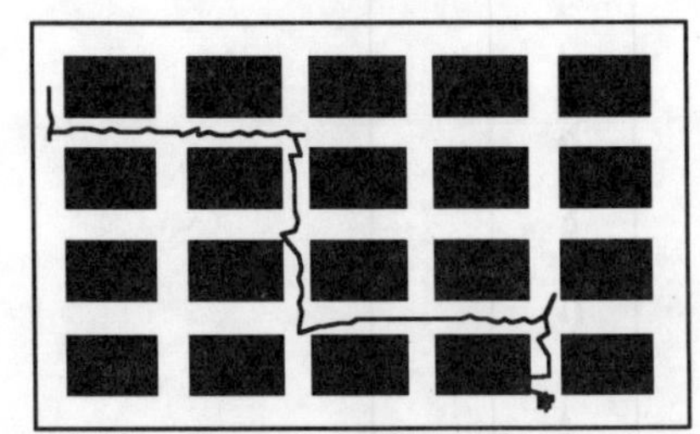

Fig.8 (b) IAR model only

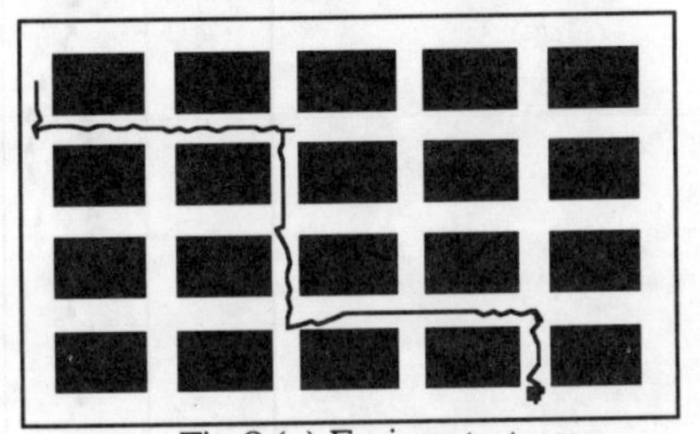

Fig.8 (c) Fusion strategy

Table1 Average Prediction error (x,y)

Time Series only: {0.808789, 0.698105}
Integrated result: {0.650625, 0.674534}

Let us consider another case. Suppose that CellA is moving along the road and CellB is passing it from right lane. CellB determines its own path considering prediction results of CellA's next position. Fig.9 shows distribution of prediction error. As fig.10 shows, we can see that error and variance of error distribution of fusion strategy is smaller than that of result by time series model only. Fig. shows trajectory of CellA and CellB. In case of (a), CellB is not confident with its prediction results. This is a case that when CellB cannot estimate behavior strategy precisely. In such a case, CellB detours not to collide with CellA. On the other hand, in case of (b), CellB could predict CellA's next position considering its behavior pattern. CellB did not have to change its course. This is an example of the predictive decision making. We human also behave similarly when we drive. It is very important to estimate uncertainty of prediction results when the robot tries to make a decision based on that. If uncertainty is large, robot should confirm the information through further observation and communication if possible. Predictive dicision making will enhance the adaptability of the robot and it results in cooperative task execution which is less dependent on communication. From view point of the information acquisition, communication reduction is possible by considering when and why it is required. In this mean, uncertainty is an important measure in decision making.

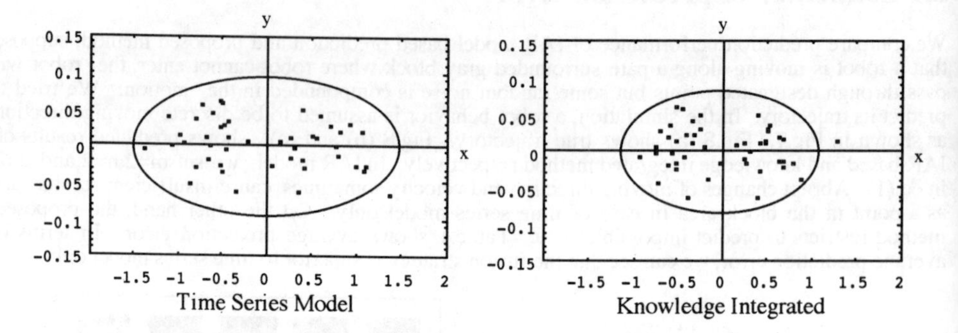

Fig.9 Distribution of prediction error

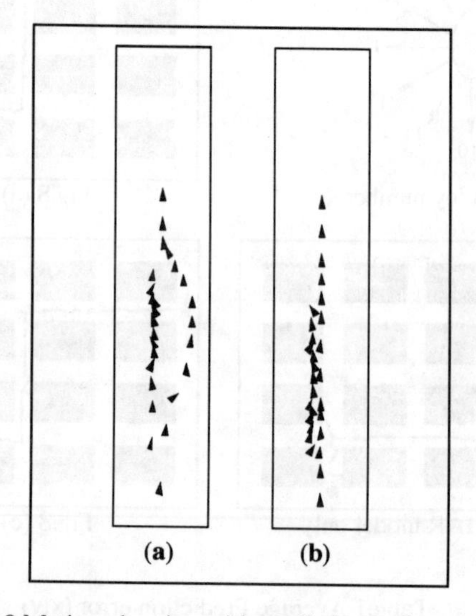

Fig. 10 Motion planning with prediction and uncertainty

5.2 Effect of the Learning equation

Since this algorithm largely depends on estimation accuracy of the action strategy $P(S_k|\theta_i)$, if we have wrong knowledge on the strategy, it is difficult to obtain proper prediction result. We will repeat the same mistakes as long as the "wrong knowledge" is improved. In previous chapter, we described learning algorithm of the strategy to avoid such a situation. In this chapter, we will evaluate the effect of this learning algorithm. Situation of the simulation is the same as previous example. Strategy is defined as its moving direction again. As initial condition, we teach the wrong information on the behavior strategy. That is, action 2,3,4,5, and 6 which generates movement forward, is almost rejected in spite of the fact that the cell actually takes strategy 4 mainly. As to the other strategy, since no information is given, they are determined by eq.(13). Fig.12 (a) shows initial estimated behavior strategy. At first, prediction error was large, although time series prediction contributed to better results. However, as step goes, learning algorithm reinforce the probability of the strategy which is influential state changes. Fig.11 shows change of the average prediction error. Fig.12 (b),and (c) shows $P(S_k|\theta_i)$ at step 9 and 17 where behavior pattern almost converged. From these result, it is considered that improvement of the prediction error is brought by learning algorithm and it efficiently contributed to better prediction.

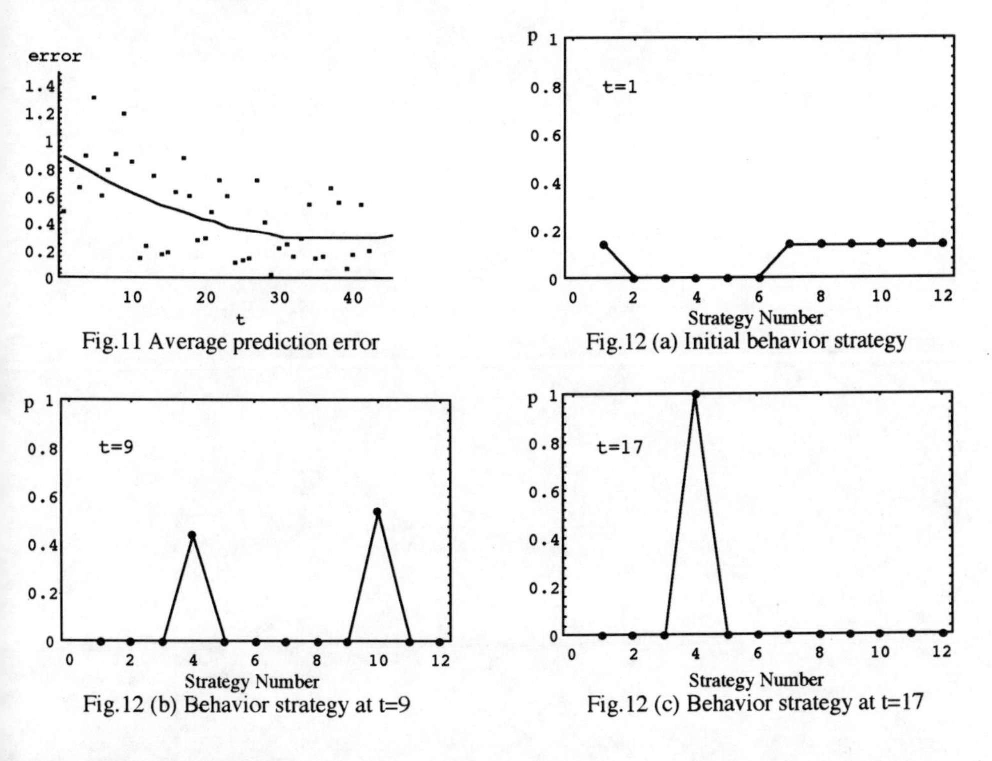

Fig.11 Average prediction error

Fig.12 (a) Initial behavior strategy

Fig.12 (b) Behavior strategy at t=9

Fig.12 (c) Behavior strategy at t=17

6. Conclusions

We described an on-line algorithm to integrate knowledge and time-series prediction by expelling impossible states and adopting conditionally dependent knowledge, environmental constraints, and behavior patterns. We can improve intention reasoning by considering not only qualitative knowledge but also numerical time series model based prediction. The method has an adaptability by inductive learning of the behaviors from the observed posterior state. In prediction, we should consider the uncertainty of the result and possibility of the prediction, that is, to what extend prediction is available.

References

[1] T.Fukuda and K.Sekiyama, Communication Reduction with Risk Estimate for Multiple Robot System, Proc. of IEEE International Conference on Robotics and Automation, pp.2864-2869,1994.
[2] A.C. Harvey, TIME SERIES MODELS, Philip Publishers Limited, Oxford, 1981.
[3] Ueyama, T., and Fukuda, T., "Cooperative Search Using Genetic Algorithm Based on Local information - Path Planning for Structure Configuration of Cellular Robot -," Proc. 1993 IEEE International Conference on Intelligent Robots and System (IROS'93), pp.1110-1117, 1993.

[4]Kawauchi, Y., Inaba, M. and Fukuda, T. "Self-Organizing Intelligence for Cellular Robot System (CEBOT) with Genetic Knowledge Production Algorithm," Proc. of 1992 IEEE International Conference on Robotics and Automation, pp.813-818, 1992.
[5]K.J. Kyriakopoulos and G.N.Saridis, Collision Avoidance of Movile Robots in None-stationary Environments, Proc. of IEEE International Conf. on Robotics and Automation, pp.904-909,1991.

On a Deadlock-free Characteristic of the On-line and Decentralized Path-planning for Multiple Automata

HIROSHI NOBORIO and TAKASHI YOSHIOKA

Dept. of Precision Eng., Osaka Electro-Communication Univ., Osaka, 572 Japan

Abstract

In this paper, we propose an on-line and decentralized path-planning algorithm for multiple automata and then discuss its deadlock-free characteristic in an infinite 2-d world. In this research, we consider many automata with a finite number in the world without any static obstacle. Each automaton with the same circular shape can move for omni-directions to arrive at the goal. An automaton basically does not see any information except its present position in an on-line manner, and therefore usually goes straight to the goal and finally stops at it. However an automaton exceptionally knows a behavior of another colliding one by its ring of tactile sensors, and in a real time way, it processes the present own and partner's behaviors to determine its next own behavior by the common sense.

By reason of this, each automaton merely determines its behavior by its surrounding information without considering any global perspective of the world. Therefore the algorithm does not always ensure any deadlock-free characteristic of all automata toward the goals. To overcome this drawback, each automaton never circulates around its goal by the clockwise order in our algorithm. The counter-clockwise action in an on-line avoidance is carefully designed under the common sense located on all automata in advance. Under topological characteristics implicitly derived from a sequence of counter-clockwise actions, each automaton never generates any periodic circulation around an arbitrary point. In result, our algorithm ensures its deadlock-free characteristic in the world.

Key words: Deadlock-free characteristic, Path-planning algorithm, On-line algorithm, Decentralized algorithm, A robot world by multiple automata

1 Introduction

Recently, many kinds of path-planning algorithms have been aggressively investigated for constructing an intelligent world with multiple robots [1-3]. By two key words "on-line" or "off-line" and "centralized" or "decentralized", such path-planning algorithms are frequently categorized into four combination types as follows [3]:

(1) An off-line and centralized path-planning algorithm
(2) An off-line and decentralized path-planning algorithm
(3) An on-line and centralized path-planning algorithm
(4) An on-line and decentralized path-planning algorithm

Firstly the authors are hardly to deal with all algorithms in the first and second types as the following reason: In such off-line algorithms, all automata communicate their conditions to a computer or a set of computers, and in a short time, the computer set gives the automata their next behaviors. However in the meanwhile, every condition around each automaton is drastically changed in a dynamic world running by multiple automata and consequently the given behaviors are usually to be out-of-date by the time delay. Moreover mechanical automata usually lead self-positioning errors in a real world by their uncertainties. As a result, a computer world always differs from its corresponding real world and consequently an off-line algorithm based on the computer world cannot ensure any deadlock-free characteristic constitutionally in a real and dynamic world with some uncertainties.

Table 1: An usual comparison between a centralized and off-line path-planning and a decentralized and on-line path-planning.

	Optimality	Computation	Safety
Cent. & Off.	○	×	×
Decent. & On.	×	○	○

In the algorithms with the third type, each automaton independently communicates an own relation for its surrounding automata to a central computer, and then the computer coordinates every complex relation from all automata to determine their new behaviors, and in succession, it communicates the new behaviors to the automata individually. Needless to say, a set of the behaviors should be selected so as for all automata not to join all types of deadlocks in a whole world. In general, it is hard for a present computer to determine all deadlock-free behaviors in a real time manner. This tendency becomes remarkable as the number of automata increases in a whole world. In addition to this, we must notice that all automata cannot maintain their robot world immediately if only a central computer is crashed.

On the basis of the above observations, the authors are very interested in the on-line and decentralized path-planning algorithms [4-6]. By the on-line algorithms, a set of behaviors is selected under real conditions in a dynamic world. Therefore the automata have some possibility to maintain their deadlock-free characteristics by a real-time avoidance. Moreover the decentralized characteristic reinforces its on-line one because of no explicit communication between many automata. Also under the decentralized characteristic, each automaton determines its behavior by processing its surrounding condition by an own computer. Thus even though a few automata or its computers are crashed, the other automata keep the senses and in result they robustly run in a dynamic world while keeping their deadlock-free characteristic safely.

In general, an on-line and decentralized path-planning algorithm has included the following two problems: (1) A deadlock-free characteristic of each automaton: In the decentralized path-planning, each automaton independently acts in a 2-d world by local sensor information. In other words, each automaton cannot consider any global perspective about its whole world. Because of the locality, some automaton may take part in a deadlock and consequently does not arrive at the goal in the world. To overcome this defective point, we must consider a global perspective to eliminate all kinds of deadlocks in the world. (2) An optimal property of each deadlock-free path: Even though an automaton gets a deadlock-free path toward the goal, the path is not always coincident with the optimal one, e.x., the shortest path toward the goal. In this paper, we propose an on-line and decentralized path-planning algorithm with the deadlock-free characteristic in a 2-d infinite world without any static obstacle. However because of the locality, it is difficult for us to pick up the optimal path in the world. Thus all deadlock-free paths selected by the proposed algorithm are not always to be the optimal ones in an unknown dynamic world.

2 An Automaton and Its Robot World

In this research, a robot world is of infinite and also does not include any static obstacle as illustrated in Fig.1. In the 2-d world, all automata have circular frames with the same scale and also each automaton puts a ring of tactile sensors around it. By a face contact on the sensor ring, an automaton can feel some force and moment against another touching automaton. In the on-line path-planning, an automaton does not usually know any action of another automaton, but if and only if an automaton touches another one, it can identify an action of the another one by analyzing its own action and sensor information. Especially in this research, the automaton restricts its action into three types of behaviors, that is, stopping, going straight, and drawing a circle in an uncertain world, and consequently an automaton stably distinguish several kinds of contacts against another automaton. In result, the automaton easily recognizes its partner's behavior (Fig.2). This is a kind of an implicit communication.

For example, if an automaton with the line behavior touches another one with the stop behavior, the former automaton cannot feel any passive force by excepting its active force from its sensor information as shown in Fig.2(a). If two automata touch each other by the line behaviors, each feels a passive force by excepting its active force from its sensor information as illustrated in Fig.2(b). Furthermore if an automaton with the line behavior touches another one with the circular one, the former automaton can find some passive force and moment, and on the other hand, the latter automaton can see another passive force (Fig.2(c)). Note that an automaton always finds a combination of active force/moment and passive force/moment by its sensor ring since its contact is formed by a face. Finally if two automata have a contact by the circular behaviors, each one feels a passive force and moment to know its partner's behavior as the circular one (Fig.2(d)).

Thus after processing own and partner's behaviors on the basis of common rules in Table 2, an automaton individually determines its new behavior by a personal computer with a few memory.

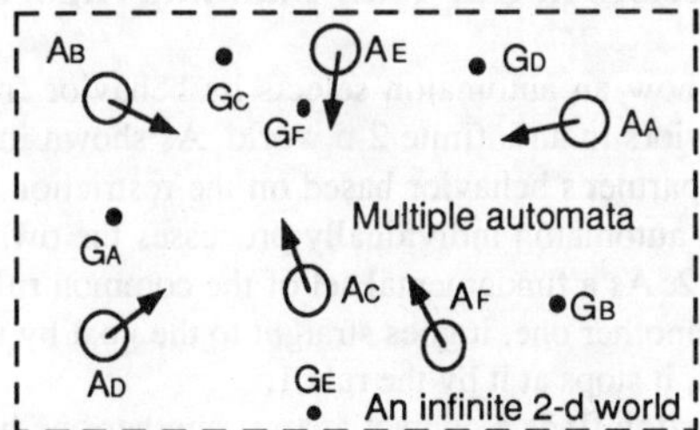

Figure 1: A robot world constructed by our decentralized and on-line path-planning.

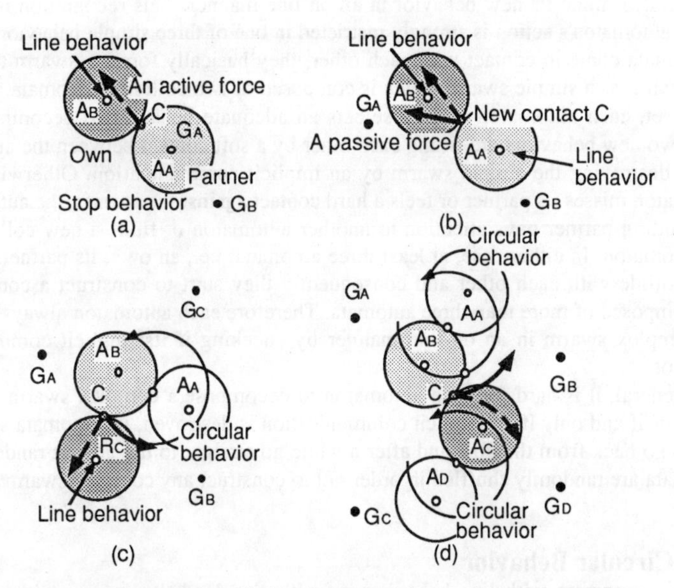

Figure 2: An automaton finds a behavior of its colliding partner by processing its own behavior and sensor information.

Table 2: A basic set of common rules to lead each automaton to its goal and to decompose a simple swarm.

No.	A present state		A next state
	Own beh.	Partner beh.	Own behavior
1.	Stop(G)	None	Stop(G)
2.	Stop(G)	Line	Circle(New)
3.	Stop(G)	Circle	Line(Bw)
4.	Line(Fw)	None	Line(Fw)
5.	Line(Fw)	Stop	Circle(New)
6.	Line(Fw)	Line	Circle(New)
7.	Line(Fw)	Circle	Line(Bw)
8.	Circle	None	Line(Bw)
9.	Circle	Stop	Circle(Cont.-*1)
10.	Circle	Line	Circle(Cont. -*1)
11.	Circle	Circle	Circle(New)

*1 Since the circular behavior has the highest priority, we present an additional condition to finish it.

3 Three Types of Behaviors in Our Path-Planning Algorithm

In this section, we explain how an automaton selects its behavior from three types, i.e., the circular, line, and stop behaviors in an infinite 2-d world. As shown in the previous section, an automaton stably knows its partner's behavior based on the restriction by its own behavior and its sensor information. Each automaton individually processes the own and partner's behaviors under a lot of rules in Table 2. As a fundamental set of the common rules, if an automaton does not find any contact against another one, it goes straight to the goal by the rule 4, and then if the automaton arrives at the goal, it stops at it by the rule 1.

Moreover if an automaton finds a contact against another one by its tactile ring, each of them recognizes own and partner's behaviors by its sensor information and then processes the behaviors to determine its new behavior in an on-line manner. This recognition is to be stable because an automaton's action is strongly restricted in one of three simple behaviors. In general, if two automata come in contact with each other, they basically forms a swarm to avoid each other. We call it as a simple swarm which is composed of two colliding automata. Based on the given rule set, each automaton usually selects an adequate behavior to decompose a simple swarm. If two new behaviors are fitting each other by a soft contact between the automata, they succeed in destroying the simple swarm by an implicit communication. Otherwise, that is, if each automaton misses its partner or feels a hard contact against the partner, the automaton finds that its colliding partner pays attention to another automaton or finds a new collision against another automaton. In either case, at least three automata, i.e., an own, its partner, and another automata collide with each other and consequently they start to construct a complex swarm which is composed of more than three automata. Therefore each automaton always notices itself touch a complex swarm in an on-line manner by checking if its implicit communication is broken or not.

In general, it is hard for each automaton to decompose a complex swarm in an on-line manner. Thus if and only if an implicit communication is destroyed, its automata select the line behavior to go back from the goal and after a while go straight to the goal at random. In result, such automata are randomly shuffled in order not to construct any complex swarm in an infinite 2-d world.

3.1 The Circular Behavior

If two automata with line behaviors or with circular behaviors or with stop and line behaviors touch each other, they adopt new circular behaviors to avoid each other under the rules 2, 5, 6, 11 in Table 2. In our algorithm, since the circular behavior has the maximum priority, such an automaton elbows its way through the world by the rules 9 and 10. Thus our algorithm misses decomposing a swarm, unless it compels at least one of its automata to leave the swarm by an additional condition. Thus in a swarm, each automaton always checks if it goes out of the swarm without colliding with its partner. In general, an automaton can go straight to the goal without the collision if its smaller angle $\angle CAG$ is larger than or equals to $\pi/2$ (Fig.3(a),(b)). This is called as a leave condition. Note that both automata never accomplish the leave condition in a swarm if and only if their goals are located within a circle traced by each center. For example, if and only if both automata A_A and A_B have their goals within a center circle as shown Fig.4(a), the smaller angles $\angle CA_AG_A$ and $\angle CA_BG_B$ are continuously to be smaller than $\pi/2$. However two automata in a simple swarm always have their goals whose distance equals to or is larger than the diameter of the center circle, i.e., each automaton, or else they miss their goals physically. As this contraposition, we can see that at least one of them always keeps the leave condition and consequently destroys the swarm by leaving it as illustrated in Fig.4(b). In addition to this, a left automaton cannot follow its partner adequately and then goes back from the goal by the line behavior according to the rule 8 in Table 2.

Finally based on the leave condition, both automata leave a simple swarm to go straight to the goal or go back from it as soon as they make a round the goals by the clockwise order. Thus in a swarm, each automaton always circulates around its goal by the counter-clockwise order as shown in Fig.5. Based on the counter-clockwise circulation, an automaton does not make any deadlock and arrives at its goal surely. These details are explained in the section 4.

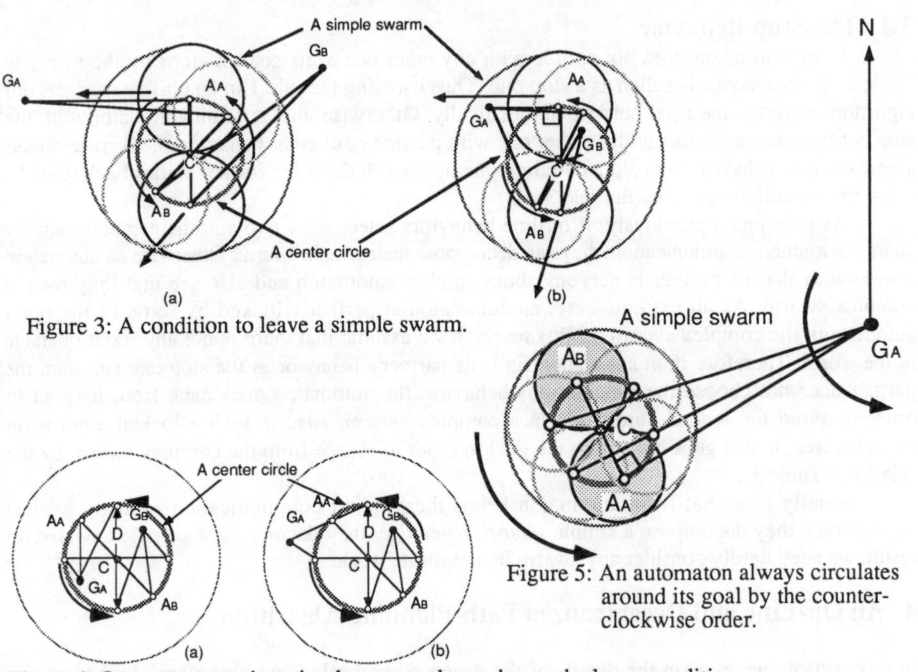

Figure 3: A condition to leave a simple swarm.

Figure 5: An automaton always circulates around its goal by the counter-clockwise order.

Figure 4: One automaton in a simple swarm always keeps the leave condition.

3.2 The Line Behavior

Basically if an automaton does not collide with any one outside the goal, it goes straight to the goal by the rule 4 in the Table 2. Thus an automaton arrives at the goal surely as long as it is not obstructed by any automaton in an uncertain world. In addition to this, an automaton with the leave condition in a swarm goes straight to the goal, and simultaneously its partner without the leave condition goes back from the goal. These are a basic usage of the line behavior. Moreover while an automaton goes back from the goal, the automaton changes its direction from Bw(Backward) to Fw(Forward) if it finds a collision newly by the rule 14 in Table 3. Otherwise, after a while, an automaton randomly reverses its movement direction so as to go straight to the goal again by the rule 13 in Table 3.

Finally an automaton always moves on a semi-line from the goal to a direction by the line behavior, and therefore the automaton does not circulate around its goal by the clockwise or counter-clockwise order. Furthermore the circular behavior explained previously has a priority over the line behavior, and thus an automaton with the line behavior always gives way to another automaton with the circular behavior by the rule 7 in the Table 2. In result the line behavior does not concern to any deadlock mentioned later.

Table 3: An additional set of rules to eliminate a complex swarm.

No.	A present state		A next state
	Own beh.	Partner's beh.	Own behavior
12.	Stop(Ng)	All	Line(Bw)
13.	Line(Bw)	None	Line(Bw- *2)
14.	Line(Bw)	Other	Line(Fw)

*2 After a while , the automaton changes its direction from BW(Backward) to FW(Forward).

3.3 The Stop Behavior

If an automaton does not collide with any other one at its goal, it stops by the rule 1 in Table 2. This behavior is called as a stop one. Thus by using the rule 1 in an on-line manner, our algorithm achieves the final purpose automatically. Otherwise, that is, if an automaton with the stop behavior has a contact with another one with the line or circular behavior, the former selects a new circular behavior to avoid a colliding one by the rule 2, or the former comes back to make room for the colliding one by the rule 3.

As mentioned previously, if on-line behaviors selected by two automata synchronously attain the implicit communication, they can decompose their simple swarm. Otherwise an automaton always sees that its partner is nervous about another automaton and also see that they form a complex swarm. As an extreme case, an automaton is perfectly locked by some of the other automata in the complex swarm. In this research we assume that there is not any fixed obstacle in the world. Therefore if an automaton finds its partner's behavior as the stop one and then the partner does not choose the synchronous behavior, the automaton goes back from its goal in order to avoid the locked automaton in a complex swarm. Also if such a locked automaton becomes free, it also goes back from the goal in order to escape from the complex swarm by the rule 12 in Table 3.

Finally note that if an automaton keeps the implicit communication robustly against another one, they decompose a simple swarm, otherwise, they decompose a complex swarm. In result, we need hardly consider any swarm in an infinite world.

4 An On-Line and Decentralized Path-Planning Algorithm

In this section, we explain the details of the proposed path-planning algorithm. An automaton always goes straight to its goal by the line behavior, and after arriving at the goal, the automaton stops forever as long as it is not touched by any automaton.

If an automaton finds a contact against another one, they independently determine their behaviors based on a basic set of common rules in Table 2. As a typical example of rules, if two automata with the stop or line behavior collide with each other, they newly select circular behaviors to avoid each other locally by the counter-clockwise order. Keeping the partner's right is also included in the example. Then if the selected behaviors are fitting each other, the automata feel a soft touch while tracing each other, and consequently their communication comes into existence implicitly without any explicit communication devise. The implicit communication easily decomposes a simple swarm composed of two colliding automata.

Moreover as a by-product of the basic rule set, an automaton implicitly finds to join the other swarms by checking if its soft contact is continuously kept. That is, if the soft contact is suddenly released or is inadequately transformed to a hard one, an automaton naturally recognizes that its partner does not perform any synchronous behavior by obstruction of the others. In either case, more than three automata construct a complex swarm. In order not to join the complex swarm, the automaton goes back from the goal by the posterior reasoning. The suspension of the implicit communication decomposes a complex swarm composed of more than three automata in wide free space of the infinite world. After escaping from such a complex swarm, each automaton is usually controlled under an additional set of common rules in Table.3. The basic and additional sets of rules are completely memorized into all automata in advance as a common sense. Based on the common sense, all automata decomposes all kinds of swarms in such a dynamic world by an on-line manner.

The proposed algorithm consists of five parallel procedures. If and only if all automata are stopping at their goal correctly, the algorithm keeps the condition as the success forever, otherwise, some of them individually act according to one of the following five procedures.

1. If an automaton does not find any contact, the automaton stops forever at its goal G by the rule 1 in Table 2, and except the goal, the automaton goes straight to it by the rule 4 in Table 2.

2. If two automata with the stop or line behavior touch each other, each automaton synchronously selects the circular behavior to trace its partner in the counter-clockwise order as illustrated in Fig.6, and consequently they form a simple swarm. In the swarm, if each automaton can go straight to the goal without any collision of the partner, the automaton selects the line behavior to go straight to the goal from the swarm. The leaving is judged by checking if the smaller angle $\angle CAG$ is larger than or equals to $\pi/2$. On the other hand, the partner automaton selects the line behavior to go back from the goal and after a while goes straight to it by the rule 8 in Table 2. In result, both automata finally go straight to the goal by the line behavior after avoiding each other. Needless to say, at least one of the automata is always left from the simple swarm after keeping the leave condition.

3. If an automaton A with the stop or line behavior touches another one B with the circular behavior, the automaton A turns back the way it has come by the rule 3 or 7, and also the automaton B continues to keep the circular behavior by the rule 9 or 10 as shown in Fig.7. The reason is that our algorithm is designed under the priority: (the circular behavior) > (the line behavior) > (the stop behavior). Then after a random interval, the automaton A reverses its moving direction to go straight to the goal by the rule 13 in Table 3.

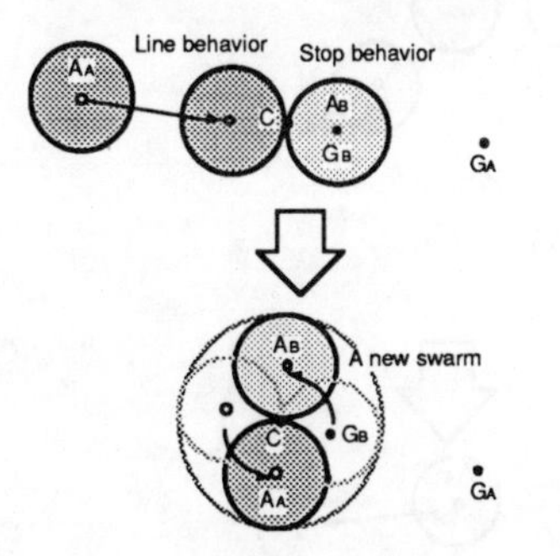

Figure 6: An implicit negotiation between two colliding automata.

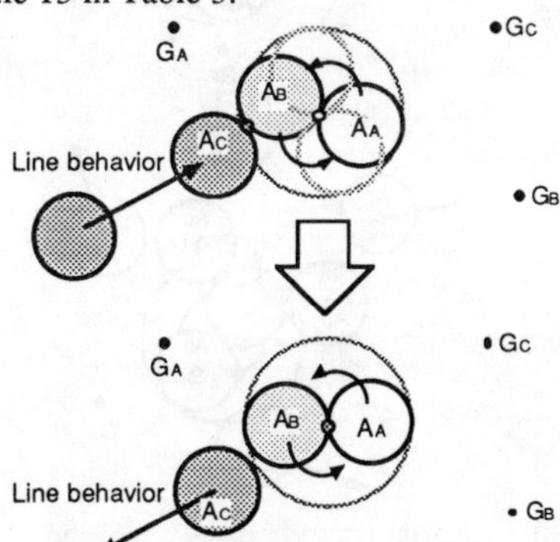

Figure 7: An implicit negotiation between colliding automaton and swarm.

4. If two simple swarms collide with each other, each colliding automaton can find an own and partner's behaviors as the circular ones. In this case, they synchronously form a new swarm centered by a new contact point C by the rule 11 as illustrated in Fig.8. Furthermore since the other automata lose their contacts or feel hard contacts, the automata go back from the goals without obstructing the new swarm by the rule 8 and after a while they go straight to the goals again by the rule 13. The direction is randomly reversed in order not to construct the same condition.

5. Even if above three types of collisions occur simultaneously, each automaton automatically selects its behavior according to one of the previous three procedures. However in a complex swarm constructed by the simultaneous collisions, almost all behaviors are not suitable for each other and consequently they miss their partners or feel hard contacts against them.

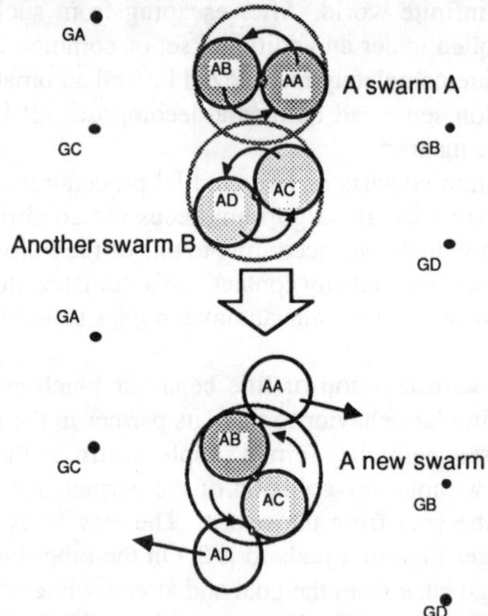

Figure 8: An implicit negotiation between two colliding swarm.

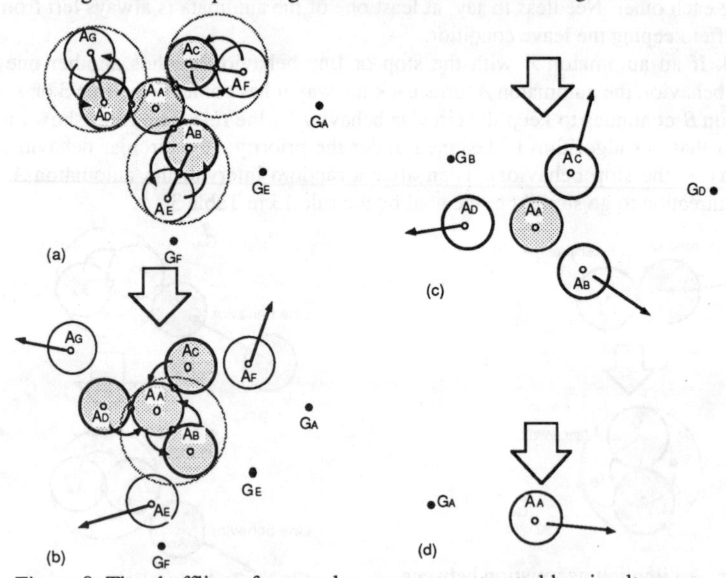

Figure 9: The shuffling of a complex swarm constructed by simultaneous collisions of three or more automata and swarms.

Because of the failure of the implicit communication, each automaton individually goes back from its goal by the rule 8 if and only if it is not locked by the others. Then after a while such an automaton goes straight to the goal by the rule 13. The interval is usually selected at random and consequently all automata in a complex swarm are perfectly shuffled to escape from it as illustrated in Fig.9.

4 On a Proof of the Deadlock-free Characteristic

In this section, we theoretically show that our on-line and decentralized path-planning algorithm has the deadlock-free characteristic by the following five theorems. Each theory is based on several topological characteristics formed implicitly by a set of automata running under the common sense. In our algorithm, each automaton always circulates around its goal by the counter-clockwise order as shown in the Theorem 1. By reason of this, we can show that each automaton never generates all kinds of deadlocks, that is, periodic circulation around an arbitrary point as denoted in Theorems 2 and 3. In result, even though an automaton is obstructed by another one, the number of the obstruction is perfectly bounded as a finite number in the Theorem 4. In addition to this, on the assumption that the number of automata is of finite in the whole world, the obstruction number of an automaton against the others is exactly limited as a finite number. As a result, each automaton always arrives at its goal at worst because the automaton always goes straight to the goal if it is not interfered by any automaton. Even though our algorithm is to be on-line and decentralized path-planning one, it always ensures the deadlock-free characteristics of all supervising automata implicity in an infinite 2-d world without any fixed obstacle (Theorem 5).

Definition: In this research, we call periodic circulation of an automaton around its goal point and another point except it as global and local deadlocks as illustrated in Fig.10 and 11, respectively. In general, a deadlock is defined as a local one if the angle $\angle AGN$ does not increase at all while its automaton *A* goes round the deadlock. Otherwise, that is, if the angle $\angle AGN$ regularly increases by 2π, a deadlock is defined as a global one in an infinite 2-d world. Here the vector *GN* is defined as a semi-line from the goal *G* to the infinite north *N*, and also the point *A* is defined as a center of an automaton. Moreover we define that a path-planning algorithm has its deadlock-free characteristic if each supervising automaton individually arrives at the goal at worst. ○

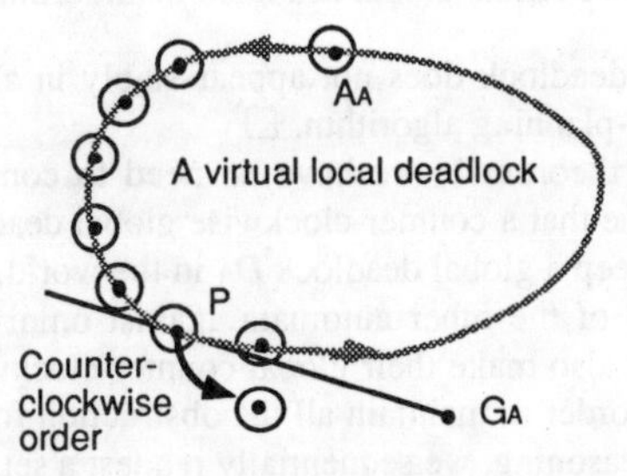

Figure 10: A virtual local deadlock in an infinite 2-d world.

Theorem 1: An automaton supervised by our algorithm never circulates around its goal by the clockwise order in an infinite 2-d world. □

Proof: In our algorithm, each automaton selects three behaviors, that is, the stop behavior, the line behavior, and the circular behavior. When an automaton uses the former two behaviors, its angle $\angle AGN$ is to be constant in the world. On the other hand, if an automaton adopts the latter one behavior, the automaton is always managed under the leave condition mentioned previously. In general, the angle $\angle AGN$ increases by the clockwise order if and only if the angle $\angle CAG$ is always larger than $\pi/2$ Thus on the basis of the leave condition, an automaton never circulates around its goal G by the clockwise order in the circular behavior. Consequently as long as an automaton is running in the world by our algorithm, its angle $\angle AGN$ usually increases by the counter-clockwise order. In result, the proposition is perfectly satisfied in an infinite 2-d world (Fig.5). ■

THEOREM 2: A local deadlock does not appear at all in an infinite 2-d world by our on-line and decentralized path-planning algorithm. □

PROOF: Because of the theorem 1, we do not need to regard any clockwise local deadlock. Thus on the assumption that a counter-clockwise local deadlock occurs in the world as shown in Fig.10, we find its following contradiction: In order to trace the deadlock, an automaton sometimes travels round its goal by the clockwise order. Therefore by the theorem 1, the proposition is perfectly satisfied in an infinite 2-d world. ■

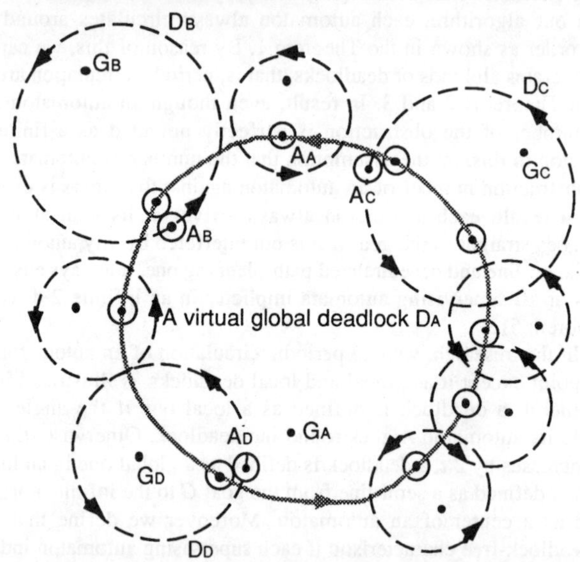

Figure 11: A virtual global deadlock in an infinite 2-d world.

THEOREM 3: A global deadlock does not appear stably in an infinite 2-d world by our on-line and decentralized path-planning algorithm. □

PROOF: Based on the theorem 1, we have no need to consider any clockwise global deadlock. Thus we first assume that a counter-clockwise global deadlock occurs in the world as shown in Fig.11. In order to keep a global deadlock D_A in the world, an automaton A_A should be perfectly obstructed by some of the other automata against omni-directions. In other words, some automata A_B, A_C, A_D, ... also make their global counter-clockwise deadlocks D_B, D_C, D_D, ... outside the deadlock D_A in order to maintain all the obstruction for the global deadlock D_A in the world. By this inductive reasoning, we sequentially request a set of outside global deadlocks in order to keep a set of inside ones. In this research, since the number of automata is of finite in the world, we finally find a set of outermost global deadlocks in the world. However in reality, an outermost deadlock is not stable since it always has an open direction which is not obstructed by any automaton in the world. For example, an automaton A_F is not obstructed by the other ones, i.e., A^* as described in Fig.12, and consequently its outermost deadlock D_F immediately disappears in the world. In succession, an automaton A_B also arrives at the goal and its global deadlock D_B disappears soon in the world. Thus by the reverse reasoning of the inductive one, all outermost deadlocks disappear sooner or later and consequently their automata arrive at the goals in regular order. In result, the proposition is perfectly ensured in an infinite 2-d world. ■

On this observation, even if some automata make their virtual global deadlocks in the world, they escape the deadlocks from the outside, in other words, they arrive at their goals from the outside. This character is suitable for the dynamic and uncertain path-planning since some automaton is not obstructed by all automata stopping at their goals.

THEOREM 4: Even though an automaton is obstructed by another one any number of times in an uncertain 2-d world, the number is of finite by our algorithm. □

PROOF: First of all, if the segment between an automaton A_A and its goal G_A does not intersect the segment between an automaton A_B and its goal G_B, the automaton A_A never collides with the automaton A_B in the world, otherwise they may collide with each other. In the latter case, two automata circulate around their goals independently by the counter-clockwise order after parting from each other in their swarm. Therefore they miss meeting each other in course of time unless they forever keep global deadlocks simultaneously around their goals (Fig.13). However based on the theorem 3, both the deadlocks are always destroyed in time and consequently their automata never touch each other at worst. In result, the proposition is perfectly kept in an infinite 2-d world. ■

THEOREM 5: Each automaton arrives at its goal at the worst and therefore its deadlock-free characteristic is perfectly ensured in an uncertain 2-d world by our algorithm. □

PROOF: As shown in the theorem 4, even though each automaton is obstructed by another one many times, the number of obstruction is always limited as a finite one. Furthermore in this research, since the number of the other automata is also of finite, the number of their obstruction is evaluated by a finite number.

Moreover in our algorithm, an automaton goes straight to its goal as long as the automaton is not obstructed by any automaton. In result, each automaton finally arrives at the goal without being obstructed by the other automata in an infinite 2-d world. As a result, the proposition is perfectly kept in an infinite 2-d world. ■

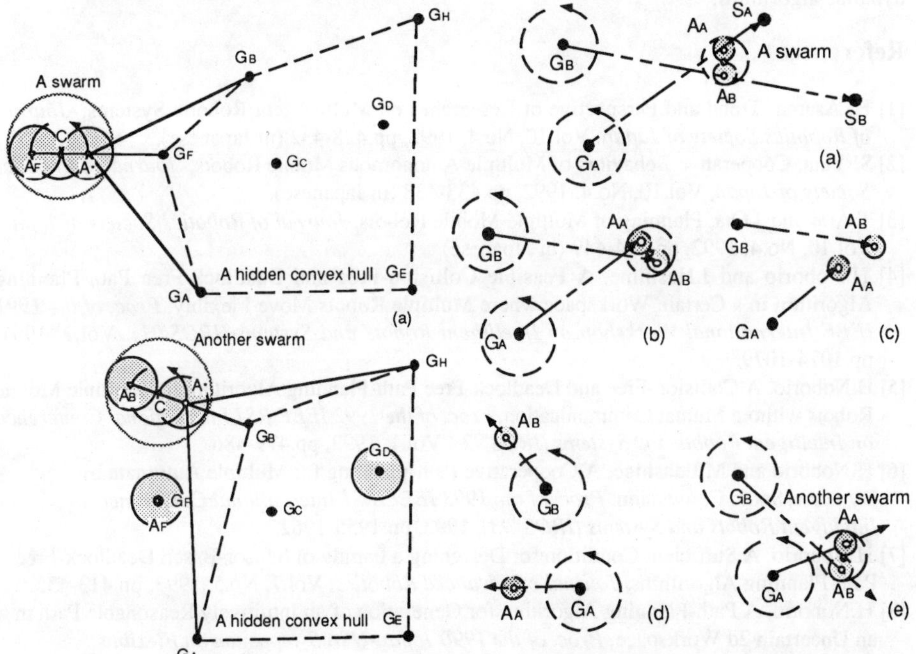

Figure 12: Outermost global deadlocks D_F and D_B always disappear under the leave condition.

Figure 13: An automaton A_A touches another one A_B after generating counter-clockwise global deadlocks around their goals G_A and G_B.

5 Conclusions

In this paper, we proposed an on-line and decentralized path-planning algorithm. The algorithm always leads all automata to their goals surely in order to ensure its deadlock-free characteristic in an infinite 2-d world without any static obstacle.

In general, there are a lot of static obstacles with free shape in a finite 2-d world. To extend the proposed uncertain and dynamic path-planning algorithm in order to deal with such a real world, we firstly consider how to identify the difference between a locked automaton and a fixed obstacle. Generally speaking, if an automaton explicitly communicate with them, the automaton obviously distinguishes their difference. Thus on this assumption, an automaton can arrive at its goal in an on-line way while avoiding static and uncertain obstacles according to some sensor-based path-planning algorithms [7,8] and simultaneously while avoiding all automata including some locked ones according to our proposed one. However two kinds of deadlock-free properties of the algorithms are ensured under different characteristics, i.e., distance and topological ones. Therefore in future, we should examine how the distance characteristic is coincident with the topological one so as to combine the static and dynamic sensor-based path-planning algorithms. As an exception of this, we already addressed the sensor-based path-planning algorithm running in a finite uncertain 2-d world with several kinds of circular obstacles [9]. In this algorithm, each automaton always arrives at the goal while tracing circular uncertain obstacles by the counter-clockwise order under the topological characteristic. Thus this algorithm is easily connected with our proposed one while ensuring their deadlock-free characteristics. In general, an uncertain world with several circular obstacles is born if some automata are stopped by a mechanical accident in our framework, but even then, the other automata safely keep their deadlock-free characteristics in such a world with stopped automata by the connection between the static and dynamic algorithms.

References

[1] H. Asama, Trend and Perspective of Researches on Multi-Agent Robotic Systems, *Journal of Robotics Society of Japan*, Vol.10, No.4, 1992, pp.428-432 (in Japanese).

[2] S. Yuta, Cooperative Behaviors of Multiple Autonomous Mobile Robots, *Journal of Robotics Society of Japan*, Vol.10, No.4, 1992, pp.433-438 (in Japanese).

[3] T.Arai and J.Ota, Planning of Multiple Mobile Robots, *Journal of Robotics Society of Japan*, Vol.10, No.4, 1992, pp.444-449 (in Japanese).

[4] H.Noborio and J.Hashime, A Feasible Collision-Free and Deadlock-Free Path-Planning Algorithm in a Certain Workspace where Multiple Robots Move Flexibly, *Proc. of the 1991 IEEE International Workshop on Intelligent Robots and Systems (IROS'91)*, Vol.2, 1991, pp.1074-1079.

[5] H.Noborio, A Collision-Free and Deadlock-Free Path-Planning Algorithm for Multiple Mobile Robots without Mutual Communication, *Proc. of the 1992 IEEE/RSJ International Conference on Intelligent Robots and Systems (IROS'92)*, Vol.1, 1992, pp.479-486.

[6] H.Noborio and M.Edashige, A Cooperative Path-planning for Multiple Automata by Dynamic/Static Conversion, *Proc. of the 1993 IEEE/RSJ International Conference on Intelligent Robots and Systems (IROS '93)*, 1993, pp.1955-1962.

[7] H.Noborio, A Sufficient Condition for Designing a Family of Sensor-Based Deadlock-Free Path-Planning Algorithms, *Journal of Advanced Robotics*, Vol.7, No.5, 1993, pp.413-433.

[8] H.Noborio, A Path-Planning Algorithm for Generation of an Intuitively Reasonable Path in an Uncertain 2d Workspace, *Proc. of the 1990 Japan-USA Symposium on Flexible Automation*, Vol.3, 1990, pp.477-480.

[9] H.Noborio, A Relation between Workspace Topology and Deadlock Occurrence in the Simplest Path-Planning Algorithm, *Proc. of the 1992 Second International Conference on Automation, Robotics and Computer Vision*, 1992, pp.RO-10.1.1-RO.10.1.5.

Distributed Strategy-making Method in Multiple Mobile Robot System

JUN OTA, TAMIO ARAI, and YOICHI YOKOGAWA

Dept. of Precision Machinery Eng., The Univ. of Tokyo, Tokyo, 113 Japan

Abstract

A distributed algorithm for making strategies in multiple autonomous system is presented in this paper. Here, every robot in the system has several *tactics* to be selected, and a set of probabilities to select each tactic is called a *strategy*. The problem is "how does each robot get one's proper strategy independently by trying tactics, getting responses from the environment around itself and revising each strategy?" From the viewpoint of simplicity of modeling and analysis, authors adopt reinforcement-learning-like approach. Each robot memorizes "evaluated pay-off values for every tactic" and the "strategy," and revises those values asymptotically by iterating four steps as follows: (1) select a certain tactic by reflecting the present strategy of its own and try it, (2) get pay-off value for the applied tactic from the environment, (3) revise the estimated pay-off value of tactics, (4) revise the strategy by using (3). Path-selecting simulation of multiple mobile robots, in which fifty robots move between two areas through two paths, is made to verify effectiveness of the proposed method. By evaluating turnaround time of robots with variety of path widths' ratios, the proposed algorithm is shown to be superior to other algorithms such as one-way-traffic-strategy. Convergence to global optimal solution derived by the proposed method is discussed for simplified situation by using pay-off matrix in theory of game.

Keywords: Multiple mobile robot system, strategy-making, reinforcement leaning, theory of game

1 Introduction

Cooperation among multiple autonomous robots is important to realize high quality tasks. The cooperation can be divided into two from the tasks for each robot[1]:

(1) cooperation when the task is common for each robot (cooperative problem solving)

(2) cooperation when the tasks are independent for each robot (negotiation and balancing)

This paper deals with one of the latter problem. Here, every robot in the system has several selective ways of action (these are called *tactics* in this paper), and a set of probabilities to select tactics (this is called a *strategy* in this paper). We develop an algorithm to learn proper strategies independently.

Many studies were presented with the topics. Parker proposed the algorithm with heuristics approach and applied to cleaning tasks by several robots[2]. Ogasawara et al. realized transportation of a large object by several robots[3]. In this paper, each robot selects each action by using Bayesian statistics. Mikami et al. solved navigation problem of mobile robots by combining distributed reinforcement learning and genetic algorithm[4]. Numaoka realized autonomous transition among tactics of robot groups by using Huberman model[5].

We adopt reinforcement learning approach[4] from the viewpoint of simplicity of modeling and analysis. Strategy-making for multiple autonomous system has characteristics below: (1)it is difficult to evaluate responses from the environment correctly because those values are influenced by other robot's strategies. Let us think of a path-selecting problem for multiple mobile robots. Here, it is impossible to judge which is better between the case (i)when it takes two minutes to go through in a vacant situation, and the case (ii)when it takes three minutes to go through the same path with (i) in a crowded situation. It is necessary to establish framework for evaluating such a situation. (2)Although it is important to analyze behaviors of robots as a whole, it is difficult to discuss optimality in a global sense while each robot act independently.

We propose strategy-making algorithm that can overcome above two problems. Concretely, asymptotic strategy-making algorithm with is presented against the former one. Pay-off matrices in theory of game are used against the latter one. In chapter 2, asymptotic strategy-making algorithm is proposed. In chapter 3, path-selecting simulation is made to verify significance of the proposed algorithm, and convergence to global optimal solution by the proposed algorithm is discussed for a simple case. We conclude this paper in chapter 4.

2 Proposal of asymptotic strategy-making algorithm

2.1 Assumption for the environment and conceptual design for strategy-making

We present strategy-making algorithm in this chapter. First, we assume the following conditions.

- There are several numbers of robots in a given environment.
- Communication between robots isn't assumed. This is due to avoid communication costs while the number of robots increases.
- Each robot i ($1 \leq i \leq n$, n: the number of robots) has more than one tactics.
- A probability r_{ij} is defined for robot i to select tactic j ($1 \leq j \leq m$, m: the number of tactic for each robot). Vector $\mathbf{r_i} = (r_{i1}, r_{i2},..., r_{im})^T$ is called "a strategy of robot i."
- When a certain tactic is applied, each robot can get quantitative response from the environment. This value is changed depend on other robots' tactics.

The strategy-making problem in this paper can be defined for each robot to learn $\mathbf{r_i}$ autonomously and independently. The proposed algorithm in this paper consists of four steps as follows, and strategy-making is accomplished by iterating these steps (Fig. 1).

(STEP 1) Determination of tactic to be applied by reflecting the present strategy
(STEP 2) Acquisition of the pay-off value for the applied tactic
(STEP 3) Revision of estimation of the applied tactic
(STEP 4) Revision of the strategy

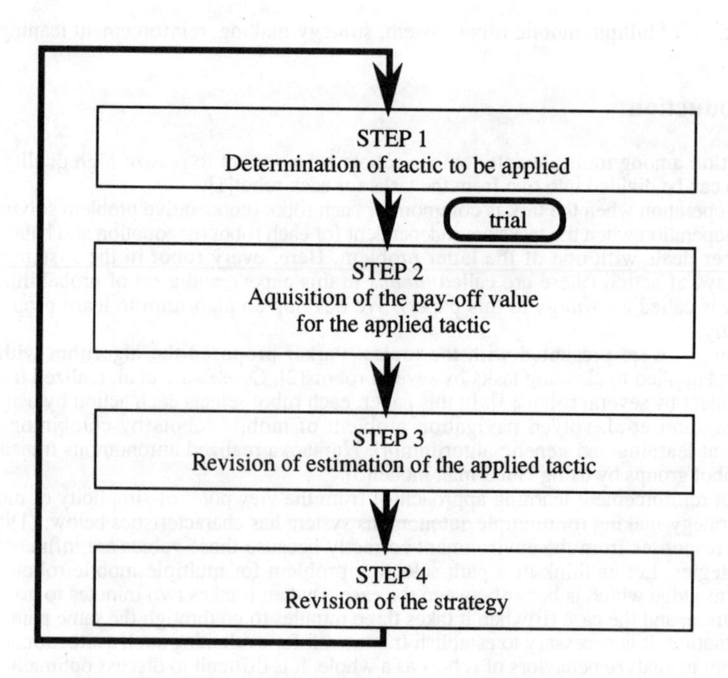

Fig. 1 The asymptotic strategy-making algorithm

In the above algorithm, each robot maintains the estimated values for tactics and the strategy r_i and revise the values in STEP 3 and STEP 4. Initial estimated values of tactics and a strategy are set to equal for all tactics.

2.2 Proposal of asymptotic strategy-making algorithm

The proposed algorithm is explained in detail in this paragraph.

2.2.1 STEP 1: Determination of a tactic to be applied next time

Each robot determines a tactic to be applied next time based on the present strategy of itself, and apply it.

2.2.2 STEP 2: Acquisition of the pay-off for the applied tactic

After applying the tactic, each robot gets pay-off from the environment.

2.2.3 STEP 3: Revision of estimated value of the applied tactic

Estimated value of the applied tactic is revised with the use of the pay-off value. When robot i ($1 \leq i \leq n$) applies a tactic j ($1 \leq j \leq m$), Estimated value of the applied tactic p_{ij} is revised as eq. (1).

$$p_{ij} = K_f\, p_{ij\_old} + (1 - K_f)\, p_{ij\_now} \qquad (1)$$

p_{ij_now}: pay-off value of the applied tactic j for robot i
p_{ij_old}: evaluated value of tactic j before revision for robot i
K_f: memory coefficients ($0 \leq K_f < 1$)

Eq. (1) makes it possible to revise estimation of the applied tactic. Moreover, eq. (1) has advantages as follows:

- errors of evaluated value of the tactic caused by variety of other robot's tactics can be absorbed with numbers of trials, and
- relative evaluation among tactics can be made.

When K_f is set to be large, older estimated values of tactics are more weighted.

2.2.4 STEP 4: Revision of the strategy

The strategy is revised in STEP 4 with eq. (2) and eq. (3).

$$r_{ij\_tmp} = \max\left(r_{ij\_old} + K_d\left(p_{ij} - \frac{\sum_{k=1}^{m} p_{ik}}{m}\right), r_{min}\right) \qquad (2)$$

$$r_{ij} = \frac{(r_{ij\_tmp} - r_{min})(1 - m r_{min})}{\sum_{k=1}^{m} (r_{ik\_tmp} - r_{min})} + r_{min} \qquad (3)$$

r_{ij_old}: r_{ij} before revision
K_d: changeability coefficients ($K_d > 0$)
r_{min}: minimal value for selecting each tactic ($\frac{1}{m} > r_{min} > 0$)

The meaning of eq. (2) is as follows:

- the higher estimated value of a certain tactic becomes, the more probability to select the tactic becomes,
- the more K_d becomes, the more the probabilities become easy to change, and
- r_{min} is the parameter that assures to apply all the tactics with more than a certain probability, by which each robot can adjust itself to dynamic changes of the environment.

With eq. (3), normalization for the probabilities is made with satisfying $\sum_{j=1}^{m} r_{ij} = 1,\ r_{ij} \geq r_{min}$.

By introducing the above algorithm to each robot, every robot can learn strategies autonomously. The proposed strategy can be said as distributed gradient method. The detail explanation is made at paragraph 3.3.

3 Path-selecting simulation

Path-selecting problem, in which many robots select each own path from the start area to the goal area without colliding each other, is very important for multiple mobile robot system. Then several numbers of studies are made on the topic (for example, [7]). In this chapter, efficiency of the proposed algorithm is verified by applying the proposed algorithm to the problem.

3.1 Problem settlement of the simulation

Simulation environment is assumed as Fig. 2. Here,
- There are two paths (called path 1 and path 2) between two areas (called area A and area B).
- Paths are expressed as grid lines as Fig. 2. Each robot precedes one grid at one step count.
- Each robot comes and goes from area A to area B by selecting and passing through path 1 or path 2. When a certain path is selected, a grid line to pass is randomly selected.

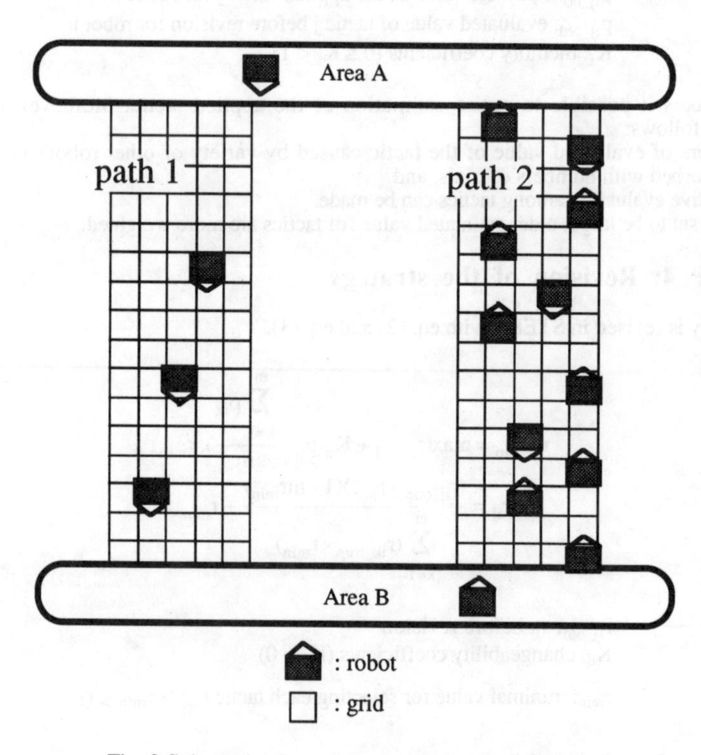

Fig. 2 Schematic view of the simulation environment

- When a certain robot enters the grid where another robot already exists, the robot takes additional time to go to next grid. This assumption models time consumption while avoiding another robot.
- Each robot makes the strategy from area A to area B and that from area B to area A independently.

By considering that each path is tactic and pay-off is $-1.0 \times$ "step counts to move from a start area to a goal area," the algorithm in chapter 2 can be applied to the simulation.

3.2 Simulation results and evaluation

Simulation is made with changing ratio of the width of path 1 and that of path 2 to 5:5, 6:4, 7:3, 8:2 and 9:1. We compare four algorithms as follows:

(a) Random select algorithm: the algorithm in which each robot selects two paths with randomly. This means paths are selected with the same probability.
(b) Proportional-to-width algorithm: the algorithm in which each robot determines probabilities to select two paths respecting the ratio of path's width. For example, when the width ratio is 9:1, probabilities to select each path become 90% and 10%.
(c) One-way-traffic algorithm: the algorithm in which all the robots select the same path. For example, every robot selects path 1 while moving from area A to area B, selects path 2 from area B to area A, and vice versa.
(d) Asymptotic algorithm: the algorithm in which each robot adopt the proposed algorithm in chapter 2.

The condition for the simulation is shown in Table 1. Although K_f, K_d and r_{min} can be set independently for each robot, they are set to common for all robots with cut and try method in this simulation. First, strategy-making process with the algorithm (d) is shown. Time series of a probability to select path 1 of a certain robot is shown in Fig. 3, where the width of path 1 and path 2 is 5:5. Fig. 3 shows the probability is converged to a certain constant value after about 50 times' trial, where the probability to select path 1 is equal to $(1 - r_{min})$(= 99 %), and the one for path 2 is r_{min} (= 1 %). Almost the same results can be seen for other robots and other environments. The final situation of the algorithm (d) for each environment becomes as follows:

- When the ratio is 5:5 or 6:4, every robot selects one path with probability of $(1 - r_{min})$ while moving from area A to area B, and selects the other path with the same probability while moving from area B to area A. This result means every robot converges to one-way algorithm.
- When the ratio is 7:3 or 8:2 or 9:1, every robot selects only wider path while moving from area A(B) to area B(A), and the ratio of "the number of robots that selects the wider path" and "the number of robots that selects the narrower path" converges to 13:37, 24:26, 36:14 respectively while moving from area B(A) to area A(B), .

To evaluate each algorithm quantitatively, we select "average step counts to move between two areas" for performance index. Comparison of each strategy is shown in Fig. 4. Here,

- (d) is results after convergence of strategies.
- (e) is the solution for evaluation. This is derived by changing robots' strategies and searching for the ones that minimize the performance index. Then this solution can be called sub-optimal solution. Please note that this is calculated by using global information and cannot be derived by distributed algorithm.

Table 1 Conditions for the simulation

The number of robots	50
K_f	0.9
K_d	0.1
r_{min}	0.01
Total number of grid lines	10
The number of grids for one grid line	13

We can derive the following results from Fig. 4.

- Although random select algorithm (a) is the simplest one, it is not adequate for it takes more step counts to move compare to others.
- Proportional-to-width algorithm (b) is valid when the width ratio between two paths is large. Results for this algorithm are constant because the probability for selecting each grid line is the same for all width's ratio.
- One-way-traffic-algorithm (c) is valid when the width ratio is small. However, the algorithm becomes inefficient when the ratio becomes large because of traffic jam on a narrower path.
- The asymptotic algorithm (d) is valid for all the ratios. The reason the algorithm (c) is better than (d) when the ratio is 5:5 or 6:4, is because in (d) it doesn't converge to perfect one-way-traffic situation because of r_{min}. r_{min} is, as written in chapter 2, to be adapted to dynamic changes of environments and is effective for the situation of temporal obstacle is put on the path and width of the path changes dynamically. Comparison of (d) and (e) shows the result of (d) is nearly sub-optimal values.

Hence, effectiveness of the proposed method is verified.

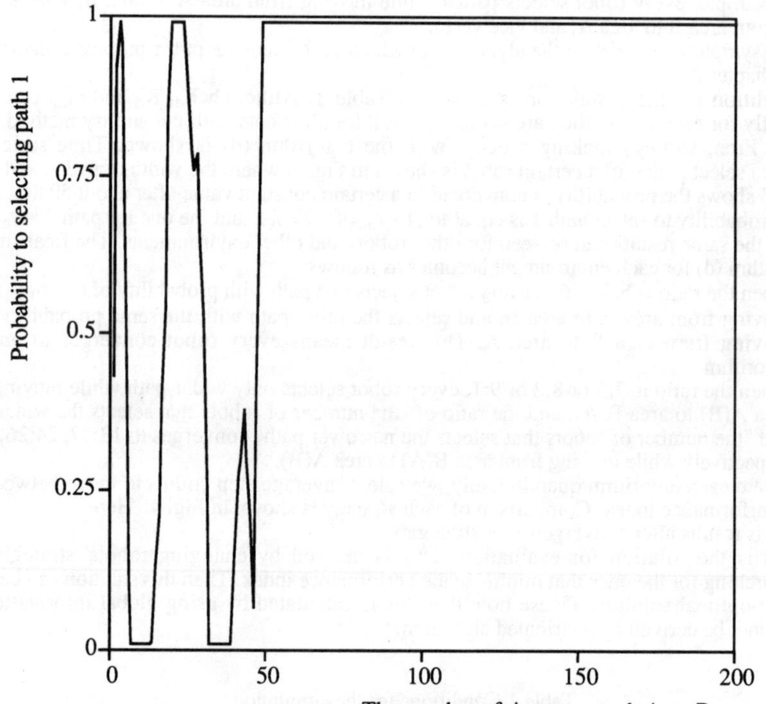

Fig. 3 Time series of a probability to select path 1 of a certain robot

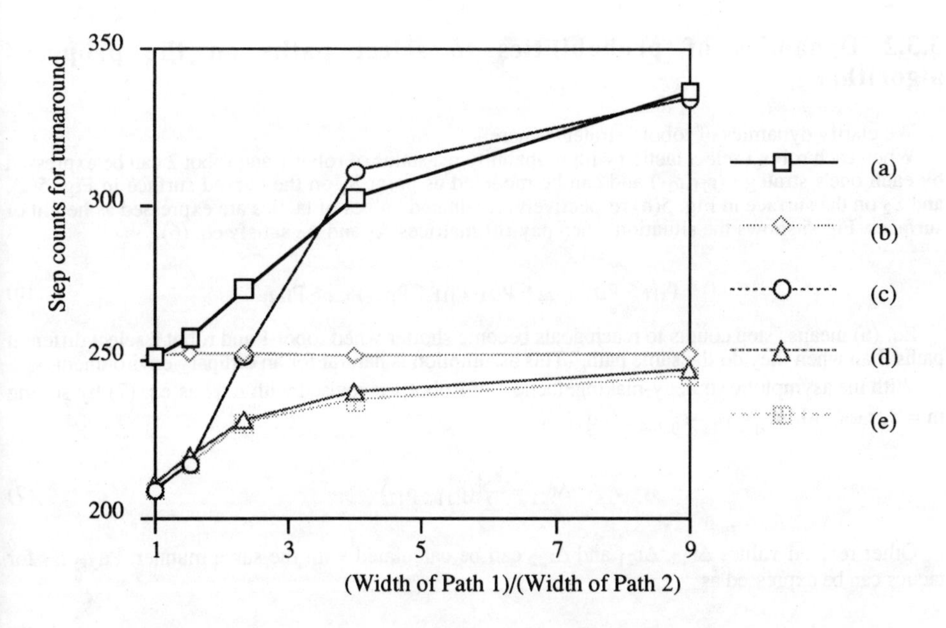

Fig. 4 Comparison of effects of algorithm (a) to (e)

3.3 Analysis of the solution of the proposed algorithm

In this paragraph, we discuss the convergence of strategies of robots with the proposed algorithm and relationship between the converged value and global optimal solution.

3.3.1 Model of the path-selecting problem

The problem dealt with in this paper can be called competitive problem of multiple agent system. Then, we introduce theory of game to analyze the problem.

For simplicity, the following assumption is set in this analysis:

- errors in evaluating pay-offs of tactics are neglected, which can be realized by setting K_d small,
- effects of r_{min} is neglected, and
- there are two robots named Robot 1 that goes from area A to area B and Robot 2 that goes from area B to area A in the environment.

With above assumption, the problem can be modelled as non-zero sum game where the number of player is two and each player has two kinds of tactics (that is called tactics 1 and tactics 2, here). Pay-off matrices of Robot 1 (A_1) and that of Robot 2 (A_2) are as follows:

$$A_1=\begin{bmatrix} p_{111} & p_{112} \\ p_{121} & p_{122} \end{bmatrix} \quad (4)$$

$$A_2=\begin{bmatrix} p_{211} & p_{212} \\ p_{221} & p_{222} \end{bmatrix} \quad (5)$$

p_{ijk}: a pay-off of robot i when robot 1 selects tactic j and robot 2 selects tactic k.
Concretely, it means -1 × (step counts for robot i to move between two areas when robot 1 selects tactic j and robot 2 selects tactic k)

Path-selecting problem can be modeled as eq. (4) and eq. (5).

3.3.2 Dynamics of probabilities to select paths on the proposed algorithm

We clarify dynamics of robots' strategies here.

When each robot i select tactic j with probability r_{ij}, status of robot 1 and robot 2 can be expressed by each one's strategy (r_{11}, r_{21}) and can be modeled as point S_1 on the curved surface in Fig. 5(a), and S_2 on the surface in Fig. 5(b) respectively. Evaluated values of tactics are expressed as height of surfaces. Fig. 5 shows the situation when pay-off matrices A_1 and A_2 satisfy eq. (6).

$$p_{i11} < p_{i21}, p_{i22} < p_{i21}, p_{i11} < p_{i12}, p_{i22} < p_{i12} \tag{6}$$

Eq. (6) means "step counts to reach goals become shorter when robot 1 and robot 2 select different paths than when they do the same path." This assumption is natural for an ordinary environment.

With the asymptotic strategy-making, tactic 1 of Robot 1 is revised with Δr_{11} as eq. (7) by setting $m = 2$, $r_{min} = 0$, $\Delta r_{ij} = r_{ij} - r_{ij\_old}$ in eq. (2).

$$\Delta r_{11} = \frac{K_d}{2}(p_{11} - p_{12}) \tag{7}$$

Other revised values Δr_{12}, Δr_{21} and Δr_{22} can be calculated with the same manner. Pay-offs for tactics can be expressed as

$$p_{11} = r_{21} \cdot p_{111} + r_{22} \cdot p_{112} \tag{8}$$

$$p_{12} = r_{21} \cdot p_{121} + r_{22} \cdot p_{122} \tag{9}$$

From eq. (8) and eq. (9), we can understand p_{11} and p_{12} correspond to the height of point A and point B respectively. This result and eq. (7) show that the asymptotic strategy-making algorithm makes robots' probabilities to select paths change in the direction of gradient of pay-offs. In other words, each robot changes its strategy to which the arrows in Fig. 5 show. When we describe the arrows for all the status, revision process of status of robot 1 and robot 2 can be expressed as Fig. 6. Note that the direction of arrows in Fig. 6(a) is only vertical direction, on the other hand, that of arrows in Fig. 6(b) is only horizontal direction. This is because each robot can only change each own strategy. The black stripe in Fig. 6 is the place where gradient is equal to zero.

By integrating Fig. 6(a) and Fig. 6(b), strategies of robot 1 and robot 2 converge to local maximum points of pay-offs (black points in Fig.7).The meaning of black points are "the situation when robot 1 always selects path 1 and robot 2 always selects path 2 (point L_1 in Fig. 7)" and "the situation when robot 1 always selects path 2 and robot 2 always selects path 1(point L_2 in Fig. 7)."

We discuss optimality in global sense. Here, we define a performance index that expresses global optimality as summation of a pay-off of robot 1 and that of robot 2. Usually, only one of the two points is considered as a global optimal solution and a converging point depends on the initial status of each robot. Therefore the proposed algorithm does not assure optimality in a global sense. However, in the path-selecting problem dealt with in this paper, a pay-off while passing through path 1(2) from area A to area B and that while passing through path 2(1) from area B to area A can be considered to be the same, in other words, $p_{112} = p_{221}$, $p_{121} = p_{212}$. Above assumption is adequate in ordinary environments. In this situation, performance indexes in a global sense become equal for both local maximum points. Then we can say the solutions given in the algorithm converge to global optimum points.

Above discussion can be applied to the case when width ratio is 5:5 or 6:4 in the simulation of the former paragraph, by considering "robots from area A to area B" as robot 1 totally, and "ones from area B to area A" as robot 2. In the case width ratio is 7:3, 8:2 or 9:1, structure of pay-offs is not as Fig. 6 and Fig. 7 but as Fig. 8 and Fig. 9. However, because the pay-offs of two local maximum points can be considered to be the same with the same discussion of Fig. 7, the proposed algorithm assures global optimality.

Although we dealt with the basic problem with the proposed algorithm in this paper, this algorithm can be applied to other examples.

4 Conclusion

Strategy-making problem in multiple mobile robot system is discussed in this paper.

- Asymptotic strategy-making algorithm by using reinforcement learning is proposed.
- The proposed method is applied to path-selecting simulation of multiple mobile robots, and effectiveness of the method is verified.
- Optimality in a global sense is discussed by using pay-off matrices in theory of game.

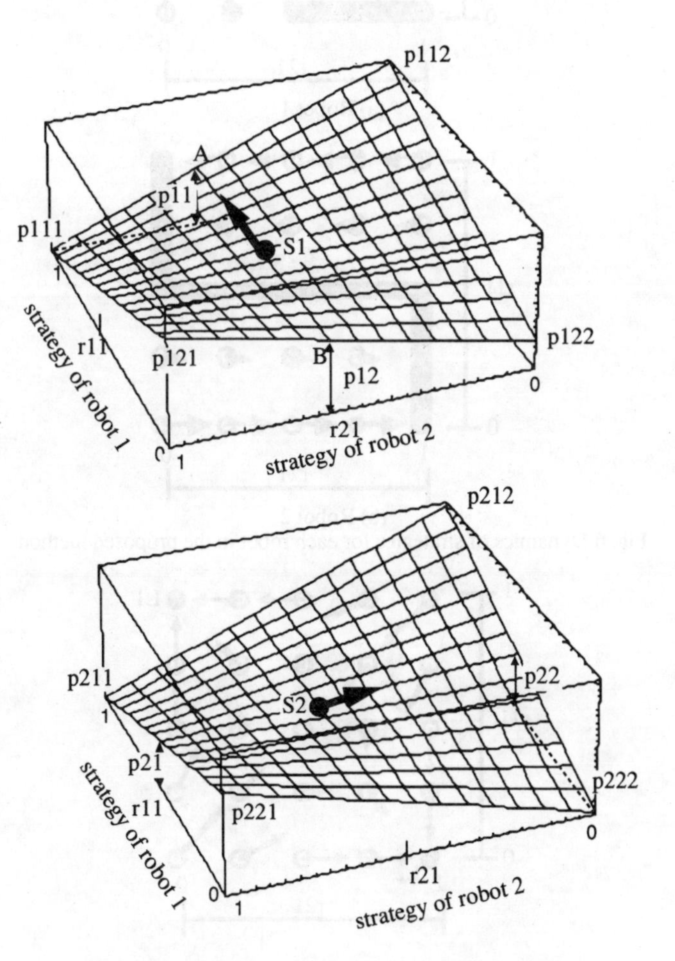

Fig. 5 Relationship between pay-offs and strategies

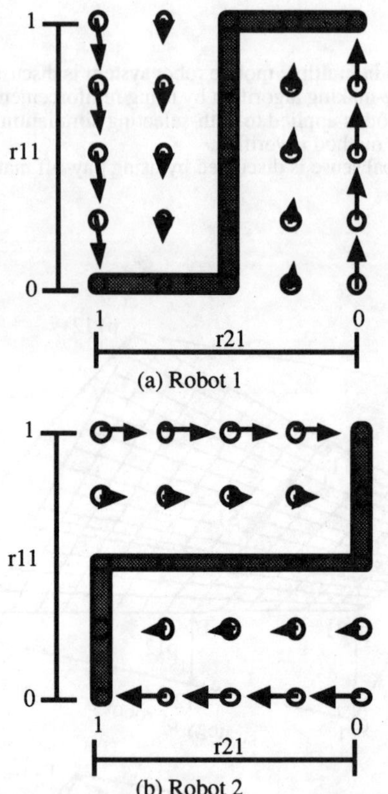

(a) Robot 1

(b) Robot 2

Fig. 6 Dynamics of strategies for each robot in the proposed method

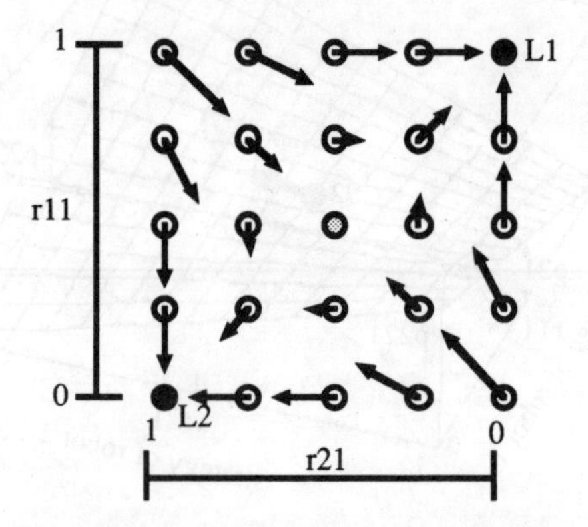

Fig. 7 Convergence of strategies of robots

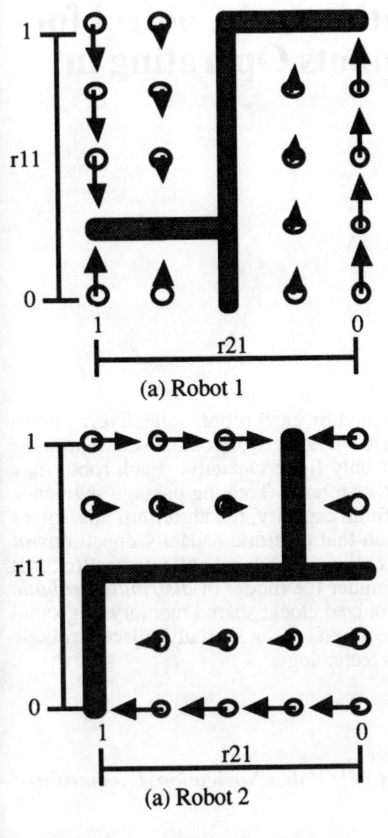

Fig. 8 Dynamics of strategies for each robot in the proposed method (When the width ratio of the two paths' width is large)

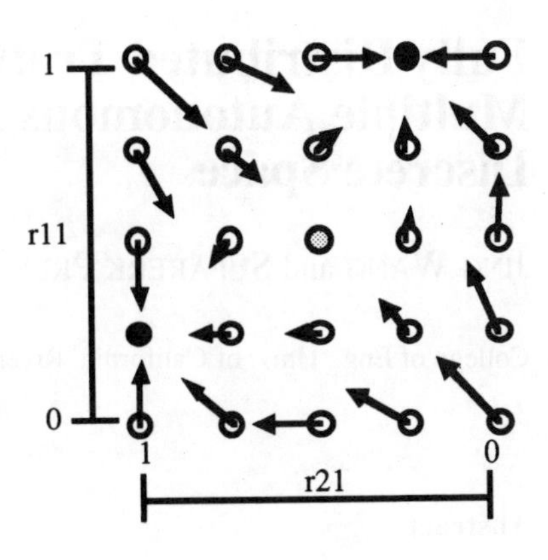

Fig. 9 Convergence of each robot's strategy (When the width ratio of the two paths' width is large)

References

[1] K. Kuwabara, T. Ishida, Distribute Artificial Intelligence (2): Negotiation and Balancing, *Journal of Japan Society for Artificial Intelligence*, Vol. 8, No. 1, 1993, pp. 17-25(in Japanese).
[2] L. E. Parker, Adaptive Action Selection for Cooperative Agent Teams, *Proc. 2nd. International Conference on Simulation of Adaptive Behavior*, 1992, pp. 442-450.
[3] G. Ogasawara, T. Omata, T. Sato, Multiple Movers using Distributed, Decision-theoretic Control, *Proceedings of Japan-USA Symposium on Flexible Automation*, Vol. 1, 1992, pp. 623-630.
[4] S. Mikami, Y. Kakazu, Conflict Resolution of Navigation Problems by Reinforcement Learning, *Proceedings of the Japan Society of Mechanical Engineers Annual Conference on Robotics and Mechatronics*, 1993, pp. 907-912(in Japanese).
[5] C.Numaoka, Collective alternation strategic types with delayed global information, *Proceedings of 1993 IEEE/RSJ Int. Conf. Intel. Robots and Systems*, 1993, pp.1077-1084.
[6] R. S. Sutton, *Reinforcement Learning*, Kluwer, 1992.
[7] T. Shibata, T. Fukuda, Coordinative Behavior in Evolutionary Multi-agent-robot System, *Proceedings of 1993 IEEE/RSJ Int. Conf. Intel. Robots and Systems*, 1993, pp.448-453.

Fully Distributed Traffic Regulation and Control for Multiple Autonomous Mobile Robots Operating in Discrete Space

JING WANG and SUPARERK PREMVUTI

College of Eng., Univ. of California, Riverside, CA 92521, USA

Abstract

A fully distributed algorithm is presented, which when executed by each robot, collectively allows multiple autonomous mobile robots to travel through a discrete traffic network composed of *passage segments*, *intersections*, and *terminals*, all of which are of only finite capacity. Each robot may establish dynamically its own desired route not known to other robots. Treating passage segments, intersections and terminals as shared, discrete resources of finite capacity, the algorithm guarantees multiple robots traveling in the discrete traffic network such that (i) finite capacity constrains of passage segments and terminals are always enforced; (ii) no collision occurs at any intersection; (iii) deadlocks are detected and resolved. The system operates under the model of *distributed robotic systems* (DRS), assuming no centralized mechanism, synchronized clock, shared memory or ground support. It is shown that inter-robot communication is only required among spatially adjacent robotic units, and the algorithm is readily implementable with today's technology.

Key Words

Distributed Robotic Systems, Autonomous Mobile Robots, Mobile Robot Navigation, Decentralized Traffic Regulation and Control, Distributed Resource Sharing.

1. Introduction

1.1 Background

Coordinated Motion Among Multiple Robots

It is generally agreed upon that the population of autonomous mobile robots will grow rapidly in the near feature, which will force robotic researchers to study cooperation and coordination strategies among autonomous robotic units under distributed control. In particular, many applications will require robots to share operating spaces of discrete nature (Fig. 1).

Coordination and cooperation among multiple autonomous mobile robots have been studied in recent years with various degrees of success. Notable contributions include Fukuda's CEBOT [1], Asama's experimental system ACTRESS [2], Arkin's motion schema [3], Beni and Wang's swarm systems [4] and Brooks' reactive based systems [5].

We nevertheless have noticed that some existing multi-agent experimental systems, which are supposed to be fully distributed, in fact inadvertently employ centralized mechanism. On the other hand, many proposed models under investigation assume unrealistic sensing, actuation and communication capabilities so far beyond that implementable with today's technology, and such research may only be demonstrated with computer simulations (not real robots).

The purpose of this paper is to demonstrate a system that (i) employs only fully distributed operating primitives; (ii) is implementable with today's technology; and (iii) exhibits impressive cooperation and coordination behavior of a group of autonomous mobile robots. Notice that the correctness of each individual protocol, as well as the overall algorithm is rigorously provable.

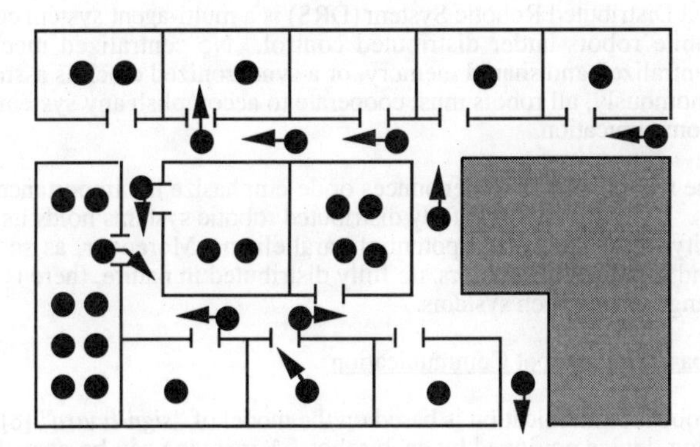

Fig. 1 Shared discrete workspace for autonomous mobile robots

Distributed Algorithms and Yamabico Experiment

One of the authors (Wang) has studied fully distributed algorithms for multiple robots to share discrete resources under the model of Distributed Robotic Systems (DRS) [6]. The other (Premvuti) was the main contributor of the demonstrated experiment for coordinated motion of multiple autonomous mobile robots (Yamabico) operating on a discrete field [7].

In the Yamabico experiment, a number of autonomous mobile robots travel through a discrete traffic network consisting of *passage segments* and *intersections* (Fig. 2). Each robot has its own route not known to others. Collisions at intersections are avoided through inter-robot communication.

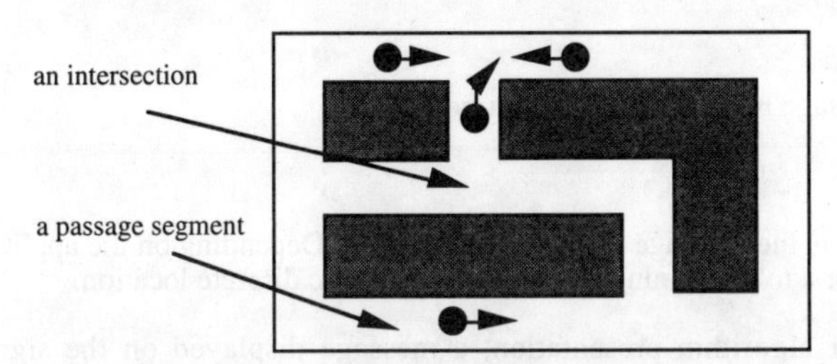

Fig. 2 The Yamabico experiment (1991)

Because the number of robots involved in the Yamabico experiment was relatively small, several potentially important problems in such a system were not considered:

- The capacity of each passage segment is trivially defined as 1, the situation of more than one robot traveling through the same passage segment was not considered.
- Deadlock detection and resolution is trivialized as robot operations at intersections.
- There is no place in the operating field for robots to stay for finite, but unspecified amount of time. (These places may correspond to work cells on manufacturing floors or stations in roadway/railway systems)

This paper generalizes the model assumed in the Yamabico experiment, and presents a rigorous, effective, robust and efficient solution readily implementable with today's technology.

1.2 Mobile Robot based Distributed Robotic Systems

A mobile robot based Distributed Robotic System (DRS) is a multi-agent system consisting of a group of autonomous mobile robots under distributed control. No centralized mechanism, such as a centralized CPU, centralized and shared memory, or a synchronized clock is assumed. Each mobile robot operates autonomously, all robots must cooperate to accomplish any system-wide task through limited inter-robot communication.

By no means that the research on DRS denounces or de-emphasize the importance of centralized and hierarchical systems. It is our belief that fully distributed robotic systems holds its ground because its robustness, reliability, redundancy and potential parallelism. Moreover, as so many human (and animal) activities and intelligent behaviors are fully distributed in nature, there is certainly a need of understanding and engineering such systems.

1.3 Sign-board based Inter-robot Communication

In this paper, inter-robot communication is based on the model of "*sign-board*" [8]. A sign-board is a *conceptual* displaying device equipped by each robot. A message can be posted on the sign-board only by this robot, and be read by robots in its neighborhood.

The sign-board model differs from that of *message passing*. Information displayed on a sign-board may or may not be seen by any specific robot during any specific time window. If a message is designated to a specific robot, handshaking has to be explicit spelled out in the algorithm.

It should be emphasized that the sign-board model is fundamentally different than that of *blackboard* commonly studied in artificial intelligence. The former is a fully distributed system, as messages on sign-boards are physically carried by robots in the operating field. Whereas the latter is a centralized mechanism, and a message displayed on a blackboard can be read by all agents.

Basic operations for the sign-board model are:

- *post*(*m*)

  display message *m* on the robot's sign-board.

- *m* := *read*(*a*)

  retrieve into *m* the message displayed by agent *a*. Depending on the application, an agent can be a robot, or a robot running on behalf of a specific discrete location.

For the purpose of algorithm presentation, a message displayed on the sign-board is usually abstracted as being divided into several *sections*. Each section is associated with a name, called an *instance*, indicating for what purpose this section of messages is posted.

Each section is further divided into a number of *logic fields*. The content of each field may be static, such as a robot's identification or other intrinsic/physical characteristic; or dynamic reflecting the robot's current operating state.

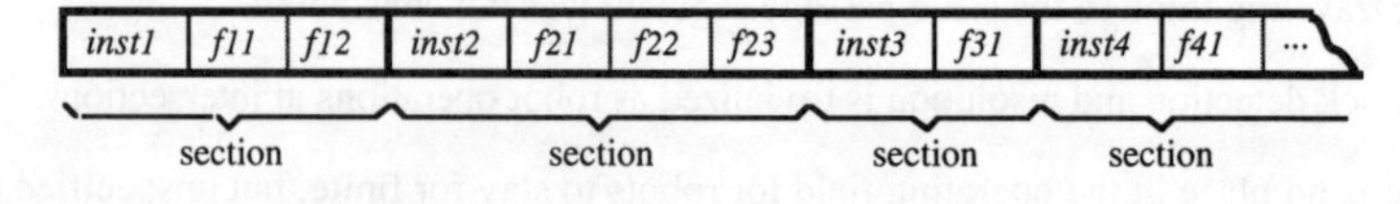

Fig. 3 Logic structure of a sign-board

The concept of *multi-section* sign-board allows clear presentation of distributed algorithms involving nested or concurrent protocol calls.

In this paper, the following two operating primitives are employed:

- *post_sect*(i; f_1: V_1; ...; f_n: V_n)

  Set field f_k to value V_k (for $k = 1, ..., n$) in section indicated by instance i. Other sections or logic fields in this section but not mentioned in this field list are not affected.

- *erase_sect*(i)

  Erase from the sign-board the entire section specified by instance i.

- *msg* := *read_sect*(i; a)

  Read the section, specified by instance i, of the message shown on the sign-board of agent a.

In this paper, an agent is always a discrete location. NULL is returned by function *read_sect* if no robot is present at the specified location a, or the sign-board of the robot at location a does not show a section identified by the specified instance *inst* (Fig. 4).

(a) returns NULL (no robot at location *a*) (b) returns sign-board of robot at location *a*

Fig. 4 Return value of function *read_sect*(i; a)

Note that sign-board is only a mathematical abstraction used for describing DRS algorithms. For a detailed explanation on the concept and implementation of the sign-board model, please consult [8].

1.4 Fully Distributed Traffic Regulation and Control

In this paper, we consider multiple, autonomous mobile robots operating in a 2-D space consisting of *intersections*, *terminals* and *passage segments*, all of which are of finite capacity (Fig. 5).

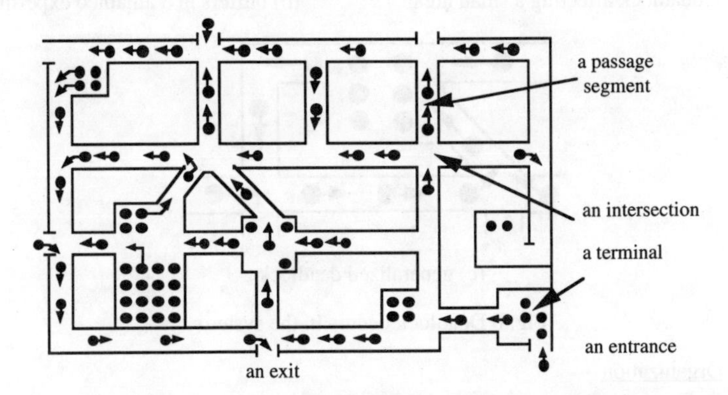

Fig. 5 Fully distribute traffic regulation and control in a 2-D field

It is should be emphasized that this system is under fully distributed control, and no centralized mechanism is to be used to regulate the traffic. For such a system to operate properly, a number of problems must be considered:

(a) Possible collision at any intersection among mobile robots must be avoided. An intersection must be passed through by robots one at a time.

(b) The finite capacity constrain of any passage segment must be enforced. It must be controlled in conjunction with the traffic regulation at its entrance.

(c) The finite capacity constrain of any terminal must be enforced in conjunction with the traffic regulation at its entrances and exits.

(d) Deadlocks must be detected and then resolved.

While the importance of (a) through (c) is easy to realize, (d) deserves more explanation.

- Deadlock does occur in this kind systems because all passage segments and terminals are of finite capacity; and the sequence of events leading robots to reserve, to move into, and to depart from various passage segments/terminals are not known *a prior*.
- Deadlock does occur even in a system of extremely low density of robot population -- sometimes a deadlock occurs affecting only a tiny portion of the operating field (Fig. 6a).
- Extra buffering areas may be required to resolve a deadlock while guaranteeing positive system progress. In fact, a buffering area at the end of each passage segment provided by the Yamabico experiment was for this purpose (Fig. 6b) [7].

This paper deals with the most generalized deadlock problem (Fig. 6c), for which a rigorous deadlock detection and resolution algorithm is presented.

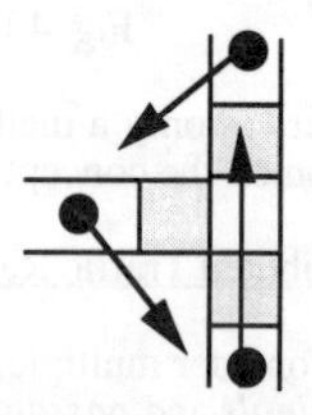

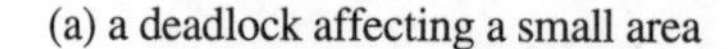

(a) a deadlock affecting a small area

(b) buffers in Yamabico experiment

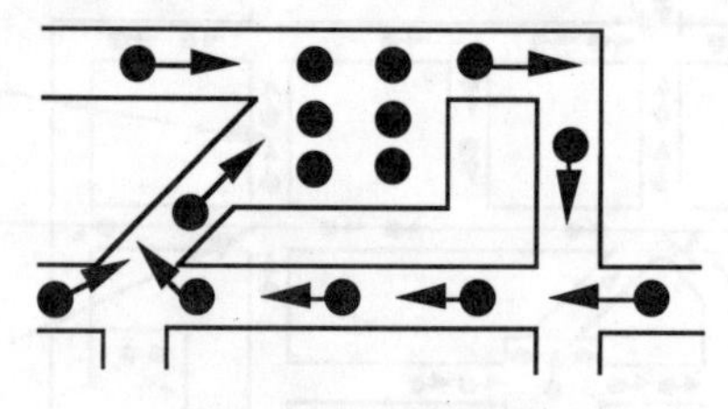

(c) generalized deadlock

Fig. 6 Deadlock occurs in the system

1.5 Paper Organization

Section 2 briefly reviews the basic operating primitives to be used in the paper. The distributed traffic control strategy (ideas and algorithms) is presented in Section 3. Section 4 briefly addresses implementation issues. Merits and shortcomings of the algorithm is the discussed in Section 5, along with current and future research directions on this topic.

2. Basic Operating Primitives

In this section, two operating primitives, are briefly reviewed. They are crucial to understand the overall algorithm described in the next section. Both of them are related to distributed resource sharing, which is important to any multi-agent system -- distributed robotic system is not an exception. The focus of this paper prohibits a detailed description for these primitives, reader should consult [9] and [10] for a detailed explanation.

2.1 1 out of N

In this problem, up to N ($N \geq 1$) agents compete for a resource of capacity 1 (Fig. 7).

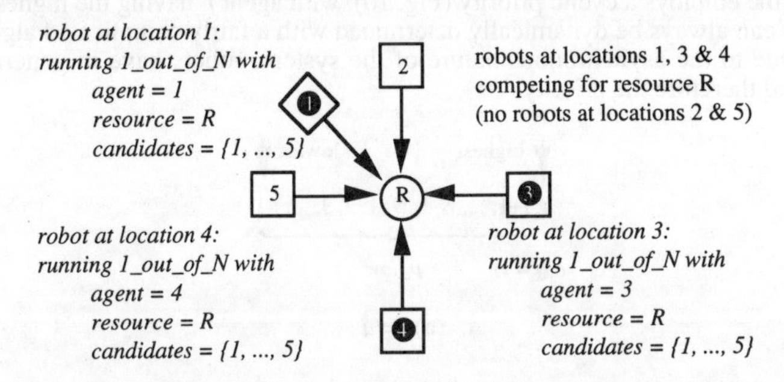

Fig. 7 Robots compete for resource R on behalf of locations: 1 out of N with N = 5

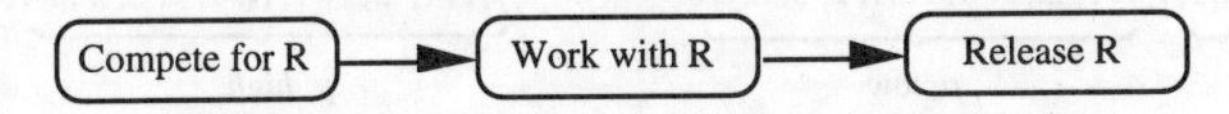

Fig. 8 One operating cycle of a robot (w.r.t. resource R)

Assumptions

It is assumed that each agent has an identification (id), and a total relation exists for the collection of agent ids. For instance, *{1, 2, ..., N}* may be used to denote the set of agent ids with relation "less than" (<) serving as the total relation. Note that *agent* is a different concept than that of a *robot*. For instance, a set of agents may be robots at entrances incident to a specific intersection. Although the system contains arbitrary number of robots, the number of potential agents competing for passing through this intersection is always a finite (and usually a quite small) number. In fact, robots (as agents) are in this case competing on behalf of entrances incident to the intersection.

Priority Schemes

Two priority schemes based on agent ids are introduced. For a specific agent, set *p_high* contains others agents of higher priority, whereas set *p_low* includes those of lower priority.

The first scheme simply sets the priority directly according to agent ids (Fig. 9). This works fine for most applications, but does not guarantee *lockout free* for low priority agents. Since the scheme is absolute and stationary, all agents in the system have a consistent view towards this priority.

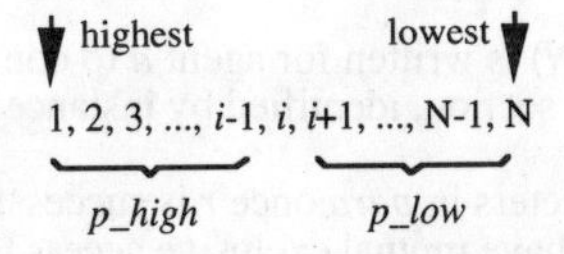

Fig. 9 Priority is set according to agent ids

Function *get_priority* returns pair <*p_high*, *p_low*>. If this absolute priority scheme is used, this function is simply,

function *get_priority*(C, a) **is**

/* C: set of all candidate agents
       a: agent id for the robot
*/

       return($<\{i \in A \mid i < a\}, \{i \in A \mid i > a\}>$)

end;

The second scheme employs a cyclic priority (Fig. 10), with agent f having the highest priority. For agent with id i, f can always be dynamically determined with a fairly complicated algorithm (details ignored) [9]. Due to the asynchronous nature of the system, robots have in general inconsistent views about f, and therefore the priority.

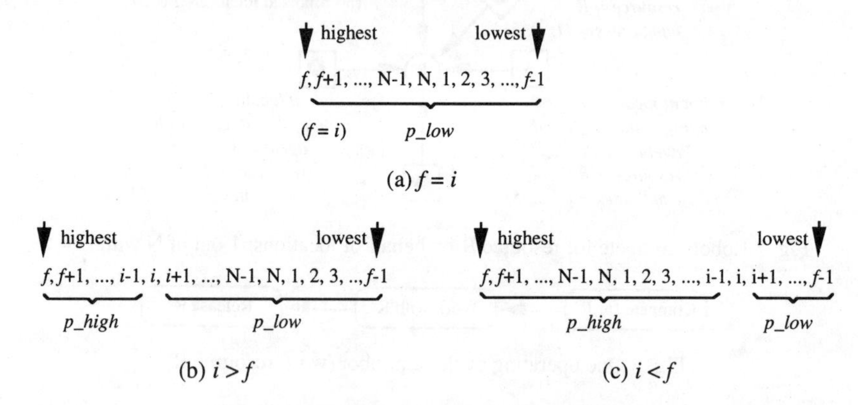

Fig. 10 A dynamic and cyclic priority scheme

Contrary to the intuition, priority schemes are not essential to guarantee mutual exclusive access to the resource. This effect is achieved through proper setting of operating states at proper times by involved agents. Priority schemes does ensure deadlock freedom for the "1 of out N" algorithm.

Sign-board Format

The sign-board used for this algorithm contains three fields (Fig. 11). The first (*agent*) is the agent id, which never changes during the period of algorithm execution. The second (*resource*) indicates which resource the agent is competing for. The last field indicates the operating state of the agent while executing this algorithm. Possible states for this algorithm are REQUEST, TAKEN, YIELD and NONE. All three fields belong to the same section identified by instant *inst*.

inst	*agent*	*resource*	*state*

Fig. 11 Sign-board used by the 1 out of N algorithm

Algorithm

The following function (*1_of_out_N*) is written for agent a to compete with others in candidate set C for resource r of capacity 1. A new section, identified by instance i, is established on the sign-board.

Routine *act* is executed with parameters in *para* once r is successfully *reserved* -- when the state of a becomes TAKEN. This allows a to have mutual exclusive access to resource r. The function assumes the value returned by executing *act*.

```
function 1_out_of_N(i,  a, r, C, act, para) is

/*      i: instance specification
        a: identification of the agent executing this program
        r: the resource (of capacity 1)
        C: the candidate set, i.e., the set of potential competitors
        act: a routine to be executed when resource r is TAKEN
        para: a list of parameters to be used by function act when it is called upon
*/

(1)  post_sect(i; agent: a; resource: r; state: REQUEST);
     <p_high, p_low> := get_priority(C, a;

(2)  for each  r ∈ p_high do begin
     lp1:  s := read_sect(i; r);
           if s = NULL or s.resource  ≠ r then s.state := NONE;
           if s.state ≠ NONE then begin
                 post_sect(i; state: YIELD); goto lp1
           end else if state = YIELD goto (1)
     end;

(3)  for each  r ∈ p_low do begin
     lp2:  s := read_sect(i; r);
           if s = NULL or s.resource ≠ r then s.state := NONE;
           if s.state = REQUEST or s.state = TAKEN then goto lp2
     end;
(4)  post_sect(i; state: TAKEN);

     s := act(para)      /*work with resource r */

(5)  erase_sect(i);

(6)  return(s)

end;
```

It can be shown that this algorithm guarantees mutual exclusive access to the resource, and it is deadlock free. The algorithms is also lockout free if the cyclic priority scheme is employed. Note that it is not required for involved robots to start running this algorithm at the same time. In particular, if an agent does not hold a robot, its "sign-board" *state* is interpreted as NONE.

2.2 Deadlock Detection

Deadlock

In [10], a fully distributed algorithm for deadlock detect in a traffic network composed of passage segments and intersections is presented, where all passage segments are assumed unidirectional and of capacity 1.

Recall in the distributed traffic control problem , a robot can *reserve* and *release* passage segments in an order not known *a priori*. Moreover, a robot must reserve its *next destination* (a passage segment) while occupying another. If the sequence of these reservations is not controlled, a *deadlock* may occur. Until the deadlock is resolved, all the robots involved are blocked indefinitely.

In Fig. 12, the desired next destinations for robots *a*, *b*, *c*, *d*, *e* and *f* form a *closed chain*. Thus nobody in this chain can proceed, because of the *deadlock*. In addition, robot *g* can not move, even though it is not in the closed chain. Similarly, robots *x* and *y* form another deadlock, which also blocks robots *w* and *z*.

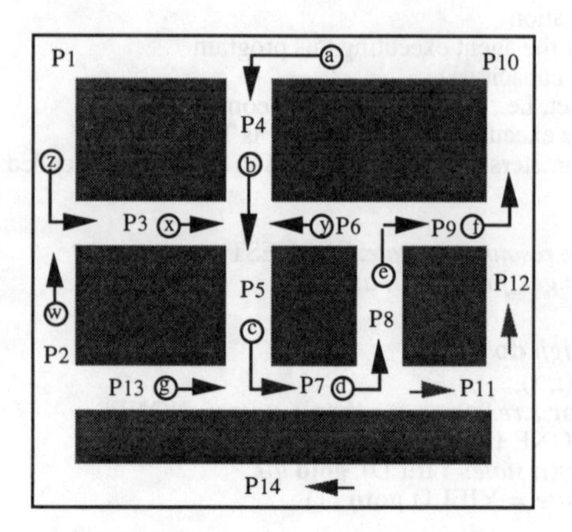

Fig. 12 Deadlocks present in a discrete operating field

Terminology

Ignoring intersections and the coordinated intersection passing problem, the operating field such as the one shown in Fig. 12 can be abstracted as a directed graph in which each passage segment is a node. If the exit of passage segment *p* is incident to an intersection connecting to the entrance of passage segment *q*, then $<p, q>$ is an edge in this directed graph. This graph is referred, in this paper, as the *Segment Connectivity Graph* (SCG) of the operating field.

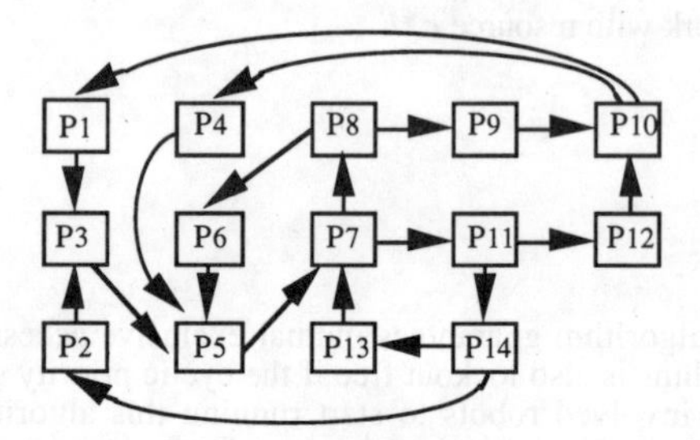

Fig. 13 SCG for the operating field shown in Fig. 12

The *next destination* for a robot at passage segment *p* is a passage segment *q* determined according to the robot's task mission. Note that s *q* must be immediately incident to *p* in the SCG. The robot at passage segment *p* is said to be *blocked by* robot at passage segment *q* if the robot at *q* currently occupies the next destination of the robot in *p*. This relation is denoted by $p \rightarrow q$.

A *Moving Dependency Graph* (MDG), as a snapshot of the system, is a directed graph $\mathbf{G} = (\mathbf{V}, \mathbf{E})$ where $\mathbf{V}$ is the set of passage segments, and $\mathbf{E} = \{<p, q> \mid p, q \in \mathbf{V} \text{ and } p \rightarrow q\}$. MDG is a subgraph of the SCG.

Note that the MDG of a system changes dynamically. Let *K* be the set of passage segments along which a cycle is formed in the MDG. *K* is called the *kernel* of the deadlock. A deadlock occurs if and only if the current MDG of the system contains a cycle. The goal of any deadlock detection algorithm is to detect cycles in the MDG that span several passage segments.

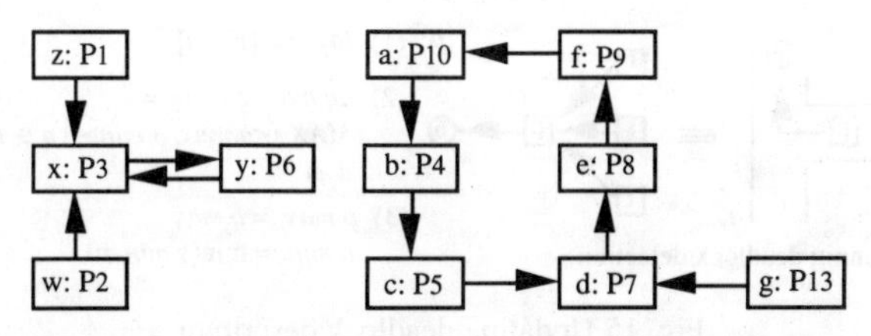

Fig. 14 The MDG of the system snapshot shown in Fig. 12

Assumptions

It is assumed that

(a) each passage segment has an identification, and there is a total relation on the set of passage segment ids.

(b) only finite number of robots ever operate in the system; or equivalently, the operating field is finite.

(c) each robot is equipped with a map of the operating field represented internally as a SCG. Moreover, each robot knows its current position in the SCG.

(d) a robot has at most one next destination at any time. This implies that a robot can belong to at most one (1) deadlock cluster. Or equivalently, all the nodes in a MDG must have out-degree one (1) or zero (0).

(e) a robot starts to run this deadlock detection as soon as it realizes that its next destination is blocked.

Idea

On behalf of the passage segment it currently occupies, a robot runs deadlock detection because it is being blocked by someone residing in its next designation.

A *deadlock descriptor*, to be displayed on the sign-board of a robot occupying passage segment p, is a pair $\langle p.max, p.min\rangle$, where $p.max$ and $p.min$ are ids of passage segments. Let p and q be two passage segments. $\langle p.max, p.min\rangle$ is said to *dominate* $\langle q.max, q.min\rangle$ if $p.max > q.max$ or $p.max = q.max$ and $p.min < q.min$. Note that *dominate* is a total relation on the set of all possible deadlock descriptors. It is therefore meaningful to talk about the *most dominating* one for a given set of deadlock descriptors.

Let p be a passage segment, its deadlock descriptor $\langle p.max, p.min\rangle$ is initiated, and updated according to the following rules (Fig. 15):

(0) Initially, $p.max = p.min = p$.

(1) Let $B(p) = \{q \mid q \rightarrow p\}$ (passage segments at which robots are blocked by the robot in p).

Select a passage segment $q \in (B(p) \cup p)$ such that the robot in q is displaying the most dominating deadlock descriptor.

(2) Set $p.max := q.max$ and $p.min := \min(q.min, p)$.

Steps 1 and 2 are repeated for robot at passage segment p, until either the deadlock detection is abolished because the robot in p is no longer being blocked, or a deadlock is declared.

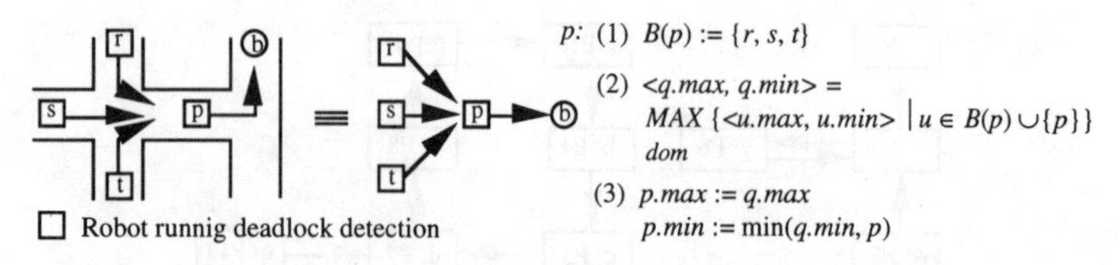

Fig. 15 Updating deadlock descriptor

The robot in passage segment p may declare the existence of a deadlock (with p in its kernel), if

- $\exists_q$: $q \rightarrow p$ and $q.min = p$ (Fig. 16) or
- $\exists_s \exists_t$: $s \rightarrow p$, $t \rightarrow p$, and $s.max = t.max$ (Fig. 17).

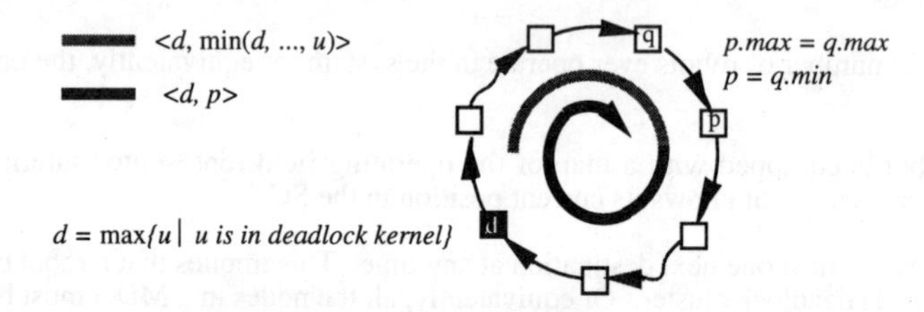

Fig. 16 Deadlock descriptor propagation when d is in the deadlock kernel

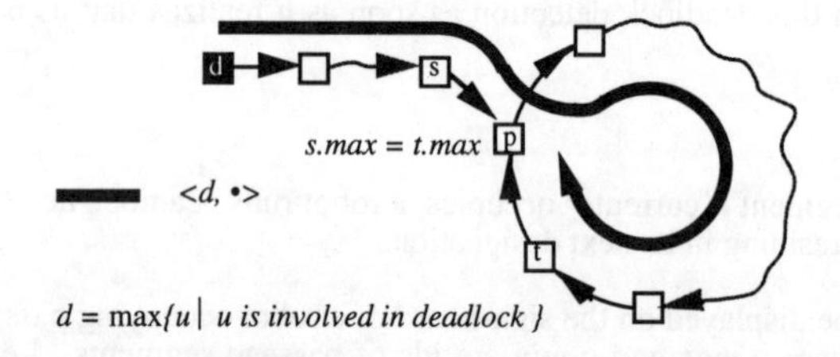

Fig. 17 Deadlock descriptor propagation when d is out side of the deadlock kernel

It has been shown rigorously in [11] that for this algorithm: (i) a robot in the system realizes the existence of a deadlock, if and only is there is one; and (ii) there is exactly one robot initially realizes the deadlock, of which it is a member of its kernel. This robot will serve as the *initiator* for deadlock resolution. The news about the deadlock is propagated along the deadlock kernel, whose members become *followers* for the resolution process.

Sign-board

The logic fields of the sign-board used in this algorithm is shown in Fig. 18.

inst	*c_pid*	*d_pid*	*state*	*max*	*min*

Fig. 18 Section of sign-board used for deadlock detection

where *c_pid* is the passage segment that the robot currently resides, and *d_pid* is the robot's next designation. Fields *max* and *min* are used to propagate deadlock descriptors. Possible states for field *state* are DETECTION, DEADLOCK and NONE. The section is identified by instance *inst*.

Algorithm

In the following, parameter *c* is the robot's current segment id, whereas *d* indicates its next designation. The function returns NO_DEADLOCK if the robot is no longer being blocked; INITIATOR, if the robot first realizes the deadlock; or FOLLOWER if the robot becomes aware of the deadlock by observing the others.

```
function deadlock(c, d) is

/*      c: id of passage segment currently occupied by robot
        d: the robot's next destination
*/
(0)     post_sect(DEADLOCK; c_pid: c; d_pid: d; state: DETECTION; max: c; min: c);
        max := min := c;                  /* set max and min to current passage segment id */

(1)     if pass_avail(d) = AVAIL then begin        /* no longer being blocked */
            erase_sect(DEADLOCK);
            return(NO_DEADLOCK)
        end;
(2)     max_count := 0;
        for each p ∈ pred(c) do begin
            s := read_sect(DEADLOCK; p);
            if s = NULL then continue;       /* continue -- in the sense of C language */
(3)         if s.state = DEADLOCK then begin       /* a follower */
                post_sect(DEADLOCK; state: DEADLOCK);
                erase_sect(DEADLOCK);
                return(FOLLOWER)
            end;

(4)         if s.state ≠ DETECTION or s.max < max then continue;

(5)         if s.max = max then begin
                max_count := max_count + 1;

(6)             if s.min  < min then min := s.min;

(7)             if s.min = c then  begin            /* an initiator */
                    post_sect(DEADLOCK; state: DEADLOCK);
                    goto(12)
                end
            end;
(8)         if s.max > max then begin
                max := s.max; min := MIN(s.min, c); max_count := 1
            end
        end;

(9)     post_sect(DEADLOCK; max: max; min: min)        /* deadlock descriptor update */

(10)    if max_count = 2 then begin                   /* an initiator */
            post_sect(state: DEADLOCK);
            goto (12)
        end;

(11)    goto (1);
```

```
(12)    for each p ∈ pred(c) do begin                    /*for the initiator */
              s := read_sect(DEADLOCK; p);
              if s.state = DEADLOCK then begin
                    erase_sect(DEADLOCK);
                    return(INITIATOR)
              end
        end;

        goto (12)

end;
```

Function *pred*(*c*) returns the set of passage segments whose exits are incident to the entrance of *c*. Function *pass_avail*(*p*) returns AVAIL if *p* is no longer blocked, and UNAVAIL otherwise.

2.3 Deadlock Resolution

Idea

A traffic system is said to have made a *positive progress* if at least one robot has advanced to its next destination, while the rest staying in their respective current passage segments. A number of deadlock resolution strategies have been proposed in the past [10]. Nevertheless none of them can guarantee positive progress for the system.

The deadlock resolution strategy proposed in this paper requires a reserved buffering area at each intersection (Fig. 19). Noted that this buffer is only logical. Physical implementation of this buffering area depends on application. It can be shown that positive progress of the system is impossible unless buffering areas specially reserved for deadlock resolution are introduced.

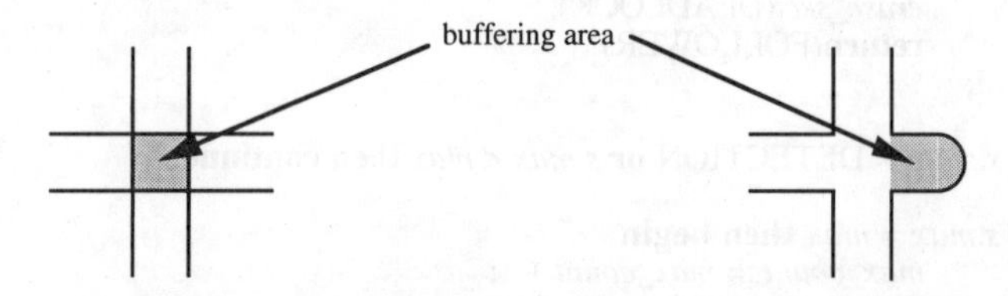

Fig. 19 Specially reserved buffering area at intersections

The resolution process is proceed as the following:

(0) The news about the deadlock is first announced by the initiator (Fig. 20a), and then propagated along the deadlock kernel (Fig. 20b).

(1) The initiator moves into the buffering area of the intersection in front of it (Fig. 20c) as soon as it gets out of the detection routine. The followers move successively into their respective next destinations along the deadlock kernel (Fig. 20d).

(2) The initiator moves into its next designation from the buffering area (Fig. 20d).

For those robots involved in the deadlock, but not in the deadlock kernel, this algorithm does not make them aware of the deadlock (although this is certainly possible). In fact, these robots may still be running the deadlock detection algorithm even after the deadlock has been announced and/or resolved. These robot swill quit the deadlock detection as soon as they are no longer being blocked, which will eventually happen as the deadlock kernel has been resolved. Fig. 21a and Fig. 21b are the MDGs corresponding to Fig 20b and Fig. 20d, respectively, with the deadlock shown in Fig. 20a (and Fig. 21a) being resolved.

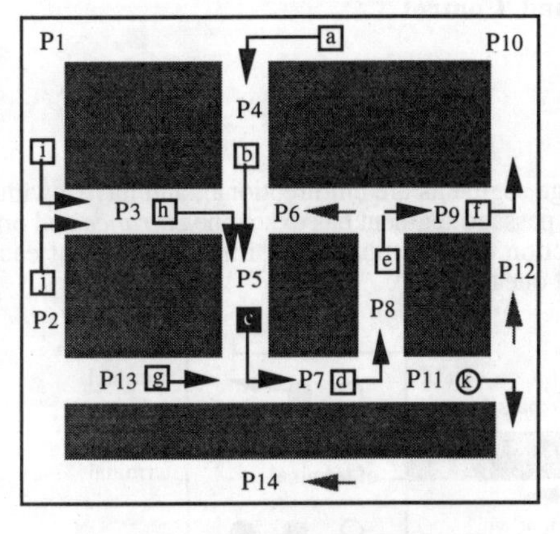

(a) Robot *c* at P_{13} is the initiator

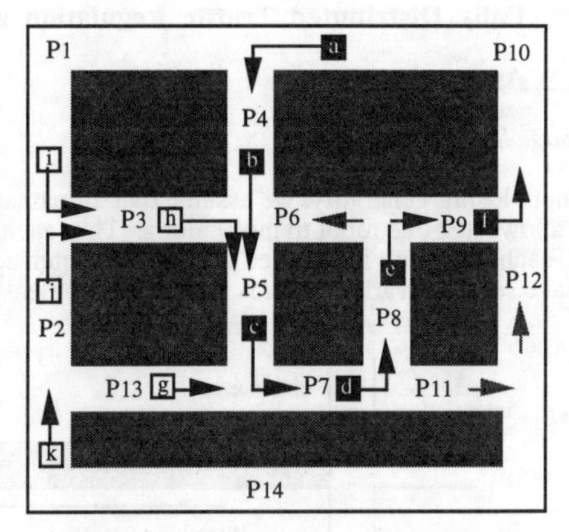

(b) DEADLOCK is propagated along the kernel

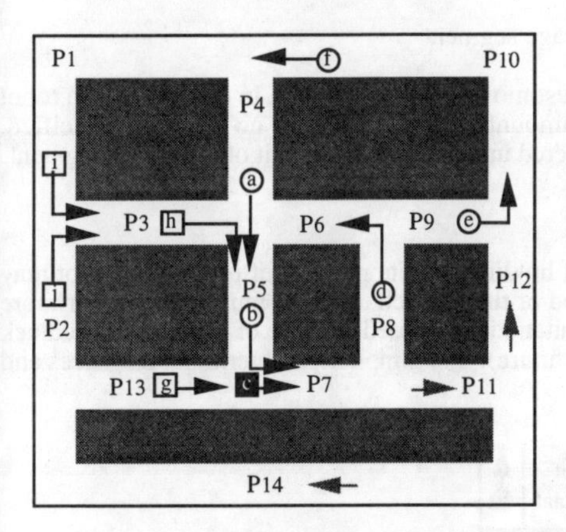

(c) Resolution of deadlock kernel

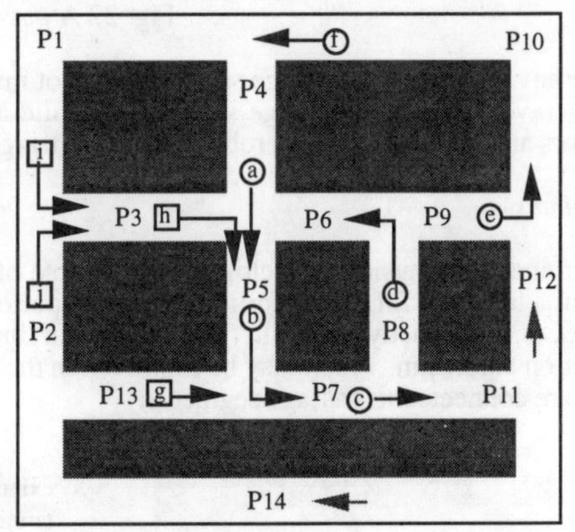

(d) Deadlock is resolved

Fig. 20 Deadlock resolution

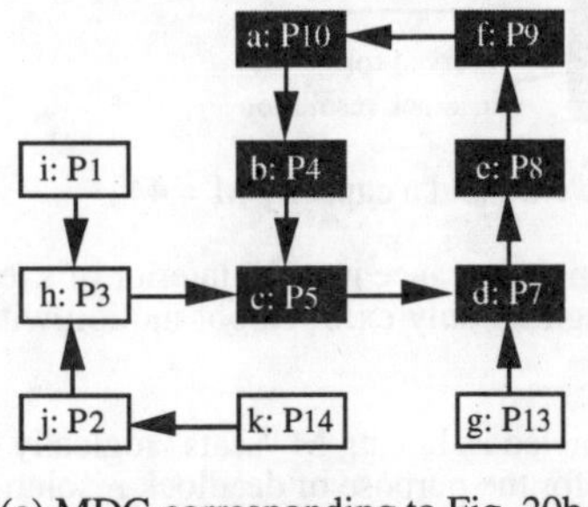

(a) MDG corresponding to Fig. 20b

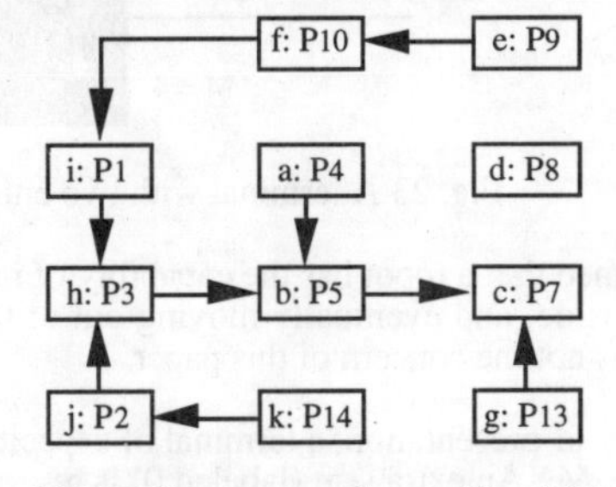

(b) MDG corresponding to Fig. 20d

Fig. 21 MDGs during deadlock resolution

3. Fully Distributed Traffic Regulation and Control

3.1 Assumptions

Passage Segments

Without losing generality, we assume that all passage segments are unidirectional, and have a width that allows just one robot to move along. Thus each passage segment has exact one *entrance* and one *exit*, each of which is exposed to either an intersection or a terminal. The finite capacity of each passage segment varies, but it is known to be at least one (Fig. 22).

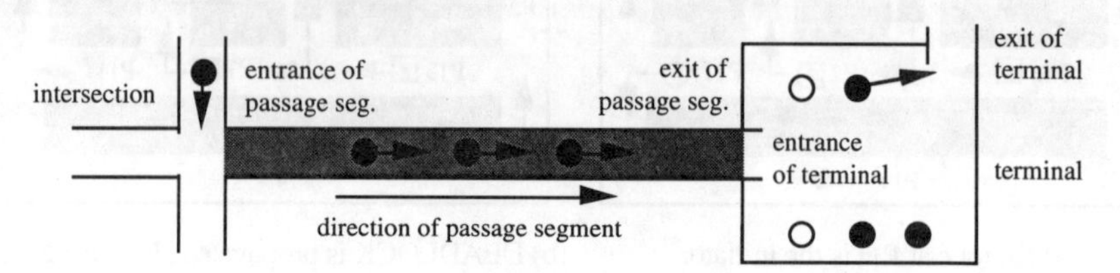

Fig. 22 A passage segment

Once having entered a passage segment, a robot must move towards its exit. In other words, a robot must travel through a passage segment in finite amount of time, although no bound is specified. Sensors are equipped on each robot, which are triggered upon arrival at the exit of a passage segment.

Terminals

A terminal is a physically enclosed area capable of holding a finite number of robots. A robot may stay at a terminal for a unspecified, but finite period of time. Each terminal must have one or more entrances, and exactly one exit. (This single exit limitation is imposed because of the current deadlock detection algorithm, which may be improved in the future). For simplicity, all terminal entrances and exits are connected to passage segments.

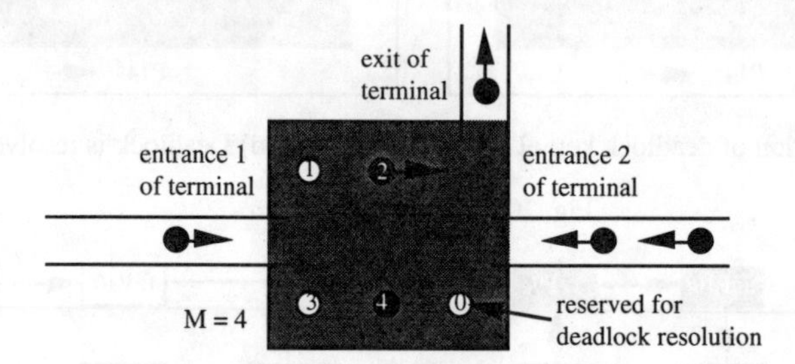

Fig. 23 A terminal with two entrances, one exit, and a capacity M = 4

It is assumed that a robot has the capability of moving from an entrance into the interior of a terminal, staying inside, and eventually moving out of the it through its only exit. Robot motion within the terminal is not the concern of this paper.

For clarity of presentation, a terminal of capacity M is depicted as having M "seats" logically labeled with 1, ..., M. An extra seat (labeled 0) is reserved solely for the purpose of deadlock resolution.

Intersections

Incident to two or more passage segments (with at least one being incoming, and one outgoing), an intersection must be passed by mobile robots one at a time. Thus a robot at the exit of a passage segment must first compete with others for the right of passing through the intersection before entering its next destination. It is assumed that a robot is capable of guiding itself through an empty intersection based on specifications given in the field map.

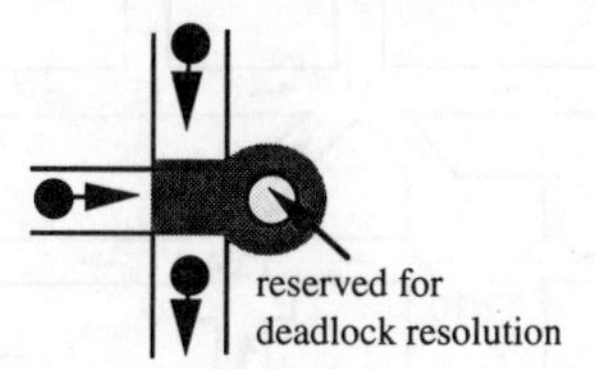

Fig. 24 A 3-way intersection

An extra buffering area with the capacity of holding one robot is reserved at each intersection solely for the purpose of deadlock resolution. The specific implementation of this buffering area (at an intersection) differs depending on application.

System Entrances and Exits

A *system entrance* is used to introduce robots into the system. And robots may withdraw from the system via a *system exit*. System entrances and exits are in fact passage segments incident to intersections or terminals.

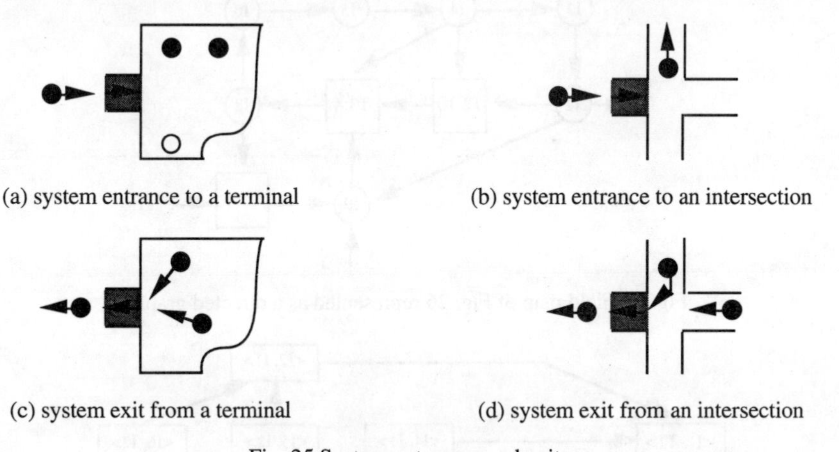

(a) system entrance to a terminal

(b) system entrance to an intersection

(c) system exit from a terminal

(d) system exit from an intersection

Fig. 25 System entrances and exits

Field Map

The operating field consists of terminals and intersections connected by passage segments. A detailed map of this operating field is equipped by each robot, which always
keeps track of its current location.

The operating field can be represented intuitively as a directed graph, in which terminals and intersections are vertices, and passage segment are edges. The attributes associated with a terminal vertex indicates its finite capacity. An edge in this graph is identified with an ordered pair of vertices (*lhs*, *rhs*>, where *lhs* is the intersection and terminal incident to the entrance of the passage segment, and *rhs* is that incident to its exit. An operating field exemplified with Fig. 26 is represented by a directed graph shown in Fig. 27. A segment connectivity graph (SCG) can be derived from this field map, and visa versa. The SCG for the operating field of Fig. 26 is given in Fig. 28.

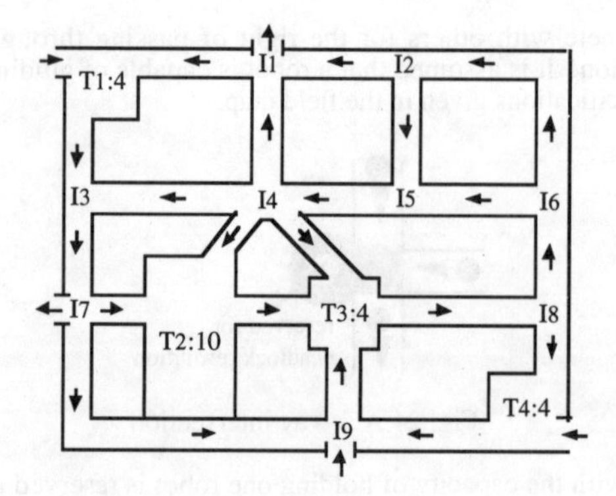

Fig. 26 Field map of an operating field

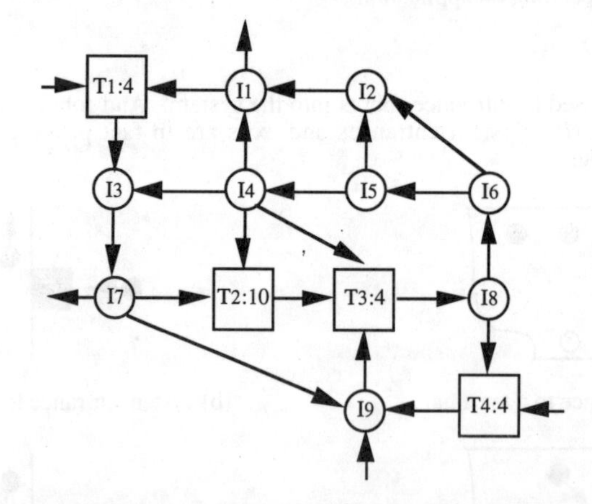

Fig. 27 Field map of Fig. 26 represented as a directed graph

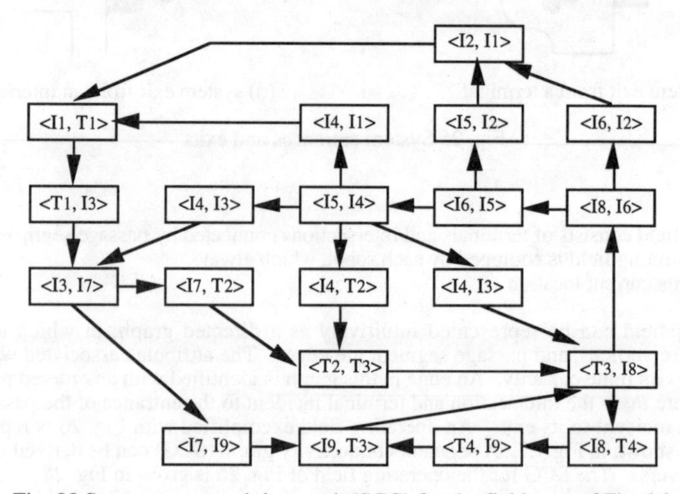

Fig. 28 Segment connectivity graph (SCG) for the field map of Fig. 26

The field map is consulted from time to time during the overall algorithm execution.

3.2 Idea

Passing Through an Intersection

To avoid collision, a robot must compete with (potential) others for the right of passing through the intersection. Once the this right is obtained, the robot travels through the intersection to enter its next destination.

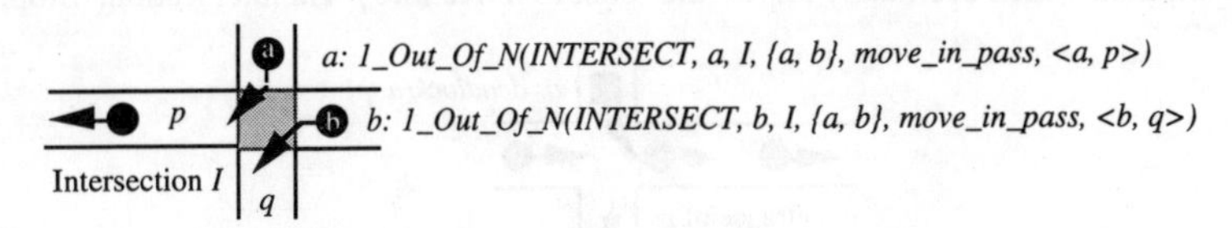

Fig. 29 Robots compete for passing through an intersection

This is clearly a "1 out of N" problem, with the resource being the intersection. Robots are competing on behalf of a finite, small number of incoming passage segments, over which inter-robot communication is required. Since no robot can stay at the middle of an intersection forever, a robot eventually gets its turn to pass through the intersection.

For the robot at passage segment *a* having reservation on passage segment *q*, as shown in Fig. 29, it needs to invoke

1_out_of_N(INTERSECT, *a*, *end_of*(*a*), *pred*(*p*), *move_in_pass*, *<a, p>*)

where INTERSECT is an constant instance; *pred*(*p*) returns set {*a*, *b*}; *end_of*(*a*) returns intersection *I*; and routine *move_pass_in*(*<a, p>*) guides the robot from *a* to *p* via intersection *I*.

Entering a Passage Segment

To enforce the finite capacity constrain of a passage segment, the traffic at its entrance must be regulated. A robot must compete with potentially others for the right to enter its next destination.

This is again a "1 out of N" problem involving finite, small number of robots at exits of passage segments incident to the same intersection. The here resource is an outgoing passage segment. Once a robot has successfully made a reservation for its next destination, it then competes for the right of passing through the intersection.

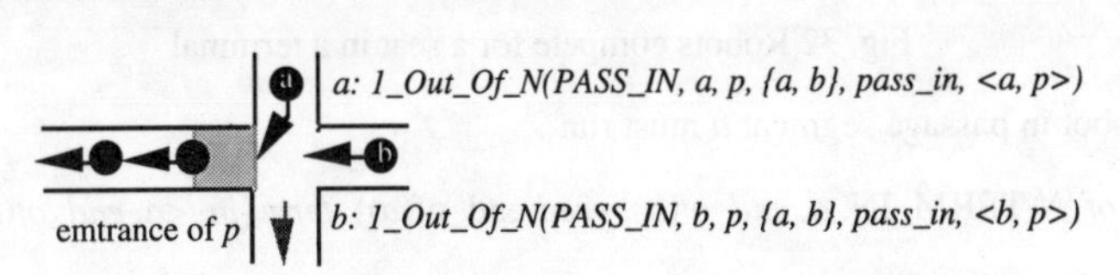

Fig. 30 Robots compete for entering a passage segment

In Fig. 30, the robot at passage segment *a* should invoke

1_out_of_N(PASS_IN, *a*, *p*, *pred*(*p*), *pass_in*, *<a, p>*)

where PASS_IN is a constant instant, and *pred*(*p*) again returns set {*a*, *b*}. Routine *pass_in*(*<a, p>*) is called upon when state TAKEN is assumed for the instance. Its main function is as follows:

- If p is available, (i.e., there is sufficient room at the entrance of p), compete for the right o passing through the intersection, and then move into p via the intersection. Stop.
- If p is blocked, (i.e., there is not sufficient room at the entrance of p), run deadlock detection.
- If deadlock detection results in NO_DEADLOCK, (i.e., p is now available), compete for th right of passing through the intersection, and move into p via the intersection. Stop.
- If deadlock detection results in INITIATOR or FOLLOWER, participate in deadlocl resolution, which eventually allows the robot to move into p via intersection. Stop.

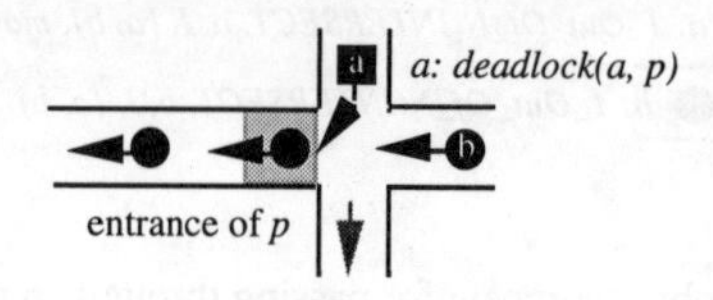

Fig. 31 A robot must run deadlock detection if it is being blocked

In Fig. 31, robot at passage segment a has the exclusive reservation for entering passage segment p Since p is not available, it must run *deadlock*(a, p) in routine *pass_in*. Note that routine *pass_in* call *1_out_of_N* recursively for passing through the intersection.

Entering in a Terminal

To enforce the finite capacity constrain for a terminal, robots must be admitted seriously (one at a time). Thus robots at multiple entrances of the terminal must compete for this logic right to enter the terminal. This is again a "1 out of N" problem.

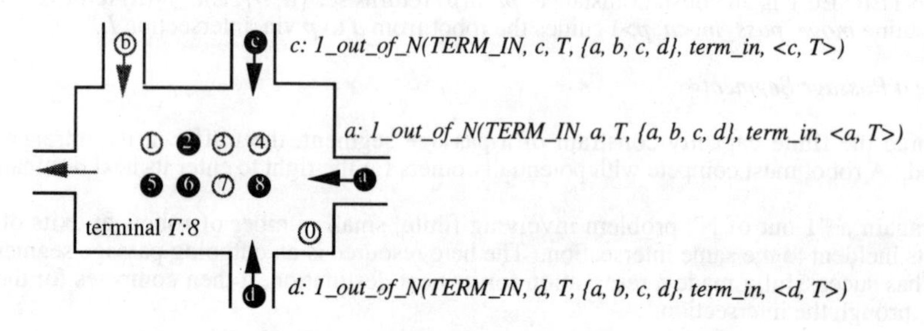

Fig. 32 Robots compete for a seat in a terminal

In Fig. 32, a robot in passage segment a must run

1_out_of_N(TERM_IN, a, *end_of*(a), *pred*(end_of(a)), *term_in*, <a, *end_of*(a)>)

where TERM_IN is a constant instance; *end_of*(a) returns the id of the terminal; *pred*(*end_of*(a)) returns the set of passage segments whose exits are entrances of the terminal. Routine *term_in* is called upon when *state* becomes TAKEN for the instance. It is designed to do the following:

- If there is a seat available in the terminal, move into that seat and stop.
- If no seat is available, run deadlock detection.
- If there is no deadlock, (i.e., the exit of the terminal is available), wait until one of the seats becomes empty. Move into that seat and stop.

If there is indeed a deadlock, participate in deadlock resolution, which eventually makes the exit of the terminal available. This in turn will make one of the seats inside the terminal available. Move into that seat and stop.

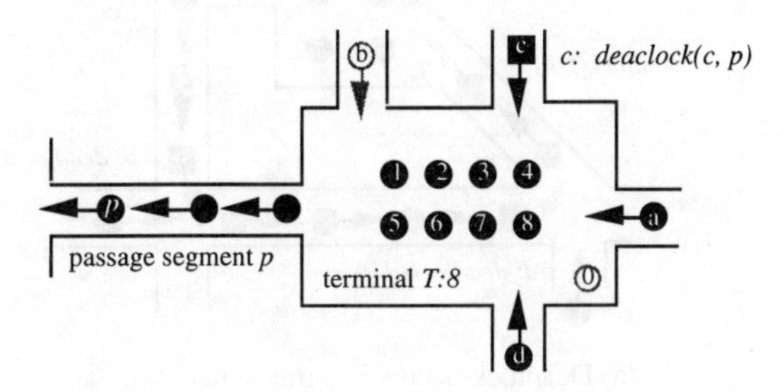

Fig. 33 Run deadlock detection at the entrance of a terminal

Exiting from a Terminal

Once having finished its work within the terminal, a robot must compete with others inside for the right of exiting from the terminal through its only exit, which is the entrance of a passage segment (Fig. 34). If the exit is not immediately available, the robot simply waits. Since it there is no deadlock, the exit will eventually become available; if there is indeed a deadlock, the exit will be available after deadlock resolution.

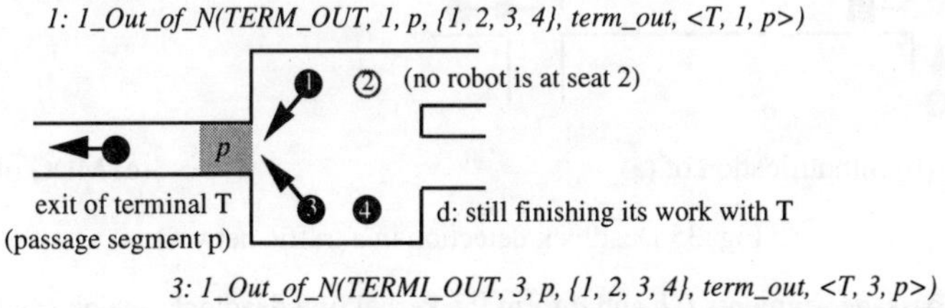

1: 1_Out_of_N(TERM_OUT, 1, p, {1, 2, 3, 4}, term_out, <T, 1, p>)

(no robot is at seat 2)

p

exit of terminal T
(passage segment p)

d: still finishing its work with T

3: 1_Out_of_N(TERMI_OUT, 3, p, {1, 2, 3, 4}, term_out, <T, 3, p>)

Fig. 34 Robots inside a terminal compete for the exit

This is again a "1 out of N" problem. In Fig. 34, a robot occupying seat 1 invokes

1_out_of_N(TERM_OUT, 1, *p*, *term_seat*(*T*), *term_out*, <*T*, 1, *p*>)

where TERM_OUT is a constant instance; *p* is the passage segment at the exit of the terminal; and *term_seat* returns the set of seats of the terminal. Routine *term_out* is called up once TAKEN is assumed by *state* for this instance. It simply moves the robot from its seat inside the terminal into the terminal's exit as soon as the exit is available.

Deadlock Detection

Deadlock occurs if a loop is formed along a series of passage segments/terminals that have all reached their respective finite capacity (Fig. 35). To reduce the communication overhead and the complexity of the overall algorithm, each passage segment is represented by the robot exposed at its exit. Note that this principle applies not only to deadlock detection, but also for all other operating primitives employed in this paper.

Thus only those robots which are representatives of passage segments need to participate in deadlock detection. A terminal is treated as if it were an intersection. Thus the traffic illustrated in Fig. 35(a) can be simplified as that shown in Fig. 35(b), which in turn has a MDG of Fig. 35(c).

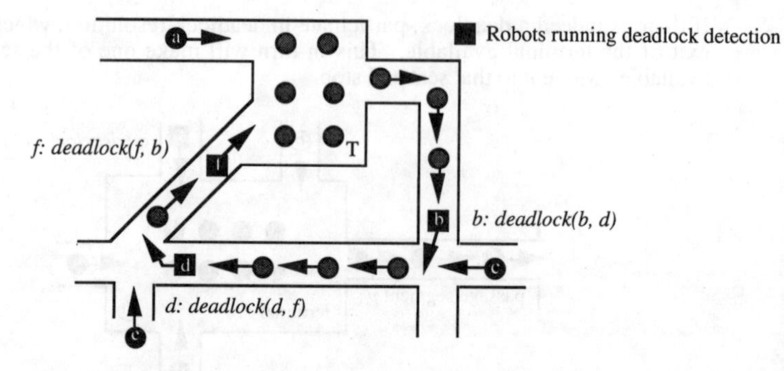

(a) Deadlock occurs in a traffic network

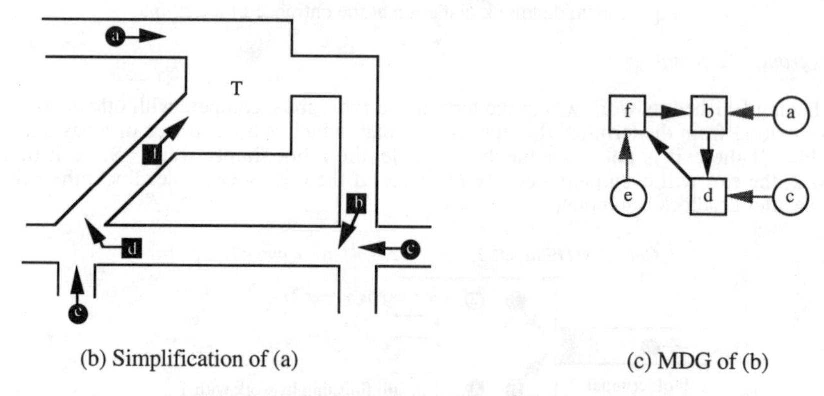

(b) Simplification of (a)

(c) MDG of (b)

Fig. 35 Deadlock detection in a traffic network

In Fig. 35(c), passage segments *f*, *b* and *d* form the kernel of a deadlock, which can be detected with the algorithm described in Section 2.

Deadlock Resolution

Deadlock resolution involves passage segments and terminals are essentially the same as that described in Section 2. Consideration has been made for those robots being blocked at entrances of a terminal:

- If deadlock detection indicates that the robot is the resolution initiator --

  Enter the specially reserved buffering area associated with the terminal.
  Wait until a seat is available, and then enter the seat.

- If deadlock detection indicates that the robot is a follower for the resolution --

  Wait until a seat is available, and then enter the seat.

It is extremely important to realized that

- No one will be able to take a seat just becomes available except the robot participating in the deadlock detection and resolution, because primitive *deadlock* is invoked by routine *term_in* under state TAKEN in *1_out_of_N*.

- A seat will eventually become available because the terminal exit will be opened after deadlock resolution, and no robot can stay inside a terminal for infinite period of time.

3.3 Description of the Overall Algorithm

The following is written for robot being introduced into the system at system entrance *e*.

Main Task

```
task dist_traffic_control(e) is

/*      e: system entrance through which the robot is introduced */

(0)     sys_in(e);                    /* initialization */
        p := e;                       /* p: passage segment currently occupied by robot */

(1)     move_along(p);                /* moving inside p towards its exit */
        t := end_of(p);               /* t is a terminal, intersection or system exit at the end of p */

(2)     if t = EXIT then begin        /* t is a system exit */
            sys_out(p);               /* house keeping before retiring from p */
            stop
        end;

(3)     if capacity(t) > 0 then begin         /* t is a terminal */

            /* reserve and move into seat s of terminal t */
            <t, s> := 1_out_of_N(TERM_IN, p, t, pred(t), term_in, <p, t>);

            /* work with terminal t while occupying seat s */
            work_term(t, s);

            /* move out of seat s of terminal t, exit to a passage segment p*/
            p := term_exit(t);
            p := 1_out_of_N(TERM_OUT, s, p, term_seat(t), term_out, <t, s, p>)
        end;

        goto (1);

(4)     d := next_dest(p);            /* t is an intersection -- determine next destination */

        /* reserve d, and move from p into d passing through intersection t */
        p := 1_out_of_N(PASS_IN, p, d, pred(d), pass_in, <p, d>); /* set p to d */

        goto (1)

end;
```

Sub-tasks

The following are the subtasks performed under various instances of *1_out_of_N*.

```
function term_out(<t, s, d> is

/*      t: terminal the robot currently in
        s: seat the robot currently occupies
        d: passage segment as exit of t
*/
```

```
(1)   while pass_avail(d) = UNAVAIL;          /* wait until exit d is available */

(2)   move_out_term(t, s, d);                 /* move from seat s to exit d */

(3)   return(d)

end;

function pass_in(<p, d>) is

/*    p: passage segment currently occupied by robot
      d: next destination
*/

(1)   if pass_avail(d) = AVAIL then goto (6); /* passage segment d is available */

(2)   r := deadlock(p, d);            /* run deadlock detection if d is unavailable */

(3)   if r = NO_DEADLOCK then goto (6); /* d is now available */

(4)   if r = INITIATOR then           /* move into buffer of the intersection */
            move_in_buf(p, end_of(p));

(5)   while pass_avail(d) = UNAVAIL;  /* wait until d is available */

(6)   r := 1_out_of_N(INTERSECT, p, end_of(p), pred(d), move_in_pass, <p, d>);

(7)   return(d)

end;

function term_in(<p, t>) is

/*    p: passage segment currently occupied by robot
      t: destination terminal
*/

(1)   s := seat_avail(t);
      if s ≠ UNAVAIL then goto (6);      /* if seat available */

(2)   r := deadlock(d, term_exit(t));    /* run deadlock detection if no seat is avail */

(3)   if r = NO_DEADLOCK then goto (6) /* terminal exit is now available */

(4)   if r = INITIATOR then move_in_buf(p, t);   /* move into buffer area in t */

(5)   do s := seat_avail(t) until s ≠ UNAVAIL; /* wait until a seat (s) is available in t */

(6)   move_in_term(p, t, s);

(7)   return(<t, s>)

end;
```

Key Operating Primitives

The routines in this section are as presented in Section 2. They are mentioned here for their external functionality.

- *r := deadlock(p, d)*

  returns NO_DEADLOCK if passage segment *d* is available; or INITIATOR, if the robot in *p* first realizes the fact of deadlock; or FOLLOWER if robot in *p* becomes aware of the deadlock by observing a DEADLOCK state raised by others.

- *r := 1_out_of_N(inst, c, d, C, act, paras)*

  executed by robot at *c* to compete with others in candidate set *C* for resource *d*. Once *d* is exclusively TAKEN, function *act* is called upon with parameter list *para*. Mutual exclusive access to *d* is effective until the end of execution of *act*, at which time the value of *act* is returned. Parameter *inst* indicates the purpose for which "1 out of N" is performed.

Robot Operations

Routines described here are all application and implementation dependent.

- *move_in_pass(p, d)*

  move from passage segment *p* to passage segment *d* via intersection.

- *move_in_term(p, t, s)*

  move from passage segment *p* into seat *s* of terminal *t*.

- *move_out_term(t, s, d)*

  move from seat *s* of terminal *t* into exit passage segment *d*.

- *move_along(p)*

  move inside passage segment *p* towards its exit.

- *move_in_buf(p, d)*

  move from passage segment *p* into the deadlock resolution buffer associated with *d*, which can be either an intersection or a terminal.

- *work_term(t, s)*

  work with terminal *t* while occupying seat *s*.

- *sys_in(e)*

  robot initialization when being introduced at system entrance *e*.

- *sys_out(e)*

  robot house keeping before retiring from system exit *e*.

- *d := next_dest(p)*

  returns the next designation for a robot currently at *p*.

- *b := pass_avail(d)*

  returns AVAIL if there is sufficient space at the entrance of *d* to accommodate one more robot, or UNAVAIL otherwise.

- $s := seat\_avail(t)$

  returns UNAVAIL if no seat is available in terminal t, or an available seat number otherwise.

Map Consulting Functions

All map functions can be easily derived based on the SCG representation of the operating field.

- $S := term\_seat(t)$

  returns the entire set of seat ids of terminal t.

- $p := term\_exit(t)$

  returns the passage segment connecting to the exit of terminal t.

- $P := pred(d)$

  returns the set of passage segments whose exits are incident to the entrance of d, if d is a passage segment. Otherwise, (d is a terminal), the function returns the set of passage segments that are entrances of d.

- $t := end\_of(p)$

  returns constant EXIT if passage segment p leads to a system exit; otherwise, it returns the id of the intersection, or that of the terminal at the end of p.

- $n := capacity(t)$

  returns the capacity of terminal t, or 0 if t is in fact an intersection.

4. Implementation

4.1 Robots

Since no ground support is assumed, a robot must have the capability to navigate itself through the discrete traffic network. Since the operating filed is highly structured, the required functionality can be readily implemented with technology well developed for autonomous mobile robots.

4.2 Sign-board based Inter-robot Communication

Messages logically displayed on the sign-board may be physically broadcasted repetitively by the robot via a radio communication channel (Fig. 36).

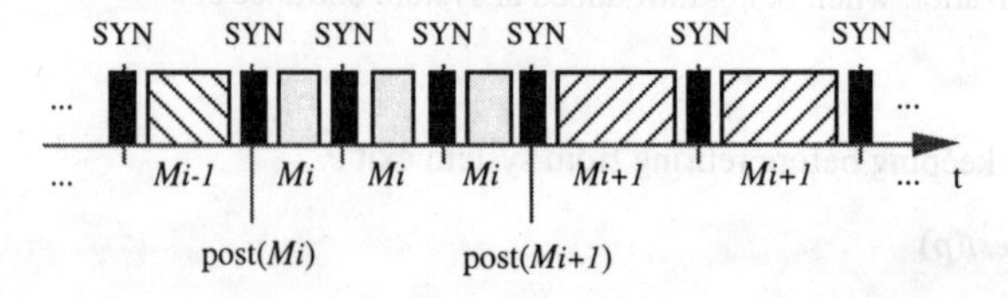

Fig. 36 Sign-board implemented as broadcasting messages

This simple and straight forward implementation of the sign-board model requires allocation of a distinct radio communication channel to each *site*, which is either a passage segment or a seat inside a terminal. In this fully distributed traffic control algorithm, a robot always operates on behalf of a site. Thus whenever a robot enters a site, it assumes the site's radio frequency. This implies that the radio transceiver equipped on each robot must be able to operate on different frequencies covering the entire communication bandwidth employed by the system.

With this implementation, the usable communication bandwidth certainly limits the complexity (number of sites) of the discrete traffic network. It is fortunate to note that the inter-robot communication is only required among robots in immediate neighborhoods. Thus if the operating field is large enough, one communication channel can be re-assigned to multiple sites that are sufficiently further apart (Fig. 37).

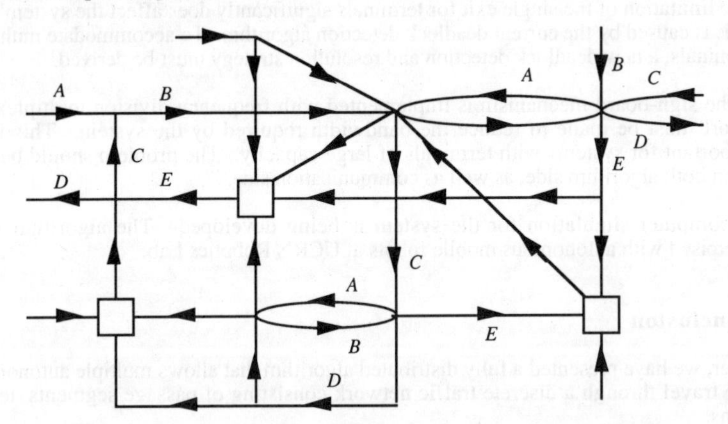

Fig. 37 Multiple use of some of the communication channels

More sophisticated wireless communication systems (such as wireless Ethernet) can be employed to implement the sign-board model, resolving the problem caused by the limited communication bandwidth. But this adds significant cost and complexity to the system which may also deteriorate the overall system performance.

5. Discussion

5.1 Merits of the System

(a) The algorithm is fully distributed. No centralized mechanism, ground support or synchronized clock is assumed. There is no requirement on the relative execution speed among robotic units. This is consistent with the model of DRS.

(b) Inter-robot communication is only required among finite (usually quite small) number of robots in immediate neighborhoods. The worst case happens during deadlock detection, where two robots may have to communication over the distance equal to the length of a passage segment plus the dimension of a terminal. Moreover, the message sizes employed in the algorithm are all extremely small.

(c) The system can accommodate arbitrary number of robots, subject to its total capacity. Robots may be introduced into the system, or withdrew from the system dynamically.

(d) Deadlock can be detected and resolved effectively. Positive system progress is guaranteed.

(e) The overall algorithm is based on a few well designed, simple and fully distributed operating primitives, which are likely to be supported by future robotic operating systems. Moreover, the correctness of these primitives are rigorously provable.

(f) The system is readily implementable with today's technology.

5.2 Current and Future Work

(a) Since deadlock detection and resolution is only called upon by robots who have successfully made reservations (*state* = TAKEN in *1_out_of_N*) for their respective next destinations, all nodes in a MDG must have in-degree and out-degree one or zero. This implies that a deadlock detection algorithm much simpler and efficient than the one presented in Section 2 can be employed.

(b) The limitation of the single exit for terminals significantly does affect the system's generality. This is caused by the current deadlock detection algorithm. To accommodate multiple exits for terminals, a new deadlock detection and resolution strategy must be derived.

(c) If the sign-board mechanism is implemented with frequency division multiplexing (FDM), effort must be made to reduce the bandwidth required by the system. This is especially important for systems with terminals of large capacity. The problem should be approached from both algorithm side, as well as communication side.

(d) A computer simulation for the system is being developed. The algorithm will also be exercised with autonomous mobile robots at UCR's Robotics Lab.

6. Conclusion

In this paper, we have presented a fully distributed algorithm that allows multiple autonomous mobile vehicles to travel through a discrete traffic network consisting of passage segments, terminals and intersections.

Collisions at intersection are avoided, and finite capacity constrains of passage segments and terminals are enforced. This is achieved by a simple distributed operating primitive *1_ouf_of_N*.

A fully distributed algorithm is used to detect deadlocks. It is shown that as far as deadlock detection and resolution is concerned, a terminal is equivalent to an intersection. (In other words, an intersection is a terminal of capacity zero). By introducing a specially reserved buffering area of capacity 1 at each intersection and as well at each terminal, deadlocks can be effectively resolved and positive system progress is guaranteed.

The algorithm employs the sign-board as its sole inter-robot communication mechanism. No communication is needed while a robot is traveling inside a passage segment. A robots competes for resources only when it exposed to the exit of a passage segment. At that time, it needs to communicate with only a finite number of robots in its immediate neighborhood.

It is demonstrated that relative impressive cooperation and coordination behavior is available employing only a small collection of simple, fully distributed operating primitives. The system is expected to have applications in controlling multiple AGVs on flexible manufacturing floors, as well as in regulating traffic for automated roadway/railway systems.

7. References

[1] T. Fukuda and S. Nakagawa, "Approach to the Dynamically Re-configurable Robotic System", *Journal of Intelligent and Robotic Systems*, Vol. 1, 1988, pp. 55-72.

[2] A. Matsumoto, H. Asama, Y. Ishida, K. Ozaki and I. Endo, "Communication in the Autonomous and Decentralized Robot System ACTRESS", *Proceedings of IEEE International Workshop on Intelligent Robots and Systems (IROS-90)*, July 3-6, 1990, Tsuchiura, Japan.

[3] R. Arkin, "Cooperation without Communication: Multi-agent Schema Based on Robot Navigation", *Journal of Robotic Systems*, Vol. 9, No. 3, April 1992, pp. 351-364.

[4] G. Beni and J. Wang, "Swarm Intelligence in Cellular Robotic Systems", Robotics and Biological Systems: Towards a New Bionics, edited by P. Dario, G, Sandini and P. Aebischer, *NATO ASI Series, Series F: Computer and System Science*, Vol. 102, pp. 703-712.

[5] R. Brooks, "A Robust Layered Control System for a Mobile Robot", *IEEE Journal on Robotics and Automation*, RA-2, April, 1986.

[6] G. Beni and J. Wang, "Theoretical Problems for the Realization of Distributed Robotic Systems", *Proceedings of the 1991 IEEE International Conference on Robotics and Automation*, Sacramento, CA, April 1991, pp. 1,914-1,920.

[7] P. Premvuti, "Cooperative Behavior of Multiple Autonomous Mobile Robots", *Ph.D. Dissertation*, Department of Computer and Information Science, University of Tsukuba, Japan, March 1992.

[8] J. Wang, "On Sign-board based Inter-Robot Communication in Distributed Robotic Systems", *Proceedings of 1994 IEEE International Conference on Robotics and Automation*, May 7-13, San Diego, CA, pp. 1,045-1,050.

[9] J. Wang, "DRS Primitives based on Distributed Mutual Exclusion", *Proceedings of the 1993 International Workshop on Intelligent Robots and Systems (IROS-93)*, June 26-30, Yokohama, Japan, pp. 1,085-1,090.

[10] J. Wang, "Deadlock Detection and Resolution in Distributed Robotic Systems", *Proceedings of International Symposium on Distributed Autonomous Robotic Systems (DARS-92)*, September 21-22, 1992, Wako-shi, Saitama, JAPAN, pp. 93-101.

[11] J. Wang, "Theory and Engineering of Cellular Robotics Systems", *Ph.D. Dissertation*, Department of Computer Science, UCSB, June 1990.

Chapter 5

Coorperative Operation

A Medium Access Protocol (CSMA/CD-W) Supporting Wireless Inter-Robot Communication in Distributed Robotic Systems

SUPARERK PREMVUTI and JING WANG

College of Eng., Univ. of California, Riverside, CA 92521, USA

Abstract

A communication media access protocol, CSMA/CD-W (Carrier Sense Multiple Access with Collision Detection for Wireless) is proposed to support both broadcasting and point-to-point communication in mobile robot based Distributed Robotic Systems (DRS). Differing from many existing experimental systems based on off-the-shelf wireless communication products for computers, no centralized mechanism or "ground support" is used, which is consistent with basic principles of DRS. The proposed protocol supports wireless data communication among mobile robots on a shared communication channel. It distincts itself from CSMA and its variations with the capability of detecting, in a wireless network, collisions of broadcast (undesignated) messages for which acknowledgments are not expected.

Key Words

Distributed Robotic Systems, Inter-robot Communication, CSMA/CD, Wireless Communication, Autonomous Mobile Robots.

1. Introduction

1.1 Background

The research on mobile robot based Distributed Robotic Systems (DRS) has received a lot of attention in recent years [1]. It is generally agreed upon that each robotic unit under the DRS model should operate autonomously, while all robots must cooperate to accomplish any system-wide (global) task [2]. Note that the principle of DRS does not allow any centralized mechanism be employed in the system. As the only resource in a DRS, mobile robots must interact through either localized broadcasting (sign-board) [3], or point-point communication (message passing) not depending on any centralized mechanism or "ground support".

Many existing mobile robot based DRS test beds are implemented with off-the-shelf wireless communication systems designed for computers [4]. Those wireless systems employ a centralized hardware to indicate the network status or a centralized communication server, (for instance, the wireless Ethernet needs a centralized controller), physically violating the principles of DRS.

DRS researchers have realized that for a fully distributed system, centralized hardware or server should not be used. Such an experimental system is used now just to "simulate" the effect of localized broadcasting or point to point communication among mobile robots. However, DRS experiments based on these systems are valid only if the cooperation and coordination algorithms exercised do not depend, either explicitly or implicitly, on the centralized mechanism. These algorithms will otherwise fail immediately once ported to a fully distributed multiple mobile robotic system with the centralized communication server removed.

Research and development on wireless communication, such as satellite communication, cellular phone systems and mobile workstation networks are progressing very fast. However, the research on wireless network systems and protocols for mobile robots operating under DRS is still far behind.

1.2 Motivation

For a distributed robotic system, broadcasting and point-to-point communication are the two basic mechanisms to convey information among robots. It is crucial to support and assure the quality and reliability of both types of communication. Due to limitations on radio bandwidth, a "flat" frequency division multiplexing (FDM), (one robot per communication channel), may not be feasible for a system containing a large number of robots (such as swarm systems [5]). Thus, Communication channels must be shared in general by a group of mobile robots.

Token ring and CSMA are two basic types of TDM (Time Division Multiplexing) protocols to resolve collisions in a shared communication channel. Wireless implementations based on the former include on Yamabico robot [6,7]. The DRS research community has expressed great interest of introducing CSMA type protocols into a fully distributed multiple mobile robotic system [8].

Since the operating environment and communication mechanism among mobile robots in a fully distributed system differ from that of an ordinary wireless networked computer system, especially with the concern of not allowing any centralized mechanism and ground support, existing variations of CSMA (Carrier Sense Multiple Access) [9] relying on a centralized mechanism to detect collision, can not be used.

In this paper, we propose a medium access protocol, CSMA/CD-W (Carrier Sense Multiple Access with Collision Detect for Wireless), specially designed for wireless network nodes to be implemented on mobile robots in a distributed robotic system.

1.3 CSMA/CD-W Protocol

A mobile robot is a *node* in the wireless communication network. A single communication channel is shared, as the common medium, for a group of robots. Only one robot should transmit during any time period. Simultaneous transmission from more than one robot results in a *collision.*

CSMA and ALOHA [9] can not be used to resolve collision among broadcasting messages. Under CSMA, a collision is detected by waiting for an acknowledge from the destination node (until time-out). A broadcasted message, however, has no specific destination, and no node (robot) will therefore respond to a broadcast message.

It is practically impossible to use CSMA/CD (Carrier Sense Multiple Access with Collision Detect) in a wireless network. A node is equipped with a transmitter and a receiver, and both of them have to be operated simultaneously to detect a collision. While a node is transmitting, its receiver will sense very strong radio signal from its own transmitter, and can not tell if there is any simultaneous transmission from other robots, as the antenna for the transmitter and that for the receiver can not be located so far apart from each other on the robot body -- as a matter fact , they may have to share the same antenna. Thus, it is in general not practical for a mobile robot to detect collision while transmitting.

This paper proposes a protocol based on CSMA. With an augmented capability of detecting collisions among broadcasting messages, the protocol (CSMA/CD-W) does not rely on any centralized mechanism or ground support, which is consistent with the basic principles of DRS.

Similar to CSMA, the proposed protocol reduces the probability of collision by introducing a random waiting time before each transmission. It differs from CSMA on the method of collision detection, which is accomplished by monitoring the state of the shared channel immediately *after* (not during) each transmission.

1.4 Paper Organization

Section 2 of the paper presents the proposed protocol in detail, which comments in Section [3], The implementation strategies for the protocol is discussed in Section 4. In Section 5, current and further research on the protocol and inter-robot communication based on this protocol is addressed.

2. CSMA/CD-W Protocol

2.1 Assumptions

We assume that radio equipment is installed in each autonomous mobile robot, serving as a node in the wireless communication network, and the network consists of only nodes installed on robots. The radio equipment onboard a robot should operate while the robot is moving while the radio communication is maintained among all robots, even if there are obstacles lie between them.

Consistent with the principles of DRS, the system does not contain any centralized mechanism such as master server/robot/CPU, globally synchronized clock or shared memory. Moreover, the protocol operation can not relying on any specific node. There is no ground support of any type. For instance, making marks on the operating field, is not allowed.

The protocol should operate with arbitrary number of nodes (robots), and the number of robots involved may change dynamically. A robot needs to be involved in this protocol only if it intends to communicate with others.

2.2 Basic Idea

A single communication channel is used as a medium for all nodes. To avoid collision, a node should check the status of the communication channel before a transmission is attempted. If the channel is busy, the node waits for a random period of time . This greatly reduces the probability of collision.

There is nevertheless still a small chance for two or more nodes to start transmit at almost the same time resulting a collision. Due to strong radio energy emitted by the transmitter, it is impossible for a node to realize the collision until the transmission is completed. The protocol is designed such that the length of a messages generated by a node is always distinct from that of others. Thus if a collision occurs, which means a few nodes started to transmit at almost the same time, the nodes involved in the collision will end their respective transmissions at different times.

The protocol instructs each node to check the status of the channel immediately after each transmission. If the channel is still busy, it means that a collision was occurred, as some other nodes are still broadcasting their messages.

Thus all nodes involved in the collision are able to realize the collision, except the one which sent the longest message. This node has to be informed by others.

To solve this problem, a node is instructed to check, right after each transmission, the validity of the carrier in addition to the status of the channel. (The validity of the carrier can be obtained from the modem.) The received carrier is valid if modulated signal from the transceiver can be demodulated by the modem, which means that the received radio signal must be emitted from a single node. The difference in message lengths, the data transmission rate and the speed of protocol execution are specified in such a way that a node detects a valid carrier if and only if it has involved in a collision with the second longest message. This node is responsible to generate a collision report [CR] message to inform the node with the longest message.

For a node involved in the collision with the third longest or shorter message, its received carrier at the time of checking is not valid. Because at the moment when it completed its transmission, more than one nodes (at least, nodes with the longest and the second longest message) are still transmitting.

The protocol asks each node to wait for a fixed period of time (Tcr) for possible [CR], even if the channel is not busy after the transmission. If a [CR} is received during the waiting period, a collision must have occurred. Otherwise, the transmission is considered successful.

Thus if a collision occurs, all nodes involved are able to realized the collision,. Re-transmission will be attempted after a random delay.

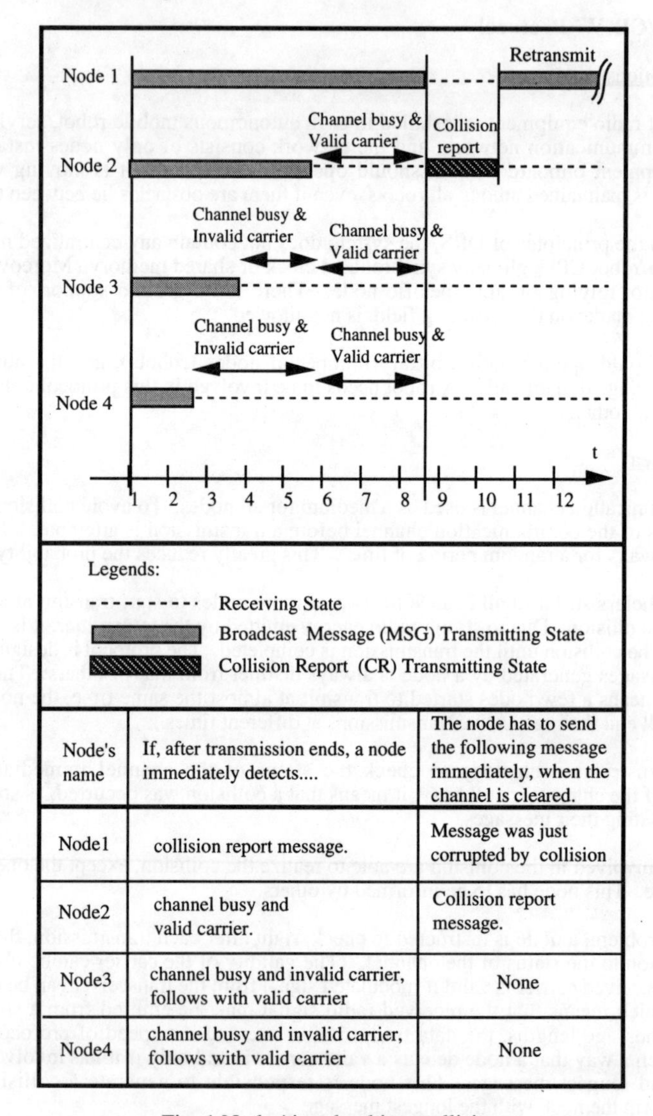

Node's name	If, after transmission ends, a node immediately detects....	The node has to send the following message immediately, when the channel is cleared.
Node1	collision report message.	Message was just corrupted by collision
Node2	channel busy and valid carrier.	Collision report message.
Node3	channel busy and invalid carrier, follows with valid carrier	None
Node4	channel busy and invalid carrier, follows with valid carrier	None

Fig. 1 Nodes' involved in a collision

The proposed protocol CSMA/CD-W consists of two layers: the *physical layer* and *medium access sublayer*.

2.3 Physical Layers

Layer name	Involved elements
Medium Access Sublayer	- Procedures - Frame
Physical Layer	-Low level signal requirements - Transceiver - Medium

Fig. 2 Layers of CSMA/CD-W

Medium

A bare radio communication channel is used as a multi-access medium shared for all nodes in the network. The protocol distributedly allocates this medium to nodes attempting transmission. The simultaneous transmissions of two robots or more nodes will cause a collision, and information involved will be garbled.

Unlike a local area network for computer systems, the medium does not contain any controller or special hardware to indicate its operating status (idle, busy or collision). This is consistent with principles of mobile robot based distributed robotic systems.

Transceiver

A transceiver tuned in the common communication channel is used to transmit and receive data. It operates under the receiving mode, except during the transmission of a message.

Controller

For the purpose of detecting collision, the controller of each node must provide the following signals to the medium access sublayer.

- [RF] (Radio Frequency Detected), a Boolean signal. RF =1 *iff* the channel is busy indicating some nodes are transmitting.
- [DCD] (Data Carrier Detected), a Boolean signal. DCD =1 *iff* the carrier enter the modem is valid indicating a single node is transmitting to the shared channel.

2.4 Medium Access Sublayer

This layer controls the access to the medium, detects collision and fulfills the task of receiving and transmitting messages for the robot. Two processes, *Receiving* and *Transmitting*, operate on data frames, logic signals and other devices to accomplish the task.

Frames

A unit of message sent through the network is called a *frame*. Two types of frames, the MSG (message) frame and the CR (collision report) frame, are employed in the protocol (Fig. 3). The former is used to hold data messages. The latter is used to broadcast a message indicating a collision. The format of these frames are similar to HDLC (High Level Data-link Control) frame [10].

01111110	Synchronization flag
DESID	Destination node (1111111 if global)
SORID	Source node ID
FRMTYP	Frame type (00000000 = MSG)
DAT1	Data #1
DAT2	Data #2
...	
DATn	Data #n
DUM1	Dummy #1
DUM2	Dummy #2
...	
DUMn	Dummy #3
CRC	Error detection code
CRC	Error detection code
01111110	Synchronization Flag

(a) MSG frame

01111110	Synchronization Flag
DESID	Destination node (1111111 if global)
SORID	Source node ID
FRMTYP	Frame type (11111111 -- CR)
CRC	Error detection code
CRC	Error detection code
01111110	Synchronization flag

(b) CR frame

Fig. 3 Frame structure for CSMA/CD-W

Random Number Generator

A random number generators is called upon to whenever a node discovers that the channels is busy, or it has just experienced a collision. It determines the waiting period before a re-transmission (Twi) attempt.

Timer

A timer is used to time the random waiting period for re-transmission and the fixed waiting period for possible CR.

Receiving Process

Radio signals received by the transceiver operating under receive mode are demodulated by the modem. Digital serial signals out of the modem are streamed into the HDLC controller, which generates an interrupt for every detected frame. The interrupt handler analyzes the type of frame. For a MSG frame, the interrupt handler extracts the data message into a message buffer, and interrupts the robot for the incoming message. For a CR message, a variable (CR_REC), also accessible by the Transmitting process, is set to 1.

The receiving process runs all the time, except when the transceiver is set to the transmitting mode. No messages originated from other nodes are to be missed.

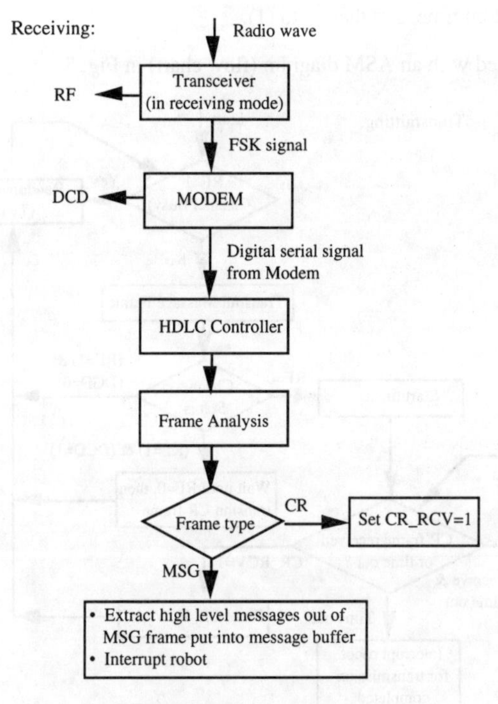

Fig. 4 Receiving process

Transmitting Process

This service is called upon only when the robot intends to send a message out. The process is described as follows.

(1) Check the status of the channel (signal RF) before a transmitting attempt. If the channel is busy, go to (4).

(2) if the channel is not (clear) busy, transmit the message.

(3) Check logic signals RF (channel busy) and DCD (carrier valid).

(3.1) If the channel is busy (RF=1) and no carrier is detected (DCD=0), there is a collision. Goto (4).

(3.2) If the channel is busy (RF =1) but carrier is detected, there is a collision. The node is also responsible to inform the node which sent the longest message. Send a CR message and goto (4).

(3.3) If the channel is not busy (RF = 0), watch logic signal CR_REV for a fix period of time (Tcr).

(3.3.1)if CR_REV becomes 1 during this time period, a collision occurred. Reset CR_REV to 0 and goto (4).

(3.3.2)if CR-REV remains 0 till time-out of Tcr, the transmission was successful. Interrupt the robot and stop the execution of the process.

(4) Wait for a random time, and then goto (1).

The process is illustrated with an ASM diagram (flow chart) in Fig. 5.

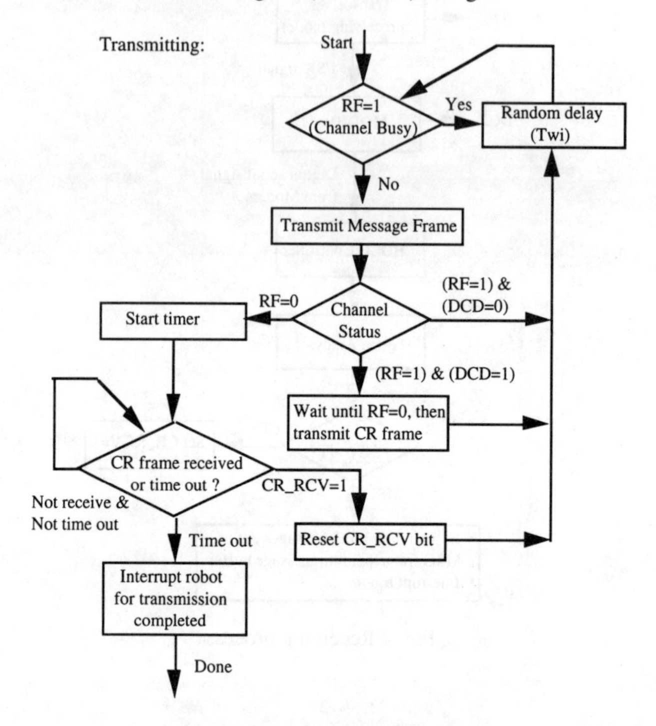

Fig. 5 The Transmitting process

3. Discussion

3.1 Protocol Design Considerations

The following must be specified for a valid implementation of the protocol.

(1) The length of data messages (MSG) originated by each robot. It is predetermined so that it is distinct from that generated by other robots. The minimum difference between any two data messages must be specified according to the transmission rate, execution speed of the protocol.

Note that this practically restricts the number of robots that can be assigned to a communication channel. We expect this number to be not large (< 10), and the real data is relatively long to minimize the effect of length variation of messages generated by the group of robots.

(2) The distribution of the random waiting time (Twi) before a re-attempt.

(3) The fixed waiting time for the collision report message (CR).

3.2 Other Comments

The proposed protocol has the following advantages over conventional CSMA type protocols. It is especially suitable for wireless communication among autonomous mobile robot based distributed robotic systems.

(1) Differing from CSMA, this protocol can detect collisions among broadcasted (undesignated) message for which acknowledgments are not expected.

(2) Differing from LAN system such as Ethernet, the proposed protocol does not rely on any centralized hardware or mechanism to detect collision. This supports autonomous mobiles under DRS environment.

(3) Since no centralized hardware is employed, the protocol can be implemented with relatively low cost with current technology.

(4) It solves the problem of sharing a common communication channel by detecting collision on radio based wireless networks, which is practically not possible using common CSMA/CD due to strong radio signal generated by the node.

(5) It is expected that this protocol will result slower throughput than other CSMA/CD based protocols (such as Ethernet. As the collision can not be detected immediately after it occurs. The authors believe that this is probably acceptable for highly autonomous mobile robots operating under distributed control where very limited inter-robot communication is expected.

(6) Many relatively frequently used high level operating primitives for distributed robotic systems, such as distributed mutual exclusion, can be effectively and efficiently implemented (taking advantages) of this protocol.

4. Implementation

The hardware of the system can be implemented with a transceiver, a modem, a HDLC control along with a micro-controller system (Fig. 6).

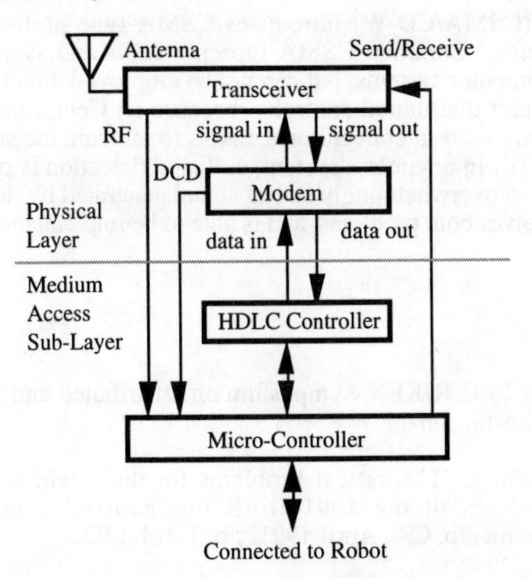

Fig. 6 Hardware implementation

Control procedures the Medium Access Sublayer is implemented with a micro-controller based system. It supplies model switch control through output digital ports, and accepts RF and DCD signals from input ports. Modem is connected through the built-in serial port.

The Transmitting process and the software portion of the Receiving process is implemented as interrupt handlers.

The RF signal for monitoring radio transmission is easily derived from by the transceiver. The DCD for monitoring carrier is equivalent to Data Carrier Detect output from the modem. HDLC frames may be caught and assembled by an HDLC controller chip available on the market.

5. Current and Future Research

Current and future research on the proposed protocol include

(1) Formal specification and verification on the correctness of the protocol.

(2) Performance analysis and discrete event simulation for the protocol.

(3) Implementation of wireless communication subsystems on autonomous mobile robots based on this protocol.

(4) Design and implementation of higher level protocols to support point-to-point communication (message passing) and broadcasting (sign-board).

(5) Higher level DRS operating primitives based on the protocol.

6. Conclusion

The DRS research communication has expressed great interest of implanting CSMA type protocols to a wireless communication network consisting of only autonomous mobile robots.

The proposed protocol (CSMA/CD-W) introduces CSMA type protocols to mobile robot based distributed robotic systems. Common CSMA type protocols and systems, such as ALOHA and Ethernet, are good for computer systems, but can not be employed directly for multiple autonomous mobile robot systems under distributed control -- because (i) Centralized hardware, node or other ground support, commonly used in conventional LANs to indicate the status of the shared channel, are not allowed in DRS; (ii) In a wireless system, collision detection is practically impossible at the time of transmission due to overwhelmingly strong signal generated by the robot itself. The protocol proposed in this papers solves both problems, and is able to be implemented effectively and efficiently with existing technology.

7. References

[1] Proceeding of the 1992 RIKEN Symposium on Distributed and Autonomous Robots, Sept. 21-22, 1992, Wakoshi, Japan.

[2] G. Beni and J. Wang, "Theoretical Problems for the Realization of Distributed Robotic Systems, Proceedings of the 1991 IEEE International Conference on Robotics and Automation, Sacramento, CA, April 1991, pp. 1,914-1,920.

[3] J. Wang, "DRS Operating Primitives Based on Distributed Mutual Exclusion", Proceedings. the IEEE /RSJ 1993 International Conference on Intelligent Robots and Systems, Yokohama, Japan, July 26-30, 1993, pp. 1085-1090.

[4] K. Ozaki, H. Asama, Y. Ishida, A. Matsumoto, K. Yokota, H. Kaetsu and I. Endo, "Synchronized Motion by Multiple Mobile Robots using Communication", Proceedings. the IEEE /RSJ 1993 International Conference on Intelligent Robots and Systems, Yokohama, Japan, July 26-30, 1993, pp. 1,164-1,170.

[5] G. Beni and J. Wang, “Swarm Intelligence in Cellular Robotic Systems”, Robotics and Biological Systems: Towards a New Bionics, edited by P. Dario, G, Sandini and P. Aebischer, NATO ASI Series, Series F: Computer and System Science, Vol. 102, pp. 703-712.

[6] S. Premvuti, S. Yuta and Y. Ebihara, "Radio Communication Network on Autonomous Mobile Robots for Cooperative Motions", Proceedings. IECON'88, Singapore, October, 1988, pp. 32-37.

[7] S. Yuta, S. Premvuti, "Consideration on Cooperation of Multiple Autonomous Mobile Robots --Introduction to Modest Cooperation--", 5th International Conference on Advanced Robotics ('91 ICAR), Pisa, Italy, June, 1991.

[8] R. Brooks, private communication.

[9] D. Bertsekas, R. Gallager, "Data Networks (Second Edition)", Prentice-Hall, Inc., New Jersey, 1992, (Chapter 4).

[10] ISO-3309, "Data Communication -- High Level Data Link Control Procedures -- Frame Structure".

The Design of Communication Network for Dynamically Reconfigurable Robotic System

SHIN'YA KOTOSAKA[1], HAJIME ASAMA[2], HAYATO KAETSU[2], HIROMICHI OHMORI[2], ISAO ENDO[2], TOSHIO FUKUDA[3], FUMIHITO ARAI[3], and GUOQING XUE[3]

[1]Graduate School of Science and Eng., Saitama Univ., Saitama, 338 Japan
[2]The Inst. of Physical and Chemical Research (RIKEN), Wako, 351-01 Japan
[3]Dept. of Mechano-Informatics and Systems, Nagoya Univ., Nagoya, 464-01 Japan

Abstract

An advanced Self-Organizing Manipulator System (SOMS) is introduced. The SOMS is a manipulator with modularized architecture, which is structured with modules called cells. This system is designed so as to reconfigure its own structure by itself depending on tasks and an environment. The SOMS is more task-oriented than other dynamically reconfigurable robotic systems, and characterized by its structuring strategy utilizing another specific manipulator for the SOMS assembly.

In this paper, the second version of the SOMS with a passive coupling mechanism and communication system with an optical connection is mentioned. At first, we introduced several types of cells for structuring the manipulator. Each cell has a distributed control system that can communicate with each other through the optical connection. These distributed control systems enhance intelligence of the cell. The new passive coupling mechanism is developed for down size of cell. Next, an optical connection for communication between cells to achieve improved reliability is described. Then, we discuss how to construct a communication network on the reconfigurable and distributed robotic system. It is found that packet filtering and routing mechanism is indispensable for the communication network to reduce traffic on the network, to achieve robustness, and to deal with changeable network topologies.

Keywords: Dynamically Reconfigurable Robotic System, Communication, Network, Cellular Robotics, Distributed Manipulator

1. Introduction

In recent years, various dynamically reconfigurable robotic systems have been proposed [1] [2]. The dynamically reconfigurable robotic system is characterized by adaptability and robustness. It is designed so as to reconfigure its own structure depending on tasks and an environment. We have been developed a kind of dynamically reconfigurable robotic system, which is called self-organizing manipulator system (SOMS) [3]. The SOMS is a manipulator with modularized architecture, which is structured with modules called cells. We can construct a structure of manipulator from any number, in any order of cells. This system is more task-oriented than other dynamically reconfigurable robotic systems, and characterized by it structuring strategy utilizing another specific robot for cell assembly. Figure 1 shows the concept of self-organizing manipulator. In this paper, a design of second version of the SOMS is mentioned. The second version of SOMS has passive coupling mechanism for more down sizing, distributed control system for intelligence enhancement and an optical communication system for reliability improvement. And, how to construct a communication network on the distributed and reconfigurable robotic system is discussed. Then, the new communication system for SOMS is introduced.

2. Design of SOMS mechanisms

The first version of the SOMS needed an actuator to combine two cells. This actuator is needed only for coupling two adjacent cells, and is not used to actuate joints of the SOMS. The first version of the SOMS needs the centralized control system and physical connectors with many contact pins for communication and power supply. For that reasons, the SOMS has problems in autonomy, reliability and robustness. Therefore, the second version of the SOMS is developed to solve the problems with new passive coupling mechanism, distributed control system and optical connection for communication between cells. Only power source is supplied by physical connection. We have developed several kinds of cell: Base cell, Bending cell, Rotation cell, Sliding cell, Gripper cell and Branch cell.

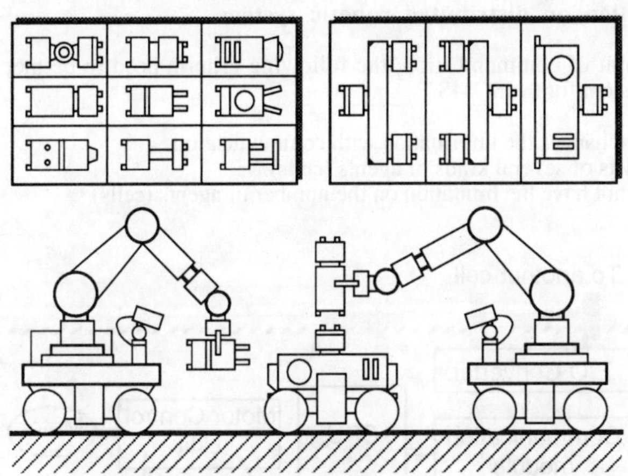

Fig. 1 Concept of self-organizing manipulator

The second version of the SOMS is adopted the passive coupling mechanism without an actuator for coupling, as shown figure 2. When the grasping points are grasped by an assemble robot, the coupling hook is open, and the assemble robot can assemble and disassemble the cell. When the coupling points are released, the coupling hook is closed by springs, and then the cell is fixed to adjacent cell. Guide pins and communication ports for optical connection are located on the contact surface.

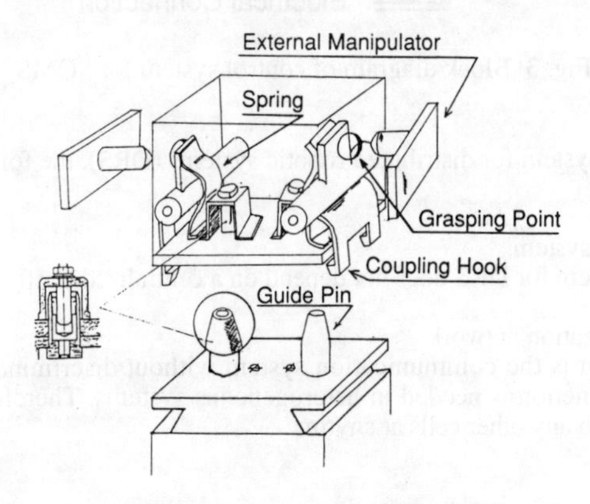

Fig. 2 Passive coupling mechanism

Each cell has a distributed control system which consists of a one-chip CPU, a motor control unit and an optical communication system. Figure 3 shows that the block diagram of the control system for the SOMS. The one-chip CPU can communicate with other cells via the optical communication system and control an actuator. The optical communication system provides communication on distributed system. Full particulars will be mentioned later. Only the base cell has a wired communication port to communicate with external systems, and the branch cell has extra communication ports.

3. Communication system for SOMS

3.1 Communication on distributed robotic system

From the view point of communication, the following criteria need to be met for designing the communication system for the SOMS.

• The each agent exchanges the information with communication.
• The SOMS consists of several kinds of agents (cells).
• The system does not have the limitation on the number of agents (cells).

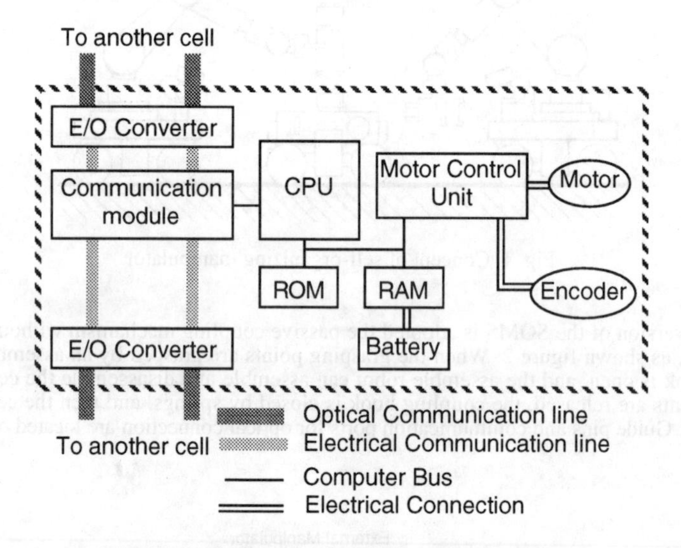

Fig. 3 Block diagram of control system for SOMS

On the communication system for distributed robotic systems (DRS), the following characteristic is indispensable.

• No centralized control system.
The communication system for DRS does not depend on a centralized traffic control system.

• Peer-to-Peer communication network.
The peer-to-peer system is the communication system without discrimination, because it is not predetermined which function is needed in heterogeneous systems. Therefore, any cell should be able to communicate with any other cells at anytime.

• Robustness.

For effective utilization of the distributed system, the communication system also should have robustness. For example, the system should have ability to prevent a fault from leading to the complete break down of the entire system.

The communication systems for DRS are investigated recently [4] [5] [6]. For realization of a communication system which can satisfy such requirements mentioned above, the following problems should be resolved. However, any discussions on the following problems have never discussed so far.

• Which the equipment and protocol should be used for communication?

• How to ensure the connectability between multi-communication media?
In heterogeneous systems, various communication media are possibly used, such as wired communication media and wireless communication media.

• How to deal with dynamic reconfiguration of communication network topology?
This problem in distributed robotic system has not been recently investigated. In the SOMS, the topology of communication network is frequently changed, and a loop topology may exist.

• How to prevent the communication traffic jam?
Traffic estimation also dose not yet considered. If the network becomes so big, the increase of communication traffic should be considered. In the field of computer science, the traffic estimation on the various networks was explored [7].

• How to realize the robustness on the communication system?
To solve these problems, the network technology in the computer science is almost applicable. However, we should give attention to the important difference that the communication network for DRS is more varying network than the computer network.

3.2 Design of communication system for SOMS

Based on previous consideration, we have been developing a communication system (in figure 4) for the second version of the SOMS.

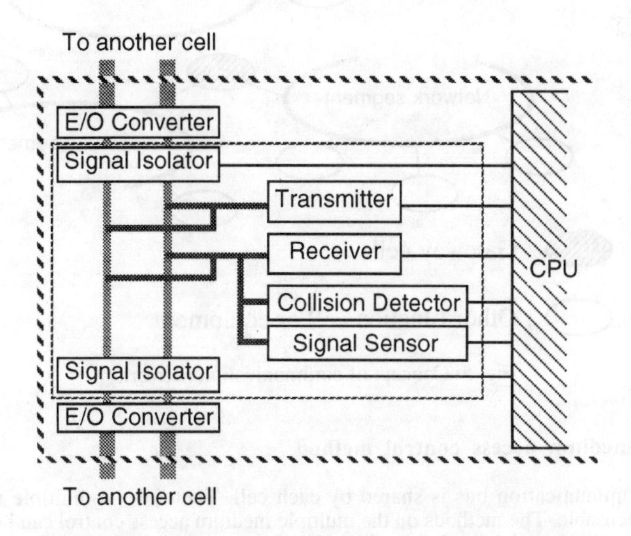

Fig. 4 Block diagram of communication system for SOMS

The communication system dose not needs a centralized traffic control system. Each system has a distributed traffic control system which can realize the peer-to-peer communication. To prevent adverse influence to the others, the system can isolate the faulty communication bus with signal isolator.

For the communication system, the duplex-single direction bus system is adapted. Each cell shares the communication bus and can communicate by packet's exchange. To achieve improved reliability, the optical connection is applied for communication between cells. This communication system is composed of transmitter, receiver, carrier sensor, collision detector, electrical/optical signal converter and signal isolator. The carrier sensor can detect the availability of the communication bus. If anybody uses the communication bus, it detects the communication signal that indicates the communication bus is occupied by another cell. The conflict of the communication is detected by the collision detector. The electrical/optical signal converter is used to transform electrical signal into optical signal. The signal isolator is employed to intercept external ray when the cell is not assembled.

4. Communication architecture of SOMS's network

4.1 Concept of SOMS's communication network

The communication system for the SOMS is equipped with a gateway function in order to solve these problems mentioned above. Figure 5 illustrates the concept of SOMS's communication network. The communication network is composed of some network segments that are connected by gateway cells. The function of gateway cell are a relay station to connect the network segments and multiple communication media [8]. In the current implementation, the branch cell is equipped with getaway function.

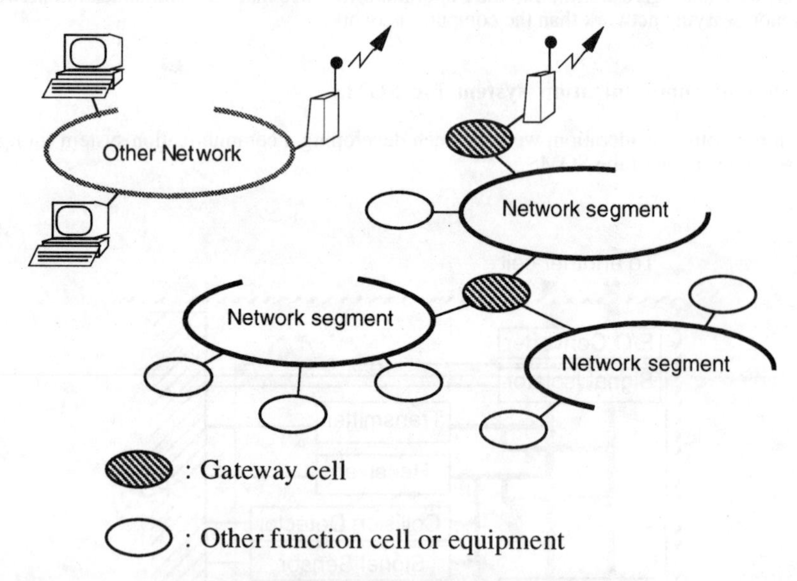

Fig. 5 Concept of communication network

4.2 Multiple medium access control method

The SOMS's communication bus is shared by each cell. Therefore, a multiple medium access control is indispensable. The methods on the multiple medium access control can be classified into the following three approaches, namely polling & selection method, token passing and contention method. The polling & selection method is required for centralized access control mechanism, thus it is not acceptable for SOMS's network. The token passing method has advantages that the communication performance is more efficient than contention method and the communication delay is predictable. However, this method has disadvantage of the complicated procedures for connecting and disconnecting of the cell from the network. The token management is also complicated. Especially, to detect token loss, centralized management is needed or the information on whole network structure needs to be know to each cell. It is undesirable aspects. The contention method has no centralized access control mechanism, and it is easy for a cell connect and disconnect from the network. Therefore, we have applied a contention method, CSMA/CD method, to SOMS's network. The CSMA/CD method [9] can divide a chance of transmitting the communication packet equally between cells. This method also does not need a centralized control system and can easily cope with reconfiguration of the cells.

4.3 Cell addressing system

Each cell has a unique Medium Access Control (MAC) address and a network address. The MAC address is used to identify the cells and also indicates the function of the cells. It is a unique number assigned to each cell and it is not changeable. The network address is composed of a network segment address that identifies network location in the whole network and a node address that is a unique number in the network segment. The network address is supplied from a gateway cell when a cell connects to a network segment. When a cell transmits a packet, the cell should include both addresses in the packet. Figure 6 illustrates the cell address system on the SOMS's network. When a cell in the segment A is transferred to the segment B, the MAC address dose not change but the network address is changed to reflect the new network segment address. By using this addressing system, the gateway function is realized as follows. As the first step, we have implemented gateway function: "packet filtering" and simple "routing" function on the branch cell. These functions are effective technique to reduce traffic on the network, to achieve robustness, and to deal with changeable network topologies.

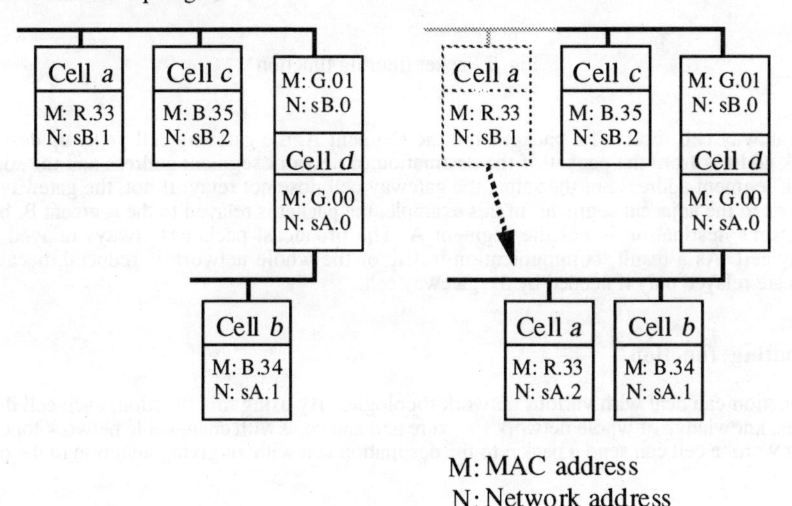

Fig. 6 Cell address system

4. 4 Gateway function

(1) Packet filtering mechanism

The packet filtering function is a part of gateway function. The gateway cell collects the information of network addresses from packets on each network segment that is connected to itself. On the basis of this information, the gateway cell decides whether the communication packet should be relayed to the other network segments or not.

An example of the packet filtering is illustrated in figure 7.

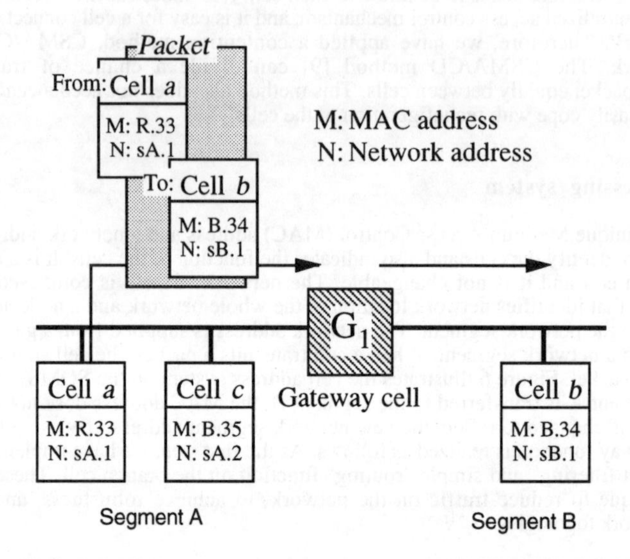

Fig. 7 Packet filtering function

If the gateway cell received a packet from the segment A, the gateway cell picks up destination network address from the packet. If the destination of network segment address and the source of network segment address are the same, the gateway cell dose not relay. If not, the gateway relays the packet to the adjacent segment. In this example, the packet is relayed to the segment B, because this packet's destination is not the segment A. The broadcast packet is always relayed by the gateway cell. As a result, communication traffic of the whole network is reduced, because the packets are relayed only if needed by the gateway cell.

(2) Routing function

This function can deal with various network topologies. By using this function, each cell dose not needs the knowledge of whole network structure and can cope with changeable network topologies. In other word, a cell can send a packet to the destination cell without giving attention to the packet's route.

The gateway cell always collects the routing information as to how to send the communication packet to destination through which network segments by exchanging information with other gateway cell (in the current implementation, this information is given from the cell assembly planner) [10]. The routing information is sets of destination segment address and the first of gateway's MAC address which is passed through which gateway cells to the destination. On the basis of the routing information, the gateway cell calculates the minimum route which reaches the each destination segment with spanning tree algorithm [11]. This function is also effective to solve the infinite loop of the packet. To prevent the infinite loop, the gateway cell calculates the minimum route information. Then, the gateway cell broadcasts the result to the all cells which are connected to own segments. When a cell wants to send the packet to the segment which is not connected itself, the cell determines the gateway cell which can relay the packet, by using this calculated result.

The gateway cell listens to the all communication packets, if it finds a packet whose destination network segment and source network segment is not the same, it relays the packet to the destination network segment. Figure 8 demonstrates the routing function briefly. We assume that the cell 'a' wants to send a packet to the cell 'b' and the cell 'a' knows the network address of the cell 'b'. (The way of finding the network address of other cell is realized by another function.) The cell 'a' can knows that the cell 'b' is not connected to its own segment from the network address of the cell 'b'. Then the cell 'a' sends the packet with destination address which is composed of the gateway cell's MAC address and the cell 'b's network address. When the gateway cell 'G1' receives the packet from the cell 'a', it recognizes that this packet should be relayed, because its destination network address and its own network address are not the same. The gateway cell 'G1' resends the packet to the destination segment by using the routing information. In this example, the gateway cell 'G1' sends the packet to the segment 'B'. If destination segment is the segment 'D', the gateway cell sends the packet to the next gateway cell 'G2'. In this manner, the packet is delivered to the destination with relayed by a series of gateway cells.

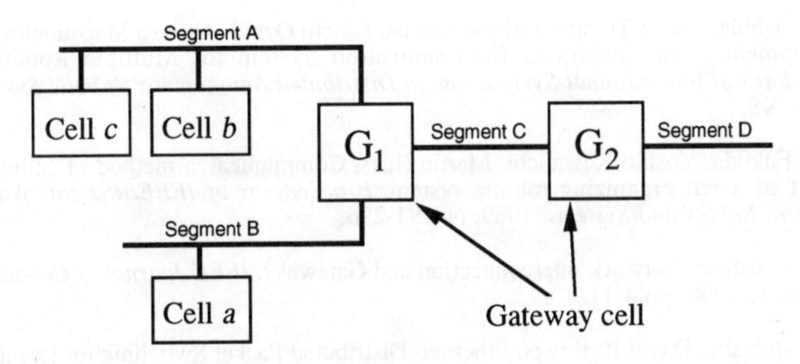

Fig. 8 Routing function

The routing function can realize robustness on the SOMS network. The gateway cell also has the function of isolating failed network segment.

The routing function includes the packet filtering function. The routing function is heavier process for CPU than the packet filtering function. The reaching range of broadcast packet is determined by the location of gateway which carries out the routing function. Therefore, the arrangement of packet filtering and routing function to the gateway cell should be determined by the consideration of network size, communication traffic and the other criteria.

5. Conclusions

(1) The new version of self-organizing manipulator system is introduced.

(2) The communication system for distributed and reconfigurable robotic system is discussed.

(3) The gateway function which are packet filtering and routing control apply to the communication network is proposed.

Acknowledgment

The authors many thank to Dr. K. Yokota and Dr. Y. Ishida for discussions and reading manuscript.

References

[1] Toshio Fukuda, Seiya Nakagawa, Approach to the Dynamically Reconfigurable Robotic System, *Journal of Intelligent and Robotic Systems*, Vol.1 No.1, 1988, pp.55-72.

[2] D. Schmitz and T. Kanade, Design of a Reconfigurable Modular Manipulator System, *Proceedings of the workshop on space telerobotics*, Vol. 3, 1987, pp. 171-178.

[3] Toshio Fukuda, Guoqing Xue, Fumihito Arai, Hajime Asama, Hiromichi Ohmori, Isao Endo, Hayato Kaetsu, A Study on Dynamically Reconfigurable Robotic Systems Assembling, Disassembling and Reconfiguration of Cellular Manipulator by Cooperation of Two Robot Manipulators, *Proceedings of IEEE/RSJ Int. Workshop on Intel. Robots and Systems*, 1991, pp. 1184-1189.

[4] Jing Wang, On Sign-board Based Inter-Robot Communication in Distributed Robotic Systems, *Proceedings of IEEE Int. Conf. on Robotics and Automation*, 1994, pp.1045-1050.

[5] Yoshiki Ishida, Shogo Tomita, Hajime Asama, Koichi Ozaki, Akihiro Matsumoto, Isao Endo, Development of an Integrated Communication System for Multiple Robotic Agents, *Proceedings of International Symposium on Distributed Autonomous Robotic Systems*, 1992, pp.193-198.

[6] Toshio Fukuda, Yoshio Kawauchi, Martin Buss, Communicaion method of cellular robotics CEBOT as a self organizing robotic system, *Proceedings of IEEE/RSJ Int. Workshop on Intelligent Robots and Systems*, 1989, pp.291-296.

[8] Carl A. Sunshine, Network Interconnection and Gateways, *IEEE Journal of Communications*, vol.8, no.1, 1990, pp.4-11.

[7] Robert Metcalfe, David R. Boggs, Ethernet: Distributed Packet Switching for Local Computer Networks, *Communication of the ACM*, vol. 26 no. 1, 1983, pp.90-95.

[9] ISO-8802/3, CSMA/CD Access Method and Physical Layer Specifications, 1984.

[10] RFC 1058, Routing Information Protcol, 1988.

[11] IEEE 802.1, High-Level Layer Interface, 1984.

A Multi Agent Distributed Host to Host to Robot Real-Time Communication Environment

FRANCESCO GIUFFRIDA, GIANNI VERCELLI, and RENATO ZACCARIA

DIST-Dept. of Commjnication, Computer and System Sciences, Univ. di Genova,
I-16145 Genova, Italy

Abstract

In this paper we present a communication environment developed for host and robot cooperation.
It is a socket based communication and message exchange protocol, that allows the dialogue between hosts, manipulators, mobile bases, and any other device connected on the network. The protocol allows the transmission of a wide range of messages: from high level messages as target positions, to the closure of the servo control loops of the manipulators.
The network is completely TCP-IP based: the mobile bases are seen as standard units on the network and the radio-links are kept using the Serial Line Internet Protocol (SLIP). Any unit connected is characterized by an unique internet address. The choose of tcp-ip is founded on it's diffusion and it's simplicity of use. Like many packet communication protocols, different speed of communication (high speed on coaxial ethernet cable, low rate on serial radio link) can coexist on the same net. Any host can exchange messages with any robot, and also robot to robot communication is allowed .

Keywords Multi-host, multi robot communications, real-time control, mobile robots

1 Introducing the multi host - multi robot world

In order to operate with different robots (mobile bases and manipulators), cooperating and real-time controlled, a high level of integration between robots and computation units is required. Complex tasks as manipulation, grasping, navigation in unknown space and cooperation between manipulators and mobile bases imply:

- **real-time control:** if the area of operation is partially or totally unknown (as in a nuclear plant emergency operation); off-line programming is not allowed, and an effective real-time control is required.
- **heavy computations:** real-time control means high speed closed loop control and it requires not only the control of the single task, but also the integration of various cooperating units.
- **real-time task planning:** the resolution of complex tasks in unknown space claims for task organization during the task execution: for example the automatic control of a valve using a manipulator and a vision system needs continuos grasping and regrasping of the valve, basing on the variation of the incoming data from a vision system.

All these points means the need of an high rate data exchange between the units, hence a powerful communication system between the units is required.
If the computer system for robot control is separated from the development system, is more difficult to communicate between them while the robot is working. The observation of the physical robot motion does not make possible to determine the program statement being executed on the robot, and to evaluate the next robot motion [4]. Real time task planning is the key for the solution of complex tasks [5]: a single real time planner can coordinate different robots for the realization cooperative tasks, but a powerful interface unit between the different modules is needed. The adoption of a powerful communication system and a flexible protocol permits high speed data transfer between the units, cooperation, collaboration and distributed computation [3].

2 Robot communications

Robot have been working for many years only as "stupid playback machines", working only on preprogrammed paths and tasks. In this way robots had to communicate only for the exchange of simple commands, like the set of the starting points or the selection of a program. Changing the role of robots operation to an active and interactive behavior, the real time planning of a complex task and the need of interaction with host computers became fundamental.
Collaboration between different robots means high speed exchange of message between robot controllers and hosts, even if the controllers are more than simple servo-loops closures; this includes direct communication between the controllers (e.g.: intrinsic collision detection). When multiple computers execute a coordinated task, the motion of one robot depends on the motion of its peers [2], and therefore not only data transfer for control, but also a high level of integration between the units is needed.
Real time control means continuos messages exchange among the robot controller, the host computer that plans and controls the tasks, and, obviously, the environment.
Error messages must be propagated among all the units involved in the task (broadcast).
The system developed allows high speed communication between different units (hosts, manipulators, mobile bases).

3 Low level protocol

Like other robot and hosts communication environments (like ACTRESS [3] [1]) we choose to develop a LAN (tcp-ip) based protocol to coordinate complex robotized areas composed of manipulators and mobile bases, but different solutions have been found. Two kinds of connections to the bus are used: manipulators and computing units are directly connected to the bus, mobile bases are wireless connected to computing units that work also as gateways. Mobile bases are connected by radio links using wireless modems, so the communication module must be able to treat high speed bus communications and low speed wireless point-to-point communications in a complete transparent way for the units involved.
Our system defines a set of base functions, including those necessary basic functions for the control of manipulators (real-time position control, point-to point), and mobile bases (point to point, jog and speed, real-time position control). Also host-to-host and synchronization primitives are implemented.
The protocol is based on tcp-ip sockets and supports stream and datagram (udp) mode; blocking and non-blocking options are included.
Since mobile bases have to move freely in the workspace, they need to be wireless connected.

In our system each mobile base is connected through a serial radio link to a gateway. No limit exists in the number of the mobile bases in the same network, and also units from different sub-networks can cooperate. The disadvantage of this method is obviously that each mobile base uses a different radio link. As the mobile bases are slip connected, no limit

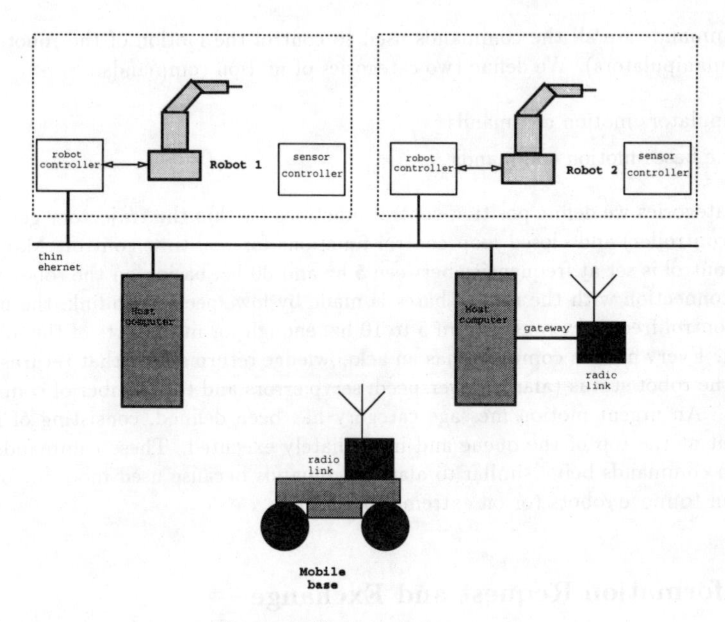

Figure 1: System architecture

in the speed of the communication is imposed by the protocol; only the limitations imposed by the radio modems (9600 baud). The synchronization between the different connection speeds (radio link, coaxial cable) is intrinsically solved by tcp-ip.
Each host computer can also operate transparently as a gateway for one or more mobile base. The network is unique: there is no distinction between the types of the messages which are sent/received from the mobile bases and the other messages, (the wireless connections are always point to point). The gateways are here used as *bridges* for the wireless units and not as links between different networks: each gateway can controls many wireless units.

4 High level protocol

A set of high level communication and control functions completely based on the low level communication protocol have been developed. We have divided the functions into the following categories:

- Motion control commands
- Information request and transfer
- Data transfer

- Synchronization and coordination
- Alarms

4.1 Motion control commands

Motion commands are all the commands used to control the motion of the robots (mobile bases and manipulators). We define two categories of motion commands:

- Manipulators motion commands
- Mobile bases motion commands

For both categories we define position control functions (within the trajectory generated by the robot controller) and closed loop control functions for real-time control. Manipulators real-time control is set at frequencies between 5 hz and 30 hz, basing on the robot typology. Since the connection with the mobile bases is made by low speed radio link, the maximum real-time control frequency ranges from 5 to 10 hz, enough for most parts of the jobs we are working on. Every motion command has an acknowledge return status that returns the host computer the robot status (alarms, overspeed, servo errors and the number of commands in the queue). An urgent motion message category has been defined, consisting of messages that are put at the top of the queue and immediately executed. These commands are not real motion commands being similar to alarm commands because used more to control the motion than to move robots (or on extreme conditions).

4.2 Information Request and Exchange

This message family is used most for message exchange (status, I/O channels read and write, controller programming). Sensors control is also made by this category of messages; stream and handshake protocol are provided.

4.3 Data Transfer

For long data transfer (more than 4096 bytes), as images or special sensor data we developed a special function category. Long data streams are divided into little packets and sequentially sent through channels. This category works also for mobile bases through the radio links, but with bottleneck side effects on the link.

4.4 Synchronization and coordination

To coordinate many cooperating robot units we developed a synchronization message category. These are urgent messages, not queued together with the motion messages on the robot, but executed immediately. A test-and-set basic semaphore control has been implemented for motion coordination tasks.

4.5 Alarms

An alarm message family that works on channels and also as a broadcast transmission has been developed.

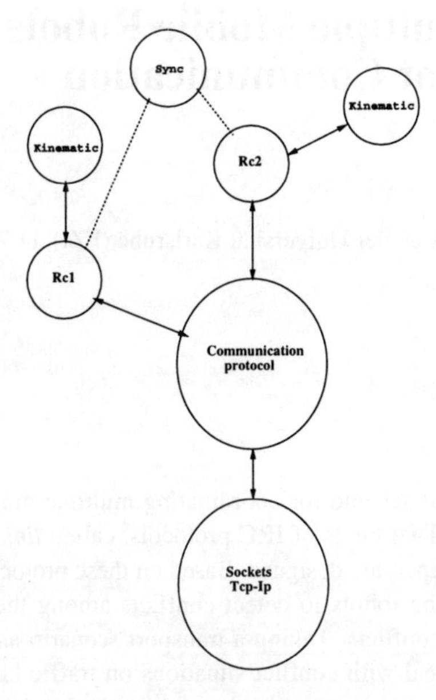

Figure 2:

5 Experimentation and conclusions

We experimented this communication protocol in our lab, that includes a mobile base, two 6 axes manipulator and eight computation units (Unix workstations), with all robots real-time controlled (see fig. 1). The computation units cooperate in the control of the robots with a load distribution: one host works mainly for the reconstruction of the environment and the real-time graphic simulation of the actions; two units work for the real-time control of the manipulators and the last unit as control unit for the mobile base and also as communication gateway. The manipulators are controlled in real-time at 20 hz frequency. No synchronization problems have been found; presently we are planning to substitute the radio links with infrared optical links, allowing to have higher rates of transmission.

References

[1] Asama H. Masumoto A. and Ishida Y. Design of an autonomous and distribuited robot system: Actress. In *Proc. of the IEEE/RSJ (IROS 89')*, pages 283–290, 1989.

[2] Yin Q. Zheng F. Performance analysis oof a token bus lan in coordinating multiple robots. In *Proc. of the 1992 IEEE Internationale Conference of Robotics and Automation*, pages 455–460, Nice, FRANCE, May 1992.

[3] Ishida I. Asama H. and Ozaki K. A communication system for a multi-agent system. In *Proc. of the 1993 JSME*, pages 424–428, Tokyo, JAPAN, August 1993.

[4] Yuta S. Suzuki S. A programming language and software tools designing a sensor based behaviour for autonomous mobile robots. In *Proc. of the 1992 IEEE International Workshop on ETFA*, Melbourne, Australia, 1993.

[5] Morasso P. Vercelli G., Giuffrida F. and Zaccaria R. Hybrid architecture for intelligent robotics systems. In *Proc. of the 1992 IEEE International Workshop Robot and Human Communication*, Tokyo, Japan, 1992.

Coordinating Multiple Mobile Robots Through Local Inter-Robot Communication

SHENG LI

Forschungszentrum Informatik an der Universität Karlsruhe (FZI), D-76131 Karlsruhe, Germany

Abstract

This paper presents a practical scheme for coordinating multiple mobile robots based on inter-robot communication (IRC). Two kinds of IRC protocols, called *link protocol* and *conversation protocol* respectively in this paper, are designed. Based on these protocols, a coordination protocol is developed which enables the robots to detect conflicts among their motion plans in a local region and then to solve the conflicts. Taking a transport scenario as an example, we show the process of our algorithm to deal with conflict situations on traffic lanes and at intersections. A wireless, asynchronous inter-robot communication system is developed using infrared wave as transmission medium. A test system with two laboratory mobile robots and the planned tests are described.

Key words: multi-robot coordination, coordination protocol, inter-robot communication (IRC)

1. Introduction

Figure 1 shows an application scenario where several autonomous mobile robots transport containers in a structured road network. In this example, a certain robot R* is assigned with a task: "Transport a container from position P1 to position P2". R* plans its path and starts to move to P1. The task for R* now is to carry out its plan while avoiding collisions and coordinating its motion with other robots in conflict cases.

This is a scenario of the European Commission ESPRIT III project MARTHA. MARTHA stands for "Mobile and Autonomous Robots for Transportation and Handling Applications". The ultimate goal of this research project is to develop a prototype system for autonomous transportation of containers at airports, rail shunting yards and harbour quays. The system consists of two main parts, a central station and distributed mobile robots. The central station serves as a man-machine interface. Via a so-called central communication system, the central station issues tasks for robots and monitors the termination of the tasks. Due to the limited bandwidth of the central communication, the central station knows only *roughly* about the states and positions of robots so that a kind of central global control for every robot is difficult and sometimes even impossible. The control at the robot level is distributed. Each robot is autonomous and plans its own path according to its static road map without taking the existence of other robots into account. This implies that conflicts among the individual plans of robots on the common system resources, in our case the free traffic roads, are inevitable. These conflicts should be dealed with between involved robots in a distributed way.

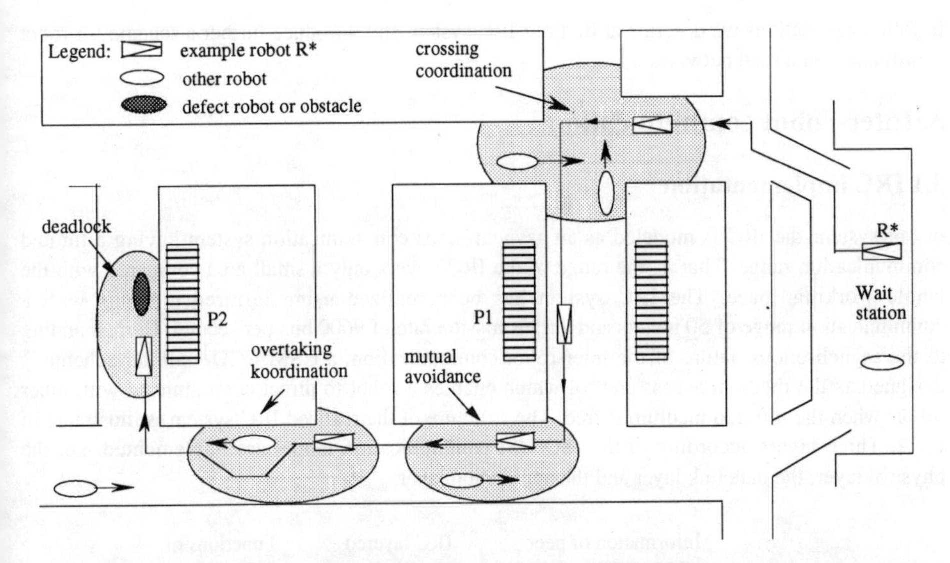

Figure1: A scenario of multi-robot transportation. The grey circles indicate possible conflict cases between robots (by intersections, mutual avoidance, overtaking and deadlocks, etc.).

In this paper we introduce a motion coordination protocol based on local inter-robot communication and describe how it works in a multi-robot system for applications in a transportation domain.

The paper is organized as follows. In section 2, a short review of the related work is given. Section 3 describes the inter-robot communication system and its communication protocols. Section 4 outlines the environment model and the coordination protocol used by robots. The test set-up with two laboratory robots is presented in section 5 together with the discussions to this paper.

2. Related work: Motion coordination of mobile robots based on IRC

Previous studies on motion coordination problems are based mainly on inter-robot communication (IRC), on common traffic rules [1], on sensor information [2] or on the combination of those [3]. On one hand, IRC can be regarded as a special kind of sensor. It provides a robot information about other robots and vice verse. On the other hand, IRC has many advantages compared with conventional active sensors. For example, IRC has normally a larger functional area and greater range than that of active sensors. The information from IRC is usually explicit and needs no further processing work. Consequently, IRC has been taken explicitly into account in most algorithms (e.g. [4], [5]).

Generally, the common process of the IRC-based coordination algorithms is that

- robots communicate their plans as a kind of requests on common resources (e.g. free space, road, etc.) in order to detect and prevent possible conflicts, and
- they negotiate according to predefined protocols to solve the conflicts. The coordination solutions respect to the motion plans for robots are normally slowing down, waiting, accelerating or changing routes.

In following sections we describe at first our IRC system and introduce further a scheme for robot coordination in a road network.

3. Inter-robot communication

3.1 IRC implementation

In our system the IRC is modeled as an asynchronous communication system having a limited communication range. That is, the range of the IRC covers only a small area compared with the whole working space. The IRC system has been realized using infrared modems with a communication range of 50 meters and a transmission rate of 9600 bits per second. Corresponding to the asynchronous nature of the inter-robot communication, a CSMA/CD[1] similar scheme is designed as the medium access control which enables a robot to direct communicate with other robots when the infrared medium is free. The structure of the realized IRC system is illustrated in Fig.2. Three layers according to the ISO/OSI communication model are implemented, i.e. the physical layer, the data link layer and the application layer.

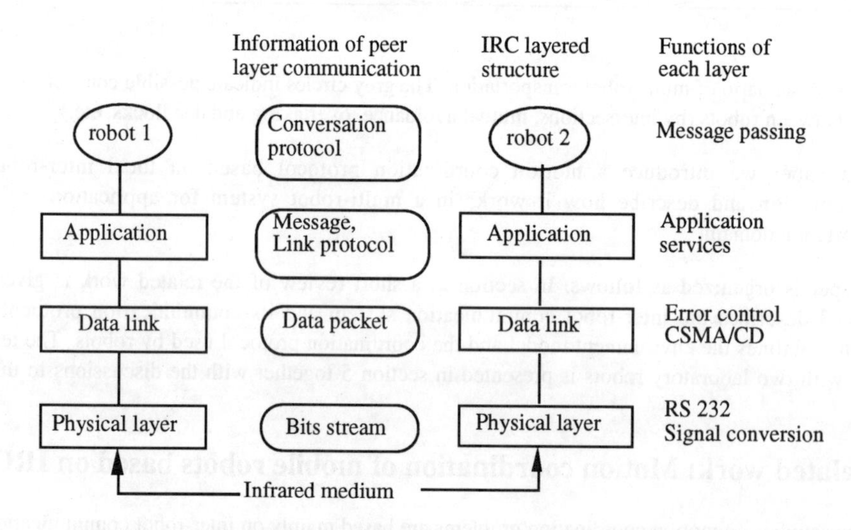

Figure 2: IRC layered structure and peer layer communication

3.2 Local inter-robot communication: Why do we need it?

The main reasons for applying an IRC with a limited transmission range are follows:

- Inter-robot communication activities occur normally only among a small number of spatially neighboured robots which have common interests on communication at a certain moment. We say that these robots have formed a local communication group.
- There is only one shared communication channel with limited transmission capacity. The limited transmission range makes possible that the communication of different independent local groups can take place at the same time with little interferences, as if the communication channel is dedicated only to every single group. So the communication

[1] CSMA/CD stands for "Carrier Sense Multiple Access with Collision Detection".

capacity can be better utilized. Otherwise, the capacity of a global communication system, for example a radio system whose range can be several hundred meters or several kilometres, would be exhausted if the number of robots is large.

3.3 Communication protocols

Two types of communication mechanisms, i.e. broadcast and point to point (PtP) communication, were implemented. The Broadcast communication is often used by a robot to announce its motion plans to all robots who may have interests to it. With PtP two robots can exchange messages directly, for example during a negotiation process. Based on these two mechanisms, two kinds of communication protocols, called conversation protocol and link protocol, have been developed.

3.3.1 Conversation protocol

The conversation protocol is a set of predefined telegrams with functions of *announcement*, *inquiry*, *answer* and *negotiation* [2]. The structure of these telegrams is as following:

Structure Conversation_Telegram
{function_type: announcement/inquiry/answer/negotiation
from: identifier of the source robot (myrobot_ID)
to: all (broadcast)/ destined robot_ID (PtP)
content: message text }

3.3.2 Link protocol

Independent of the conversation protocol, an activity done by the IRC is to broadcast *periodically* a predefined link telegram called Link_T. A Link_T contains only the self identifier (myrobot_ID). Through Link_Ts the IRC of each robot establishes and updates a list of robot IDs called Comm_list which tells the robot which other robots are currently within its IRC communication range. Based on this information, the robot can decide its communication activities, for example, to communicate with the robots which are *always* in its communication range to see whether there will be a conflict. The Comm_list can also be used to check whether a destined robot_ID in a PtP message is possible. Besides of these, the Comm_list shows also the density of robots in an area. This information could be used for predicting traffic jams in a part of the working area.

To avoid communication costs, the Comm_lists can also be maintained during message passing processes between robots without using the Link_T, for example by applying our coordination protocol described in 4.2.

4. Coordination of multiple robots

4.1 Environment model and safety zone

The working environment of robots is modeled simplified through traffic lanes and intersections (Fig.3). The environment map for path planning is presented as a directed graph in which each intersection is a node and each lane a directed arc.

Each lane is assumed to have two intersections as its entrance and exit respectively. So a lane is defined through an assigned number (Lane_No) and the numbers of its entrance and exit

[2] Similar message passing protocol as [6].

intersections. An intersection is presented through a given number (Int_No) and all the lanes connected to it. For instance, the lane L3 and the intersection C2 in Fig.3 are presented as:

L3(Lane_No, entr_C2, exit_C1) and
C2(Int_No, L9, L10, L4, L3).

We use the concept of *safety zone* as the resource request of a robot. Two types of safety zones are defined for a robot. While moving on a lane, the safety zone is defined as:

Sz_lane (position, speed),

where "position" and "speed" are the current position and moving speed of the robot respectively. We assume that all the robots are of the same type so that through these two parameters the width and the length of the safety zone can be calculated (the grey field on L3 in Fig.3).

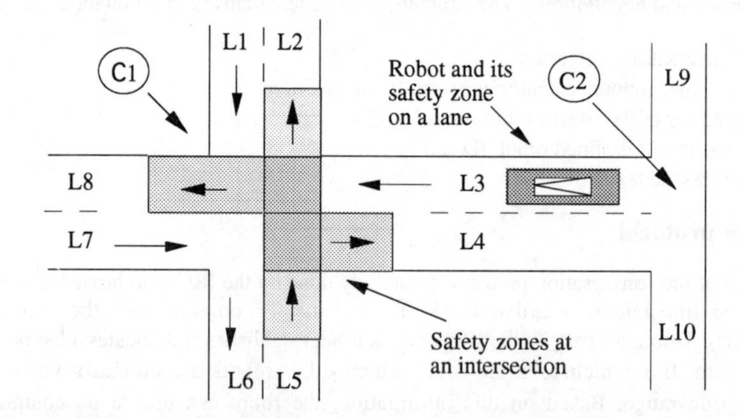

Figure 3: A simplified environment model and safety zones

At an intersection, the safety zone is specified according to the moving directions of a robot. For example, at the intersection C1 in Fig.3 three safety zones are depicted with dotted fields. The corresponding moving intentions are turning left from L5 to L8, straightforward from L5 to L2 and turning right from L5 to L4. The safety zone for the last case is presented as:

Sz_int (C1, L5, L4, Time, Duration),

where "Time" is the time point when the robot enters the safety zone and "Duration" is the time the robot needs to pass this zone.

Supposing that two robots want to cross C1 during overlapping time intervals, there will be a conflict if the two safety zones overlap too. For example, a conflict is considered occurred in the case:

Robot1_Sz_int_C1 (C1, L5, L2, 12:10:35, 10) and
Robot2_Sz_int_C1 (C1, L3, L6, 12:10:40, 15).

4.2. Coordination protocol

A state transition diagram is used to describe the coordination protocol. Figure 4 illustrates a simplified giagram with six states (presented by circles) of a robot and their transition conditions (presented by arcs with arrows).

According to the coordination protocol, a certain robot R works as follows (where S means a state and Tr a transition):

1) R waits for a task (S_waiting).

2) R obtains a task from the central station (Tr_task).

3) R plans a path and starts to move (S_planned).

4) R broadcasts its intended safety zone Sz_lane before entering a new lane and its current safety zone periodically while moving on a lane.

5) R broadcasts its intended safety zone (Sz_int) before entering an intersection and periodically while passing the intersection.

6) Upon receiving broadcasts of safety zones from other robots, R verifies the current situation to see whether there will be a conflict (S_verifying). The "inquiry" and "answer" may be used by the involved robots to exchange more information.

7) At the coordinating state, the involved robots try to solve the detected conflict through negotiation (see 4.3). If no conflict solution can be made between the robots, the robots involved stop their current actions and a deadlock message is sent to the central station (the assumed high level deadlock solver in our system).

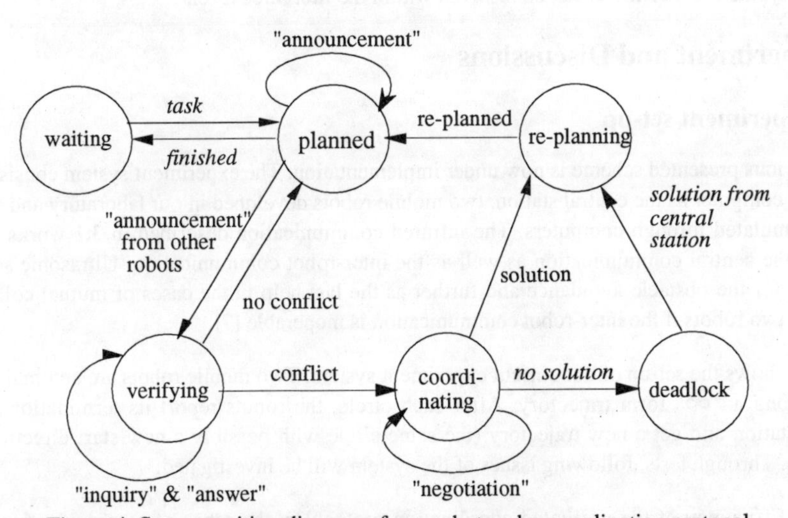

Figure 4: State transition diagram of one robot under coordination protocol

One case which is not shown in Fig.4 is the possible breakdown of conversation or negotiation processes due to IRC errors at the verifying and coordinating states. This fault, however, can be detected by IRC lower layer protocols and the robot's current state is transferred to S_planned.

This is a simple, but a practical solution to cope with this problem because it is often difficult and sometimes impossible to find out the reason of the breakdown and to resume the communication link. To precaution collisions due to this error, the robots would slow down and rely mainly on sensor information.

4.3. Negotiation for conflict solving

The aim of the negotiation is to make a common agreement between involved robots so that each of them can *re-schedule* (for instance delay or advance) its prior planned motion accordingly. That means, in our case the original paths of robots will not be changed at this stage. A deadlock between robots in a local area is said occurred if no one can continue its motion without changing the routes of some robots. To simplify the functions of robots, the deadlock solving function including changing route and task of a robot is resided at the central station.

In order to reach an agreement, robots exchange information in a multi-level way within a negotiation loop. All the robots obey the predefined rules to supply the others with its information, to make proposal and to try to settle an agreement at a possible lower level. The following five levels are defined (from low to high):

Level_1: task priority of the robots; Robots currently without tasks have the lowest priority.

Level_2: velocity assignment; The faster robot has the priority over the slower one.

Level_3: traffic rules such as "keeping distance" and "right before left";

Level_4: Coordination within a group of robots by a group leader;

Level_5: deadlock solution from the central station.

Obviously, the most conflicts can be resolved within the first three levels.

5. Experiment and Discussions

5.1. Experiment set-up

The previous presented scheme is now under implementation. The experiment system consists of a personal computer as the central station, two mobile robots developed in our laboratory and virtual robots emulated through computers. The infrared communication described in 3.1 works in our tests as the central communication as well as the inter-robot communication. Ultrasonic sensors are used for the obstacle avoidance and further as the last help in the cases of mutual collisions between two robots if the inter-robot communication is inoperable [7].

Figure 5 shows the set-up of our current experiment system. Two mobile robots are commanded to move along a "∞" form trajectory. After each circle, the robots report its termination to the central station and get a new trajectory (the same circle with possibly a new start direction and velocity). Through tests, following issues of the system will be investigated:

- Efficiency of the presented coordination protocol in the sense of communication costs and some related parameters, such as communication periodic versus moving speed or population density of robots;

Interaction between the central station and robots for global time synchronization and deadlock solving, etc.;

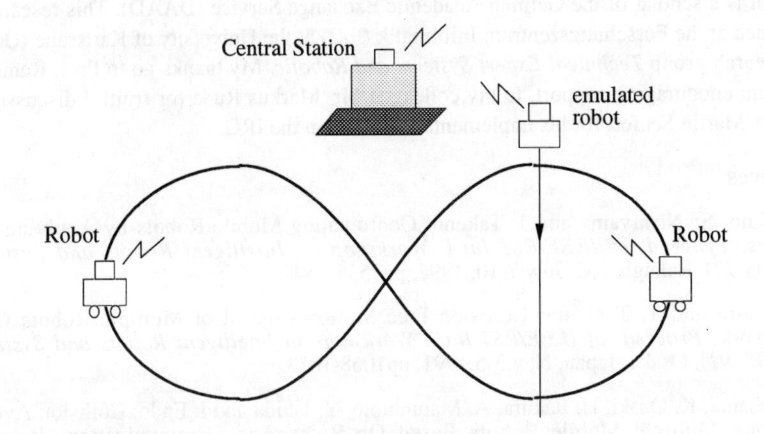

Figure5: Set-up of multi-robot tests

5.2 Discussions

The introduced coordination protocol is application oriented and based on the *exact* time and locations of robots. One assumption has been made by the coordination protocol that all robots have the same time base. In practice and particularlly in our case, the time base could be organized in a *centralized* way that the central station broadcasts a global system time periodically to all the robots. But the time bases of robots are by no means exactly the same. There are different communication delays by different robots and there are further differences between the clocks of on-board computers of robots. In our system, the robots are equipped with digital radio clock which is similar as a centralized system clock. Another possible way is to use the IRC to synchronize the robots within a group during the negotiation process (see also [8]).

Problems may occur with the current coordination protocol if one robot has at the same time for example two different types of conflicts with two robots respectively, i.e. different states in the state transition diagram. Furthermore, the random access nature of CSMA/CD protocol is sometimes unfavorable during the process of negotiation between several robots. We currently work on designing a more efficient IRC protocol. The robots within the same negotiation process would be organized explicitly as a *local communicating robots network* so that the communication can be carried out more effectively (for instance using token passing protocol).

In a real application, a central station (at least a man-machine interface) is always needed. So it is important to investigate a combined scheme of centralized and distributed control of multi-robot systems. With the help of central information, some drawbacks of distributed control, for example information inconsistency between robots, could be better countered with. Some results have been presented before [9], [10]. It is to emphasize that the goal of robots is to carry out tasks and not to coordinate with each other. In this sense, the coordination process should be as simple as possible.

Acknowledgements

The author is a scholar of the German Academic Exchange Service (DAAD). This research work is performed at the Forschungszentrum Informatik (FZI) at the University of Karlsruhe (Germany) in the research group *Technical Expert Systems and Robotic*. My thanks go to Prof. Rembold for his constant encouraging support, to my colleague Mr. Markus Rude for fruitful discussions and also to Mr. Martin Seufert for his implementation work on the IRC.

References

[1] S. Kato, S: Nishiyama and J. Takeno: Coordinating Mobile Robots by Applying Traffic Rules, *Proceed. of IEEE/RSJ Int'l. Workshop on Intelligent Robots and Systems '92 (IROS'92),* Raleigh, NC, July 7-10, 1992, pp1535-1541.

[2] M. Saito and T. Tsumura: Collision Free Motion Control of Multiple Robots On Path Network, *Proceed. of IEEE/RSJ Int'l. Workshop on Intelligent Robots and Systems '91 (IROS '91),* Osaka, Japan, Nov.3-5,1991, pp1068-1073.

[3] H. Asama, K. Ozaki, H. Itakura, A. Matuumoto, Y. Ishida and I. Endo: Collision Avoidance Among Multiple Mobile Robots Based On Rules And Communication, *Proceed. of IEEE/RSJ Int'l. Workshop on Intelligent Robots and Systems '91 (IROS '91),* Nov. 3-5, 1991, Osaka, Japan, pp1215-1220.

[4] F. R. Noreils: Coordinatied Execution of Trajectories by Multiple Mobile Robots, *Proceed. of IEEE/RSJ Int'l. Workshop on Intelligent Robots and Systems '91 (IROS '91),* Nov.3-5, 1991, Osaka Japan, pp1061-1067.

[5] S. Yuta and S. Premvuti: Consideration on Coopera-tion of Multiple Autonomous Mobile Robots -- Introduction to Modest Cooperation--, *Proceed. of the 5th Int'l. Conf. on Advanced Robotics (ICAR'91)*, Pisa, Italy, June 1991, pp545-550.

[6] H. Asama, M. K. Habib and I. Endo: Functional Distribution among Multiple Mobile Robots in An Autonomous and Decentralized Robot System, *Proceed. of the 1991 IEEE Int'l Conf. on Robotics and Automation* , Sacramento, California, April 1991, p1921.

[7] M. Rude: Cooperation of Mobile Robots by Event Transforms into Local Space-time, *to be published in the Proc. of IEEE Int'l. Workshop on Intelligent Robots and Systems '94 (IROS '94)* , 1994, Munich, Germany.

[8] K. Ozaki, H. Asama, Y. Ishida, A. Matsumoto, K. Yokota, H. Keatsu and I. Endo: Synchronized Motion by Multiple Mobile Robots Using Communication, *Proceed. of IEEE/RSJ Int'l. Workshop on Intelligent Robots and Systems '93 (IROS'93),* Yokohama, Japan, July 26-30, 1993. pp1164-1170.

[9] H. Asama, K. Ozaki, Y. Ishida, M.K.Habib, A. Matsumoto and I. Endo: Negotiation between Multiple Mobile Robots and An Environment Manager, *Proceed. of the 5th Int'l. Conf. on Advanced Robotics (ICAR '91)* , Pisa, Italy, June 1991, pp533-538.

[10] S. Yuta and S. Premvuti: Coordinating Autonomous and Centralized Decision Making to Achieve Cooperative Behaviors Between Multiple Mobile Robots, *Proceed. of IEEE/RSJ Int'l. Workshop on Intelligent Robots and Systems '92 (IROS'92),* Raleigh, NC, July 7-10, 1992, pp1566-1574.

Negotiation Method for Collaborating Team Organization among Multiple Robots

KOICHI OZAKI[1], HAJIME ASAMA[2], YOSHIKI ISHIDA[3], KAZUTAKA YOKOTA[4], AKIHIRO MATSUMOTO[5], HAYATO KAETSU[2], and ISAO ENDO[2]

[1]Graduate School of Science and Eng., Saitama Univ., Saitama, 338 Japan
[2]The Inst. of Physical and Chemical Research (RIKEN), Wako, 351-01 Japan
[3]Computer Center, Kyushu Univ., Fukuoka, 820 Japan
[4]Faculty of Eng., Utsunomiya Univ., Tochigi, 321 Japan
[5]Faculty of Eng., Toyo Univ. Saitama, 350 Japan

Abstract

In this paper, an organization method for a collaborating team in a multi-agent robotic system is discussed. In order to realize a flexible and robust intelligent robotic system, we have been developing a distributed autonomous robotic system, ACTRESS, which is composed of multiple and heterogeneous robotic agents. In ACTRESS, every robotic agent aids other agents to achieve a common target, which cannot be carried out by a single agent. The agent requiring the collaboration organizes a collaborating team using communication, which consists of a leader agent (called *coordinator*) and some follower agents (called *cooperators*). For organizing the collaborating team, we developed an efficient negotiation method with a learning mechanism based on histories of past negotiations, and evaluated the method by an implemented simulator. Firstly, this paper presents teams and negotiation for organization in distributed autonomous robotic systems. Secondly, an organization strategy using negotiation with learning mechanism is discussed. Finally, the implementation and evaluation of this method are described, and experimental results are shown.

Keywords: heterogeneous multi-agent robotic system, collaborating team, organization, negotiation, learning mechanism

1. Introduction

In order to realize a flexible and robust intelligent robotic system for high-level tasks such as maintenance task in nuclear power plants or flexible automation in manufacturing plants, we have been developing a distributed autonomous robotic system (DARS), ACTRESS (ACTor-based Robots and Equipments Synthetic system) [1]. ACTRESS is a multi-agent robotic system, which consists of many types of robotic agents such as robots, computing systems and equipment with various functions. The design of this system is based on the concept of functional distribution and cooperation. In order to validate its concept, we have developed a prototype system [2] composed of three autonomous mobile robots and some stationary agents. The stationary agents are workstations whose roles differ from the mobile robots. Every agent in the system is equipped with an integrated communication system [3] utilizing radio modems for mobile agents and LAN for stationary agents. These agents cooperate mutually for a collaborating task if necessary. Figure 1 shows an image of collaborating tasks, which identify examples of handling, monitoring and management by some robotic agents through communication. In a previously developed strategy for collaborating tasks [4], the roles of cooperative agents are classified into a single *coordinator* and several *cooperators*. The coordinator should carry out motion planning of his own and of the cooperators, and assigns them subtasks. It was necessary to organize a team consisting of the coordinator and the cooperators, in order to achieve the collaborating task based on the planning.

In previous works, Noreils [5] or Parker [6] has developed an example of collaborating motions by multiple mobile robots. However, they have not sufficiently discussed a problem of how to organize a team for collaboration. In order to solve the cooperative problems, Hackwood and Beni have proposed the self-organization of sensors [7], and Osawa and Tokoro have suggested the collaborative plan construction [8], in which optimality of the organized team and efficiency of negotiation process have not been addressed. Ueyama, *et. al.* have developed the structural organization basedonrandom walk by mobile agents [9], and have evaluated efficiency of the organization in this system by an entropy of structure configuration [10]. However, this investigation assumed that a

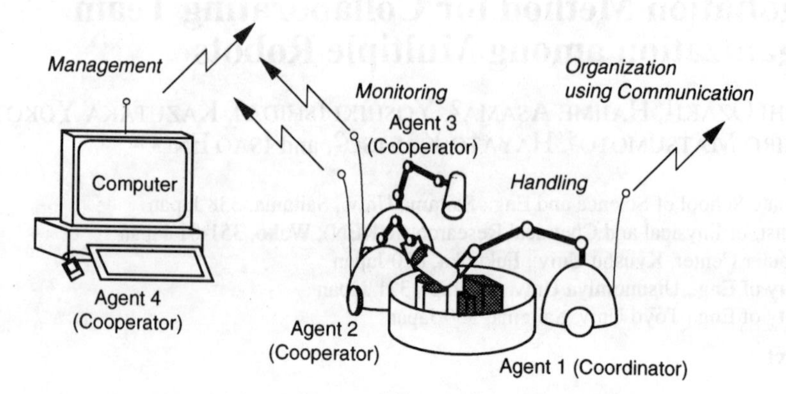

Fig. 1 Image of cooperative task by collaboration

goal of structure is given. A learning mechanism for negotiation among multiple robots has been validated from the viewpoint of communication [11], where the functionality of agents required for tasks have not been taken into account.

In order to address the above issues, we have developed an efficient negotiation method for team organization with a learning mechanism through communication, and evaluated its efficiency among multiple robotic agents with generating a list of required functions. In this paper, we discuss the negotiation method, and show the experimental results carried out with the implemented simulator.

2. Collaborating team organization

2.1 Collaborating team in DARS

Generally, there are two types of configurations in DARS. One is composed of homogeneous agents with the same function, while the other consists of heterogeneous agents with various functions. The functional properties of an organized group fundamentally depend on the type of DARS. In the homogeneous DARS, properties of the organized group depend on the number of agents. In the heterogeneous DARS, its properties depend not only on the number of the agents but also on functions of varied agents in the group. Therefore, the homogeneous DARS is more flexible. However, the agents have to organize dynamically various group according to collaborative tasks.

ACTRESS is a heterogeneous DARS and it allows existences of several groups in which all the agents are equipped with similar attributes. Figure 2 illustrates a concept of some group in ACTRESS. In this figure, every agent is partitioned into the group of stationary agents and another group of mobile agents. A single agent identifying a task, which is impossible to be carried out alone, tries to organize a group for collaboration. The organized group is called a collaborating team. Every agent in the organized collaborating team has a common goal. Any agents, capable of dealing with the goal required by the collaboration, may be included in the team. Thus, the collaborating team is composed of not only those of agents physically located in the area of the task but also other, remote agents. For example in fig. 2, the collaborating team consists of the stationary agents and the mobile agents. In ACTRESS, the collaboration among the mobile robots and stationary agent, a workstation with image processing function was implemented [3].

2.2 Negotiation for organization

To establish team organization, the problem of searching for suitable agents for the collaborating task must be addressed. This paper proposes a strategy for the team organization taking advantage of negotiation via communication. In Distributed A. I., inter-process negotiation is generally utilized

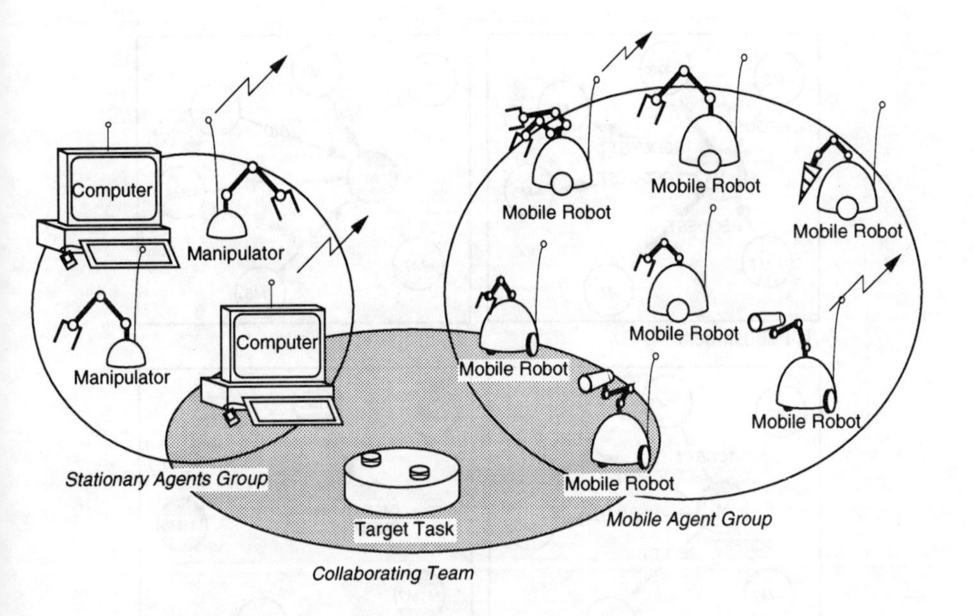

Fig. 2 Concept of collaborating team and groups

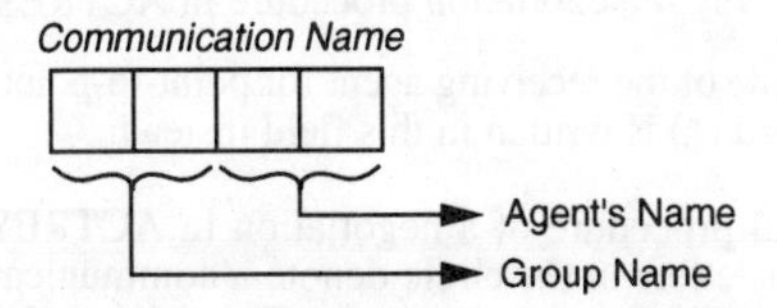

Fig. 3 Structure of communication name

for cooperative problem solving, where the cost of communication of the negotiation is almost ignored [11]. The negotiation is utilized in few studies of robotics because the communication cost is expensive for real robots. One of the problems of wireless communication is possibility of interference between transmissions from agents. Restricting the communication to local area is one way to avoid such interference [9][12]. Ueyama, *et. al.*, used local communication for team organization [9]. However, so long as communication is local, it is not possible to bring agents outside communication range into organized group. In DARS consisting of various agents in global area, the local communication is not suitable for the team organization since the agents must search for satisfied agent in this area. In ACTRESS, agents have the intelligent communication system [3]. It can avoid the interference as every agent has multi-communication channels and changes the channel dynamically according to free channels, and covers the global communication area. Therefore, the agents can search for agents with required functions for the collaborative task. In ACTRESS, a negotiation procedure was previously developed based on the contract-net protocol [13][14], in order to solve general problems in cooperation among multiple agents [15]. The team organization is involved one of this problem. Thus, the negotiation can be applied to achieve the team organization.

2.3 Negotiation procedure and distinction of network in ACTRESS

The communication system of the agent can receive a message only when the addressee's name of the reached message and the communication name accord. The communication name is composed of two fields: the name of the group and the name of the agent, as shown in fig. 3. The group name field is left blank when the receiving agent does not belong to any group. The agent name field is

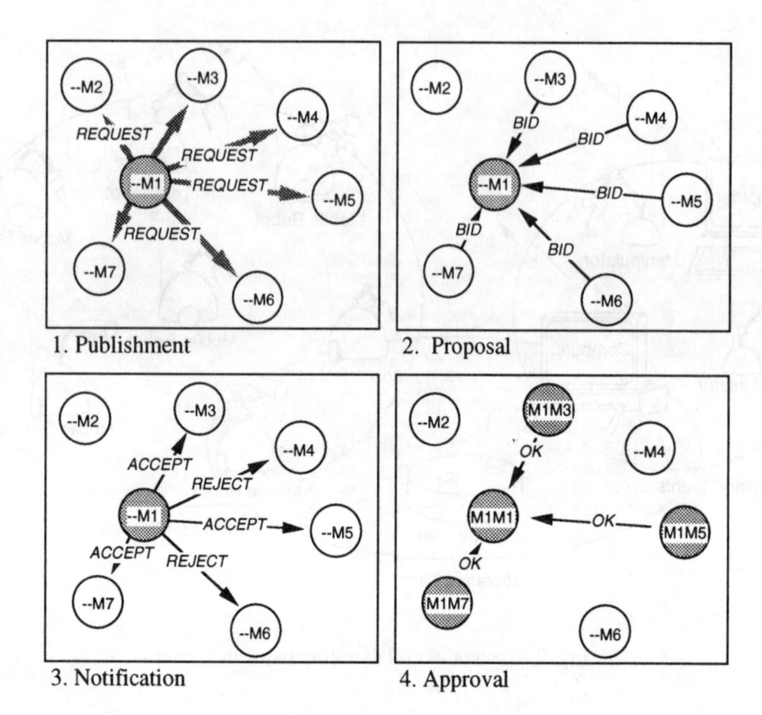

Fig.4 Negotiation procedure in ACTRESS

filled with the individual name of the receiving agent for point-to-point communication. For broadcast communication, the wild card (*) is written in this field instead.

Figure 4 illustrates a typical procedure of a negotiation in ACTRESS. In this figure, every circle identifies an agent, and four letters in the circle denote a communication name. The '--' denotes the group name meaning free attribution of any group. The 'M1', 'M2' and etc. identify respectively each agent name. The negotiation procedure is divided into four phases: *publishment, proposal, notification* and *confirmation*. The publishment means that an agent requests to other agents using the broadcast communication. The proposal means that receiving agents response against the broadcasted message. These phases are described below:

1. *Publishment*
   A coordinator, M1, sends a message of *REQUEST* by broadcasting to all agents, M2, M3, M4 and etc.

2. *Proposal*
   The agents which can comply with request, M3, M4, M5, M6 and M7, send each message of *BID* using the point-to-point communication, and became bidders in this negotiation. The agent M2 which cannot comply does not reply.

3. *Notification*
   The coordinator notifies the results to all bidders using the point-to-point communication after deciding which agents are to be members of a new group. In the specific example, the coordinator sends *ACCEPT* message to agents selected as member of M3, M5 and M7. In addition, it sends *REJECT* message to the agents no selected, M4 and M6.

4. *Confirmation*
   The M3, M5 and M7 receiving the *ACCEPT* message send a message of confirmation. After the communication each agent changes its own group name to new group name. In this case, M1, the coordinator name, is entered.

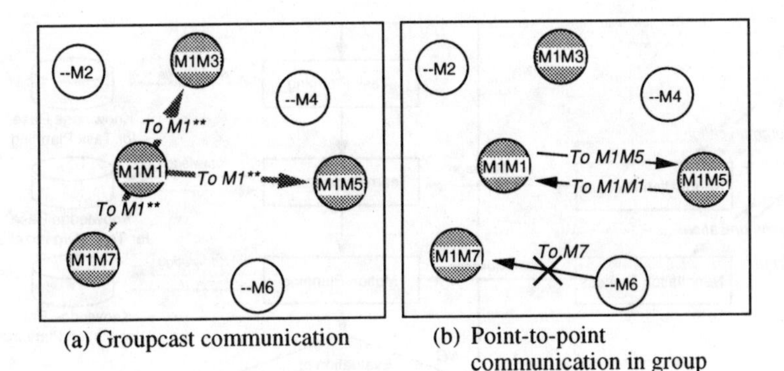

(a) Groupcast communication (b) Point-to-point communication in group

Fig. 5 Advantage of group with distinguished network

Organizing the new group is possible by this procedure, and this group distinguishes a new communication network for the group from general network. Figure 5 illustrates an advantage of the organized group with specific network. With the communication system in ACTRESS, it is possible to broadcast restrictively to agents in the group, as shown in fig. 5 (a). The coordinator, M1, sends messages only to these agents whose group name is M1. This broadcast is called the groupcast in ACTRESS. The coordinator has to adopt the groupcast when it tries to negotiate with the group in the later stages of planning in publishment. Because the agents outside the group do not receive massages, the communication cost in the whole system will be restrained. Figure 5 (b) illustrates the communication in the group. The every agent in the group recognizes its group name, and can communicates with the other agent. In this case, message passing between M1 and M2 is possible. However, M6 which is not included in the group cannot transmit messages to M7 in the group. Consequently, the communication cost in the group can also be restricted because agents outside the group cannot communicate with the agents in group.

3. Negotiation method for team organization

3.1 Task processing in collaborating task

In this section, the task processing flow of the coordinator in a collaborating task is discussed, see fig. 6. The process of task processing is divided into three phases: task planning, team organization and motion planning. Generally, some knowledge-base is necessary for planning, and every agent is equipped with specific knowledge-bases for task planning, team organization and motion planning. In the task planning phase, the coordinator decomposes a task into several task steps referring to its knowledge-base. In the team organization phase, the coordinator requests for required functions, corresponding to each task step, from other agents. Then it organizes a collaborating team as previously described. In the motion planning phase, it finally confirms agreement between the required functions and agents' functions and carries out the motion planning for whole of the team.

Information sharing is necessary in team organization and in motion planning. Every agent can obtain the information using the negotiation processes, and can update the information in knowledge-bases for the team organization and the motion planning.

Generally, precise motion plans cannot be made without knowing cooperators and their performance, and the cooperators cannot be determined without planning how to process the task. Therefore, the motion planning and team organization are done interactively. In the strategy for cooperative task processing, we proposed the coordinator backtracks and re-organized the team, only when it fails in motion planning.

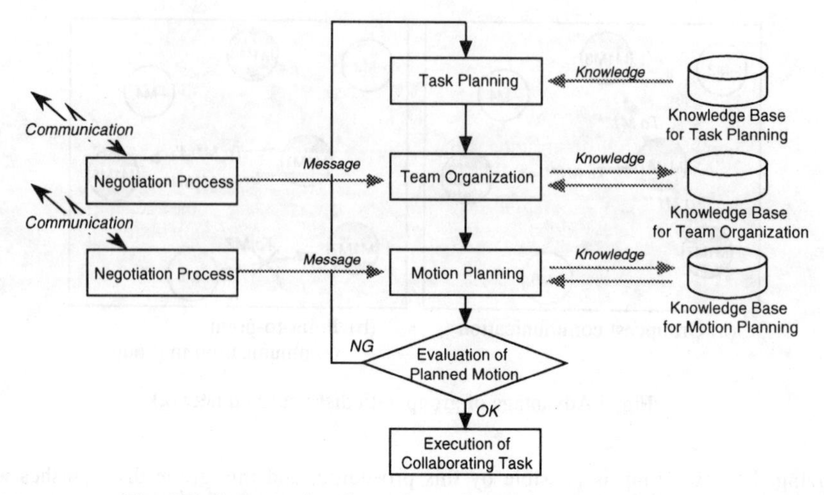

Fig. 6 Task processing flow of coordinator in collaborating task

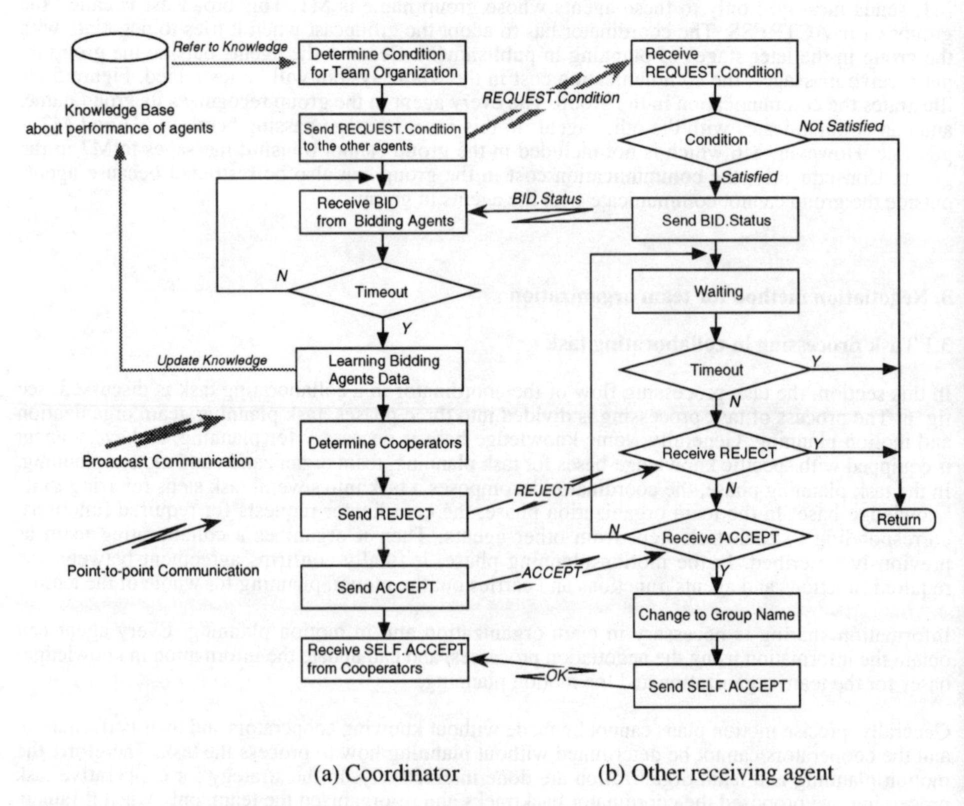

Fig. 7 Processing flow of negotiation for team organization in coordinator and receiving agents

3.2 Negotiation method for team organization

The processing flow of the coordinator and a receiving agent for the team organization is shown in fig. 7. In this figure, the flow on the left hand side denotes the process of the coordinator, and the flow on the other side denotes the process of the receiving agent. Firstly, the coordinator infers some required functions according to the collaborating task, and determines conditions of agents with those functions referring to the knowledge-base. It sends the message of *REQUEST* with the conditions of required functions by broadcast communication. Every receiving agent replies the message of *BID* with its current status or its functions such as current performance of velocity, capability, sensing ability and etc. if satisfying the conditions, then it becomes a bidder. The coordinator then updates its knowledge-base, based on the bidding message, about the current status and the functions of each bidder. It selects some agents, which are satisfactory for cooperators, from the bidders. The coordinator sends a message of *ACCEPT* to the selected agents, and also sends a message of *REJECT* to the non-selected agents. Every agent receiving the *ACCEPT* message changes its group name to the name of the coordinator, replies the message of OK, and becomes a cooperator. Accordingly, a collaborating team consisting of the coordinator and the cooperators is organized.

3.3 Learning mechanism for negotiation

For achievement of the team organization, it is necessary to record information about function or performance of other agents. However, each agent has different performance as the heterogeneous multi-agent robotic system and the performance changes dynamically. Therefore, accounting information about all agent in the system as knowledge is impossible, or is inefficient even if it is possible. We propose a learning mechanism to accumulate and update the knowledge into a knowledge-base (fig.7) through the historical record of negotiation process. This learning mechanism contributes to not only rapid organization but also optimum team organization.

The knowledge-base includes the information about history of past negotiation and potential performance of other agents, such as maximum velocity, maximum capacity, sensing ability and etc. The coordinator applies this knowledge-base for determining conditions of *REQUEST* at the publishment stage of negotiation and for selecting suitable cooperators. For example, if the coordinator is not equipped with any knowledge, it not able to determine suitable condition, and has to choose the most relaxed condition or the strictest condition. In the former case, the coordinator receives bidding messages from a large number of agents and thus the number of message transmissions will increase according to the number of agents in the system. In contrast, in the latter case, only few or no agents will reply to the request.

4. Implementation of team organization using negotiation

4.1 Model for team organization

In order to verify the team organization utilizing the negotiation method with learning mechanism, a model of a team organization was assumed. The task requirements and the performances of an agent are abstracted and represented by quantitative value, performance points, in order to discuss generally the team organization among multiple agents with diverse performance in various situations by the negotiation method.

Performance of a task requirement is represented as *task point* (*tsk_p*). Every agent (a_j; where j = the number of agents) has a *potential performance point* (*pot_p* (a_j)), which means its maximum performance, and this point is constant a parameter. However, the agents are unable to deal with the task using the *pot_p* (a_j) all the time, as they autonomously execute tasks in parallel. Therefore, the current performance of the agent is represented as the *current performance point* (*cur_p* (a_j)).

When an agent (a) obtains a task, it compares the required point (*tsk_p*) and its current point (*cur_p* (a)). If *tsk_p* $\leq$ *cur_p* (a), it recognizes the task can be carried out by itself. If *tsk_p* > *cur_p* (a), it recognizes the necessity of collaboration with other agents, and becomes a coordinator ($a \rightarrow coor$). The coordinator begins negotiation to organize a collaborating team. The coordinator has to gather some agents having required performance to carry out the collaborative task, where the *required performance point* is represented as $req\_p = tsk\_p - cur\_p\ (coor)$. It decides a condition of the collaborative task (or a collaborating team); the *conditional performance point* (cnd_p). It broadcasts a request message with cnd_p to other agents. The other agents (a_j) which receive the request message compare the cnd_p with the $cur\_p\ (a_j)$ of themselves, each agent whose $cur\_p\ (a_j)$ is larger than cnd_p replies a message of bid with its $cur\_p\ (a_j)$, and becomes a bidder (bidder; b_k, where k = the number of the bidders) in this negotiation. The coordinator waits until timeout receiving the bidding message, and then it determines some suitable agents as cooperators in the bidders by selecting agents with similar $cur\_p\ (b_k)$ so that:

$$\sum_{l=1}^{m} cur\_p\ (b_l) \geq req\_p \tag{1}$$

where m equates the number of cooperators. If the sum total of $cur\_p\ (b_k)$ of all received bidders is less than the req_p, the coordinator decides another condition (cnd_p), and tries to negotiate again.

4.2 Implementation of learning mechanism in team organization model

In the negotiation, the coordinator refers to a knowledge-base in order to decide a condition of a collaborating team. Figure 8 illustrates the structure of the knowledge-base, with some sample data filled, for the team organization. The knowledge-base consists of several specific data files of each agent. Data of agent IDs and past negotiations are recorded in these data files. In this figure, the data files of Agent 1, 2 and 3 are illustrated. The coordinator records negotiation times, conditional performance point (cnd_p_i), replied current performance point (cur_p_i) of bidders and current time ($time_i$) at the negotiation. The coordinator can infer the current performance referring to the knowledge-base. An *inferred point of the current performance* ($inf\_p\ (a_j)$) of the agent (a_j) is calculated by the following equation:

$$inf\_p\ (a_j) = \frac{1}{n} \sum_{i=1}^{n} cur\_p_i \tag{2}$$

where, if $cur\_p_i = 0$, $cur\_p_i = cnd\_p_i - 1$. In this equation, the $inf\_p\ (a_j)$ is the average of the current performance of its agent (a_j) in past negotiations. When the coordinator begins to negotiate for a team organization, it obtains inferred points ($inf\_p\ (a_j)$) referring to its knowledge-base in the current status, and calculates virtually a team based on the $inf_p\ (a_j)$s by the equation (1). This team is called a virtual team. The virtual team consists of agents recorded in its knowledge-base.

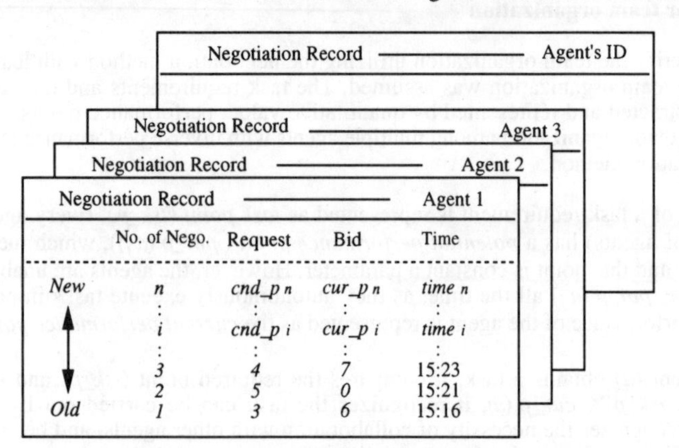

Fig. 8 Structure of knowledge-base

Therefore, the coordinator can determine a conditional performance point (*cnd_p*) based on the *inf_p* (a_j)s in the virtual team.

4.3 Implementation of negotiation procedure for experimentation

We implemented the negotiation procedure proposed in subsection 3.2, in the developed prototype system. We assume the following types of strategies for negotiation procedures:

1. Strategy I
   The coordinator has no knowledge for team organization. It sets the strictest condition (*cnd_p* = *req_p*) for request at first, then makes it relaxed gradually if any bidding messages are not received.
2. Strategy II
   The coordinator has no knowledge for team organization. It sets the most relaxed condition (*cnd_p* = 1) for request, and waits for all bidding messages to select cooperators of the highest performance point.
3. Strategy III
   The coordinator has knowledge for team organization, which is accumulated through past negotiations. By utilizing the knowledge, it determines the fittest conditions for request, for which the replies only from suitable cooperators are expected.

In the strategy I, it is predicted that if the required performance cannot be fulfilled, then the coordinator cannot organize. In the strategy II, it is predicted that the messages are increased according to a number of replying agent. In the strategy III, the coordinator can use the knowledge-base during the negotiation. Therefore, we expect then the strategy III will be more efficient than the other strategies.

4.4 Experimental results by simulation

We developed a simulator for a team organization based on this model. In this simulator, wireless communication [4] equipped with actual mobile robots is assumed for inter-agent communication, The performance of real wireless communication system are shown below:

1. Timeout for acknowledge to each message transmission: 1.0 [sec]
2. Retry in case of failure of message transmission: three times
3. Broadcast communication time: 180 ± 20 [msec]
4. Point-to -point communication time: 360 ± 40 [msec]

Simulation experiments were carried out, where a coordinator should determine suitable cooperators to deal with a given task, and the characteristics of the proposed strategies were investigated. The followings are the experimental conditions:

1. Timeout of bidding messages from receiving agents: 3.0 [sec]
2. The number of the coordinator: one agent
3. The number of receiving agents: 10 through 50
4. Potential performance point of the coordinator (*coor*): *pot_p* (*coor*) = 10
5. Potential performance point of the receiving agents (a_j): *pot_p* (a_j) = 10
6. Current performance point of the receiving agents (a_j): *cur_p* (a_j) = 0 - 10 (randomly changed)
7. Given task point: *tsk_p* = 15, 30 and 50

The experimental results of the team organization by the developed negotiation procedures are shown in fig 9, 10 and 11. The fig. 9 is the case of given *tsk_p* = 15, the fig. 10 is the case of given *tsk_p* = 30, and the fig. 11 is the case of given *tsk_p* = 50.

The *tsk_p* relates to a number of agent in organized collaborating team. In this simulation, the every agent's maximum performance point, *pot_p*, equates 10. In the *tsk_p* = 15, the team consists of two agents, which has performance of 15 through 20. In *tsk_p* = 30, it consists of three or four agents. In *tsk_p* = 50, it consists of five through seven agents. In the experimental results, the optimum organization to make a team with a small number of agents in the current status, was achieved in all strategies.

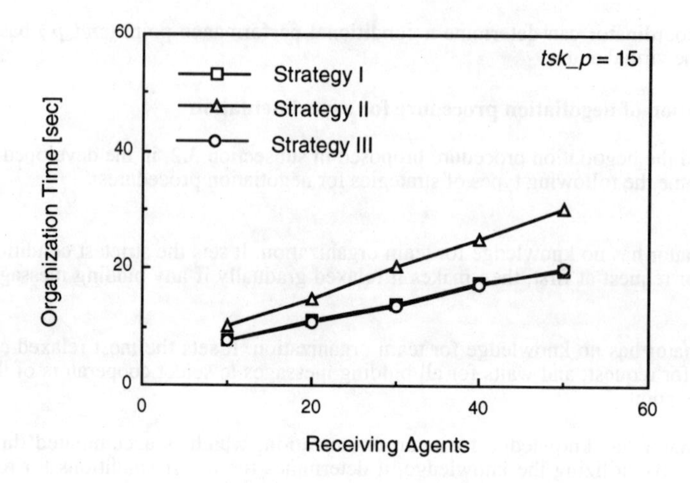

Fig. 9 Organization time against the number of receiving agents (*tsk_p* = 15)

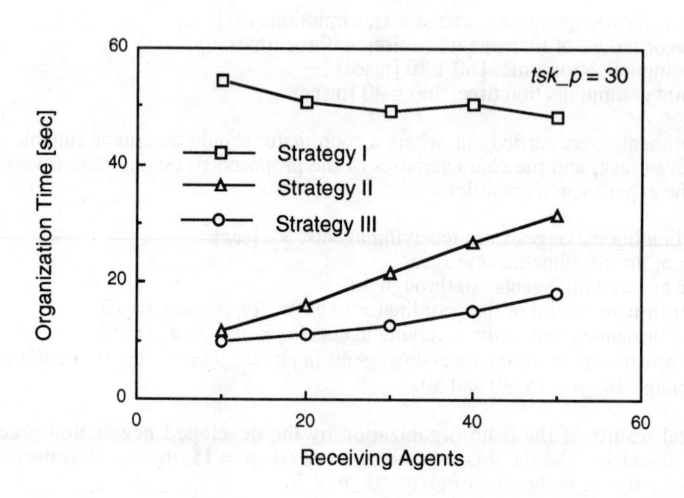

Fig. 10 Organization time against the number of receiving agents (*tsk_p* = 30)

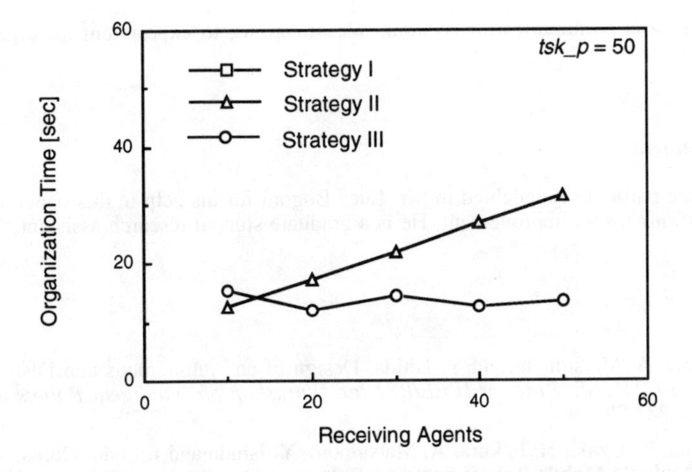

Fig. 11 Organization time against the number of receiving agents (tsk_p = 50)

The strategy I is without any knowledge-base. Therefore, the coordinator does not learn any performance of the other agents. It sets the strictest condition of required performance; cnd_p equates req_p in the first. If any bidding messages are not received, the coordinator should make its cnd_p to the relaxed point gradually as this equation; $cnd\_p = cnd\_p - 1$, in order to complete a team organization. Therefore, this negotiation is continued until the team is organized. In the negotiation procedure, timeout for bidding messages is 3.0 [sec]. The coordinator has to wait for this timeout, even if satisfying agents are not existing. Then, this strategy is not efficient as conditions. The organization time in fig. 9 was short as strategy III, because the cnd_p equates 5 and many agents with the cur_p of more 5 than exist in a large number of receiving agents. The organization time in fig. 10 and 11 was longest because the negotiation in this case is failed for some times as the cnd_p is higher than the other strategies. In fig. 11, experimental result was not represented as organization time has the average of 170 [sec].

In the strategy II, the coordinator selects sufficient cooperators from all agents without any knowledge. It sets the most relaxed condition; cnd_p = 1. Consequently, it receives many bidding messages from almost all agents (if cur_p = 0, the agent does not reply the bidding message) in this negotiation. Necessarily, the inter-agent communication was proportionally increased according to the number of agents. Therefore, this strategy cannot be applied if a large number of receiving agents exist, given the any tsk_p.

In the all results, organization times of the strategy III is shorter than the other strategies. In this strategy, the coordinator is able to determine previously appropriate conditional performance points (cnd_p) beforehand utilizing the knowledge-base. It can negotiate with only restricted number of agents. Especially, this strategy is more efficient for a number of receiving agents as validated in fig. 9, 10 and 11.

5. Conclusion

A negotiation method for a team organization in the collaborating action has been developed. In this method, it is necessary to search for and select cooperators from various agents in global area. In order to estimate this method, a simulator of the collaborating team organization has been developed. In the experimental results on simulations, the organization time was reduced by the strategy of this method using simple a recording mechanism. It was verified to be available in a heterogeneous and multiple robotic system, ACTRESS.

This negotiation method will be implemented in actual mobile robots which are equipped with developed wireless communication system. We are going to experiment an organization for collaborating team.

Acknowledgement

The authors are particularly indebted to Mr. Luca Bogoni for his help in this paper and the many useful suggestions for its improvement. He is a graduate student research assistant, University of Pennsylvania.

References

[1] H. Asama, A. Matsumoto and Y. Ishida, Design of an Autonomous and Distributed Robot System: ACTRESS, *Proc. of IEEE/RSJ Int. Workshop on Intelligent Robots and Systems*, 1989, pp. 283-290.

[2] H. Asama, K. Ozaki, H. Itakura, A. Matsumoto, Y. Ishida and I. Endo, Collision Avoidance among Multiple Mobile Robots based on Rules and Communication. *Proc. of IEEE/RSJ Int. Workshop on Intelligent Robots and Systems*, 1991, pp. 1215-1220.

[3] Y. Ishida, S. Tomita, H. Asama, K. Ozaki, A. Matsumoto and I. Endo, Development of an Integrated Communication System for Multiple Robotic Agents, *Proc. of Int. Symp. on Distributed Autonomous Robotic Systems*, 1992, pp. 193-198.

[4] K. Ozaki, H. Asama, Y. Ishida, A. Matsumoto, K. Yokota, H. Kaetsu and I. Endo, *Proc. of IEEE/RSJ Int. Conf. on Intelligent Robots and Systems*, 1993, pp. 1164-1169.

[5] F. R. Noreils, Toward a Robot Architecture Integrating Cooperation between Mobile Robots: Application to Indoor Environment, *The Int. Journal of Robotics Research*, Vol. 12 No. 1, 1993, pp. 79-98.

[6] L. E. Parker, Designing Control Laws for Cooperative Agent Teams, *Proc. of IEEE Int. Conf. on Robotics and Automation,* 1993, 582-587.

[7] S. Hackwood and G. Beni, Self-organization of Sensors for Swarm Intelligence, *Proc. of IEEE Int. Conf. on Robotics and Automation*, 1992, 819-829.

[8] E. Osawa and M. Tokoro, Collaborative Plan Construction for Multiagent Mutual Planning, E. Werner and Y. Demazeau Eds., *Decentralized A. I.-3*, Elsevier Science Publishers B. V., 1992, 169-185.

[9] T. Ueyama, T. Fukuda and F. Arai, Approach for Self-Organization – Behavior, Communication and Organization for Cellular Robotic System –, *Proc. of Int. Symp. on Distributed Autonomous Robotic Systems*, 1992, pp. 77-84.

[10] T. Ueyama, T. Fukuda and F. Arai, Distributed Structural Organization of Cellular Robots Using Random Walks, *Proc of Int. Symp. on Autonomous Decentralized Systems*, 1993, pp. 215-221.

[11] T. Ohko, K. Hiraki and Y. Anzai, LEMMING: A Learning system for Multi-Robot Environments, *Proc. of the IEEE/RSJ Int. Conf. on Intelligent Robots and Systems*, 1993, pp. 1141-1146.

[12] F. Hara and S. Ichikawa, Effects of Population Size in Multi-Robots Cooperative Behaviors, *Proc. of Int. Symp. on Distributed Autonomous Robotic Systems*, 1992, pp. 3-9.

[13] R. G. Smith, The Contract Net Protocol: High-Level Communication and Control in a Distributed Problem Solver, *IEEE Trans. on Computers*, Vol. C-29 No. 12, 1980, pp. 1104-1113.

[14] S. E. Conry, R. A. Meyer and V. R. Lesser, Multistage Negotiation in Distributed Planning, A. H. Bond and L. Gasser Eds., *Readings in Distributed Artificial Intelligence*, Morgan Kaufman Publishers, 1988, pp. 367-384.

[15] H. Asama, Y. Ishida, K. Ozaki, M. K. Habib, A. Matsumoto, H. Kaetsu and I. Endo, A Communication System between Multiple Robotic Agents, *Proc. of the Japan U. S. A. Symp. on Flexible Automation, The American Society of Mechanical Engineers*, 1992, pp. 647-654.

Resource Sharing in Distributed Robotic Systems Based on A Wireless Medium Access Protocol (CSMA/CD-W)

JING WANG and SUPARERK PREMVUTI

College of Eng., Univ. of California, Riverside, CA 92521, USA

Abstract

Resource sharing is crucial in any multi-agent systems, a distributed robotic system (DRS) is not an exception. A new, general strategy of sharing multiple, discrete resources with predetermined capacities under the model of distributed robotic systems (DRS) is proposed. It is based upon a media access protocol, CSMA/CD-W (Carrier Sense Multiple Access with Collision Detection for Wireless), which supports wireless inter-robot communication among multiple autonomous mobile robots without using any centralized mechanism. This resource sharing strategy is derived based on the fact that with the single, time-multiplexed communication channel, asynchronous events for requesting and releasing resources are effectively serialized. It is shown that the control protocol is effective, efficient, reliable and robust.

Key Words

Distributed Robotic Systems, Mobile Robots, Inter-Robot Communication, CSMA/CD, Resource Sharing.

1. Introduction

1.1 Mobile Robot based DRS

The research on Distributed Robotic Systems (DRS) based on multiple autonomous mobile robots has received a lot of attention in recent years. It is generally agreed upon that each robot under the DRS model should operate autonomously, while all robots must cooperate to accomplish any system-wide (global) task through limited inter-robot communication [1].

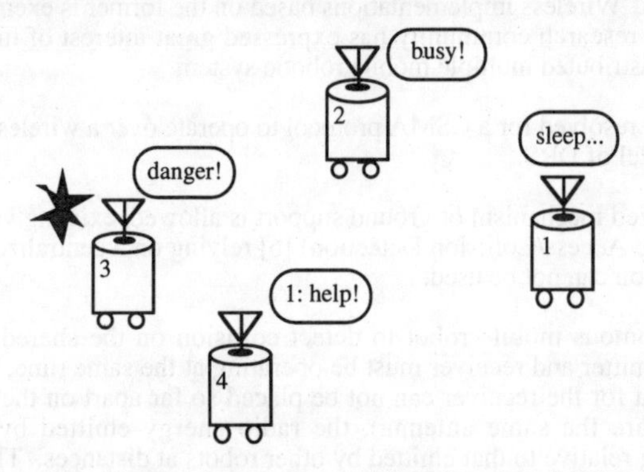

Fig. 1 Interaction among mobile robots in a DRS

Note that the principles of the DRS model do not allow any centralized mechanism, such as a centralized CPU, a centralized and shared memory, or a synchronized clock. Moreover, the system should only consists of multiple, autonomous mobile robots. No "ground support", either centralized or distributed, should be employed.

By no means that the research on fully distributed robotic system (such as DRS) denounces the importance of centralized and/or hierarchical strategies for multiple agent systems. On the contrary, the DRS is complimentary to these strategies for certain tasks with which distributed parallelism is advantageous, or under situations where the centralized mechanism may break down resulting failure of the entire system. DRS is expected to play important roles in space exploration, flexible manufacturing, and military/policing operations where high reliability, redundancy and parallelism is required.

1.2 Inter-Robot Communication

Autonomous mobile robots in a DRS interact through either localized broadcasting (sign-board) [2], or point-to-point communication (message passing) -- not depending on any centralized mechanism such as a centralized communication server, a "blackboard", or other types of ground support.

Many existing experimental DRS test-beds are implemented with off-the-shelf wireless communication systems designed for computers [3], employing either a centralized hardware to indicate the network status, or a centralized communication server to relay messages. (For instance, the wireless Ethernet has a centralized controller). The presence of any centralized mechanism physically violates the principles of the DRS model.

It should be noted that some experimental DRS systems and simulation platforms employ a centralized communication server with the intention of "simulating" the effect of fully distributed, localized broadcasting, or point-to-point communication among mobile robots. DRS experiments based on such a system are valid only if the cooperation and coordination algorithms exercised on the platform do not depend, either explicitly or implicitly, on this centralized mechanism. These algorithms will otherwise fail immediately once ported to a fully distributed multiple mobile robotic system with the centralized communication server removed.

1.3 CSMA/CD-W

Broadcasting and message passing are two basic mechanisms for autonomous mobile robots to communicate in a DRS. Due to limitations on usable radio bandwidth, a "flat" frequency division multiplexing (FDM), (one robot per communication channel), may not be feasible for a system containing a relatively large number of robotic units.

Token ring and CSMA (Carrier Sense Multiple Access) are two basic mechanisms of TDM (Time Division Multiplexing). Wireless implementations based on the former is exemplified with Yamabico robot [4,5]. The DRS research community has expressed great interest of introducing CSMA type protocols into a fully distributed multiple mobile robotic system.

Two problems must be resolved for a CSMA protocol to operate over a wireless radio communication network under the model of DRS.

First, since no centralized mechanism or ground support is allowed, existing variations of CSMA/CD (Carrier Sense Multiple Access/Collision Detection) [6] relying on a centralized mechanism to detect and to indicate a collision can not be used.

Second, for an autonomous mobile robot to detect collision on the shared radio communication channel, both its transmitter and receiver must be operating at the same time. Since the antennas for the transmitter and that for the receiver can not be placed so far apart on the mobile robot, (in fact, they may have to share the same antenna), the radio energy emitted by the transmitter is so overwhelmingly strong relative to that emitted by other robots at distances. This makes the detection on simultaneous transmissions (collision) on a shared radio communication channel practically impossible.

A medium access protocol, CSMA/CD-W (Carrier Sense Multiple Access with Collision Detection for Wireless), designed for wireless network nodes (in this case, autonomous mobile robots) has been proposed by the authors [7]. This protocol avoids these two problems allowing multiple mobile robots to share a common radio communication channel without using any centralized mechanism, in consistency with the principles of DRS.

1.4 Resource Sharing

Distributed resource sharing is an extremely important issue in any multiple agent systems, a distributed robotic system is not an exception.

The problem assumes that there are potentially infinite number of *resource types* $\{R_i\}$, each of which has a predetermined, but finite capacity $\{C_i\}$. A resource type is denoted by a pair: R_i:C_i. A typical resource type in a robotic system may be physical, such as an enclosed physical space capable of holding finite number of robotic units, or logical, such as the right to serve an emergency event which needs only finite number of robots (Fig. 2).

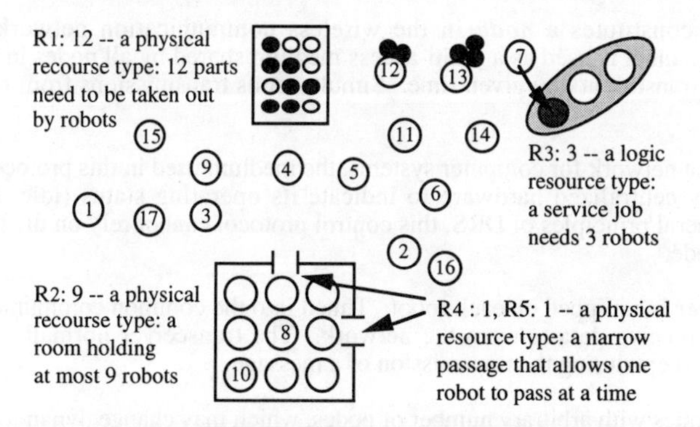

Fig. 2 Resource types in a DRS

Robots in a DRS compete for resources from a *resource pool*. Once desired resources (may be of different types) are obtained, a robot uses these resources to carry out its task. A resource should be released by the robot back the resource pool if it is no longer needed. In general, a robot should not keep any resource indefinitely.

It should be noted that the sequence of requesting and releasing resources by the collection of robots is not known *a prior*. Moreover, no centralized mechanism such as a centralized resource manager (that allocates/collects resources to/from robots) is allowed. Robots must cooperate with each other to share resources through a fully distributed protocol.

The basic requirements for a fully distributed protocol supporting sharing of multiple resource types of finite capacity are:

- The capacity of each resource type should never been exceeded.
- The capacity of a resource type must be reached, if there is sufficient interest for the resource type.
- Deadlock, if there is any, must be detected, and then resolved.

- The system should allow a robot to join and/or leave the system at any time.

Resource sharing based on message passing has been extensively studied by computer scientists in distributed computing, and remains an active topic. Study on resource sharing in DRS based on the sign-board model has also been reported in [8,9].

1.5 Paper Organization

Section 2 of the paper briefly introduces the CSMA/CD-W protocol. The strategy of resource sharing based on this protocol is presented in Section 3. In Section 4, current and future research directions on resource sharing based on this protocol are addressed.

2. CSMA/CD-W Protocol

The protocol is only briefly introduced due to the size limitation of the paper. Interested reader should consult [7] for details.

2.1 Assumptions and Requirements

A mobile robot constitutes a *node* in the wireless communication network. A single radio communication channel is used as a multi-access medium shared by all nodes in the network. Only one node should transmit at any given time. Simultaneous transmissions from more than one node cause a *collision*.

Unlike a local area network for computer systems, the medium used in this protocol is raw, and is not supported by any centralized hardware to indicate its operating status (idle, busy or collision). Consistent to general principles of DRS, this control protocol cannot rely on the functionality of one fixed "master" node.

A radio *transceiver* is equipped on each robot. Tuned into the common communication channel, it is used to transmit/receive data to/from the network. The transceiver normally operates under the receiving mode, except during the transmission of a message.

The protocol operates with arbitrary number of nodes, which may change dynamically.

2.2 Basic Idea

A single radio communication channel is used as the raw medium for all nodes. To reduce the probability of simultaneous transmission (collision), a node checks the status of the shared communication channel before a transmission is attempted. If the channel is busy, it waits for a random period of time.

There is nevertheless still a small chance for two or more nodes to start transmit at almost the same time, which results a collision. Due to strong radio energy emitted by the transmitter, it is impossible for a node to realize the collision until the transmission is completed.

The protocol is designed such that the length of a message generated by a node is always distinct from that generated by others. Thus if a collision occurs, i.e., more than one node started to transmit at almost the same time, they will end their respective transmissions at different time moments (Fig. 3).

The protocol instructs each node to check the status of the channel immediately after each transmission. If the channel is still busy, (RF=1 in Fig. 3), a collision has occurred, as some other nodes are still broadcasting their messages.

Thus all nodes involved are able to realize the collision, except the one which sends the longest message. This node has to be informed by others (about the collision).

A node is instructed to check, right after each transmission, the *validity of the carrier* in addition to the status of the channel. The received carrier is valid (DCD=1 in Fig. 3) *iff* modulated signal from the transceiver can be demodulated by the modem. In other words, the received radio signal must have been emitted from a single node. The difference in message lengths, the data transmission rate and the speed of protocol execution are specified in such a way that a node detects a valid carrier if and only if it has involved in a collision with the second longest message. This node is responsible of generating a collision report [CR] message to inform the node which sent the longest message.

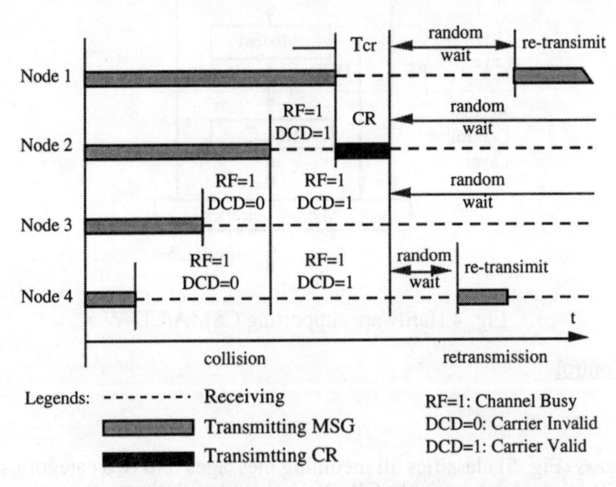

Fig. 3 Collision detection in CSMA/CD-W

For a node involved in the collision with the third longest or shorter message, its received carrier at the time of checking is invalid (DCD=0 in Fig. 3). Because at the moment when it completed its transmission, more than one nodes (at least, nodes with the longest and the second longest message) are still transmitting.

The protocol instructs a node to wait for a predetermined period of time (T_{cr}) for the possible [CR], even if the channel is not busy after the transmission. If a [CR] is received during the waiting period, a collision must have occurred. Otherwise, the transmission is considered successful.

Thus if a collision occurs, all nodes involved are able to realize the collision. A re-transmission is attempted after a random waiting period.

2.2 Hardware

The hardware of the system consists of a transceiver, a modem, a HDLC controller and a micro-controller, which operates the communication protocol, and serves as the interface between the robot and its communication subsystem (Fig. 4).

The following signals must be presented to the protocol control (running on a micor-controller) by physical hardware.

- [RF] (Radio Frequency detected): a Boolean signal. RF=1 *iff* the channel is busy indicating some nodes are transmitting.
- [DCD] (Data Carrier Detected): a Boolean signal. DCD=1 *iff* the carrier entering the modem is valid indicating a single node is transmitting to the shared channel.

The RF signal for monitoring radio transmission is easily derived from the transceiver. The DCD for monitoring carrier is the *Data Carrier Detect* output from the modem. HDLC frames is captured, and assembled by a HDLC controller.

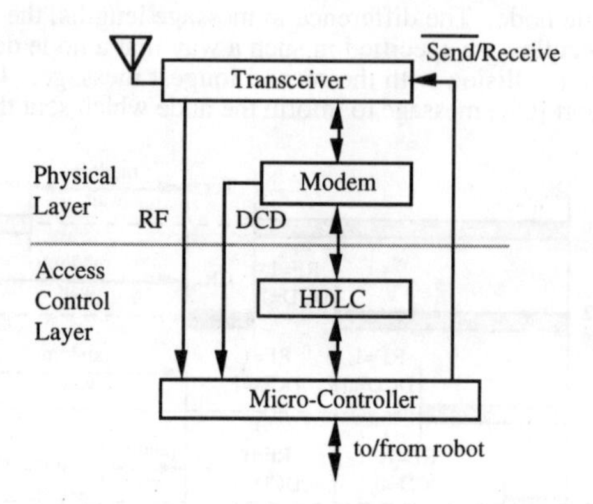

Fig. 4 Hardware supporting CSMA/CD-W

2.3 Protocol Control

Receiving

The *receiving process* (Fig. 5) classifies all incoming messages into two categories, MSG or CR. If a message of type CR is received, variable *CR_RCV*, shared with the *transmitting service*, is set to 1. Otherwise, an interrupt to robot processes is generated.

As the receiving process operates continuously except during transmission, no messages originated from other nodes are to be omitted.

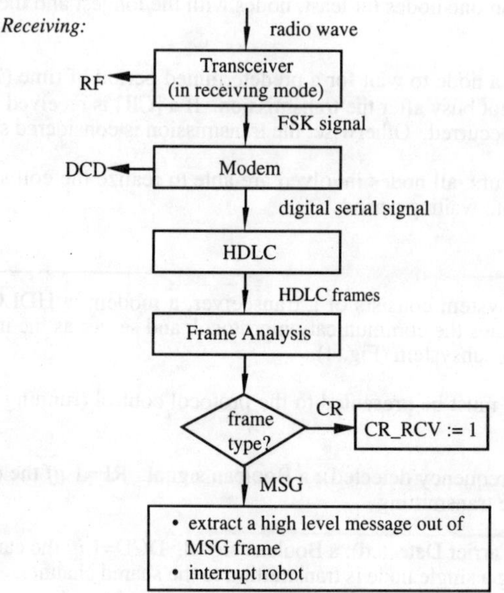

Fig. 5 Receiving process

Transmitting

Implemented according to the collision detection scheme described in Section 2, the *transmitting service* is called upon whenever a robot intends to send out a message. Notice that the service may be aborted at any time upon an *abort* interrupt.

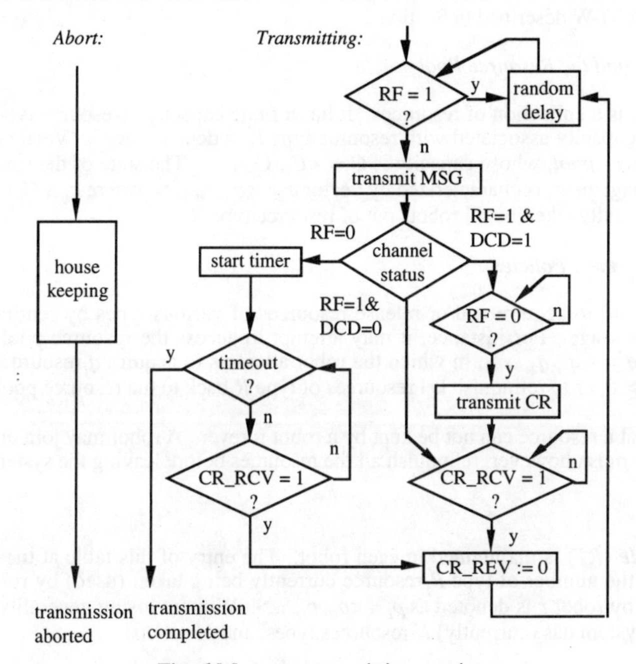

Fig. 6 Message transmitting service

3. Resource Sharing with CSMA/CD-W

3.1 Basic Idea

Since the communication channel is shared by all nodes (robots), at most one message is transmitted at any given time. In other words, all messages successfully transmitted over the shared communication channel are effectively *serialized.*

On the other hand, to enforce finite capacity constrains of respective resource types, the access to the resource pool, (taking/releasing resources from/to the pool), must be serialized. To access the resource pool, a robot is instructed to *attempt* a "resource pool access request" message, referred hereafter as a message of RESOURCE type. Because only one message can transmit at any time, the problem of competing for the right of accessing the resource pool is reduced to that of competing for transmitting a RESOURCE message to the shared communication channel.

Before a RESOURCE message is *attempted*, a robot must make sure that the requested resources are indeed available. If the right of accessing to the resource pool is obtained, i.e., the RESOURCE message is transmitted successfully, the requested resources are considered granted to the robot. However, if a robot receives a RESOURCE message generated by another robot before its own is transmitted, its message transmitting attempt must be aborted, and its "request" of accessing the resource pool is considered denied. The robot must re-evaluated the availability of the resource pool, and may re-attempt a new RESOURCE message only if its requested resources are still available.

3.2 Definitions and Notations

Robots

Robots in the system are denoted by $r_1, r_2, \ldots$. Each robot has an identification, and is equipped to support CSMA/CD-W described in Section 2.

Resource Types and the Resource Pool

A *resource type* is a collection of resources. It has a finite capacity. Resource types are denoted by $R_1, R_2, \ldots$. The capacity associated with resource type R_i is denoted by C_i. Vector $\boldsymbol{R} = \langle R_1, R_2, \ldots \rangle$ forms the *resource pool*, whose capacity is $\boldsymbol{C} = \langle C_1, C_2, \ldots \rangle$. The state of the system, with respect to resource management, is characterized by vector $\boldsymbol{c} = \langle c_1, c_2, \ldots \rangle$, where $c_k \leq C_k$ is the total number of resources currently taken by all robots out of resource type R_i.

Resource Management Policies

A robot is allowed to acquire and/or release resources of various types by sending out one single RESOURCE message. For instance, it may attempt to access the resource pool with a *resource request vector* $\boldsymbol{q} = \langle q_1, q_2, \ldots \rangle$, in which the robot attempts to acquire q_i resources out of resource type R_i when $q_i > 0$, or to relinquish $|q_i|$ resources of type R_i back to the resource pool when $q_i < 0$.

It is assumed that a resource can not be kept by a robot forever. A robot may join or leave the system at will. A robot must, however, relinquish all the resources before leaving the system.

Resource Table

A *resource table* (*RT*) is maintained in each robot. The entry of this table at the *i*-th row and *j*-th column (p_{ij}) is the number of type R_j resource currently being taken (used) by robot r_i. Resources currently taken by robot r_i is denoted as $\boldsymbol{p}_i = \langle p_{i1}, p_{i2}, \ldots \rangle$. Without losing generality, it is assumed in Fig. 7 that the system has (currently) *N* resources types, and *M* robots.

	Presence	R_1: C_1	...	R_N: C_N
r_1	z_1	p_{11}	...	p_{1N}
r_2	z_2	p_{21}	...	p_{2N}
:	:	:	:	:
r_M	z_M	p_{M1}	...	p_{MN}

Fig. 7 Resource table (RT)

Column *Presence* of *RT* indicates whether a robot is currently in the system -- z_i is TRUE if robot r_i is in the system, and FALSE otherwise. It is clear that if z_i = FALSE, then $\boldsymbol{p}_i = \boldsymbol{0}$. The reverse is not necessarily true.

The resource tables maintained by all robots in the system must be consistent, although there is no centralized location in the system to store this table.

Interface with CSMA/CD-W

The interface between the CSMA/CD-W and the *Resource Management* in each robot is illustrated in Fig. 8.

The following primitives are used in the collection of resource sharing algorithms.

- **interrupt** *msg_in*;

  Generated by the *In-Msg Handler,* it signals the *Resource Management* the arrival of an incoming message.

- **function** *msg* := *read*();

  This function is used to retrieve into *msg* an incoming message from the interface.

- **function** *result* := *send*(*msg*);

  This functions returns after the given *msg* is sent by the transmitting service under CSMA/CD-W. The transmission can not be aborted.

- **function** *result* := *attempt*(*msg*);

  As mentioned in the following, a message generated by any of the resource sharing algorithms always has its *type*. When the given *msg* is *attempted*, the transmitting service will try to send the message out under CSMA/CD-W, but the transmission attempt can be terminated by an *abort* interrupt generated by the In-Msg Handler. This happens when an incoming message of the same type is received before the attempted message is successfully transmitted.

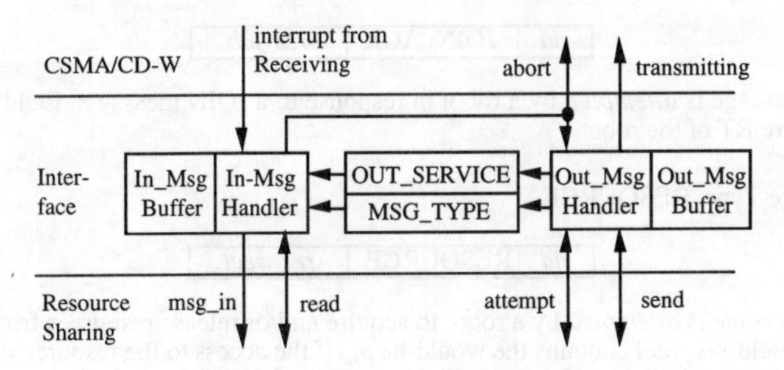

Fig. 8 Interface to CSMA/CD-W

Upon an interrupt from *Receiving* process, the In-Msg Handler checks, by examining variable *OUT_SERVICE*, which is shared with the Out-Msg Handler, whether there is currently a message ATTEMPT. If this is the case, the type of the incoming message is compared with that of the message being attempted (through variable *MSG_TYPE*). In-Msg Handler aborts the transmission attempt if they are of the same type. It always signals the resource management the arrival of an incoming message through the *msg_in* interrupt.

When the resource management unit of the robot attempts a message, its type is registered by the *Out-Msg Handler* in variable *MSG_TYPE*. Variable *OUT_SERVICE* is set to ATTEMPT. The Out-Msg Handler then calls for a Transmitting service under CSMA/CD-W, which may be aborted if a message of the same type is received before the attempted message is successfully transmitted.

When a message is being sent, (not attempted), variable OUT_SERVICE will be set by Out-Msg handler as SEND. The transmission therefore can not be aborted.

Message Formats

A message used in this paper has two parts. The first part is of fixed length and format. The length and format of the second part (message body) varies, depending on the *msg_type* field in the first part.

rid	*msg_type*	message body

Fig. 9 Message format

The following types of messages are employed by th resource sharing algorithms.

- Message Type: JOIN

rid	JOIN

This message is *sent* by a robot about to join the system.

- Message Type: LEAVE

rid	LEAVE

This message is *sent* by a robot that is about to leave the system.

- Message Type: JOIN_ACK

rid	JOIN_ACK	*res_tab*

This message is *attempted* by a robot in response to a JOIN message. Field *res_tab* contains the entire RT of the robot.

- Message Type: RESOURCE

rid	RESOURCE	*res_vect*

This message is *attempted* by a robot to acquire and/or release resources from/to the resource pool. Field *res_vect* contains the would-be $\boldsymbol{p}_{rid}$ if the access to the resource pool is granted.

3.3 Resource Sharing Algorithms

Joining the System

The following is executed when a robot is about to enter the system. Let *rid* be the identification of this robot.

(0) attempt a JOIN message;

(1) Wait until a JOIN_ACK message is received or time-out (T_{join});

(2.1) If time-out, (this indicates that no one was in the system), set $\boldsymbol{p}_i := \boldsymbol{0}$ for $i = 1, 2, ..., M$; Set $z_i = FALSE$ for $i = 1, 2, ..., M$, excluding *rid*, for which $z_{rid} :=$ TRUE. Goto (3).

(2.2) If a JOIN_ACK message is received, initialize RT according to the *res_tab* given in the JOIN_ACK message.

The robot is now ready to compete for the resource in the system.

Acknowledging a JOIN Message

Upon receiving a JOIN message, a robot broadcast its RT (which is globally consistent), as a courtesy to the robot about to enter the system.

(1) Form a JOIN_ACK message with RT as its *res_tab* field;

(2) Attempt the JOIN_ACK message formed in (1).

Reacting to a RESOURCE Message

Upon receiving a RESOURCE message by robot r_i, a robot r_{rid} should update its RT as follows:

(1) Set $\boldsymbol{p}_i$ (the resource vector of robot r_i) in RT according to the *res_vect* field of the received RESOURCE message;

Note that this is handled upon a *msg-in* interrupt.

Accessing the Resource Pool

The following is executed by robot r_{rid} when it intends to acquire and/or release resources from/to the resource pool. More resources are attempted only if they are available -- provided no other robot gains access to the resource pool before r_{rid} does.

(0) Let $\boldsymbol{C}$ be the resource capacity vector. Let $\boldsymbol{q} = \langle q_1, q_2, \ldots \rangle$ to denote the incremental resource vector required by r_{rid} to carry out its next task. If $q_i > 0$, q_i more type R_i resources are being requested. If $q_i < 0$, $|q_i|$ type R_i resources are to be returned to the resource pool. Without losing generality, we assume $\boldsymbol{q} \neq \boldsymbol{0}$.

(1) Let $\boldsymbol{c} = \langle c_1, c_2, \ldots \rangle$ be a vector indicating the state of the system with respect to resources, where $c_i \leq C_i$ indicates the number of resources currently being taken out of resource type R_i. Note that $\boldsymbol{c} := \sum \boldsymbol{p}_i$.

(2) Let $\boldsymbol{b} = \langle b_1, b_2, \ldots \rangle$, and set $\boldsymbol{b} := \boldsymbol{c} + \boldsymbol{q}$. Thus $\boldsymbol{b}$ would be the state of the system with respect to resources, if the access attempt to the resource pool is successful.

If $\exists k$: $b_k > C_k$, (C_k is the capacity of resource type R_k), goto (1).

(3) Form a RESOURCE message with $\boldsymbol{p}_{rid} + \boldsymbol{q}$ as the *res_vect* field. This indicates the would-be resource utilization of robot r_{rid}, if the access to the resource pool with this RESOURCE message is granted.

(4) Attempt the RESOURCE message formed in (3).

(5.1) If the attempt is successful, i.e., the message formed in (3) was successfully transmitted, the requested resources are considered granted/released. Replace $\boldsymbol{p}_{rid}$ with $\boldsymbol{p}_{rid} + \boldsymbol{q}$ in RT. Goto (6);

(5.2) If the attempt is not successful, goto (1).

(6) Inform the robot task process that its request $\boldsymbol{q}$ is granted.

Leaving the System

A robot may leave the system only if its has returned all the resources to the resource pool.

(1) Form a LEAVE message with the robot's id.

(2) Send the LEAVE message formed in (1).

Reacting to a LEAVE Message

(0) Let i be the id of the robot which originated the LEAVE message.

(1) Make sure that $\boldsymbol{p_i} = \boldsymbol{0}$, and set $z_i := FALSE$.

3.4 Discussion

(1) Since multiple resource types with finite capacity are involved, deadlock may occur in the system. The existence of a deadlock can be detected by each individual robots, as its TB contains the global state of system. Once the deadlock is detected. it can be resolved by forcing some robots to relinquish some of their resources.

(2) The Receiving process, the In-Msg Handler, and the handler for *msg_in* interrupt should be set with the highest priority to ensure correctness of this collection of algorithms.

(3) Resource requesting and releasing are combined in a single message type REQUEST to allow maximum flexibility in the application.

(4) If a robot realizes that its receiving process is malfunctioned, it can simply ignore its RT, and pretend that it is a new robot about to entering the system. The only change needed in the previous section is the initialization of its own resource vector, which may not be $\boldsymbol{0}$.

(5) Inconsistency among TB's can be detected, and a reconciling procedure can be invoked during which each robot reports its own resource vector, and a consistent TB is re-created inside each robot.

(6) The consistency checking can be further enforced by instructing each robot to broadcast (send) its resource vector once in a while, even if the robot has no activity with respect to the resource pool.

4. Current and Future Research

Current and future research topics and projects on CSMA/CD-W, and the resource sharing strategies based on this communication protocol include

- Formal specification and verification on the correctness of CSMA/CD-W.
- Performance analysis, evaluation and discrete event simulation for CSMA/CD-W.
- Design and implementation of inter-robot communication subsystem for autonomous mobile robots supporting CSMA/CD-W.
- Deadlock detection and resolution algorithms for resource sharing using CSMA/CD-W.
- Strategies of detecting and automatically correcting inconsistency among TBs of robots.
- Adaptation of this resource sharing strategy to a fully distributed traffic control system for autonomous mobile robots operating on discrete space [9].

5. Conclusion

The DRS research communication has expressed great interest of implementing CSMA type protocols to a wireless communication network built upon autonomous mobile robots. Without using any centralized mechanism (hardware or communication server), CSMA/CD-W detects collision on the common radio communication channel over a wireless network.

Resource sharing strategies based on CSMA-CS-W are discussed. Using the fact that under CSMA/CD-W, one message is transmitted at a time over the shared communication channel, the problem of serializing accesses to the resource pool is reduced to that of competing with other robots for sending out a RESOURCE message. A fully distributed resource sharing mechanism based on this principle is presented.

It is the belief of the authors that CSMA/CD-W is promising for multiple autonomous mobile robotic systems operating under the DRS model, and the resource sharing strategies based on CSMA/CD-W presented in this paper is effective, efficient and robust.

6. References

[1] G. Beni and J. Wang, "Theoretical Problems for the Realization of Distributed Robotic Systems, Proceedings of ICRA-91, Sacramento, CA, April 9-11, 1991, pp. 1,914-1,920.

[2] J. Wang, " On Sign-board based Inter-robot Communication in Distributed Robotic Systems", Proceedings of ICRA-94, May 12-15, San Diego, CA, pp. 1,045-1,050.

[3] K. Ozaki, H. Asama, Y. Ishida, A. Matsumoto, K. Yokota, H. Kaetsu and I. Endo, "Synchronized Motion by Multiple Mobile Robots using Communication", Proceedings of IROS-93, Yokohama, Japan, July 26-30, 1993, pp. 1,164-1,170.

[4] S. Premvuti, S. Yuta and Y. Ebihara, "Radio Communication Network on Autonomous Mobile Robots for Cooperative Motions", Proceedings of IECON-88, October, 1988, Singapore, pp. 32-37.

[5] S. Yuta, S. Premvuti, "Consideration on Cooperation of Multiple Autonomous Mobile Robots --Introduction to Modest Cooperation", Proceedings of ICAR 91, June, 1991, Pisa, Italy. pp. 545-549.

[6] D. Bertsekas, R. Gallager, "Data Networks (Second Edition)", Prentice-Hall, New Jersey, 1992, (Ch.4).

[7] S. Premvuti and J. Wang, "A Medium Access Protocol (CSMA/CD-W) Supporting Wireless Inter-Robot Communication in Distributed Robotic Systems", Proceedins of DARS-94, July 14-16, 1994, Wakoshi, Japan.

[8] J. Wang, "DRS Operating Primitives Based on Distributed Mutual Exclusion", Proceedings of IROS-93, July 26-30, 1993, Yokohama, Japan, pp. 1,085-1,090.

[9] J. Wang and S. Premvuti, "Fully Distributed Traffic Regulation and Control for Multiple Autonomous Mobile Robots Operating on Discrete Space", Proc. of DARS-94, July 14-16, 1994, Wakoshi, Japan.

An Experimental Realization of Cooperative Behavior of Multi-Robot System

SUMIAKI ICHIKAWA and FUMIO HARA

Dept. of Mechanical Eng., Science Univ. of Tokyo, Tokyo, 162 Japan

Abstract

In this paper we deal with experimental study on the possibility of emergent behaviors of multi-robot system when the robot system consists of simple-functioned mobile robots and they move essentially in a random manner. We manufacture 10 hardware robots equipped with simple mechanical functions such as locomotion and detection of obstacle by touch-sensor as well as equipped with small intelligence such as obstacle avoidance based on the information from their touch-sensor. When these mobile robots make locomotion in a given space, the trajectories made by these robots are found to fill more densely the given space with increasing the population size of the multi-robot system. We evaluate the ability in terms of 'Space Coverage Ratio' and find that the space coverage ratio increases more effectively than that in population size. Secondarily, the following function and intelligence are implemented on each individual robot: A function of communication within a limited spatial range to transmit a signal message at regular interval, and the intelligence to cease the locomotion and to return a couple of steps toward a signal message transmitter and then to become a signal message transmitter when the robot has lost the signal message transmitted from other robots. The 10 robots are initiated to move freely and communicate each other and then we find experimentally the multi-robot system to create a communication network starting from the initial location area and expanding over the given space. We experimentally evaluate the reliability of the communication network. In these technical aspects of a multi-robot system, we have experimentally demonstrated the emergent cooperative behaviors by using 10 hardware robots.

Key Words: Multi-Robot System, Emergent Cooperative Behavior, Distributed Robots, Swarm Intelligence, Local Communication

1. Introduction

Many scientists and engineers in robotics have been interested in multi-robot system since the cooperation in multi-robot system is expected to produce a number of technological advantages[1][2][3]. One of those advantages is said to increase the technical ability of the multi-robot system when it works cooperatively. It is coming from such an idea that, when a single robot has ability to accomplish a certain task, the two robots may have the ability twice as much as a single robot. In other words, the function of a single robot can be simplified when the two robots working cooperatively for a given task, or even a single robot can not achieve the given task, but the two robots can do it when they work cooperatively. The simplification in the functionality of a single robot may reduce the size of the robot system and leads to its low-cost production. Another advantage is following: As a multi-robot system consists of a number of robots for accomplishing a certain task, the system is expected to be robust against a single robot's malfunction. However when a number of robots work together in a given area, there exist some problems such as collision avoidance in locomotion of robots, communication between robots, and cooperation among robot behaviors. It also exists an unsolved problem how far the function and intelligence can be simplified to design a single robot when it is used as a member of multi-robot system.

In these situations, we are interested in the emergent ability of 'Swarm Intelligence'[4] of multi-robot system, and we consider it as essential ability to be exhibited in multi-robot system. Even if the multi-robot system consists of many robots equipped with simple mechanical function and small intelligence, they are expected to behave more intelligence when such abilities are utilized in multi-robot system. The behavior and ability of the multi-robot system are obviously dependent on the intelligence and function implemented in the individual robots. Thus we are also interested in the relationship between the level of individual robot's intelligence and the swarm intelligence generated in the multi-robot system. In other words, what kind of function and intelligence are needed for individual robots when they realize an intelligent behavior or swarm intelligence.

Along these problems mentioned above, we have undertaken computer simulation studies[5][6] and found the followings; 1) When each robot is equipped with action principle only for collision avoidance, with increasing in population size of multi-robot system, the system covered up almost every section in the given space for searching, 2) using local communication, multi-robot system could build up a communication network in the given space.

In this paper, to demonstrate experimentally these two kinds of intelligence in the multi-robot system, we manufactured ten hardware robot and implemented the most simple mechanical function and the most low-level intelligence to each of them. The functions are to perceive an obstacle, to receive an optical signal message, to transmit the optical signal message, to proceed forward, to steer itself toward right or left and to count time-lapse. We carry out the two experiments corresponding to the two kinds of intelligence exhibited in the multi-robot system and find experimentally the cooperative behaviors of multi-robot system.

2. Structure and Function of a Single Robot

2.1 Function and intelligence

In this paper a single robot is considered as an agent that perceives its outerworld situation, decides its action from the outerworld information taken in and carries out the action. We call the taken information as input information, the decided action as output information, and the transformation from input information onto output information as intelligence of robot. The input function of robot is defined by the transformation of outerworld condition onto into input information, and the output function of robot is to transform the output information onto the action of robot. **Figure 1** illustrates

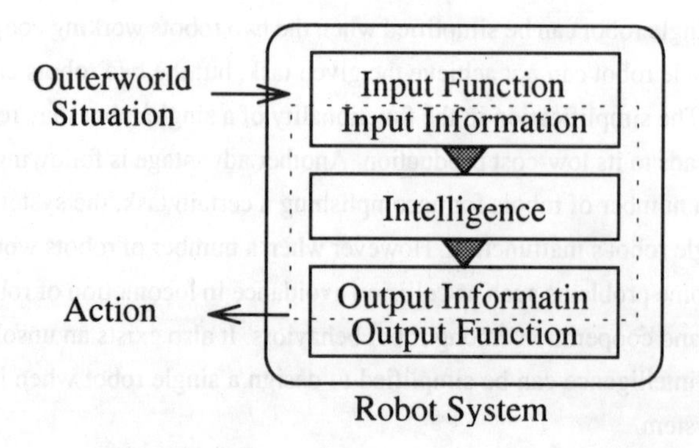

Fig. 1 Conceptual structure of a single robot's function and intelligence

this conceptual structure of a single robot's function and intelligence.

2.2 Hardware of a single robot

Ten hardware robots were manufactured and they were equipped with the following mechanical functions: locomotion by two wheels driven by DC-motors, and detection of obstacle by touch sensors, transmission and reception of infrared optical signal emitted by LED, and timer to count time-lapse and also equipped with a small quantity of intelligence to transform the input information to the output information. **Figure 2** is a schematic drawing of an individual robot used in our experiments. The robot functions mentioned above are classified into two groups: One is related to input function, and the other is to output function. The former includes 1) detection of obstacle and 2) reception of the LED optical signal, and the latter includes 1) locomotion and 2) transmission of LED optical signal. Note that the input function of robot is considered as the function of a single robot to perceive the outerworld situation around the robot and the output function as that for the robot to interact with the outerworld.

Table 1 summarizes the input function of a single robot in terms of function name, situation and input-code. The input code is constituted by binary digit code system. **Table 2** summarizes the output function of a single robot in terms of function name, action and output-code.

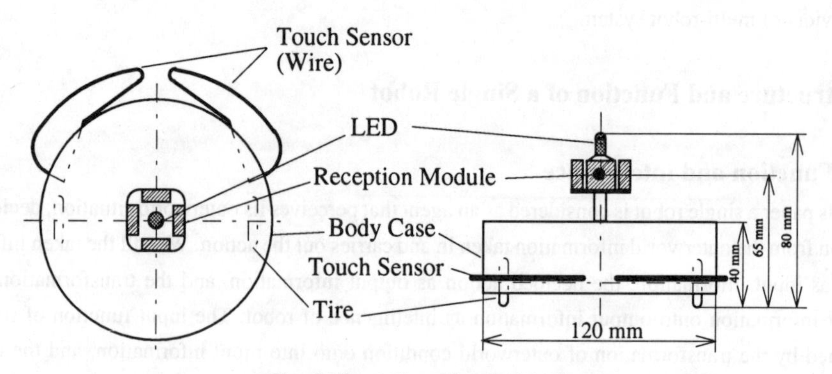

Fig. 2 Illustration of a single robot

Table 1 Transformation of outerworld situation into input information code

Function Name	Situation	Input Code
Obstacle	None Right Side Left Side Both Sides	00 01 10 11
Communication	None Reception	0 1
Regular Communication	Continuos Lost	0 1
Timer	in Time out of Time	1 0

Table 2 Transformation of output information code into robot action

Function Name	Output Code	Action
Locomotion	00 01 10 11	Proceeding Right Turning Left Turning Stop
Communication	0 1	No Action Transmitting
Timer	0 1	Disable Enable

With respect to four input functions in Table 1, function 'Obstacle' means that a single robot perceives an obstacle by two touch sensors equipped in the right hand side and left hand side of the front portion of the robot. Function 'Communication' means that a single robot receives the LED infrared optical signal from other robots by a photo-sensor installed at the head of the robot as shown in Figure 2. Function 'Regular Communication' means that a single robot maintains the reception of the LED optical signal transmitted at a regular interval of time by other robots. The function 'Timer' means that a single robot counts the time steps since the timer starts working.

In Table 2, three output functions are defined in the following manner: Function 'Locomotion' means that a single robot takes four kinds of locomotion behavior, i.e., proceeding, right-turning, left-turning, and stopping. A single robot has two wheels driven by two DC-motors and can take four modes of locomotion mentioned above by regulating rotation speed of the two DC-motors in an appropriate manner. Function 'Communication' means that a single robot transmits an optical signal of infrared ray emitted by LED mounted at the top head of the robot. Function 'Timer' is means that a single robot makes the timer to work. It is noted here that, in addition to those input and output functions, a single robot has a memory function to memorize what kind of output function is undertaken at one time step before, which is specified by three binary digits. This is, in this paper, denoted as 'Internal Information'. One binary digit of the internal information is used to specify the steering direction of a single robot, and the other two binary digits are used for identification of action mode. The internal information is feedbacked into the input information.

A single robot has a capacity of eight binary digit codes as input information and of seven binary digit codes as output information. The intelligence of a single robot is, here in this paper, defined as an appropriate one-to-one transformation from the input codes to the output codes. **Figure 3** shows a schematic diagram of information flow among inputs, outputs and internal memory, in which the intelligence of a single robot is represented as a transformation between input information and output information under taking account of internal information.

2.3 Performance of a hardware robot

Table 3 shows a performance ability of locomotion in a single robot whose two wheels and driven by DC-motors. Because of no feedback control system in locomotion function, there are some variation

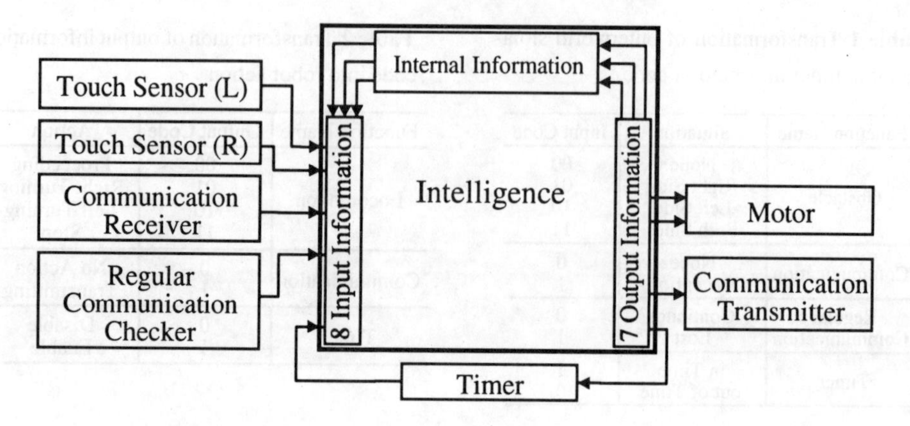

Fig. 3 Structure of function and intelligence of a single root

in velocity and turning rate among each robot and thus the table indicates the average and standard deviation for these two parameters. **Figure 4** shows an example of communication reliability between an arbitrary pair of transmitter and receiver. The reliability is evaluated in terms of the ratio of LED signal reception number to a hundred times of the LED signal transmission. When the distance between the transmitter and receiver is further than 75 cm, the ratio decreases rapidly with distance and at 125 cm of distance the ratio is almost zero. **Table 4** summarizes the reliability of the LED infrared signal communication with respect to 8 directions and 3 distances, where the values indicated are average over 10 hardware robots.

Table 3 Ten robot's abilities of locomotion

	Average	Std. Dev.
Velocity [m/sec]	0.113	0.011
Truning Rate [rad/sec]	4.61	0.44

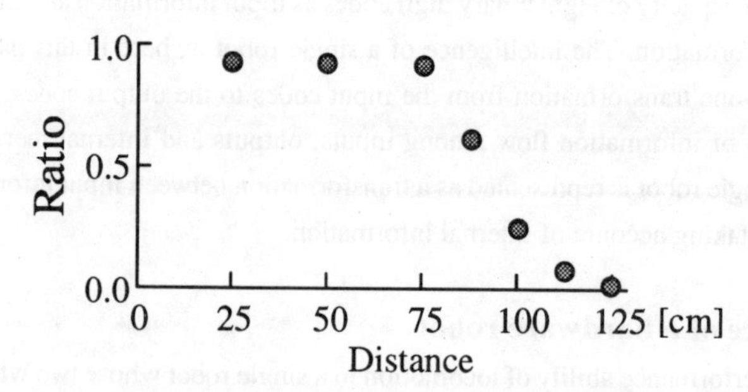

Fig. 4 Reliability of communication between transmitter and receiver

Table 4 Ten robot's reliabilities of communication in 8 directions and 3 distances

		Direction							[%]
		1	2	3	4	5	6	7	8
Distance	75 [cm]	89.1	62.8	94.6	69.5	90.6	71.4	90.4	67.9
	100 [cm]	49.3	19.1	35.2	0.9	50.9	7.5	39.6	9.5
	125 [cm]	10.7	0.8	5.2	1.0	5.4	1.1	7.8	0.7

Ref. for Direction

Front

8 1 2
7 ○ 3
6 5 4

3. Ability to Cover-up the Area for Searching

For the experimental study to demonstrate the ability for a group of robots to cover-up the area specified for searching an object, an individual robot was equipped with the two functions, i.e., 'Obstacle' and 'Locomotion' for input function and output function, respectively, and one internal information (denoted by internal information 1) for memory of the direction of steering undertaken. The intelligence of an individual robot is only the ability to avoid collision with each other or walls surrounding the area in question, or in other words, the intelligence is an ability for a single robot to take three modes of locomotion, i.e., proceeding, right-turning and left-turning, based on the input information obtained by two touch sensors as well as on the internal information. **Table 5** shows the transformation between the input information and the output information in terms of input code and output code systems. This table reads as follows:

i) If individual robots perceive no obstacle, they keep going forward and maintain the internal information 1 as it is (see 1st and 2nd rows in Table 5).

ii) If individual robots detect another robot or wall as an obstacle by the right or left touch sensor, they steer themselves toward the opposite direction to the obstacle and input the appropriate code into their internal information 1 for specifying the direction of robot-turning, i.e., right-turn is 1 and left turn is 0 (see the 3rd and 4th rows in Table 5).

iii) If individual robots detect an obstacle by both touch sensors, they steer themselves, based on the internal information, in the same direction as that taken one time-step before and maintain the internal information as it is (see 5th and 6th rows in Table 5).

Table 5 Intelligence map for free travelling

Input Code	Output Code
[00, 0]	[00, 0]
[00, 1]	[00, 1]
[01, x]	[10, 0]
[10, x]	[01, 1]
[11, 0]	[10, 0]
[11, 1]	[01, 1]

Input:[Obstacle, Internal Info.1]

Output:[Locomotion, Internal Info.1]

x: unspecified

The intelligence given in Table 5 is implemented into each of 10 hardware robots. The behavior generated due to the intelligence shown in Table 5 is called as 'free-traveling'.

When a single robot implemented with the intelligence of collision avoidance is placed at an arbitrary location in the area (2 m x 3 m) bounded by four walls, and initiated to move in an arbitrary direction, the trajectory of the single robot was experimentally found that it was almost along the walls as seen in **Figure 5** (a), where the speed of the robot was set at about 10 cm/s. Due to a slight variation in the power of DC-motor for both wheels of the robot and also some irregularity on the floor surface used for the experiment, the trajectory was slightly curved in the free-traveling paths.

When 10 robots implemented with the intelligence of collision avoidance were initially placed in a random manner within the circle region limited by 50 cm radius around the center point of the 3 m x 2 m area, their trajectory covered almost all portions in the area after the time lapse of 100 sec. This is shown in Figure 5 (c), where we still see the similar curved free-traveling paths to that in Figure 5 (a). From these two demonstrations, it should be noted that, although a single robot is not implemented with the intelligence to move around in a random manner for passing any sections of the area specified, the multi-robot system of 10 robots implemented only with the intelligence of collision avoidance can cover the almost any sections in the area by their trajectories. This implies that the multi-robot system can generate the ability or intelligence to search for an object located at a certain position in the area since there is, at least, one robot which happens to come cross the section. This intelligence emerged in the multi-robot system may be called one of swarm intelligence in our multi-robot system.

The swarm intelligence shown here is very primitive but should be carefully examined in the following ways: We evaluate the ability of multi-robot system to cover the area by their trajectory in terms of space coverage ratio E, which is defined as the ratio between the number of element sections passed

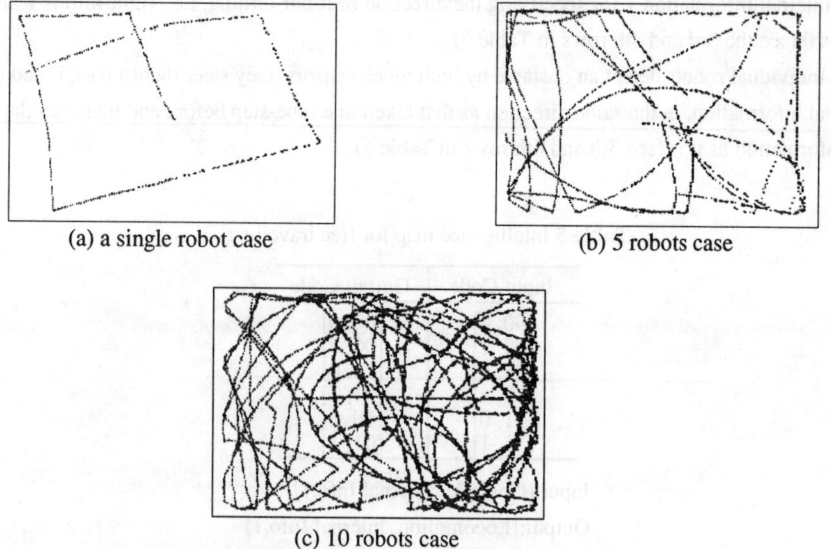

(a) a single robot case

(b) 5 robots case

(c) 10 robots case

Fig. 5 Trajectories of free travelling (100 sec)

by their trajectory and the total number of sections for the specified area (75 x 50 = 3750).

$$E=\frac{N}{75\times 50}$$

E: Space coverage ratio
N: Number of searched space units

This space coverage ratio was evaluated for a given number of the robots and also for various amounts of time-lapse specified in the experiment. The results are shown in Figure 6, where the time scale for a single robot and 5 robots cases was multiplied by 10/1 and 10/5, respectively. **Figure 6** indicates the higher ability of space coverage for the case of 10 robots than that for 5 robots, which means that the larger population size can generate a better performance in space coverage behavior than the small one does.

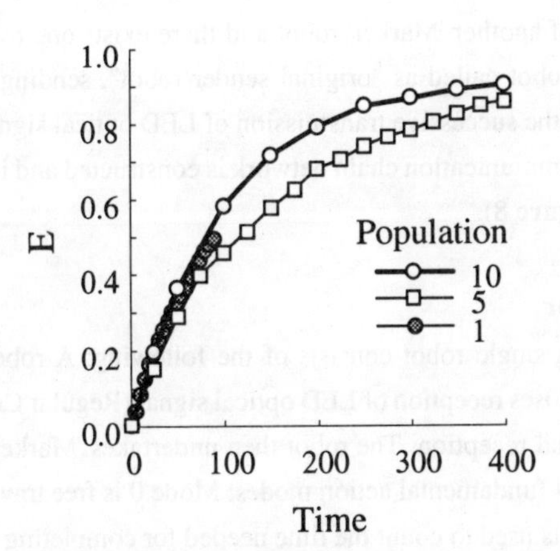

Fig. 6 Space coverage ratio E (relative time scale)

4. Building A Communication Network

4.1 Hello-call chain communication

Let's implement the following functions to an individual robot in addition to those mentioned in section 3: 1) 'Communication', 'Regular Communication', and 'Timer' for input function, and 2) 'Communication' and 'Timer' for output function. 'Communication' for input function means that a single robot receives LED infrared optical signal, when it locates within the communication range of other robots. 'Communication' for output function means that a robot transmits LED optical signal. We call this kind of communication as "hello-call-communication" since a robot happens to receive LED optical signal when it comes within a communication range. This implies that, when a robot receives LED optical signal, the robot knows that the location of itself is in the vicinity of other robots sending the LED optical signal (see **Figure 7**). 'Regular Communication' means that a robot transmits LED optical signal at a given regular interval.

We introduce a new action to robots which is called 'Marker' in this paper. The action 'Marker' means that a robot ceases its locomotion and transmits a LED optical signal when the robot receives LED optical signal. If a single 'Marker' robot, which is undertaking the action 'Marker', is located within

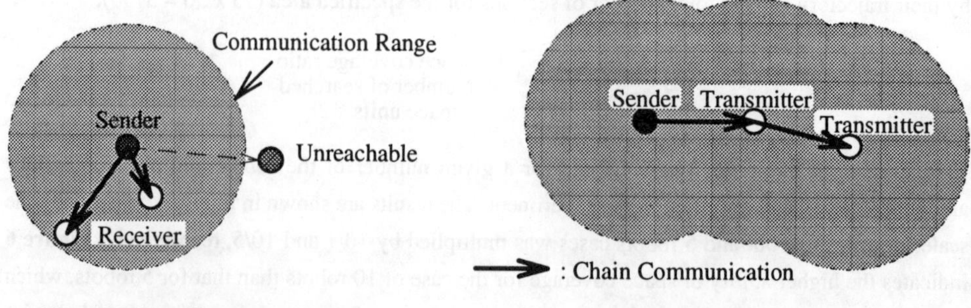

Fig. 7 Hello-call communication

Fig. 8 Communication chain

the communication range of another 'Marker' robot and there exists one robot undertaking 'Regular Communication', i.e., the robot called as "original sender robot", sending LED optical signal at a given regular interval, then the successive transmission of LED optical signal is achieved among the 'Marker' robots. Thus the communication chain network is constructed and it is here called "hello-call chain communication" (**Figure 8**).

4.2 Expansion behavior

The intelligence added to a single robot consists of the following: A robot in free-traveling stops locomotion when the robot loses reception of LED optical signal (Regular Communication) and steps backward to restart the signal reception. The robot then undertakes 'Marker' action. In other words, the intelligence consists of 4 fundamental action modes: Mode 0 is free traveling, mode 1 is turning-backward in which 'Timer' is used to count the time needed for completing the turning action, mode 2 is proceeding and mode 3 is 'Marker' action. The change in action mode is done according to the reception of LED optical signal. **Figure 9** shows the condition for changing the action mode. **Tables 6 to 9** show the details of the transformation between the input code and output code in each action mode. A single robot undertakes different actions according to the input information at each mode, and needs two bits internal information for distinguishing which action is taken. This intelligence of a single robot can emerge a new behavior in our multi-robot system to be shown later and we call it as 'expansion behavior'.

4.3 Demonstration of expansion behavior

To demonstrate the building-up of communication network in a specified space (3 m x 2 m), the 10 robots were randomly allocated within the initial section at the beginning of experiment and started the free-traveling and the communication mentioned above. **Table 10** shows the parameters in the communication between individual robots used in experiments. When robots cannot receive the regular communication for the time duration set by 'Regular communication check time' in Table 10, they judge the regular communication lost.

The experiment has revealed that they build up a communication network initiated at the start section in a certain time lapse. **Figure 10** shows the location of 'Marker' robots undertaking the so-called hello-call chain communication, in which the start section is indicated by a hatched circle and the

Table 6 Intelligence map for mode 0

Input Code	Output Code
[00, 0, 00, 0]	[00, 00, 0, 0]
[00, 0, 00, 1]	[00, 00, 1, 0]
[00, 0, 01, x]	[00, 10, 0, 0]
[00, 0, 10, x]	[00, 01, 1, 0]
[00, 0, 11, 0]	[00, 10, 0, 0]
[00, 0, 11, 1]	[00, 01, 1, 0]
[00, 1, xx, x]	[01, 01, 1, 1]

Input:[Mode, Obstacle, Intrenal Info.1]
Output:[Mode, Locomotion, Internal Info.1, Timer]

Table 7 Intelligence map for mode 1

Input Code	Output Code
[01, 1, 0]	[01, 01, 1]
[01, 0, 1]	[10, 00, 0]
[01, x, 0]	[00, 00, 0]

Input:[Mode, Timer, Regular Com.]
Output:[Mode, Locomotion, Timer]

Table 8 Intelligence map for mode 2

Input Code	Output Code
[10, 0, 00]	[10, 00, 0]
[10, 0, 10]	[10, 11, 0]
[10, 0, 01]	[10, 11, 0]
[10, 0, 11]	[10, 11, 0]
[10, 1, xx]	[11, 00, 1]

Input:[Mode, Communication, Obstacle]
Output:[Mode, Locomotion, Communication]

Table 9 Intelligence map for mode 3

Input Code	Output Code
[11, 0]	[11, 0, 11]
[11, 1]	[11, 1, 11]

Input:[Mode, Communication]
Output:[Mode, Communication, Locomotion]

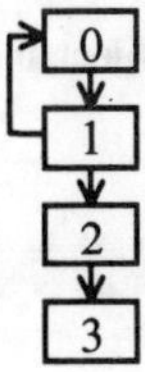

0 into 1 : for regular communication lost.
1 into 2 : for turnnig-back completed.
2 into 3 : for regular communication recieved.
1 into 0 : for regular communication recieved.

Fig. 9 Flow of mode change

Table 10 Communication parameters

Communication range	Ref. Table 4
Regular com. interval	0.6 [sec]
Regular com. check time	0.7 [sec]
Timer	0.7 [sec]

original sender robot is No. 7. **Figure 11** shows the reliability of the communication linkage between arbitrary two robots, where reliability was defined as ratio of successful reception number to 50 times of LED optical signal transmission.

5. Concluding Remarks

This paper has experimentally demonstrated the two kinds of emergent intelligence in our multi-robot

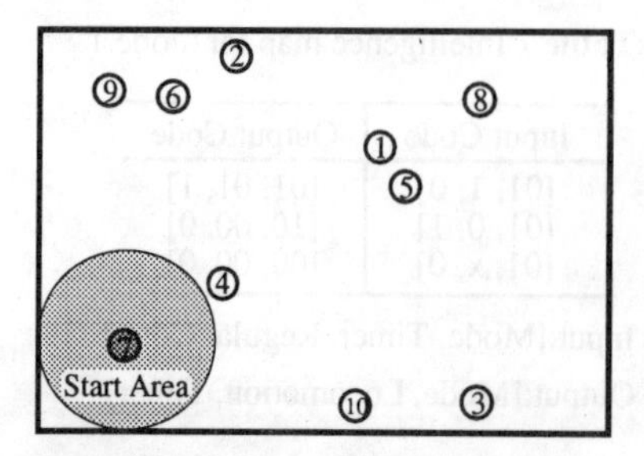

Fig. 10 Location of robot after a certain time lapse during which free-traveling and communication were performed in the multi-robot system.

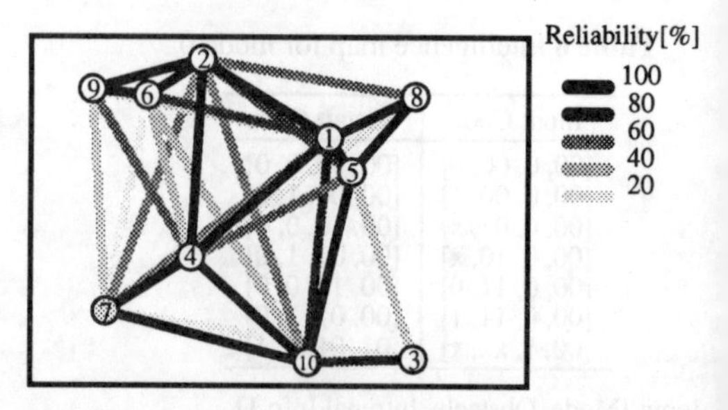

Fig. 11 Communication reliability in the network consisted of 10 robots

system: One is the ability to cover-up the area for searching an object and the other to build-up a communication network. The robots used in the experiment were implemented with certain number of input functions and output functions as well as the intelligence to transform the input information to the output one.

In near future, we will implement a 'returning behavior' to the robots in addition to the intelligence to perform the hello-call chain communication, in other words, detection of communication direction to a robot sending LED optical signal and locomotion toward the sending robot. When the returning behavior is realized, our multi-robot system may be expected to find its object and return to the initial position in a given space without the world map.

References

[1] Y. Kawauchi, M. Inaba, T. Fukuda, A Study on Cellular Robotics System (A Realization of a Robotic System Capable of Adaptation, Self-organization, and Self-evolution), *Journal of the Robotics Society of Japan*, Vol. 12, No. 1, 1994, pp. 116-132.

[2] H. Asama, etal, Functional Distribution among Multiple Mobile Robots in An Autonomous and Decentralized Robot System, *Proceeding 1991 IEEE International Conference Robotics and Automation*, Vol. 3, 1991, pp. 1921-1926.

[3] R. Brooks, et al, Lunar Base Construction Robots, *Proceeding of the 1990 IEEE/RSJ International Symposium on Intelligent Robots and Systems*, 1990, pp. 389-392.

[4] G. Beni, S. Hackwood, Coherent Swarm Motion under Distributed Control, *Proceeding of International Symposium on Distributed Autonomous Robotic Systems*, 1992, pp. 39-52.

[5] F. Hara, S. Ichikawa, Effects of Population Size in Multi-Robots Cooperative Behaviors, *Proceeding of International Symposium on Distributed Autonomous Robotic Systems*, 1992, pp. 3-9.

[6] S. Ichikawa, F. Hara, Cooperative Route-Searching Behavior of Multi-Robot System Using Hello-Call Communication, *Proceeding of the 1993 IEEE/RSJ International Conference on Intelligent Robots and Systems*, 1993, pp. 1149-1156.

Chapter 6

Self-Organization

Self Organization of a Mechanical System

SHIGERU KOKAJI, SATOSHI MURATA, and HARUHISA KUROKAWA

Mechanical Eng. Laboratory, AIST, MITI, Tsukuba, 305 Japan

Abstract

A design method is discussed for a mechanical system which is composed of homogeneous units. Two examples of the system, Fractal Machine and Fractum Machine are introduced. Fractal Machine is composed of variable length links and it can move on a plane. When it is decomposed into several subsystems, each subsystem still hold the capability to move as each of subsystems reorganizes itself. Fractum Machine is composed of units which can change their connection autonomously. It can change its form to construct a target configuration, starting from a random pattern at the beginning. A general scheme for designing this type of mechanical system is discussed and the idea of micro-rules and a global index is proposed.

Keywords: cell structured machine, distributed control, form generation, self repair, self organization.

1. Introduction

There are two types of systems which are called "distributed autonomous machine system" (Fig. 1). One type is a system which is composed of similar units. These units are connected, often physically, to realize a group function. In this type, single unit is made simple and its function is very limited when it works alone[1]. We call this type as "A type."

The other type is a system which is composed of certain number of universal units or robots. Each unit is physically independent usually and it can carry out basic functions when it works alone[2]. We call this type as "B type."

In this paper, we would like to propose a new design method for a machine system of "type A." That is, in short, to use only one kind of units to construct a machine. It may sound very strange but when we think that there are many creatures and organs which are made of similar cells only, it is not a brand new idea but a "conventional" design method in the biological world.

If it becomes possible to make a machine combining just one type of units, then, manufacturing process will be very simplified. Also maintenance process will become simple as well. If we have enough spare units then we can repair the machine by changing the malfunctioning part with a spare part.

There is, however, no guarantee that we can design this type of machine to realize useful functions. We will discuss the possibility and introduce several examples which have been designed and assembled in our study.

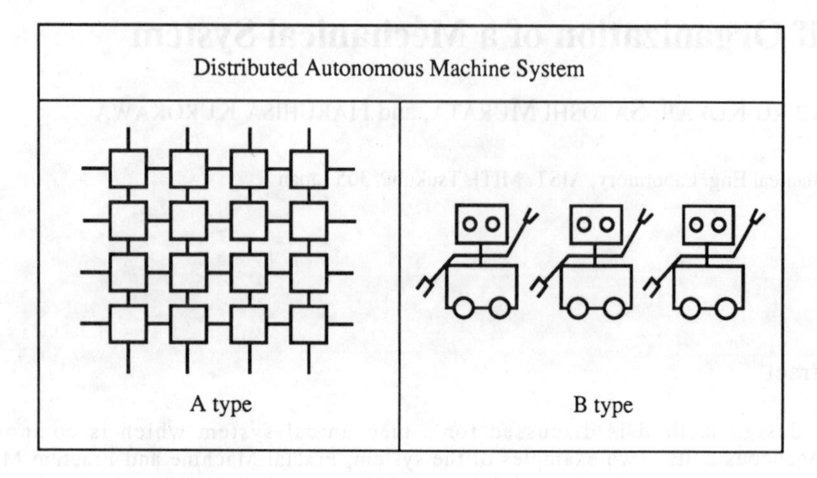

Fig. 1 Two aspects of distributed system

2. Homogeneous Unit

We would like to extend the definition of the characteristics of "type A" machine as follows.

1) The machine is composed of only one type of units.

2) Each unit has intelligence and can communicate with neighbor units.

We keep the first condition strictly. There is no supervisory control system in the machine. If the unit has software, it is assumed that the software of each unit is identical at the beginning. All the units of the machine are the same in hardware and software.

If all the units keep working in the same manner, the group function will be limited to very simple one. We try to realize that these units differentiate according to the location in the system so that useful functions can be organized with the cooperation of these units.

3. Fractal Machine

"Fractal Machine[3]" has been designed to realize the two characteristics given in the previous section. Each unit is a variable length link and connected to its neighbors at pivots. (Fig.2)

The most charming point of this machine is that there is no limitation in its size. At the beginning, the member units communicate with neighbor units and find the size of the machine themselves. They also find their location in the system and decide the operation mode accordingly. After these initialization process, the machine can walk on a plane.

Suppose that "Fractal Machine" of certain size is decomposed into several smaller parts. If one of small parts is reset, then its member units find the new size and decide their mode and can walk again. That is, each decomposed part works as a complete system (Fig. 3).

Fig. 2 Fractal Machine

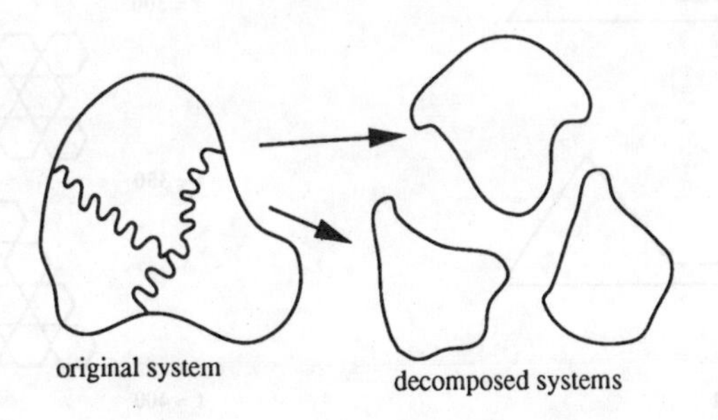

Fig. 3 Decomposition of Fractal Machine

4. Fractum

In the case of Fractal Machine, the connection between the units is done manually. If connection and disconnection are done by units themselves, the system will become very flexible.

We designed another machine to realize this flexibility. This machine system has the characteristics given in the section two and its unit can connect with other units using electromagnetic force. This unit is named as "Fractum[4]" (Fig.4). This automatic connection/disconnection capability makes following functions.

1) metamorphosis

The system can change its form by changing the connection between the units. We developed an algorithm so that the units can realize a given global form starting from any initial form. Each unit communicates with its neighbors and decides to move or not to move using local evaluation function. Once they succeed to build the target form, the form becomes stable. This stability gives useful function in several ways. "Self-assembly" or "self-organization" can be realized by this. The information about the target form is given to each unit at the beginning and from any initial condition (connection), they

Fig. 4 Fractum

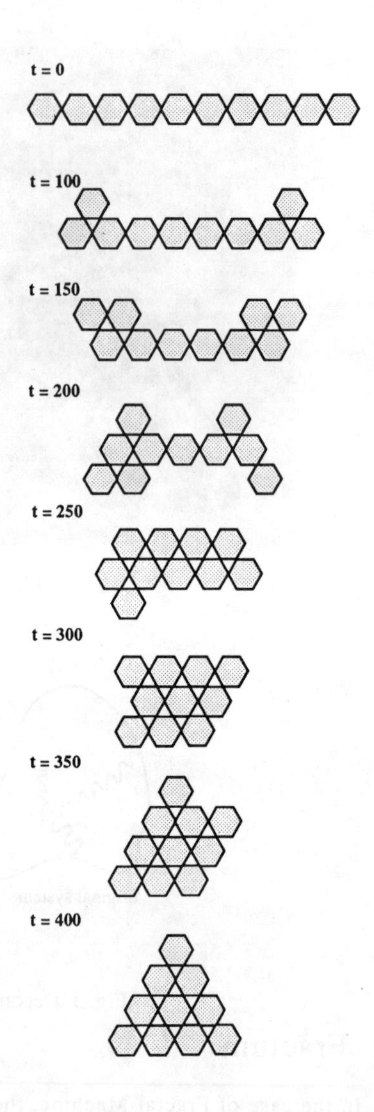

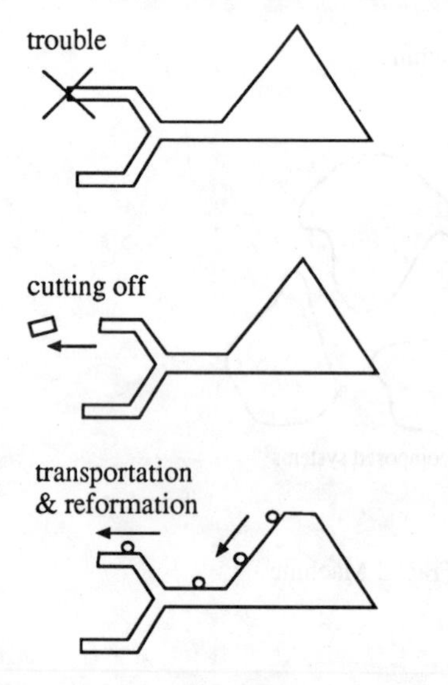

Fig. 6 Self-repairing process

FIg. 5 Metamorphosis function using Fractum

automatically change the connection and realize the target form. Fig. 5 shows the process which converges to the triangular form given as the target. Also, this principle can be applied to realize self-repairing process (Fig. 6).

2) generation of movement

The automatic connection/disconnection capability can be used to transfer a unit along the edge of the global system. By repeating this towards certain direction, the system can realize transportation of itself (Fig. 7).

3) flexible structures and actuators

If the connection strength between the fractum is set loose, flexible structures are constructed. If it is controlled in active manner, then, an actuator can be built (Fig. 8).

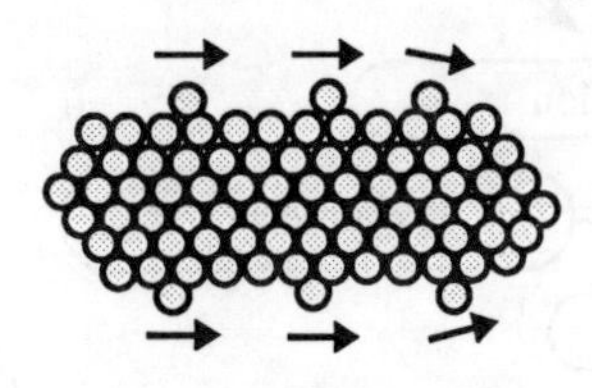

Fig. 7 Transportation process

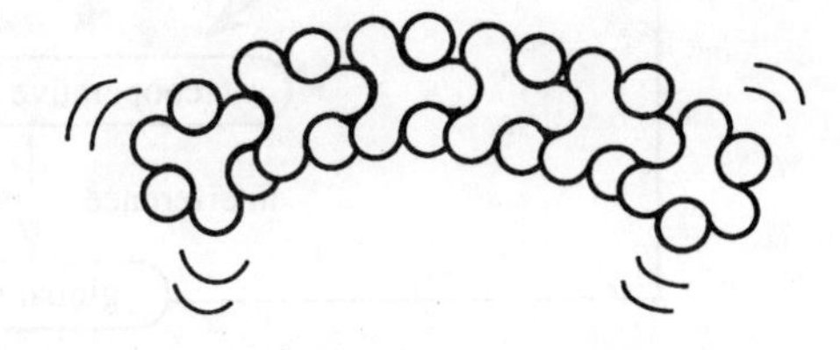

Fig. 8 Actuation using Fracta

5. Summary

We can summarize the connection of our "type A" machine as Fig. 1. We make our efforts in designing the operation of the unit. This operation is defined by a set of rules. The functions of these units as a whole (the group function) are realized through the interaction of units. In other word, the group functions are realized as the integral of the unit operation. By this, the rules of each unit can be called as "μ-rules." From its nature, these μ-rules are expressed implicitly and often difficult to be understood in explicit way. We have to develop a methodology which helps us to design the μ-rules of a unit from the group function we intend.

If we analyze the design of our example systems, it is found that a feed back mechanism

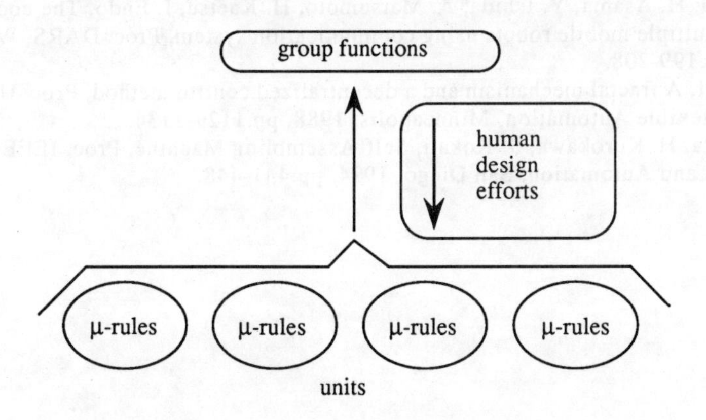

Fig. 9 Embedded μ-rules and group functions

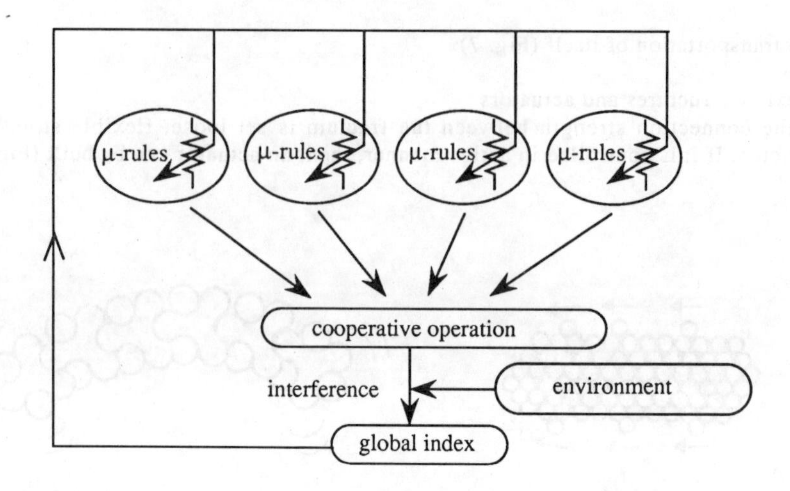

Fig. 10 Toward Robustness

is realized in the "type A" machine. Fig. 10 explains the mechanism, in which "a global index" is formed as a result of cooperative action of units and this modifies the μ-rules so that the system can realize an adaptive operation. In the case of Fractal Machine, the system size can be thought to be this global index. The adaptability of the system will be enhanced if the global index is designed to be affected by the interference between the system and the environment. We envisage that these mechanism will make the homogeneously structured system more practical and robust.

References

[1] G. Beni, S. Hackwood, Coherent swarm motion under distributed control, Proc. DARS, Wako, Japan, 1992, pp.39-52.

[2] K. Ozaki, H. Asama, Y. Ishida, A. Matsumoto, H. Kaetsu, I. Endo, The cooperation among multiple mobile robots using communication system, Proc. DARS, Wako, Japan, 1992, pp.199-208.

[3] S. Kokaji, A fractal mechanism and a decentralized control method, Proc. USA-JAPAN Symp. Flexible Automation, Minneapolis, 1988, pp.1129-1134.

[4] S. Murata, H. Kurokawa, S. Kokaji, Self-Assembling Machine, Proc. IEEE. Int. Conf. Robotics and Automation, San Diego, 1994, pp.441-448.

Evolutional Self-organization of Distributed Autonomous Systems

YOSHIO KAWAUCHI[1], MAKOTO INABA[1], and TOSHIO FUKUDA[2]

[1]Applied Tech. Dept., Toyo Eng. Corp., Chiba, 275 Japan
[2]Dept. of Micro System Eng., Nagoya Univ., Nagoya, 464-01 Japan

Abstract

This paper discusses a framework and algorithms for reconfiguration of distributed autonomous systems(hereinafter called DAS). Most of DASs have ill-structured systems which mean non-linear and complex systems. The proposed method based on Genetic Algorithms(hereinafter called GA) is applicable to optimize such a system. Many parameters of ill-structured systems are hierarchically mapped into parameters of the method as chromosomes in GA. The parameters of the system regarded as chromosomes are updated to increase(or decrease) fitness of each state of the system. As one example, the proposed method is applied to realize the self-organization of intelligent system of cellular robotic system(hereinafter called CEBOT) previously studied by authors. Both hardware and software of CEBOT consists of many functional units called "cells". In this paper, self-organization of the intelligent system of CEBOT is mainly discussed. The proposed intelligent system consists of many kinds of knowledge sources, which have simple levels of data and intelligence together with some blackboards which dynamically organize a hierarchical structure. It is assumed that each intelligent unit is allowed to locally communicate with the other units to exchange data and information. Therefore, communication load of units can be defined. It is desired that the intelligent system can reorganize in order to maximizing (or minimizing) performance index calculated by using load of each unit. The efficiency of the proposed method is shown by some simulations of maximizing (or minimizing) the performance index of the system during self-organization.

Keywords: Cellular robotic systems, evolutional programming, self-evolution, self-organization, optimization

1. Introduction

The authors have studied and presented a Cellular Robotic System(hereinafter call CEBOT), which has the capability of self-organization, self-evolution, and functional amplification [1-6]. System components called "cells" are defined as they have a simple function, database, and intelligence. A cell has the limitations of both hardware and software capacity. If many fundamental cells conglomerate to organs, the organs can perform complicated functions(refer to Fig. 1). Idea of CEBOT can be expanded to a "Robotic Society" organized by numerous autonomous robotic systems. CEBOT uses the same principle as elementary functional cells, which have a simple mechanical function and a coupling mechanism to connect with other cells. By connecting many cells to modules or large structures, difficult tasks can be executed. CEBOT system components are widely distributed, autonomous and freely exchangeable, which leads to fault-tolerance and self-repair capabilities. Therefore, it is important to study evolutional self-organizations of CEBOT to realize flexibility, adaptability and the functional amplification[6].

In the following chapters, the fundamental architecture of the cellular intelligence system of CEBOT with evolutional self-organization is proposed. The intelligent system consists of many units. The units are classified into two groups called "knowledge cell" and "blackboard cell". Algorithms installed into each blackboard cell for self-organization are also proposed. The proposed algorithms are mainly based on "Genetic Algorithms(hereinafter called GA)" learnt from mechanisms found in the natural selection. In the algorithms, system configurations which have

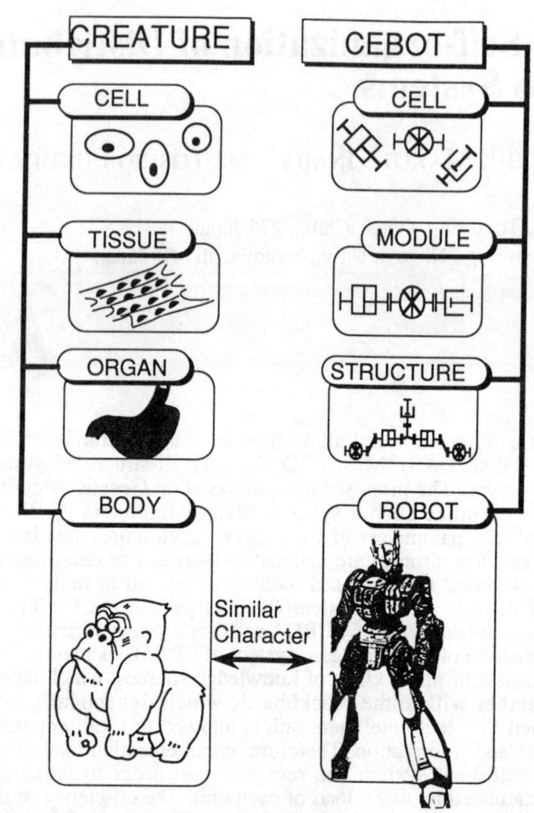

Fig. 1: Analogy of CEBOT and Biological Systems

"PATLABOR 2 the Movie"
©HEADGEAR/EMOTION/TFC/ING

hierarchical structure are encoded as chromosomes in GA. Some performance indexes(hereinafter described PI) such as communication load of each cells evaluated by entropy are defined, and the algorithms aim to maximize(or minimize) the PI of the system. According to operations of the algorithm such as Reproduction, Crossing-over, and Mutation , system configurations of the cellular intelligence are reorganized, and optimal(or semi optimal) organization is evolutionally achieved in simulational experiments.

2. Fundamentals and Models of Cellular Intelligence System

This chapter describes a fundamental configuration of the cellular intelligent system, which consists of many kinds of knowledge sources and hierarchically connected blackboards cells.

2.1. Hierarchical system architecture

As addressed in the previous chapter, CEBOT consists of many cells. It is desired that each cell can be controlled distributedly and autonomously. We have already designed a hierarchical system configuration and distributed control laws based on an idea of servo control[6]. The system configuration is expressed in Fig. 2. The system consists of many cells which form hierarchical layers. Each cell has its own goal and changes its state to satisfy the goal. We applied the architecture and the control laws to motion control of many cells which carry one object coordinatedly[6].

In this paper, we apply the proposed system architecture to intelligent system of CEBOT, which is desired to have capability of reconfiguration, knowledge amplification and evolution.

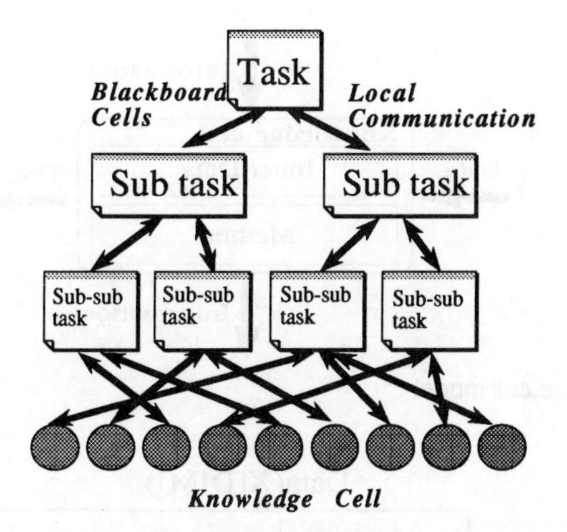

Fig. 2: A dynamically reconfigurable intelligent system composed of knowledge cells and blackboard cells

Figure 2 shows a architecture of the cellular intelligence system of CEBOT which composed of knowledge cells and blackboard cells. The blackboard cells organize a hierarchical configuration as shown in Fig. 2. The knowledge cells form the bottom layer, and the blackboard cells construct multi-layers above the knowledge cells. The knowledge cells are clustered into groups, and each group has one blackboard cell as a common working memory. In the same way as knowledge cells, the blackboard cells form similar groups, the group has one blackboard cell on the next upper layer to them. On the top layer, there is one blackboard cell as a master, which can receive a task from others. The given task is decomposed into many subtasks, the subtasks are also broken down into sub-sub-tasks. Each subtask is assigned to a blackboard. The given task is carried out as many subtask executions on each blackboard cell.

The subtasks on the blackboard cells are gradually carried out in each local group of blackboard cells and knowledge cells. The blackboard cell fills a role of "blackboard[12]" toward the lower units. The blackboard cells also play a role of "knowledge source" toward the next upper blackboard cell because each blackboard cell can read and write data on the upper blackboard cell. The definitions and functions of these units are described in the following section.

2.2. Definitions of a knowledge cell and a blackboard cell

As reported in the previous section, the proposed intelligent system of CEBOT is organized by some "knowledge cells" and "blackboard cells", and forms hierarchical architecture. The functions of knowledge cell are defined as follows (refer to Fig. 3).

The knowledge cell has its own "database" and "knowledge"(hereinafter called "inner data" and "method", respectively). The "method" is corresponded to a subroutine of a program, and the "inner data" is regarded as a local data. The knowledge cell is nputted and output both "data" and "information" as shown in Fig. 3. The "data" is regarded as global data which is displayed on the upper blackboard cell. The "data" is composed by two part, which are "value" and "attribute" as shown in Fig. 4. The knowledge cell and the blackboard cell are activated only when there are data of which attribute are same as their attribute on the blackboard cell in the next upper layer. There is no supervisor to activate the knowledge cells and blackboard cells, therefore the system is considered to be distributed autonomous system.

The blackboard cell is modeled as shown in Fig. 5. The blackboard cell mainly consists of two parts, which are "Log" and "Data Pool". The frequency of activation of knowledge cells and blackboard cells in the same group are recorded in the "Log" part of a blackboard cell in the upper layer. The "Data Pool" part is composed of two parts, which are "Required Data Pool" and

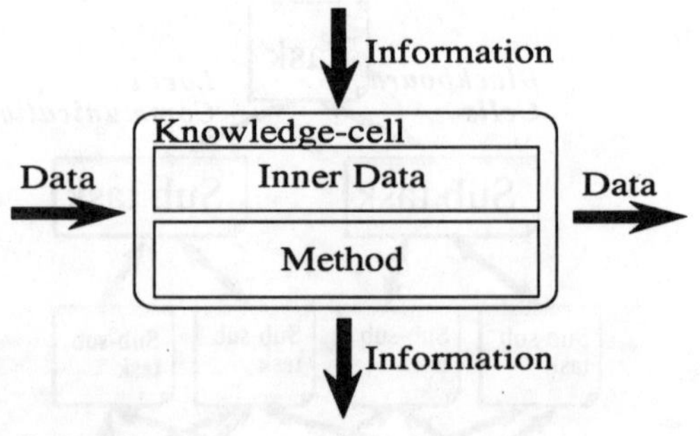

Fig. 3: A knowledge cell model

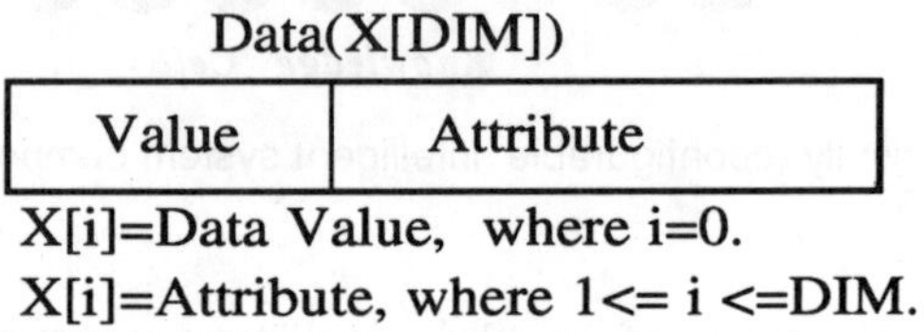

Data(X[DIM])

Value	Attribute

$X[i]$=Data Value, where $i=0$.

$X[i]$=Attribute, where $1 <= i <= DIM$.

Fig. 4: Data structur of IN/OUT of a cell

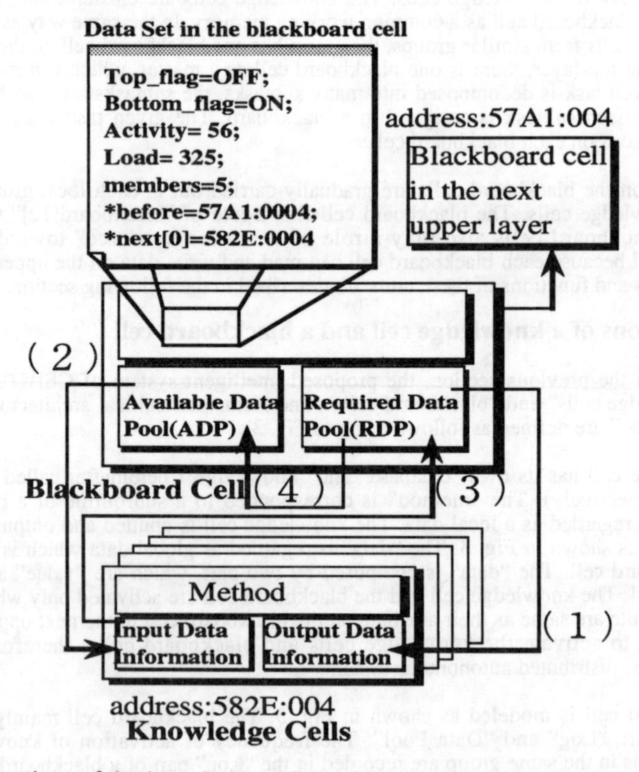

Fig. 5: Basic actions of the knowledge cells and the blackboard cells

"Available Data Pool" (hereinafter called "RDP" and "ADP", respectively). The knowledge cells and the blackboard cells in the same group can locally access to the blackboard cell in the upper layer.

Each blackboard cell locally possesses its own data as shown in Fig. 5. The following is the description of each data type.

(1)Top_flag: If the blackboard cell is in the top layer, the Top_flag is turned ON, otherwise is made OFF.
(2)Bottom_flag: If the blackboard cell is in the bottom layer, this flag is turned ON, otherwise is made OFF.
(3)Activity:This value expresses frequency of usage of the cell throughout task executions.
(4)Load:This value expresses loads of cells such as communication load, processing load and so on.
(5)members: This value is the number of blackboard cells which is connected to the knowledge cell in the lower layer.
(6)*before: The pointer of the blackboard cell in the upper layer connected to the knowledge cell.
(7)*next[members]: The pointers of the blackboard cells(or knowledge cells) in the lower layer connected to the blackboard cell. The dimension of the data type of *next[] is equal to the number of members connected to the blackboard cell denoted by "members".
(8) Others: Any kinds of indexes can be considered as inner data of cells such as correlation index, mutual information amount, and so on.

2.3. Basic actions of the knowledge cell and the blackboard cell

The basic actions of cells are shown in Fig. 5. When there are data of which attribute are same as attributes of knowledge cells or blackboard cells, they are activated autonomously, and access to "ADP" to search the proper data for them. If there is no proper data for the activated cells, the cells request the blackboard cell in the upper layer to list the data on its "RDP".
When the activated cell can get a complete kind of data from "ADP" to output the data listed on the "RDP", the data on the "RDP" is deleted and is added on the "ADP" of the blackboard cell in the upper layer. As described in the section 2.1., each blackboard cell initially has its own subtask to be carried out in its "RDP". When the subtask broken down by the blackboard cell in the upper layer is removed from the "RDP" to the "ADP", the blackboard cell recognizes that the subtask has executed.

3 Definitions of Performance Indexes and the Measures of Complexity

3.1 Activity

It is assumed that frequency of usage of cells are related to types of given tasks. For example if a given task is "Move from Tokyo to New York", it is regarded that knowledge related on "transportation" and "geography" are mainly used. It is defined that the activity of cell-i is hereinafter expressed by a(i). It is assumed that each blackboard cell can know(or measure) its own activity depending on kinds of given tasks. Activity of cell-i is defined as Eq. (1)

$$a(i) = (1/m)\cdot\sum_{j=1}^{m} a(j) \qquad (1),$$

where m is the number of cells connected to cell-i as shown in Fig. 6.

3.2 Communication load

As described in sect. 2.3, it is assumed that given tasks can be carried out by communication among cells. If a cell connects to too many cells with high activity, the blackboard cell is accessed many times by the cells as a common working memory. It is assumed that the communication amount of a blackboard cell increases in proportion to number of cells and activity of the connected cells. We define communication load of cell-i(hereinafter expressed by L(i)) can be calculated by using Eq. (2) as one of the load of cells.

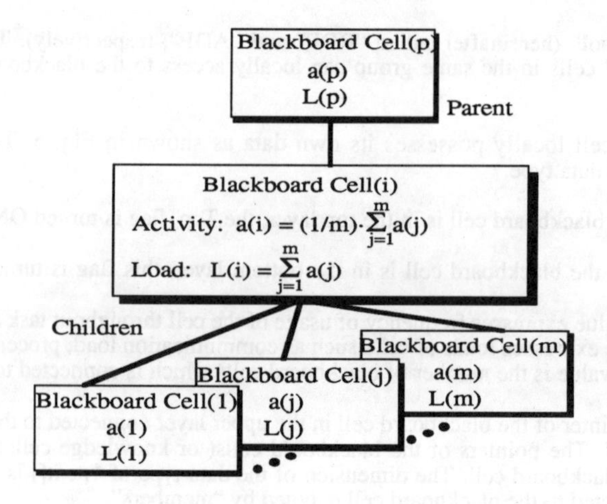

Fig. 6: Activity and communication load

$$L(i) = \sum_{j=1}^{m} a(j) \qquad (2).$$

3.3 Shannon's entropy

While cells execute given tasks by using communication on the hierarchically organized blackboard system, each cell has its own communication load expressed by Eq. (2). To evaluate the total system performance, we apply the theory of entropy to a performance index for realizing evolution of the system.

Generally, Shannon's Entropy H(A) is defined as Eq. (3),

$$H(A) = -\sum_{i=1}^{K} pi \cdot \ln(pi) \qquad \text{(nat)} \qquad (3),$$

where A is a discrete process which has K states, pi implies a probability of event(i). By using the communication amount of each blackboard cell, the entropy of the total system (hereinafter expressed by H(system)) can be defined as Eqs. (4) and(5)

$$H(\text{system}) = -\sum_{i=1}^{N} qi \cdot \ln(qi) \qquad \text{(nat)} \qquad (4),$$

$$qi = L(i) \Big/ \sum_{i=1}^{N} L(i) \qquad (5),$$

where N is the total numbers of blackboard cells in the system.

It is assumed that the blackboard cell(i) can observe the frequency of activation of blackboard cell(j) in one lower layer, therefore the cell(i) can count out a(i) by using Eq.(1).

4 Evolutional reconfiguration of systems based on Genetic Algorithms

CEBOT is required to deal with many kinds of tasks under various surroundings. Therefore, activity of each cell is changed at each task execution. To optimize the performance indexes of the system, it is desired for blackboard cells to reconfigure depending on situations. Genetic Algorithms have been studied by many researchers in order to give evolvability to systems[5,9-11]. The authors have previously studied and reported a method based on GA for distributed intelligence system[5]. In the previous paper, we have proposed Genetic Cell Production Algorithms(hereinafter call GCPA) for determining connection intensity among cells, not for dynamically reconfiguration. In this paper algorithms based on GA which is used for reconfiguration of the cellular intelligence system is newly proposed and discussed. It is regarded that GCPA is the algorithm for local evolution and organization, and algorithms proposed in this paper is used for global evolution, because the system is totally reorganized by using them.

4.1 Coding

Generally, chromosomes in GA are simple binary vectors. In these simulations, each chromosome has to express a hierarchically connected cellular intelligence as shown in Fig. 2. To express the tree architecture, data structure of each blackboard cell and knowledge cell defined in sect. 2.2 are stored in a file as a chromosome in GA. The data structure of cells possess "number of members", "address of its parent", "address of its children", and so on defined in sect. 2.2. By recording all data of cells in a file, tree of cells as shown in Fig. 2 can be reproduced. A chromosome in the proposed algorithms are corresponded to a file which records all data of cells. Therefore, the number of created files are equal to the number of chromosomes assumed in the algorithms.

4.2 Reproduction and crossing-over

A reproduction of the algorithm is corresponded to duplicating files. Crossing-over is carried on a chromosome(i) as shown in Fig. 7. Two candidates to be exchanged are randomly selected in a chromosome(i). If one of the selected blackboard cell belongs downstream of the other candidate, the selection is retried not to make ill-structured chromosomes.

4.3 Mutation

In the proposed system, the number of layers and blackboard cells are dynamically changed according to the kinds of given tasks and environment. We defined two types of organizations as follows;
(1) Horizontal organization: To change the number of blackboard cells in a layer.
(2) Vertical organization: To change the number of layers.
These two organizations are realized in the proposed algorithm as described in followings.
We assumed two modes which are "Cell Production" and "Cell Annihilation".

(1)Cell Production: In the first step, a blackboard cell is randomly selected from a chromosome. If the selected cell has more than four members as its children, the children are divided by two groups. In order to avoid making a group of which member is one, the minimum number of members of the selected cell is set at four. One group is suspended from the selected cell, and the other group is made member of a newly produced blackboard cell in a same layer of the selected cell as shown in Fig. 8.

(2)Cell Annihilation: After one blackboard cell is randomly selected, the selected cell is annihilated ,if top_flag of the cell described in sect. 2.2 is not ON. All of members of the deleted cell are grouped in the parent cell of the annihilated cell as shown in Fig. 9.

These two operations which are cell production and cell annihilation are carried out in a probability of mutation set previously.

4.4 Selection

System is evaluated by using some performance indexes such as entropy discussed in sect. 3.1. If the number of chromosomes are set at arbitrary number hereinafter expressed by NC, total number of files amount to 2NC because of reproduction. After calculation of performance indexes of all chromosomes, NC chromosomes are selected as chromosomes of the next generation in order of their value of performance indexes. Therefore number of chromosomes of next generation also start at NC.

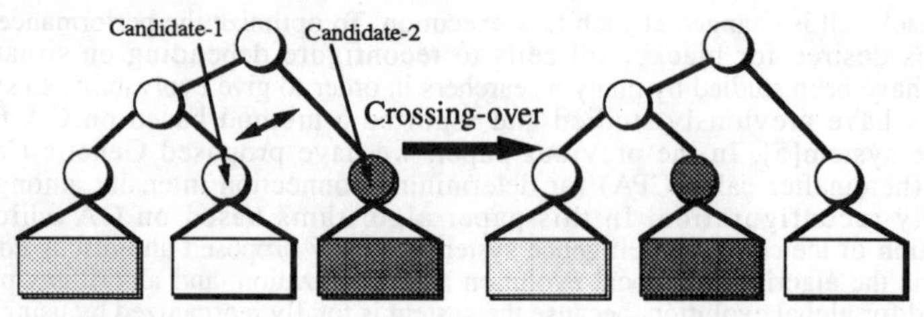

Fig. 7: Crossing-over

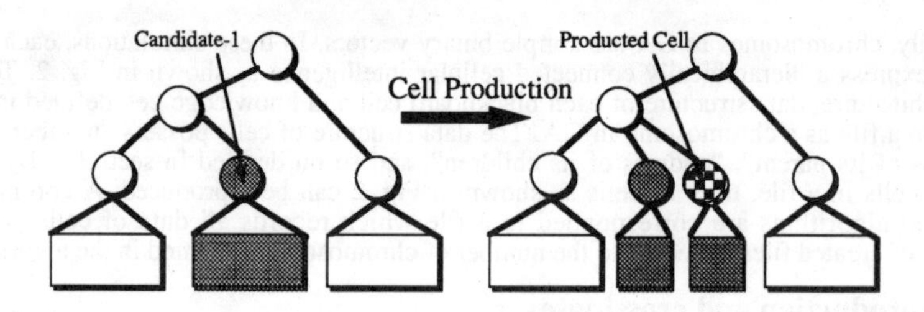

Fig. 8: Cell creation

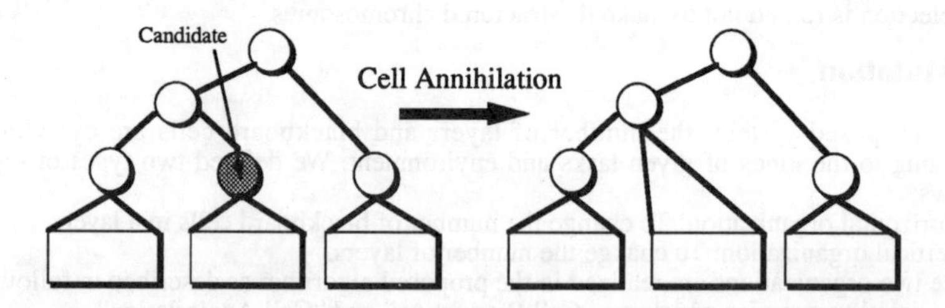

Fig. 9: Cell annihilation

5. Simulational experiments

By using a simple model of the proposed system, the system behavior is simulated, and the system performance is evaluated by using a performance index defined in sect.5.1.

5.1 Assumptions and procedure

Based on the model of cellular intelligence defined in chapt. 2, simulational experiments are carried out. In the simulations, the number of knowledge cells are fixed at 32(hereinafter expressed by NKC). Activity of each knowledge cell are randomly set at first step of the simulation. A initial organization of blackboard cells are shown in Fig. 10. As shown in Fig. 10, one blackboard cell is assigned to a knowledge cell, and two blackboard cells are allocated in one upper layer to observe the thirty two blackboard cells. One blackboard cell is set in the top layer as shown in Fig. 10. The blackboard cells in the bottom layer are randomly connected to knowledge cells in every initial state of chromosome. Performance index to evaluate the total system(hereinafter denoted by PI(system)) is defined as Eq. (6) based on entropy discussed in sect. 3.3.

PI(system) = H(system) / ln(N) (nat) (6),

where H(system) can be calculated by using Eqs.(4) and (5). N is the total number of blackboard cells in the system.

Entropy of the system defined by Eq. (4) is at maximum value which is ln(N) in every event is equally occurred in a probability(1/N)[8]. Therefore, we see for all PI(system) that

$$0 \leq \text{PI(system)} \leq 1 \qquad (7).$$

PI(system) increase, if communication of each blackboard cell is equally loaded, and vice versa. In these simulational experiments, we aim at well dispersified communication load of each blackboard cell. Therefore, chromosomes which have high performance index denoted by Eq. (6) are selected as selected chromosomes of the next generation.

We set NC, NKS, Mutation Probability and Cell Creation Probability at 25, 32, 30% and 70%, respectively. Mutation is assumed to be operated in 30%. If it is determined to operate mutation on a chromosome, Cell Creation is executed in 70%, otherwise Cell Annihilation is operated on the chromosome. We randomly make twenty five initial states of cellular intelligence system as chromosomes. One of them is expressed in Fig. 10. These chromosomes are duplicated, and some operations such as crossing-over and mutation defined in chapter 4 are executed on every chromosome. After all operations, twenty five chromosomes which have high performance index defined by Eq. (6) are selected as chromosomes of next generation. This procedure is iteratively executed. Simulational results are shown and discussed in the following section.

5.2 Results and discussions

Figure 11 shows one example of chromosome in 300th generation. In the initial state, each chromosome has thirty five blackboard cells which organize three layers expressed in Fig. 10. In Fig. 11, the number of blackboard cells and the number of layers are seventeen and eight, respectively. Value of averaged PI(system) in each generation from initial to 300th are plotted in Fig. 12. As shown in Fig. 12, averaged PI(system) tends to increase with generation. In 300th generation, averaged PI(system) is almost equal to one which is a maximum value of the index as described in the previous section. It shows that every blackboard cell has almost same communication load. To verify the dispersion of communication load, communication loads of each cell are plotted in every 100th generations in Fig. 13. In Fig. 13 the y axis is communication load of each blackboard cell calculated by Eq. (2), and the x axis is the ID number of blackboard cells in

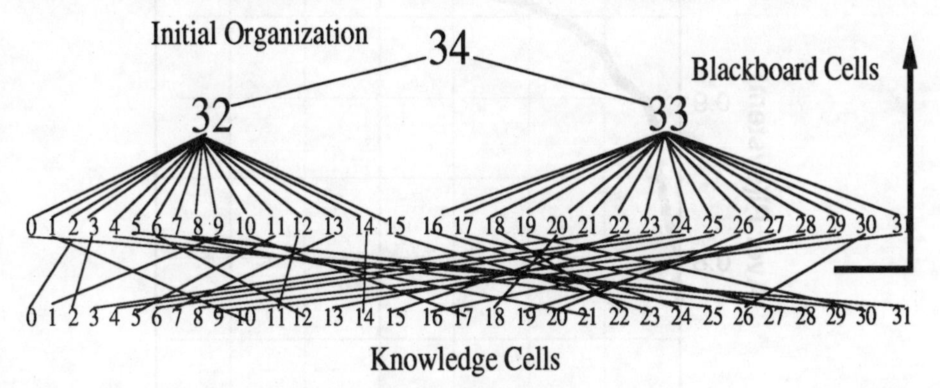

Fig. 10: One example of chromosome in initial state

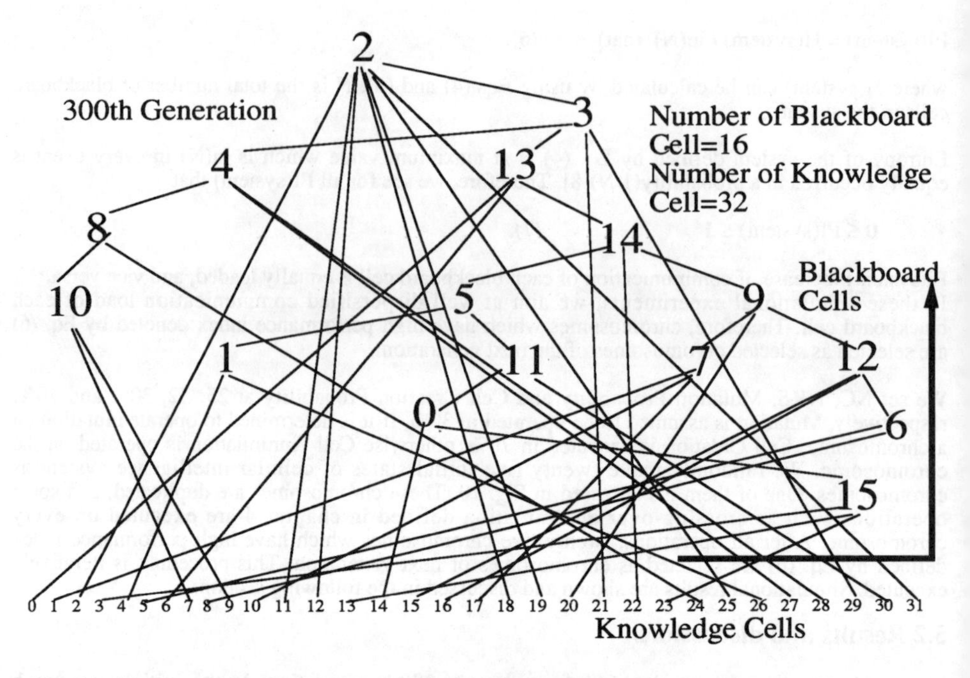

Fig. 11: One example of chromosome in 300th generation

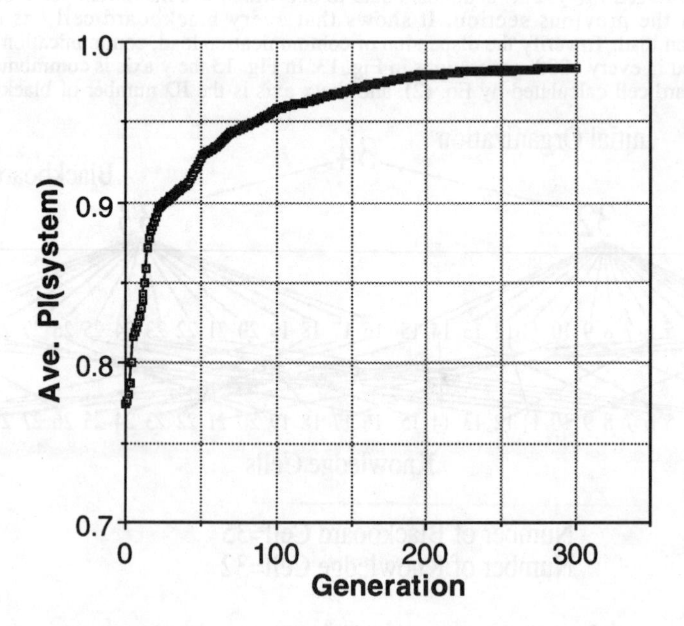

Fig. 12: Generation vs. PI(system)

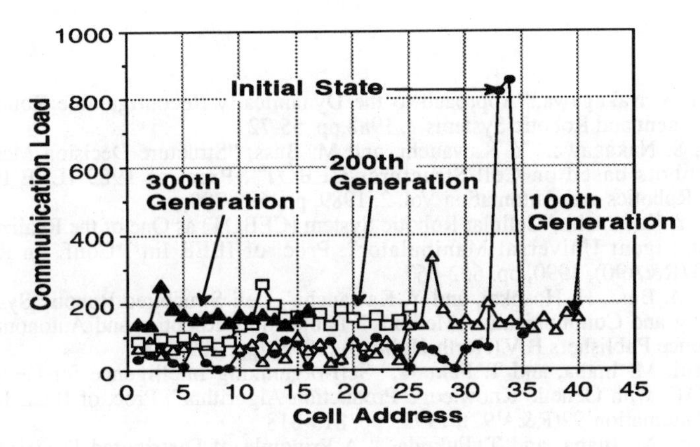

Fig. 13: Communication load of each blackboard cell in every 100 generations

Table 1: Evaluation of system in every 100 generation

	Initial	100th	200th	300th
CELLS	35	40	25	16
MAX	854.4	336.0	256.8	253.6
MIN	2.0	22.0	70.0	135.6
AVE	98.7	102.8	135.1	164.2
STDEV	187.1	59.7	46.3	28.1

each generation. In the initial state denoted by ● in Fig.13, some cells such as cell(33) and cell(34) have high communication load, and communication loads of the other cells are not equalized well. However, communication load of each cell is well equally allocated in 300th generation expressed by ▲ in Fig. 13 .Table 1 shows the total number of blackboard cells, maximum communication load, minimum communication load, averaged communication load and standard deviation of the communication loads in each generation. Table 1 statistically shows that allocation of communication load is well achieved by using the proposed algorithm based on GA.

6. Conclusions

This paper addresses a fundamental architecture and basic algorithms of the cellular intelligence system of CEBOT. The points of this paper are described as follows;

(1) The system consists of many units which are knowledge cells as knowledge sources and blackboard cells as common working memory and knowledge source.

(2) The cells configure a hierarchical architecture. Cells makes groups composed of one blackboard cell and some units(blackboard cells or knowledge cells) in the next lower layer. A given task is decomposed into many subtasks, and they are assigned to each blackboard cell. Only local communication among same group is allowed.

(3) In order to make the system flexible and adaptable to many kinds of task, it is desired to reconfigurate the hierarchical architecture dynamically and evolutionally. As one example of algorithms, we propose the algorithm based on GA. The efficiency of the algorithm is demonstrated by simulatinal experiments.

References

[1] T. Fukuda, S. Nakagawa,:"Approach to the Dynamically Reconfigurable Robotic System", Journal of Intelligent and Robotic Systems 1, 1988,pp. 55-72
[2] T. Fukuda, S. Nakagawa, Y. Kawauchi, and M. Buss, "Structure Decision Method for Self Organizing Robots based on Cell Structures-CEBOT", Proc. of 1989 IEEE International Conference on Robotics and Automation Vol. 2, 1989, pp. 695-700 .
[3] T. Fukuda., Y. Kawauchi.,"Cellular Robotic System (CEBOT) as One of the Realization of Self-organizing Intelligent Universal Manipulator", Proc. of IEEE Int'l Conf. on Robotics and Automation'90 (R&A90), 1990, pp. 662-667
[4] T. Fukuda, M. Buss, H. Hosokai, and Y. Kawauchi," Cell Structured Robotic System CEBOT Control Planning and Communication Methods", Publication, Robotics and Autonomous Systems 7, Elsevier Science Publishers B.V.(North-Holland),(1991), pp. 239-248
[5] Y. Kawauchi, M. Inaba, and T. Fukuda," Self-organizing Intelligence for Cellular Robotic System "CEBOT" with Genetic Knowledge Production Algorithm", Proc. of IEEE Int'l Conf. on Robotics and Automation'92(R&A'92), 1992, pp. 813-818
[6] Y. Kawauchi, M. Inaba, and T. Fukuda, " A Principle of Distributed Decision Making of Cellular Robotic System(CEBOT), Proc. of IEEE Int'l Conf. on Robotics and Automations,1993
[7] Christopher G. Langton," Life at the Edge of Chaos", Artificial Life II, SFI Studies in the Sciences of Complexity, vol. X, edited by C. G. Langton, C. Taylor, J. D. Farmer, & S. Rasmussen, Addison-Wesley, pp. 41-91,1991
[8] N. Abramson(Translated by miyagawa) ,"The introduction of information thory ",published by Kougakusha,1949(in Japanese)
[9] J. R. Koza, M. A. Keane, and J. P. Rice," Performance Improvement of Machine Learning via Automatic Discovery of Facilitating functions as Applied to a Problem of Symbolic System", Proc. of IEEE Int'l Conf. on Neural Networks(IJCNN'93), vol. 1, 1993, pp. 191-198
[10] J.H. Holland,"Adaptation in natural and Artificial Systems," Univ. Michigan Press,1975
[11] D.E. Goldberg,"Genetic Algorithms in Search, Optimization, and Machine Learning," Addison Wesley,1989
[12] L. D. Erman, F. Hayers-Roth, V. R. Lesser, and D. R. Reddy, "The Hearsay-II speech-understanding system: Integrating knowledge to resolve uncertainty," Compt. Surveys, vol. 12, June 1980, pp. 213-253

Intention Model and Coordination for Collective Behavior in Group Robotic System

TOSHIO FUKUDA and GO IRITANI

Dept. of Mechano-Informatics and Systems, Nagoya Univ., Nagoya, 464-01 Japan

Abstract

This paper addresses the intention model and its application to an evolution of collective group behavior on decentralized autonomous robotic systems. The decentralized autonomous robotic systems refer to multiple robotic systems including many autonomous robots, such as Cellular Robotic System(CEBOT). The CEBOT, which has been proposed by authors, consists of a number of functional robotic units called "cells." In the research on the CEBOT same as the decentralized robotic systems, it is one of the most important issues to cooperate among many robots for the group behavior and to coordinate the group behavior, since it is important to organize the behavior of the each robot in the multiple robotic systems. In order to organize the behavior in dynamic environment, we proposed a concept of the "self-recognition" for the decision making of the behavior in a robotic group. In addition to the proposed concept, this paper will show the evolutional group behavior incorporated with the intention model. The concept of the coordination of intention means that each robot coordinate with other robots' intention in its local aria by communication. Based on this idea, we present the behavioral evolution of the group robotic systems.

1. Introduction

In recent years, the research on decentralized autonomous robotic systems has been becoming active, because of the expansion of the robotic application fields and the development of computer devices with low cost. The decentralized autonomous systems are regarded as an approach method to organize complex systems with a number of robots. Additionally, the decentralized robotic systems have the ability of the flexibility, redundancy, extendibility, reconfigurablity and so on. Therefore, in laboratories, industrial facilities and so on, many researches on the decentralized robotic systems have been carried out[1-4]. On the other hand, we proposed the Cellular Robotic System(CEBOT) as one of the decentralized robotic systems [5-7]. The CEBOT is an autonomous distributed robotic system composed of a number of functional robotic units called "cells." Figure 1 illustrates the concept of the CEBOT.

The CEBOT has three characteristics, as follows. The first characteristic is that it is regarded as an autonomous distributed robotic system that can have an optimal/suboptimal configuration composed by many cells. The optimal/suboptimal configuration is organized dynamically by docking and detaching of cells with cooperation. The second one is that it is considered as a self-organizing robotic system that can reconfigure dynamically to carry out given tasks orderly depending on its environment and its given tasks. The third one refers to a "*group robotic system*" composed of many cells. The "*group robotic system*" generates and evolves its behavior with coordination and cooperation among cells. The topic dealt with in this paper focuses on how to evolve and reconfigure the group behavior consisting of multiple robots that carry out plural tasks.

As the research on the analysis of the group behavior, Huberman[8] and Parker [9] analyzed the model of the group behavior, which are closely related the behavior of living creatures[10].

Deneubourge simulated the group behavior of ant-like robots in comparison with robot-like ants[11]. Ueyama showed the structural organization of the CEBOT in dynamic environment, where the cells moved randomly with restricted sensing and communication area[7]. These are mainly researched in Artificial Life[12].

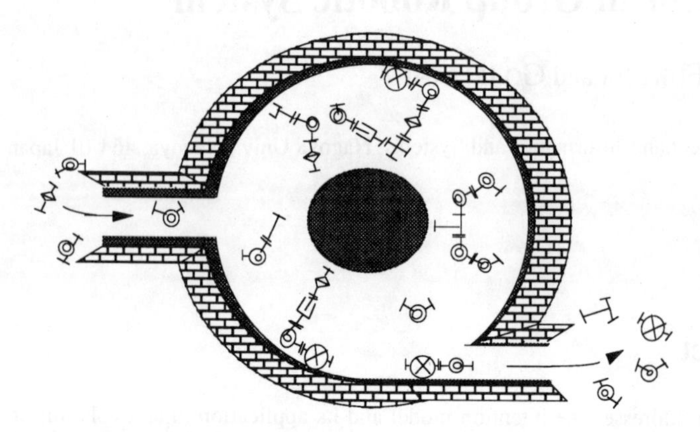

Figure 1 Concept of CEBOT

The research on the planning and decision making as the group behavior has been carried out in several research fields. Especially, the research field on distributed artificial intelligence[13] is closely related to the research on the planning or decision making as the group behavior, since several agents coordinate or cooperate to make decision. That is, the coordination or cooperation can be considered as the group behavior. As the models of planning or decision making, we can see contract net protocol[14], scientific metaphor[15], blackboard model[16], and so on.
In addition, the research on the evolution of the group behavior will be the most attractive research, since the research is closely related to the organization and the evolution of social systems or team work. As the related research, Kube and Zhang presented the evolution of the group behavior, which is based on artificial logic network[17]. On the other hand, Arkin and Balch[18] represented communication of the behavioral state in multi-agents. As the evolution of group, Shaew denoted the learning and adaptation of distributed systems in the distributed artificial intelligence[19], which is based on a genetic algorithm[20].
In this paper, we describe an evolution and organization method of the group behavior in dynamic environment. The group behavior discussed in this paper is emerged by the interaction among robots. The group behavior is carried out by the cooperation between autonomous robots. For the problem of cooperation between autonomous robots, we consider that each robot needs to recognize that the robot itself carries out tasks in the group as a part of the group. So, we proposed a concept of the "*self-recognition*" for the decision making of the group behavior as a basic strategy of group robotic systems[21]. Moreover, if the group robotic system is required to carry out various tasks, each robot has to change its behavior depending on the environment and its given tasks. In this paper we propose the concept of the intention model and its coordination with the self-recognition. The intention in this paper implies the priority of given tasks. The idea of the coordination of intention means that each robot coordinates with other robots' intention in its local area by communication. By using this idea, we present the behavioral evolution of the group robotic systems. The evolution of the group behavior is based on adaptation of each robot for the group behavior, where these strategies are based on the concept of the "self-recognition." In the following chapter we will show some simulation results for a simple task.

2. A Group Robotic System with "Self-Recognition"

2.1 Task Execution in Group Robotic System

In the group robotic system, some kind of master will be required explicitly or implicitly to carry out tasks cooperatively and coordinately. In this case, the master has to give any order to each autonomous robot to carry out given tasks. We shall consider two methods that the master gives a common goal of task to each autonomous robot. The two methods are described as follow:

(1) Sub-task
The first method is that the master decomposes the given task into sub-tasks in detail and gives them to each autonomous robot. It is necessary for master to plan from an initial state to a final state and to clear up various collisions and conflicts in the dynamic environment. If the group robotic system consists of a few robots, it can excuse the given task surely, but the larger the number of cells is, the more complex both interactions between the cells and between cells and its environment are. The physical load is also concentrated on the master or the center of the system. Moreover, the group robotic system has the less ability of the flexibility and the tolerance of breakdown.

(2) Sub-goal
In the second method, the master gives the excusing method of task and sub-goals of the task to each autonomous robot. Each robot of the group robotic system makes a plan reactivity toward the final state of the task depending on the change of its environment and the task conditions with the cooperation and the conflicts between other robots. Taking this method, the physical load of the master reduces and the group robotic system has the more ability of the flexibility and the tolerance of breakdown, but it has the problem how to plan in the group robotic system.
In the two methods, we choose the latter case. Because, we take the matter of autonomy.

2.2 Self-Recognition

When some common tasks are assigned to multiple autonomous robots, the robots should organize groups to carry out the tasks more than execute the tasks individually. Moreover, in order to carry out the required task effectively, it is necessary for the group system to organize the group behavior with cooperation. In this case, the problem how each robot should plan is occurred. To clear up this problem, it is important for each robot to recognize that it is a member of the group system. In case that each robot behaves selfishly without recognizing that it is a part of the group system, it will have bad influence upon behavior of other robots and the possibility of causing deadlock problem. As a result, it has any possibility to decrease the efficiency of the group system. So, it is important for each robot to recognize that it is a member of the group system. Moreover, we have proposed the concept of "self-recognition" that is the important strategy to organize the group behavior. When many autonomous robots execute the task simultaneously in the dynamic environment, each robot always influences other robots and the group system. It is a great issue for each robot to recognize the influence of its own behavior on other robots and the group system. So, it is necessary for each autonomous robot to plan with inferring the influence on other robots and the group systems and to make the efficiency of the group system increase.
In general, the "self-recognition" presents that *each robot infers the influence of its behavior on the group behavior, when it uses the strategy of optimizing the group behavior*.

3. Intention Model and Coordination

In our research, we have treated the case that the group robotic system is given one common task. In this case, when each autonomous robot executes the "self-recognition," each cell changes only its behavior to give the good influence to the group system, and it executes that behavior. As the result, the behavior of group robotic system is organized. On the contrary, the behavior and the task that each autonomous robot should take always change depending on the environment and the state of tasks, when many common tasks are given. So, it is necessary for each autonomous robot to have the standard which shows the priority of tasks in the group robotic system in order to infer the influence of its behavior on the group behavior, that is, the "self-recognition."
In this paper, we propose the concept of the intention model to infer the influence of its behavior on the group behavior. The value of intention shows the priority of tasks in the group robotic system. If the intention of a task is high, the robot executes the task preferentially. Each autonomous robot executes the "self-recognition" with this intention. If the same task is required and the value of intention defers, the behavior of the robot changes.
It is impossible for the group system to organize the behavior, if each robot decides the value of the intention selfishly. So, we propose the coordination of intention. The coordination of intention is

that each autonomous robot shows its intention and coordinates the intention with those of other robots which is in its local area. By the coordination of the intention, the group robotic system has the network of the intention and generates the potential of the intention. In this paper, we show the generation and the evolution with the potential of the intention. Figure 2 shows the flow chart to determine the behavior of each robot with "self-recognition" and the coordination of the intention.

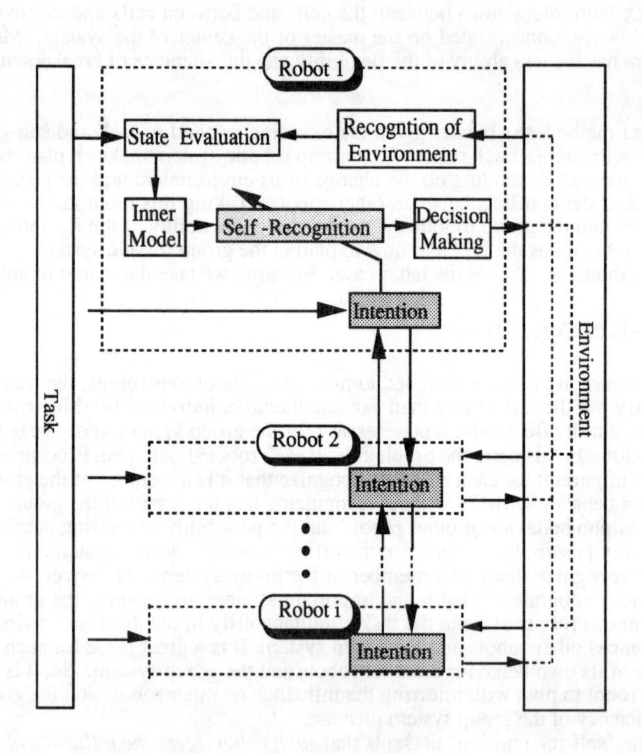

Figure 2 The "self-recognition" with the coordination of the intention

Figure 2 shows that each robot makes the inner model based on the environment and the state of tasks and decides the behavior with the "self-recognition." Each robot takes the value of intention into consideration when it executes the "self-recognition." Equation 1 shows the function of the value of the intention.

$$I_k(t) = \begin{pmatrix} i_1(t) \\ i_2(t) \\ \bullet \\ i_j(t) \end{pmatrix} \qquad i_i(t+1) = \Phi\left(\alpha \bullet i_i(t), \sum_n \beta \bullet i_n(t), M_i(t) \right) \bullet\bullet\bullet (1)$$

where

$\Phi()$: the function of the value of the intention

$I_k(t)$: the value of the intention of robot k for all tasks.

$i_j(t)$: the value of the intention of robot k for task i.

$M_i(t)$: the reward for the task i.

$\sum_n i_n(t)$: the value of the intention of robots which is in the local area of robot k

α: the coefficient of oblivion of robot k

β: the coefficient of connection with robots which is in the local area of robot k

Each autonomous robot renews the value of intention with its past intention, the intention of robots which is in the local area and the reward for the task. By the coordination of intention with local communication, the group robotic system composes the network of the intention and generates the potential of the intention. This group robotic system is the open system. If the group robotic system does not execute the tasks, the value of intention asymptotically approaches 0 by the coefficient of oblivion. As the value of the intention of the group robotic system is given the reward as the energy, it changes to the equilibrium of the energy of the reward and that of throwing away by oblivion. In result, the potential of the intention composes an order and collapses it and recomposes other order depending on the change of the environment and that of the demands of tasks. Using this character, the robots which have high priority of one of the given tasks assemble themselves to that task. So, the group behavior generates and evolves. Moreover, if we give the reward to the group robotic system appropriately, the system can learn the tasks and adapt to the change of the environment and the breakdown of a part of system. Figure 3 shows that the group behavior generates for many tasks.

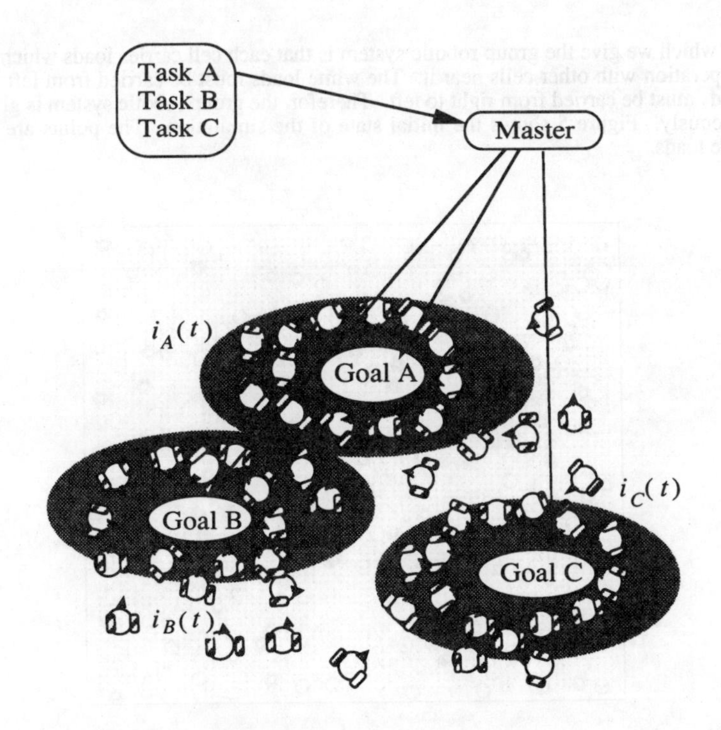

Figure 3 Generation of the group behavior

4. Simulation Environment

This section describes the environment of the simulation. Figure 4 shows the task which we give the group robotic system.

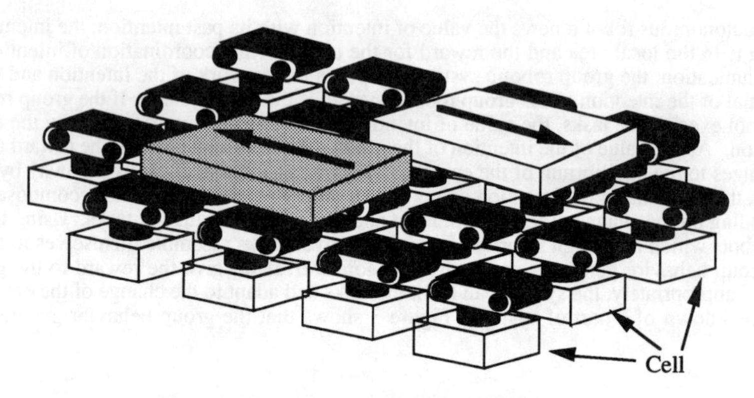

Figure 4 The environment of simulation

The task which we give the group robotic system is that each cell carries loads which move over it with cooperation with other cells near it. The white loads must be carried from left to right. The other loads must be carried from right to left. Therefor, the group robotic system is given two tasks simultaneously. Figure 5 shows the initial state of the simulation. The points are cells and the circles are loads.

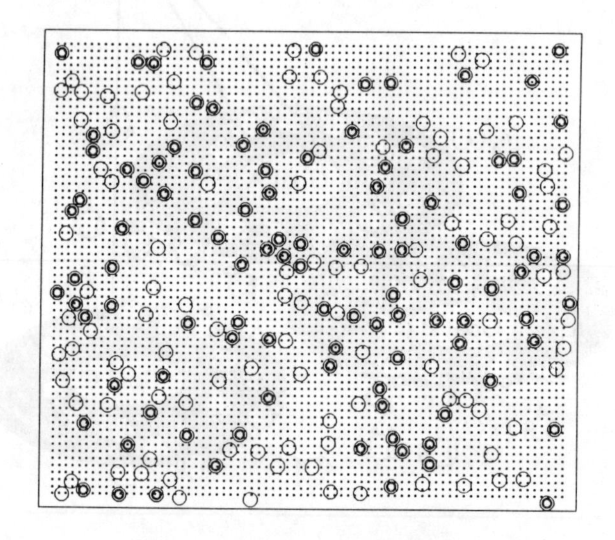

Figure 5 The initial state of the simulation.

The basic rules for each cell are as follows.
(1) Each cell can distinguish the type of the load.

(2) Figure 6 shows that each cell can communicate with four adjacent cells and learn their value of the intention. Equation 2 shows the method of renewal of the intention

$$I_i(t+1) = \alpha \cdot I_i(t) + \sum_j^4 \beta \cdot I_j(t) + M(t) \bullet\bullet\bullet (2)$$

$$-1 \le I_i(t) \le 1$$

where

$I_i(t)$: the value of the intention of cell i

$M(t)$: the reward for the task.

$\sum_j^4 \beta \cdot I_j(t)$: the value of the intention of four adjacent cells

α: the coefficient of oblivion of cell

β: the coefficient of connection with four adjacnt cells

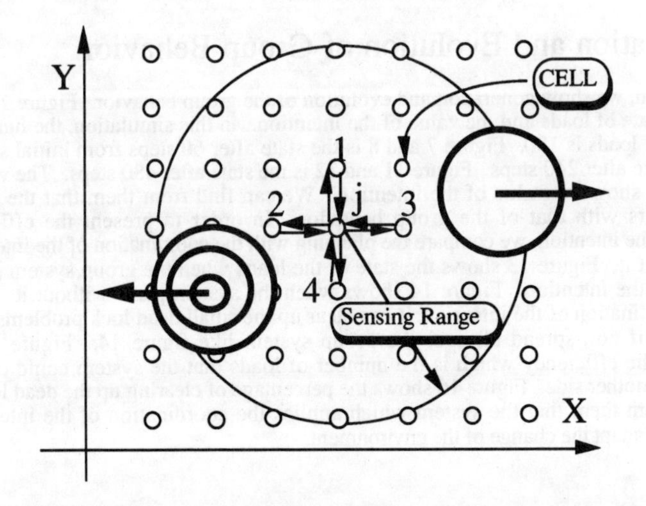

Figure 6 The condition of each cell

In this simulation, α=0.59, β=0.1, and $M(t)$=0.1, if a cell can move the white load and $M(t)$=-0.1, if it can move another kind of load. In result, if the value of the intention approaches to 1, the cell carries the white loads with higher priority and if it approaches to -1, it carries another load with higher priority.

(3) The sensing range for each cell is assumed to be 3 lattice intervals. Equation 3 and 4 show the force of cell when a load is on it.

The force to move a white load is given by this equation and

$$\begin{pmatrix} F_x \\ F_y \end{pmatrix} = \begin{pmatrix} 0.8 + I_i(t) \\ I_1(t) - I_4(t) \end{pmatrix} + \sum_{q=1}^{n} \frac{\gamma}{(x_p - x_q)^2 + (y_p - y_q)^2} \begin{bmatrix} x_p - x_q \\ y_p - y_q \end{bmatrix} \bullet\bullet\bullet (3)$$

The force to move a black load is given by this equation

$$\begin{pmatrix} F_x \\ F_y \end{pmatrix} = \begin{pmatrix} -0.8 + I_i(t) \\ -(I_1(t) - I_4(t)) \end{pmatrix} + \sum_{q=1}^{n} \frac{\gamma}{(x_p - x_q)^2 + (y_p - y_q)^2} \begin{bmatrix} x_p - x_q \\ y_p - y_q \end{bmatrix} \bullet\bullet\bullet (4)$$

where

(F_x, F_y): the force of cell

(x_p, y_p): the position of load which cell wants to carry.

(x_q, y_q): he position of load which is in the sensing area of the cell

(4)The force on the load comes from all the cells under it.
(5) A load is not allowed to move to the position already occupied.
(6) When a load arrives at the desired end, it starts moving from the other end at the same Y position.

5. Simulation Results

In this section, we show two types of simulation results.

5.1. Generation and Evolution of Group Behavior.

In this simulation, we show generation and evolution of the group behavior. Figure 7 - 12 show the change of the state of loads and the value of the intention. In this simulation, the number of cell is 3600 and that of loads is 180. Figure 7 and 8 is the state after 60 steps from initial state. Figure 9 and 10 is the state after 240 steps. Figure 11 and 12 is the state after 480 steps. The vertical axes of figure 8, 10, 12 show the value of the intention. We can find from them that the change of the intention concurs with that of the group behavior. In order to present the efficiency of the coordination of the intention, we compare the planning with the coordination of the intention with the planning without it. Figure 13 shows the state of the loads when the group system plans with the coordination of the intention. Figure 14 shows when the system plans without it. If the system applies the coordination of the intention, it can clear up the small dead lock problems, but the dead lock problems, if not, spread all over the group system like figure 14. Figure 15 shows the comparison of the efficiency which is the number of loads that the system could carry from the starting side to another side. Figure 16 shows the percentage of clearing up the dead lock problems. We can find from them that the system which applies the coordination of the intention is more efficient and can adapt the change of the environment.

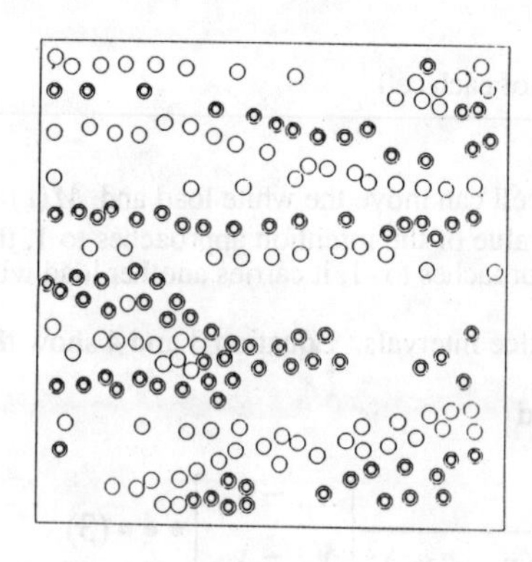

Figure 7 The state of after 60 steps

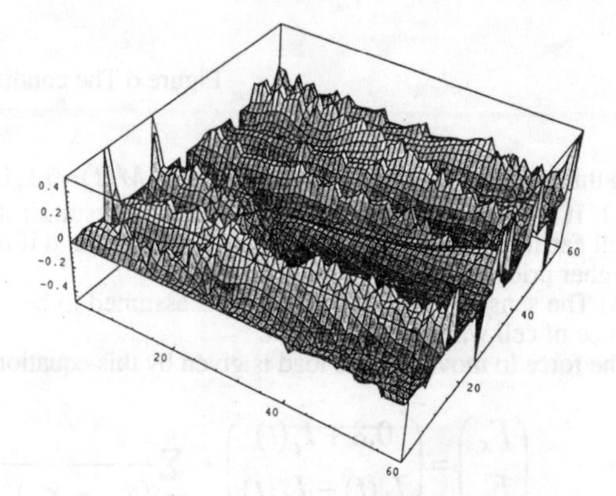

Figure 8 The intention of after 60 steps

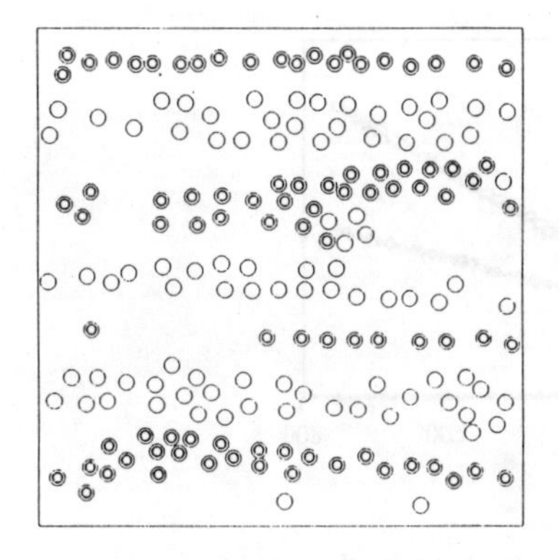

Figure 9 The state of after 240 steps

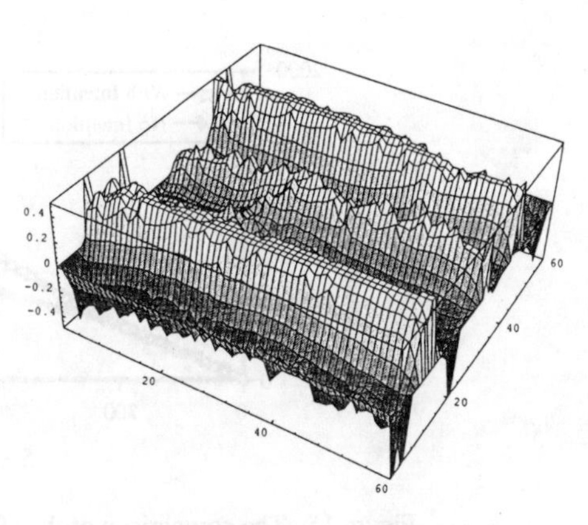

Figure 10 The intention of after 240 steps

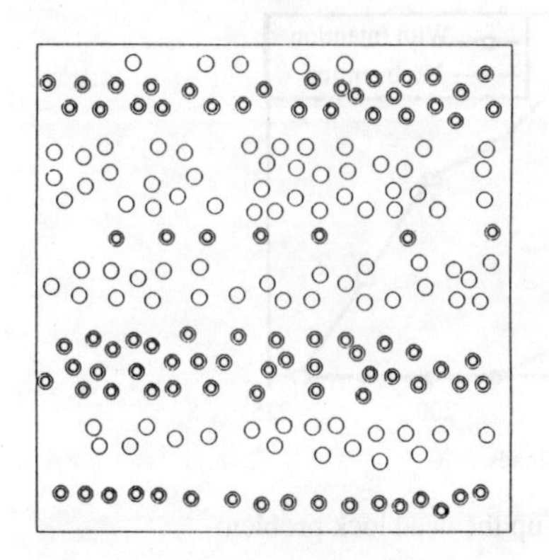

Figure 11 The state of after 480 steps

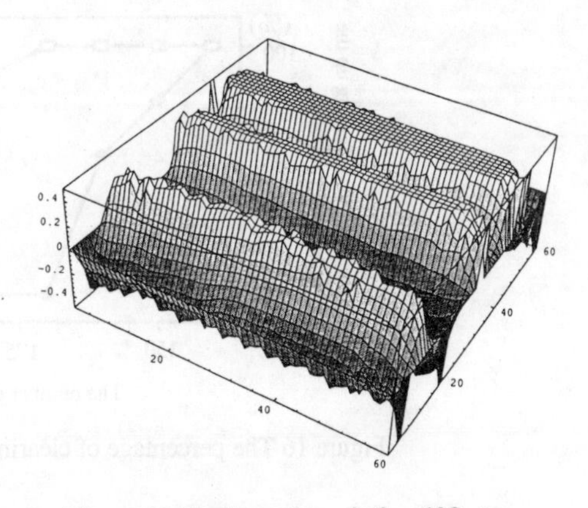

Figure 12 The intention of after 480 steps

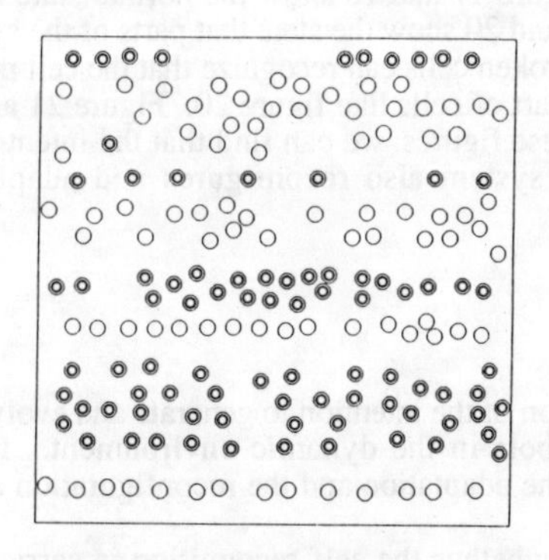

Figure 13 The state with intention

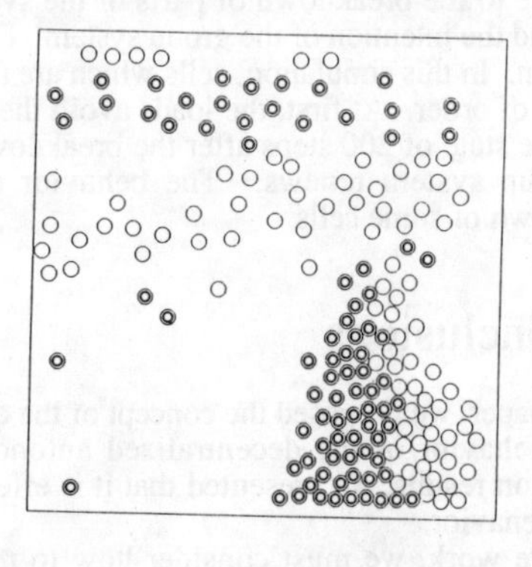

Figure 14 The state without intention

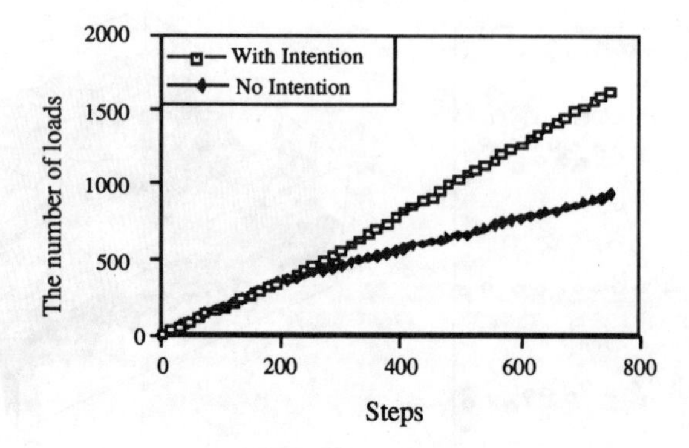

Figure 15 The comparison of the efficiency between the system with intention and the one without it

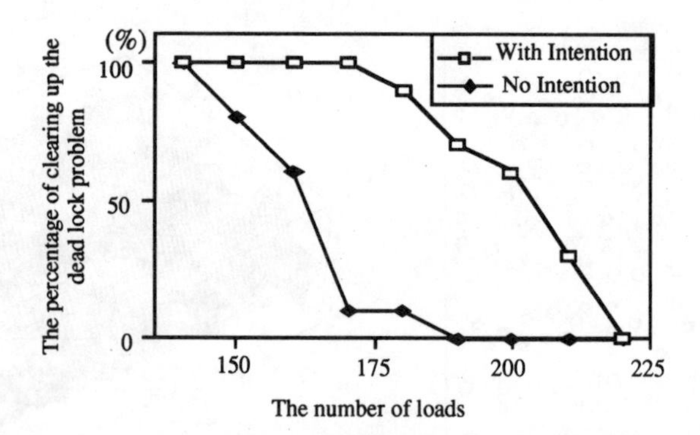

Figure 16 The percentage of clearing up the dead lock problem

5.2 Adaptation to the Breakdown of Parts of the System

In this section, we show that the group robotic system with the coordination of the intention has the tolerance to the breakdown of parts of the system. Figure 17 and 18 show the normal state of the loads and the intention of the group system. Figure 19 and 20 show the state that parts of the system is broken. In this simulation, cells which are near the broken cells can recognize that the cell next to it is out of order. At first, the loads avoid the broken part of cells like figure 20. Figure 21 and 22 show the state of 200 steps after the breakdown. In these figures, we can find that the intention of the group system renews. The behavior of group system also reconfigures and adapts the breakdown of some cells.

6. Conclusions

In this paper, we proposed the concept of the coordination of the intention to generate and evolve the group behavior of the decentralized autonomous robots in the dynamic environment. In the simulation results, we presented that it is effective in the adaptation and the reconfiguration of the group behavior.

In future work, we must consider how to recognize whether the self-recognition is carried out successfully or not.

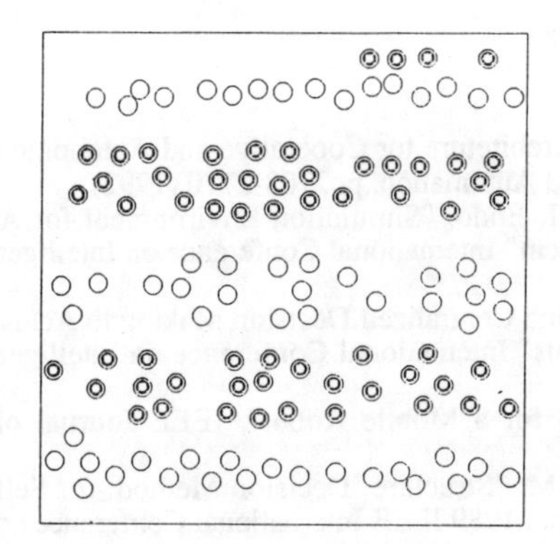

Figure 17 The normal state

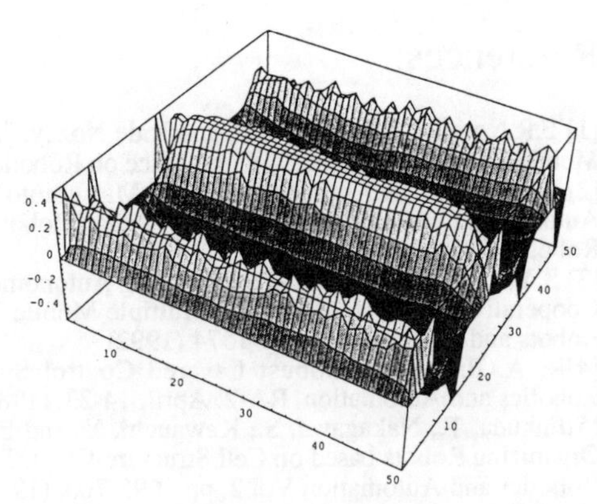

Figure 18 The normal state of intention

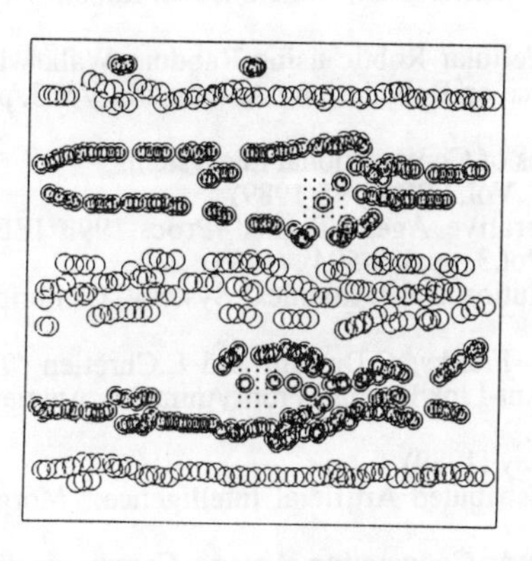

Figure 19 The state soon after the breakdown

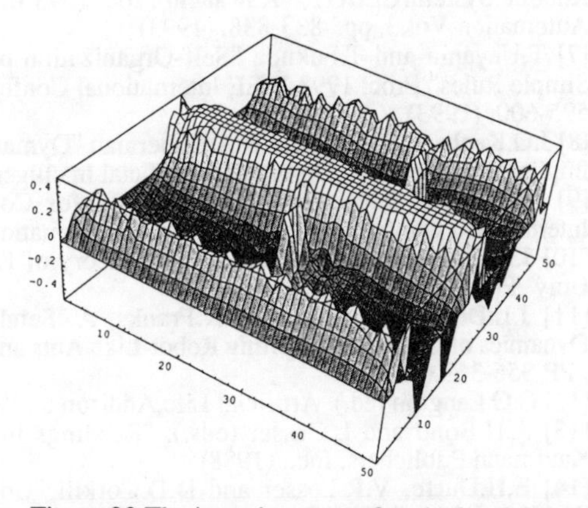

Figure 20 The intention soon after the breakdown

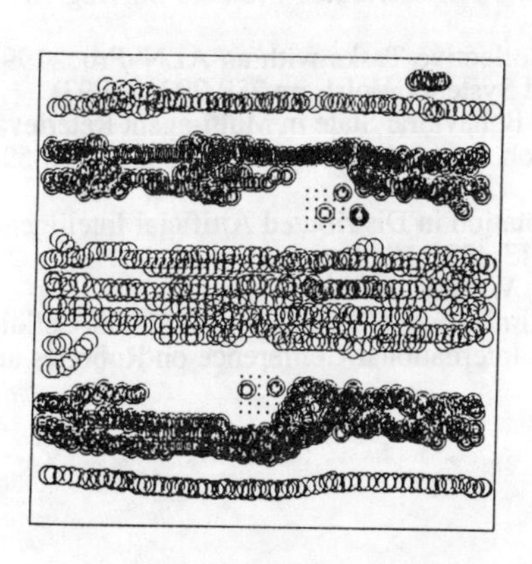

Figure 21 The state of 200 steps
steps after the breakdown

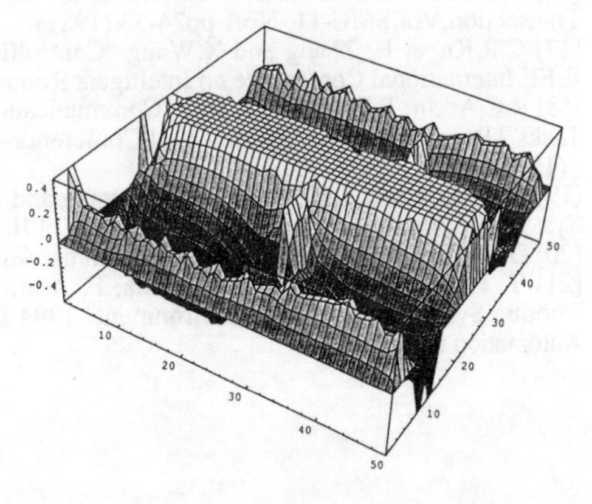

Figure 22 The intention of 200
steps after the breakdown

References:

[1] F.R.Noreils, A.A.recherche, and R.de Nozay. "An Architeture for Cooerative and Autonomous Mobile Robots" Intemational Conference on Robotics and Automation, pp2703-2710 (1992)
[2] M.K.Habib, H.Asama, Y.Isida, A.Matsumoto and I. Endo. "Simulation Environment for An Autonomous and Decentralized Multi-Agent Robot System" Intemational Conference on Intelligent Robots and Systems, pp1550-1557 (1992)
[3] S.Yuta and S.Premvti. "Coodinationg Autonomous and Centralized Decision Making to Achive Cooperative Behaviors Between Multiple Mobile Robots" Intemational Conference on Intelligent Robots and Systems, pp1566-1574 (1992)
[4]R. A. Brooks,"A Robust Layered Control System for a Mobile Robot", IEEE Journal of Robotics and Automation, RA-2, April, 14-23, (1986)
[5]Fukuda, T., Nakagawa, S., Kawauchi, Y. and Buss, M. "Structure Decision Method for Self Organizing Robots based on Cell Structure-CEBOT," Proc. 1989 IEEE International Conference on Robotics and Automation Vol.2, pp. 695-700, (1989)
[6] Y.Kawauchi , M.Inaba and T.fukuda "A Plinciple of Disributed Decision Making of Cellular Robotic System(CEBOT)" Kawauchi Proc. 1993 IEEE International Conference on Robotics and Automation Vol.3, pp. 833-838, (1993)
[7] T.Ueyama and T.Fukuda "Self-Organization of Cellular Robtic using Random Walk with Simple Rules" Proc. 1993 IEEE International Conference on Robotics and Automation Vol.3, pp. 595-600, (1993)
[8] J.O.Kephart, T.Hogg,and B.A.Huberman "Dymamics of Computational Ecosystem: Implications for DAI ",Distributed Artificial Intelligence ,Vol.2 PP.79-95(1989)
[9] L.E.Parker "Designing Control Laws for Cooperative Agent Teams" Proc. 1993 IEEE International Conference on Robotics and Automation Vol.3, pp.588-594 , (1993)
[10] J.Hofbauer and K.Sigmund "The Theory of Evolution and Dynamical System" Cambridge Univ .Press
[11] J.L.Deneubourg, S.Goss, N.Franks, A. Sendova-Franks, C.Detrain and L.Chretien "The Dynamics of Collectinve Sorting Robot-Like Ants and Ant-Link Robots" From Animal to Animats I , PP 356-363,(1990)
[12] C.G.Langton (ed.), Artificial Life,Addison on Wesley (1989)
[13] A.H.Bond and L.Gasser (eds.), "Readings in Distributed Artificial Intelligence," Morgan Kaufmann Publichers, Inc., (1988)
[14] E.H.Durfe, V.R.Lesser and D.D.Corkill "Coherent Cooperation Among Communicating Problem Solvers" IEEE Transaction,Vol.C-36, No.11, pp1275-1291, (1987)
[15] W.A.Kornfeld and C.E Hewitt "The Scientific Community Metaphor" IEEE Transaction,Vol.SMC-11,No.1 pp24-33 (1981)
[16] R.G.Smith and R.Davis "Frameworks for Cooperation in Disributed Problem Solving" IEEE Transaction,Vol.SMC-11, No.1 pp24-33 (1981)
[17] C.R.Kube, H. Zhang and X.Wang "Controlling Collective Tasks with an ALN' Proc. 1993 IEEE International Conference on Intelligent Robots and Systems, Vol.1, pp.289-293 , (1993)
[18] R.C.Arkin, T.Balch and E.Nitz "Communication of Behavioral State in Multi-agent Retereival Tasks " Proc. 1993 IEEE International Conference on Robotics and Automation Vol.3, pp.588-594 , (1993)
[19] M.J.Shaw and A.B.Whinston, "Learning and Adaptation in Distributed Artificial Intelligence Systems," Distributed Artificial Intelligence Vol.II, pp.413-428, (1989)
[20] D.E.Goldenberg (ed.),Genetic Algorithms, Addison Wesley , (1989)
[21] T. Fukuda, G. Iritani, T. Ueyama, F. Arai, "Optimization of Group Behavior on Cellular Robotic System in Dynamic Environment" 1994 IEEE International Conference on Robotics and Automation to appear

High-order Strictly Local Swarms

GERARDO BENI, SUSAN HACKWOOD, and XIAOMIN LIU

College of Eng., Univ. of California, Riverside, CA 92521, USA

Abstract

In physical realizations of Swarms, the interactions are restricted to short range in order to avoid time-delays and complexity of connections. However, short-range interactions restrict the Swarm to low-order equations. In order to take full advantage of the possibility of producing patterns generated by higher order differential equations while maintaining physical locality of interaction, and hence guarantee no communication time delays, we have recast the model so that only nearest neighbor interactions are considered. The tradeoff is that each Swarm unit updates n variables (a multi-channel process) where n is the order of the differential equation. We show by simulation and experiments on systems of microprocessors that the multi-channel algorithm converges and that the rate of convergence is not slower than for the standard Swarm algorithm. Hence, strictly local high-order Swarms appear to be practically realizable.

Keywords: Swarms, Distributed Robotic Systems, Autonomous Robots, Self-Organization, Asynchronous systems

1 Introduction

Swarms are self-organizing distributed robotic systems characterized by conservation and asynchronicity [1].

Conservation (of some physical quantity, e.g. a common resource) is a physical requirement of the Swarm. Without conservation the Swarm problem becomes trivial since the self-organizing pattern is constructed from local variables. Conservation conditions arise from self-organizing problems that depend on some *global* variable.

On the other hand, the asynchronicity of Swarms is *local*. It derives from the probabilistic self-activation behavior of each unit of the Swarm, rather than from time-delays in communications between units, as it usually happens in distributed computer systems. Communication time delays are generally neglected in Swarms because the interaction is assumed short-range. In recent mathematical models of the Swarms [2,3,4] the short-range of the interaction has been guaranteed by the restriction to low-order differential equations models.

We have found that this restriction can be lifted by recasting the linear system of equations describing the Swarm in terms of multiple variables.

In the most general mathematical model of the Swarm [5] each unit of an M-unit Swarm updates one variable, and the pattern produced by the Swarm is an M x 1 vector in these variables. There is no restriction to the order of the differential equation and hence the range. In order to take full advantage of the possibility of producing patterns generated by higher order differential equations while maintaining physical locality of interaction, and hence guarantee no communication time delays, we have recast the model so that only nearest neighbor interactions are considered. The tradeoff is that each Swarm unit updates n variables (rather than a single one) where n is the order of the differential equation.

The notion of Swarm has evolved in time from a general concept [2] to a precise model [6]. Currently the mathematical definition of the Swarm is as follows :

Def 1 : SWARM

Swarm: a set of M ordered entities (k = 1,2,...M) such that:

1) they share one resource y allocated to the units with some *conservation* constraint; (e.g., $\Sigma_k y_k = N$);

2) they are connected via linear functions G_k of the resource allocation, i.e., $G_k = y_k - \Sigma_j b_{kj} y_j$ where b_{kj} are constant ($j = 1,2,...M; j \neq k$).

3) all entities, except for a symmetry breaking one (i.e., for k = 1, 2, ...M-1), update their, and their neighbor's, resource allocation by *asynchronously* exchanging a quantum of resource according to an *identical, local* rule, as follows :

$\forall$ k: Calculate G_k

• if $G_k > \varepsilon$ then

$y_k \rightarrow y_k - 1$; $y_{k+1} \rightarrow y_{k+1} + 1$;

• if $G_k < -\varepsilon$ then

$y_k \rightarrow y_k + 1$; $y_{k+1} \rightarrow y_{k+1} - 1$;

• else do nothing;

(ε is a constant : $0 \leq \varepsilon \leq 1$).

Swarms with external inputs have been investigate in [3-4]. The swarm equation is as follows:

$$G_k = y_k - y_{k+1} + A(u_{k+1} - u_k) \qquad (1)$$

where u_k is a bounded external input to the k-th unit. The exact solution is obtained for $G_k = 0$. The experimental results confirm the finding that, under rather general condition, the Swarm can be made to converge in a time which is independent on its size, and the convergence rate of the Swarm is depended on its total configuration error.

The Swarm considered in [2] is described by a second order swarm equation as follows:

$$G_k = y_k - 2c^{-1} y_{k+1} + c^{-1} y_{k+2} \qquad (2)$$

where the constant $c > 1$ and k = 1,2...M, the number of units in the Swarm. (The exact solution is obtained for $G_k = 0$). Simulation results indicate again that, under rather general condition, the Swarm converges to a close neighborhood of the exact solution, the distance being a function of ε in Definition 1 above.

Considering both cases with or without external input, the swarm equation can be generally described by a n-th order difference equation which is most conveniently written into the form:

$$G_k = (a_0 y_k + b_0 u_k) + (\sum_{i=1}^{n} a_i y_{k+i} + \sum_{i=1}^{n} b_i u_{k+i}) \qquad (3)$$

where $G_k = 0$ for the exact solution, and: a_0, b_0, a_i, b_i are constants, y_k is the amount of resource held by the k-th unit, u_k is the external input to the k-th unit, y_{k+1} is the amount of resource held by the (k+1)-th unit, u_{k+1} is the external input to the (k+1)-th unit.

In order to evaluate Eq. (3), a swarm unit *k* must know, besides the value of the resource it holds and the external input it receives, also the resource y holdings of each of its next n nearest neighbors, i.e. the k+1, k + 2, ..., and k+n unit, as well as their external inputs u_k. This requires that the physical structure of each unit of the swarm be capable of acquiring information from its next n neighbors (e.g., by setting up n communication channels and/or n sensing channels) and of doing so simultaneously. Obviously, the building cost and complexity of such a physical swarm unit increases with the order of the swarm equations, i.e. with the number of neighbors to interact with. Besides the added complexity, the crucial limitation introduced by high order equations is on the requirement of synchronicity among n neighbors. This is contrary to the physical meaning of the Swarm model which is asynchronous but without time delays.

More precisely, the Swarm model is meant to be asynchronous, but only locally. The asynchronicity is due to the probabilistic self-activation behavior of each unit of the Swarm, rather than from time-delays in communications between units, as it usually happens in distributed computing systems. Communication time delays are generally neglected in Swarms because the interaction is assumed short-range. If, however, we model the Swarm with high order difference equations, the assumption of no-time-delay in communication between units becomes difficult to justify. And if the assumption is kept, which is necessary mathematically, the Swarm would be *pseudo-asynchronous*, in the sense of being asynchronous in the self activation of *groups* of *synchronous* units.

Thus, although in the mathematical definition of Swarm , the locality is required not in the sensing and communication (part 2 of Def.1) but in the exchange of resources (part 3 of Def. 1) yet, physically, part 2 of Def. 1 should be restricted to a short range in order to model a truly asynchronous system.

On the other hand, there is no restriction, in the Swarm definition, as to the amount of information that can be exchanged between neighbors. Generally, time delays can be introduced by shear volume of information transfer , not just by number of information sources. But, again, this information-transfer delay, which can be significant for distributed computing models, it is hardly of importance for Swarms. The reason is that the basic difficulty with multiple sources, which is the necessity to synchronize the self-activation of n units of the Swarm, is not carried over to multiple channels of information from the same unit. If only two units exchange information, all that is required is still to synchronize the self-activation of two units, regardless of the time that it is taken in transferring the information between them. This time delay also does not affect the parallel operation of other units. It simply may increase the average time of each reconfiguring step.

The latter is important in comparing the results of simulations, as we shall see below. For example, if the simulations indicated identical reconfiguration rates (inverse of the number of steps to reconfigure) still the reconfiguration rate of the multichannel system would be slower in proportion to the increase of transfer time between units. For example, a serial transfer of information from the different channels would likely lead to such an increase.

In this regard, the range and the number of channels of communication may be varied based on the task of the Swarm, but they are often limited by the physical constructions of many distributed systems. For example, the units in the Swarm we have constructed in [4] are connected in series via the two serial ports. The ability of information acquiring of each unit is restricted to between two nearest neighbors, the kth and k+1th units. The physical construction of Swarms as [4] is of special practical importance for its simplicity, which is desired in designing DRS's. But even simple Swarm models such as that described by Eq. (3), i.e. with $n = 2$, cannot be carried out on it.

However, solutions for high order systems can be carried out by the high order strictly local model of Swarm described in this paper. In Section 2, we describe the multichannel model for high order strictly local swarms. In Section 3 we describe a second order and a fourth order model. In section 4 we discuss the physical realization and in Section 5 we discuss the results of the experiments and simulations.

2 Multi-channel exchange

Swarms described by 2nd and higher order equations (Eq. 3) can be solved (i.e. made to converge to the solution of Eq. 3) by a strictly local model (i.e. a first order model) , if we allow multiple channels for exchanging resources between two neighboring units. Mathematically, this is equivalent to expressing a n-th order differential equation as a system of n 1st-order differential equations. This well known mathematical method is not obviously translated into the Swarm problem since the system of 1st order differential equations is supposed to be solved symultaneously, i.e. sysnchronously. The situation is analogous to the situation that we find in the basic Swarm problem, namely, to solve asynchronously a system of difference equations which, when solved simultaneously, give the (discrete) solution of the Swarm differential equation. Here, once again, we rely on the property of the Swarm algorithm to produce the correct solution by local asynchronous exchanges.

Thus, given a n-th order Swarm equation, we introduce n additional variables besides the resource y and the external input u, i.e. each unit k holds, besides the external input u_k, the actual resource y_k *and* n 'virtual resources' $p_{k,i}$ $(i = 1,2 \ldots n)$.

The multi-channel resource exchange process proceeds as follows:

Unit k holds a vector of n+1 resources (one real and n virtual) : $P_k = [p_{k,0}, p_{k,1}, p_{k,2}, \ldots p_{k,n}]$ where we have defined $p_{k,0} = y_k$. When unit k and k+1 activate for an exchange, they exchange a quantum of the actual resource ($y_k = p_{k,0}$) according to Def. 1 point 3, which is the normal Swarm exchange; but, in addition, they exchange the virtual resources so that $P_k \rightarrow P'_k = [\, p'_{k,0}, p_{k+1,0}, p_{k+1,1}, \ldots p_{k+1,n-1}]$, where $p'_{k,0}$ is calculated according to Def. 1 point 3.

Basically, during the multi-channel exchange between the k-th and the (k+1)-th unit, the k-th unit 'rewrites' its virtual resources according to the virtual resources of the (k+1)-th unit shifted by one unit of resource index. Hence, in particular, the n-th virtual resource, p_n, does not participate in the exchange as a 'source' (i.e. it is not copied from the (k+1)-th unit) and the actual resource, y_k, does not participate in the exchange as a 'destination' (i.e. it is not *written* from the (k+1)-th unit).

More generally, in considering also external inputs, each unit k holds also a vector of n+1 external inputs (one real and n virtual): $V_k = [v_{k,0}, v_{k,1}, v_{k,2}, \ldots v_{k,n}]$ where we have defined $v_{k,0} = u_k$, the actual external input on the k-th unit. When unit k and k+1 activate for an exchange, they exchange actual and virtual resources so that $P_k \rightarrow P'_k$ as specified above ; but, in addition, they exchange also the virtual inputs so that $V_k \rightarrow V'_k = [\, v_{k,0}, v_{k+1,0}, v_{k+1,1}, \ldots v_{k+1,n-1}]$.

We can summarize the multi-channel exchange process by amending the Def.1 of the Swarm (Section 1) as follows:

• Point 2) is rewritten in the following equivalent (for an n-th order system) form:

"they are connected via linear functions G_k of the resource allocation, i.e.

$$G_k = \left(\sum_{i=0}^{n} a_i y_{k+i} + \sum_{i=0}^{n} b_i u_{k+i}\right) \tag{3a}$$

where a_i, b_i are constants, y_{k+i} is the resource held by the (k+i)-th unit , and u_{k+i} is the external input on the (k+i)-th unit ."

•We add point 4) as follows:

"We define the virtual resources and virtual inputs on unit k as

$$p_{k,i} = y_{k+i} \; ; \; v_{k,i} = u_{k+i} \tag{4}$$

During exchange of the actual resource, the k-th unit rewrites its virtual resources and inputs so that:

$$p_{k,i} \rightarrow p_{k+1,i-1} \quad ; \; v_{k,i} \rightarrow v_{k+1,i-1} \qquad \text{for } i = 1,2,\ldots n." \tag{5}$$

Physically, the multi-channel exchange process corresponds to trading range of interaction for extra variables so that, in an actual implementation, the difficulty of building long range connections to transfer simple messages is replaced with the difficulty of building short range connections to transfer complex messages. The latter difficulty is more manageable by asynchronous systems since it requires only minimal synchronicity (between two units rather than n units) as discussed in the introduction. The trade-off in complexity of the information contained in each message is easily dealt with in robotic and computer systems in general since parallel transfer along multiple data channels between two units is generally trivial.

3 Examples

In this Section we will consider to example model for the application of the multi-channel exchange process to achieve strictly local (i.e. nearest neighbor interactions only) Swarms governed by high-order (>1) equations.

3.1 Second order model

The first model is the 2nd-order model consider in ref. [2] where the Swarm model was first presented. The system of linear, second order, homogenous difference equations with constant coefficients, which is derived from the differential equation:

$$d^2y/dx^2 + Ey = 0 \tag{6}$$

was considered in [2]. It is well known that Eq. (6) can give stationary waves solutions. Eq. (6) can be written into the difference equations as in Eq. (2) above, rewritten here for clarity .

$$G_k = y_k - 2c^{-1}y_{k+1} + c^{-1}\,y_{k+2} \quad ; k = 1,2 \ldots M \,;\, c > 1. \tag{2}$$

The cyclic boundary conditions and the conservative condition for the resource are specified respectively as :

$$y_{M+k} = y_k ; \sum_{k=1}^{M} y_k = N . \tag{7;8}$$

The asynchronous solution to Eqs. (2,7,8) can be obtained by using the Cyclic Swarm model of ref. [2].

For the same problem, the multi-channel exchange model involves two additional variables $p_{k,1}$ and $p_{k,2}$ in each k unit, besides the resource variable y_k. According to the multi-channel exchange process defined at the end of the previous section, i.e. Eq. 5, there are three (two in the case of the M-th unit) equations recalculated in each swarm units (k = 1, 2, ..., M-1) at every exchange activation, as follows:

The first equation is the basic Swarm equation,

$$G_k = y_k - 2c^{-1}p_{k,1} + c^{-1}\,p_{k,2} \quad ; k = 1,2 \ldots M\text{-}1; \tag{2a}$$

with $p_{k,1} = y_{k+1}$; $p_{k,2} = y_{k+2}$

From Eq.(2) the new value of y_k and y_{k+1} are determined according to point 3 of Def 1.

In addition, however, also the virtual variables are recalculated:

$$p_{k,1} \dashrightarrow p_{k+1,0} = y_{k+1} \;;\; p_{k,2} \dashrightarrow p_{k+1,1} \tag{9;10}$$

Note that the constraints of the cyclic boundary conditions and the conservation condition require (ref. [2]) that the symmetry breaking unit (e.g., the M-th unit) be inactive with respect to the Swarm main equation (2a). This is true also in the multi-channel exchange case, but, in the case of the multi-channel exchange, the symmetry breaking unit is still active with respect to the virtual variables, i.e., it recalculates two values as follows:

$$p_{M,1} \dashrightarrow p_{1,0} = y_{M+1} = y_1 \;;\; p_{M,2} \dashrightarrow p_{M+1,1} = p_{1,1} \tag{11;12}$$

3.2 Fourth order model

We consider now the following system of 4th-order difference equations:

$$10y_k - 2y_{k+1} + 3y_{k+2} - 2y_{k+3} + y_{k+4} = 0; \quad k = 0, 1, 2, \ldots M \tag{13}$$

a fourth order Swarm can be defined by these equations as follows:

$$G_k = 10y_k - 2y_{k+1} + 3y_{k+2} - 2y_{k+3} + y_{k+4}; \quad k = 0, 1, 2, \ldots M \tag{13a}$$

The cyclic boundary condition and the conservative condition are as in Eqs. (7,8),i.e.:

$$y_{M+k} = y_k, \tag{14}$$

$$\sum_{k=1}^{M} y_k = N. \tag{15}$$

For the same problem, the multi-channel exchange model involves four additional variables $p_{k,1}$, $p_{k,2}$, $p_{k,3}$, and $p_{k,4}$ in each k unit, besides the resource variable y_k. According to the multi-channel exchange process defined at the end of Section 2, i.e. Eq. 5, there are five (four for the M-th unit) equations recalculated in each swarm units (k = 1, 2, ..., M-1) at every exchange activation, as follows:

The first equation is the basic Swarm equation,

$$G_k = 10y_k - 2p_{k,1} + 3p_{k,2} - 2p_{k,3} + p_{k,4}; \quad k = 0, 1, 2, \ldots M-1 \tag{13b}$$

with $p_{k,1} = y_{k+1}$; $p_{k,2} = y_{k+2}$; $p_{k,3} = y_{k+3}$; $p_{k,4} = y_{k+4}$.

From Eq.(13b) the new value of y_k and y_{k+1} are determined according to point 3 of Def 1.

In addition, however, also the virtual variables are recalculated:

$$p_{k,1} \dashrightarrow p_{k+1,0} = y_{k+1} \tag{16}$$

$$p_{k,2} \dashrightarrow p_{k+1,1} \tag{17}$$

$$p_{k,3} \dashrightarrow p_{k+1,2} \tag{18}$$

$$p_{k,4} \dashrightarrow p_{k+1,3} \tag{19}$$

As in the 2nd-order case, the constraints of the cyclic boundary conditions and the conservation condition require (ref. [2]) that the symmetry breaking unit (e.g., the M-th unit) be inactive with respect to the Swarm main equation (2a). But, in the case of the multi-channel exchange, the symmetry breaking unit is still active with respect to the virtual variables, i.e., it recalculates four values as follows:

$$p_{M,1} \dashrightarrow p_{1,0} = y_{M+1} = y_1 \tag{20}$$

$$p_{M,2} \dashrightarrow p_{M+1,1} = p_{1,1} \tag{21}$$

$$p_{M,3} \dashrightarrow p_{M+1,2} = p_{1,2} \tag{22}$$

$$p_{M,4} \dashrightarrow p_{M+1,3} = p_{1,3} \tag{23}$$

4 Experimental set-up

In order to compare the behavior of the multi-channel exchange process Swarms with the standard Swarms algorithm applied to higher (> 1)order systems we have set-up a system of interconnected microprocessors. The system is similar to the one reported in [7] but with appropriate modifications as follows.

Figure 1 shows schematically the system of 16 microprocessors connected as nearest neighbors. Each Swarm unit is an Intel 286 CPU. The computers are connected in series via two serial COM1 and COM2 ports in each computer, i.e., a serial cable connects the COM1 port of a computer to its neighbor's COM2 port. A random counter is used in each unit to realize the asynchronous operation of the Swarm, as discussed in Ref. [7].

The System Connection Diagram

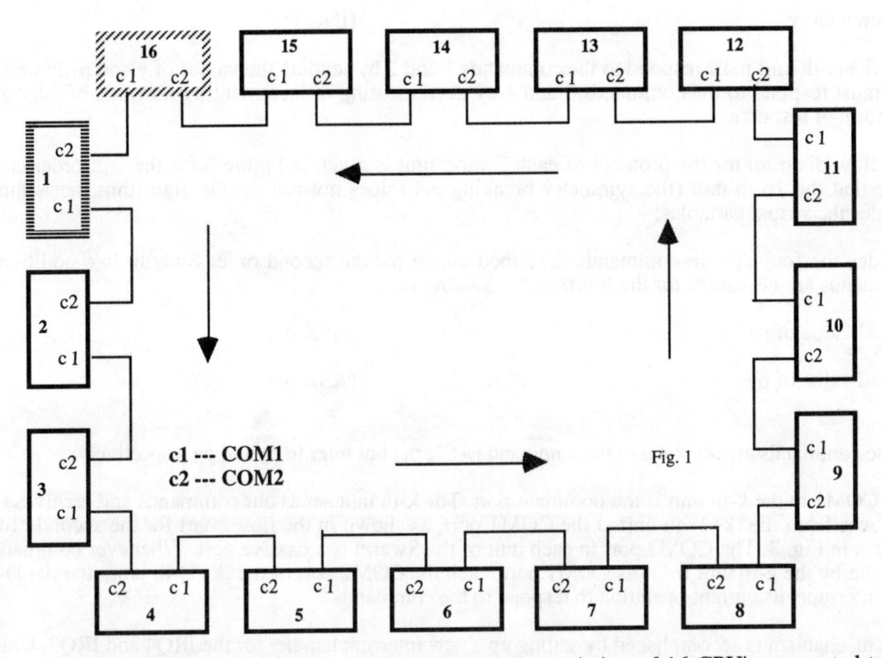

Figure 1: The Swarm experimental configuration consisting of 16 CPU's connected to the neighbors via the two serial ports c1 and c2. All units run identically except at the boundaries (1 and 16). The direction of the command flow is indicated by the arrows.

Communication Diagram Between Two Units

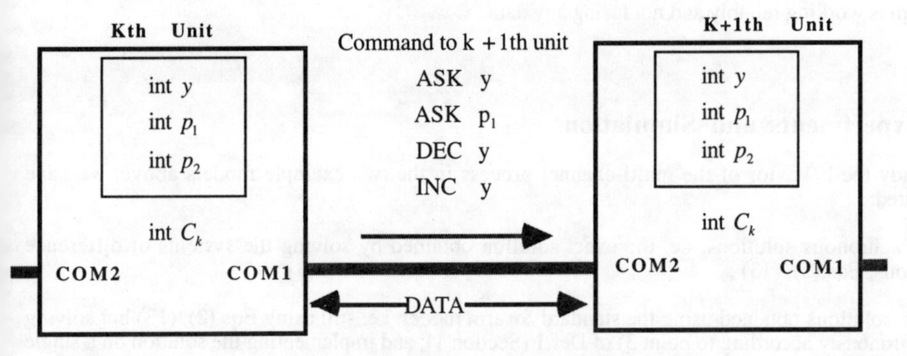

Figure 2: Command and data communication between two neighboring units. "int" refers to "integer". All symbols explained in the text.

For the 2nd-order model, Eqs. (2a,9,10), the k-th unit microprocessor must be capable of sending out 4 commands to the (k+1)-th unit microprocessor (see Fig. 2), i.e.

1. read value of y (ASK y)

2. read value of p_1 (ASK p_1)

3. decrement y (DEC y)

4. increment y (INC y)

The (k+1)-th unit must respond to the commands 1 and 2 by sending the values of y and p_1 to unit k; and must respond to the commands 3 and 4 by decrementing or incrementing its value of y by one quantum of resource.

The flow-diagram for the protocol of each Swarm unit is given in Figure 3 for the 2nd-order case. Note that the 16-th unit (the symmetry breaking one) does not run the G_k algorithm, but simply updates the virtual variables.

Besides the four system commands described above for the second order Swarm, two additional commands are necessary for the fourth order Swarm, i.e.,:

5. read value of p_2 (ASK p_2)

6. read value of p_3 (ASK p_3)

These commands are identical to the command (ASK p_1) but refer to p_2 and p_3 respectively.

The COM1 in the k-th unit is the dominant port. The k-th unit sends out commands and receives the response from the (k+1)-th unit at the COM1 port, as shown in the flow chart for the second order Swarm in Fig. 3. The COM2 port in each unit of the Swarm is a passive port. Whenever commands sent out by the k-th unit from its COM1 port reach the COM2 port of the (k+1)-th unit, the (k+1)-th unit interrupts its current operation to respond to the commands.

This mechanism is accomplished by setting up a new interrupt handler for the IRQ4 and IRQ3, which corresponds to the interrupts from COM1 and COM2, respectively. The interrupt handler's flow chart for the second order Swarm is shown in Fig. 4. If the interrupt is from the COM1 port, then the value of the (k+1)-th unit's resources, y_{k+1} or $p_{k+1,1}$, can be read from COM1 and stored in the variables $p_{k,1}$ or $p_{k,2}$ accordingly. If the interrupt is from the COM2 port, then a command can be read (see Fig. 3 for the actual commands) and the unit carries out the action corresponding to the command. Note, in particular, that whenever the y_k resource is updated, the counter C_k is incremented by one. This allows us to count the number of reconfiguring steps and compare the speed of convergence of different algorithms. Overall, this interrupt driven serial port ensures that the system is working reliably and not losing any data.

5 Experiments and Simulations

To study the behavior of the multi-channel process in the two example models above, we have compared:

(A) synchronous solutions, i.e. the exact solution obtained by solving the systems of difference equations, Eqs.(2), (13) ;

(B) the solutions obtained using the standard Swarm model, i.e. still using Eqs (2), (13) but solving asynchronously according to point 3) of Def.1 (Section 1); and implementing the solution on a single computer simulating multiple asynchronous processes.

(C) same as in (b) but implementing on 16 microprocessors each connected to its nearest neighbor.

(D) the solutions obtained using the multi-channel exchange process implemented on a single computer simulating multiple asynchronous processes.

(E) the solutions obtained using the multi-channel exchange process applied to 16 microprocessors each connected to its nearest neighbor.

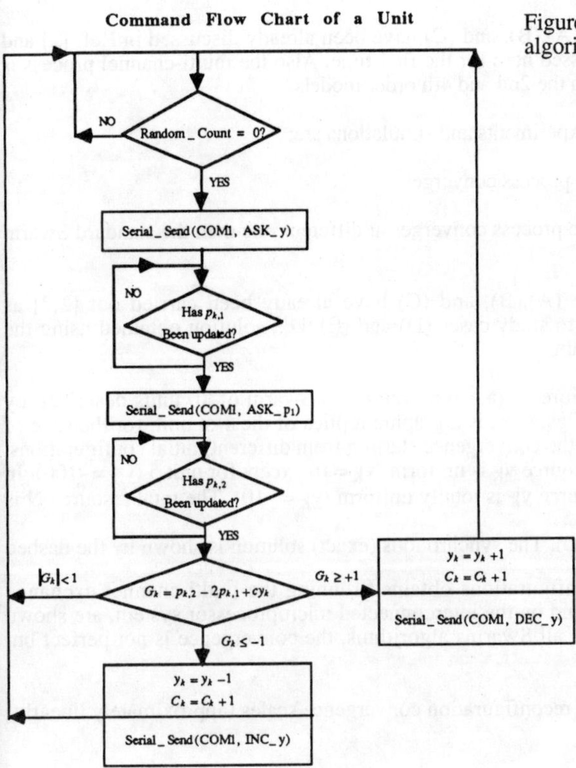

Figure 3: All units except the last one run this algorithm. See text for symbols.

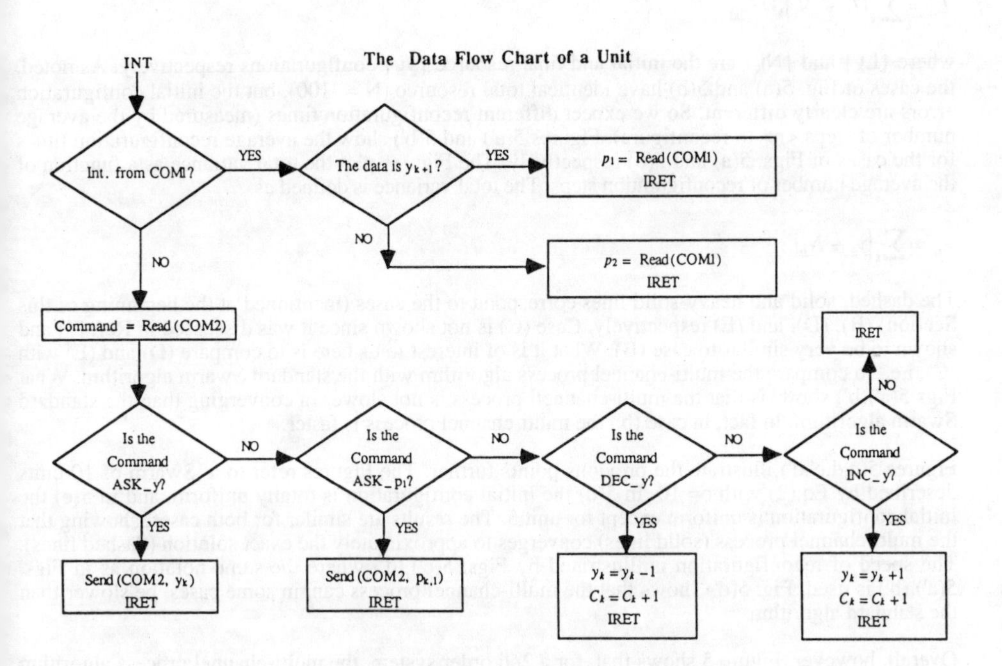

Figure 4: Interrupt handler's flow chart. IRET is "interrupt return". Other symbols in text.

For the second order model, methods (A),(B), and (C) have been already discussed in Ref. [2] and Ref. [7]. The 4th-order model is discussed here for the first time. Also the multi-channel process is discussed here for the first time for both the 2nd and 4th order models.

The basic issues to be resolved by the experiments and simulations are:

(1) whether the multi-channel exchange process converges

(2) whether the multi-channel exchange process converges at different rate than the standard Swarm algorithm.

For the 2nd-order case, Eq.(2), since (A),(B), and (C) have already been carried out [2,7] as mentioned above, we have proceeded to study cases (D) and (E), i.e., solution obtained using the multi-channel exchange process algorithm.

The results are shown in Figure 5. Figures 5 (a,a',b,b') refer to a Swarm of 10 units described by Eq.(2) with c= 1.1. The 11-th unit in Figs. (a,b) is a graphic replica of the 1-st unit, for the sake of illustration only. Figs. 5(a,a') describe the convergence starting from different initial configurations. In 5(a), the initial distribution of the resource y_k is uniform ($y_k = 0$) except for unit 5 ($y_5 = -100$). In 5(b), the initial distribution of the resource y_k is totally uniform ($y_k = -10$). The total resource N is identical in both cases ($\sum_{k=1}^{M} y_k = N = -100$). The synchronous (exact) solution is shown by the dashed lines in Figs. 5(a,b). Typical final configurations obtained running the multi-channel exchange process algorithm, both in simulation and on the interconnected microprocessor system, are shown by the solid lines in Figs. 5(a,b). As in all Swarms algorithms, the convergence is not perfect but very close, as discussed in Ref. [5].

In Ref. [7], it was found that the rate of reconfiguration convergence scales (approximately linearly) with the initial error

$$E_0 = \sum_k |L_k - N_k|,$$

where $\{L_k\}$ and $\{N_k\}$ are the initial and final resource $\{y_k\}$ configurations respectively. As noted, the cases of Fig. 5(a) and 5(b) have identical total resource (N = -100), but the initial configuration errors are clearly different. So we expect different reconfiguration times (measured by the average number of steps $<n>$ to reconfigure). Figures 5(a') and 5(b') show the average reconfiguration times for the cases of Figs. 5(a) and 5(b) respectively. The Figures plot the total variance as a function of the average number of reconfiguration steps. The total variance is defined as

$$\varepsilon_0 = \sum_k |y_k - N_k|^2.$$

The dashed, solid and heavy-solid lines correspond to the cases (mentioned at the beginning of this Section) (B), (D), and (E) respectively. Case (C) is not shown since it was discussed in Ref. [7] and shown to be very similar to case (B). What it is of interest to us here is to compare (D) and (E) with (C), i.e., to compare the multi-channel process algorithm with the standard Swarm algorithm. What Figs 5(a')(b') show is that the multi-channell process is not slower in converging than the standard Swarm algorithm. In fact, in case (b') the multi-channel process is faster.

Figures 5(c,d,c',d') illustrate the previous points further. The Figures refer to a Swarm of 10 units described by Eq.(2) with c= 10. In 5(d) the initial configuration is totally uniform, and in 5(c) the initial configuration is uniform except for unit 5. The results are similar for both cases, showing that the multi-channel process (solid lines) converges to approximately the exact solution (dashed lines). The speed of reconfiguration is illustrated by Figs. 5(c'),(d') where the same notation as in Figs. 5(a'),(b') is used. Fig. 5(d') shows that the multi-channel process can, in some cases, be slower than the standard algorithm.

Overall, however, Figure 5 shows that, for a 2nd order system, the multi-channel process algorithm is basically equally efficient in reconfiguring the Swarm as the standard algorithm.

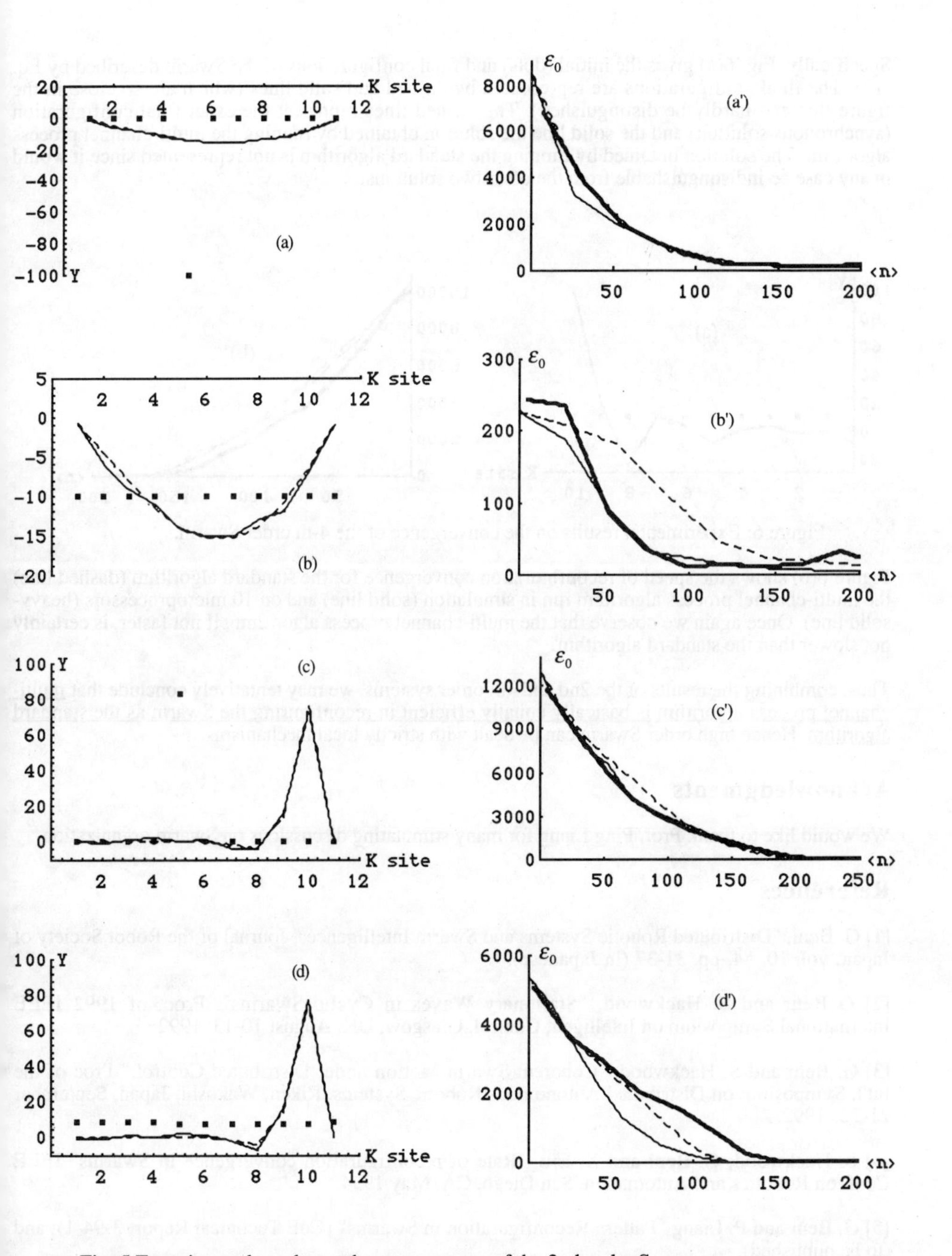

Fig. 5 Experimental results on the convergence of the 2nd-order Swarm.

It is of course of interest to check if this is still the case for higher order systems, since the main motivation for introducing the multi-channel process is dealing with significantly high order systems. Figure 6 shows the results of the experiments and simulations applied to the 4-th order case.

Specifically, Fig. 6(a) gives the initial (dots) and final configurations of the Swarm described by Eq. (13). The final configurations are represented by dashed and solid lines (which are so close in the figure that can hardly be distinguished). The dashed lines represent the exact final configuration (synchronous solution) and the solid line the solution obtained by running the multi-channel process algorithm. The solution obtained by running the standard algorithm is not represented since it would in any case be indistinguishable from the other two solutions.

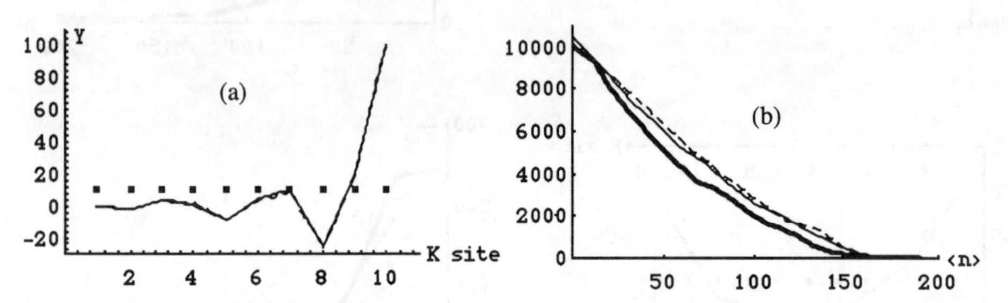

Figure 6: Experimental results on the convergence of the 4-th order Swarm.

Figure 6(b) shows the speed of reconfiguration convergence for the standard algorithm (dashed line) the multi-channel process algorithm run in simulation (solid line) and on 10 microprocessors (heavy-solid line). Once again we observe that the multi-channel process algorithm , if not faster, is certainly not slower than the standard algorithm.

Thus, combining the results of the 2nd and 4th order systems, we may tentatively conclude that multi-channel process algorithm is basically equally efficient in reconfiguring the Swarm as the standard algorithm. Hence high order Swarm can be dealt with strictly local mechanisms.

Acknowledgments

We would like to thank Prof. Ping Liang for many stimulating discussions on Swarm organization

References

[1] G. Beni, "Distributed Robotic Systems and Swarm Intelligence," Journal of the Robot Society of Japan, vol. 10, #4, pp. 31-37 (in Japanese).

[2] G Beni and S. Hackwood, "Stationary Waves in Cyclic Swarms", Proc. of 1992 IEEE International Symposium on Intelligent Control, Glasgow, UK, August 10-13, 1992.

[3] G. Beni and S. Hackwood, "Coherent Swarm Motion under Distributed Control," Proc of the Int'l. Symposium on Distributed Autonomous Robotic Systems, Riken, Wakoshi, Japan, September 21-22, 1992.

[4] S. Hackwood, G. Beni and X. Liu, "Rate of reconfiguration convergence in Swarms" IEEE Conf. on Robotics and Automation, San Diego, CA, May 1994.

[5] G. Beni and P. Liang "Pattern Reconfiguration in Swarms" (CoE Technical Report 3-94-1); and (to be published).

[6] Q. Huang "Partially and Totally Asynchronous Algorithms for Linear Swarms" Masters Thesis, Department of Computer Science, University of California, Riverside (June 1994).

[7] S. Hackwood, G. Beni, X. Liu, "Rate of reconfiguration convergence in Swarm," accepted by IEEE Conference on Robotics and Automation, San Diego, June, 1994.

Self-Organization of an Uniformly Distributed Visuo-Motor Map through Controlling the Spatial Variation

ZHI-WEI LUO[1], KAZUYOSHI ASADA[2], MASAKI YAMAKITA[1,2], and KOJI ITO[1,2]

[1]Bio-Mimetic Control Research Center, The Inst. of Physical and Chemical Research (RIKEN), Nagoya, 456 Japan

[2]Dept. of Information and Computer Sciences, Toyohashi Univ. of Tech., Toyohashi, 441 Japan

Abstract

This paper studied the distributed learning strategies for an autonomous robotic system to successfully generate its movement in a work space through constructing a visuo-motor map. Two learning algorithms inspired by Kohonen's self-organizing map algorithm using artificial neural networks have been discussed. The first is the extended self-organizing feature map algorithm and the second is the neural-gas algorithm. Based on this discussion, we propose an original approach to control the learning process by taking into account the statistical learning property (e.g. the spatial distribution of the neuron units). Computer simulations show that our approach not only derives a faster learning process but also results in a higher quality map with desired distribution of the units. Utilizing this map, we can expect the robot to achieve a more accurate movement in the work space than the other two algorithms.

Keywords: self-organization, visuo-motor map, autonomous robotic system, neural networks, learning process, spatial distribution

1 Introduction

In this paper, we study an autonomous robotic system's distributed learning strategies. If we define **intelligence** as the ability of the robot to fill a gap between the given task requirements and the task realization, the **autonomous property** in robot learning is determined according to whether this intelligence is obtained through the robot's **self-organization** or from an external teacher.

Consider a tennis game for example. To achieve the final objective of winning the game, we can give the task requirement as: (1) to win the game, directly; (2) to return the ball back to the opponent's court; and (3) to hit the ball; and (4) most simply, to move the joints of the robot arm to the given desired angles.

To realize the task following the 4th requirement, we need only to equip the robot with some robust joint feedback controllers. However, we must give the desired robot joint angles in detail from the final objective! For the task requirement of the 3rd level, it is necessary for the robot to construct its sensory-motor map from the ball position in the work space (through the camera retinas) to the joint angles. And, for the 2nd or 1st requirement, the robot should further learn the ball dynamics and the opposite player's motion behavior. If there is no external teacher that teaches the robot at the joint control level, then the robot should **learn through action**. An appropriate learning algorithm is therefore necessary for an autonomous robot to learn who to realize the task itself. This learning algorithm should not only solve the nonlinear relations between the robot's task space and the actuation space, even if there is a redundancy, but also adapt to the uncertainties of the environment and the robot's variation, and converge as fast as possible.

From an information processing point of view, Kohonen proposed a parallel distributed learning algorithm using artificial neural networks[1]~[4]. This algorithm self-organizes the weight vectors of each neurons on a given structured network in parallel such that: i) neurons have similar weight vectors are located in neighborhood; and , ii) the density of the weight vectors tends to approximate the density of the input vectors. Inspired by this algorithm, Ritter et al. proposed an extended self-organizing feature map algorithm that approximates a mapping between the input space and the output space by establishing a topology-conserving map[5]~[7], and they applied this algorithm for a robot to self-organize its visuo-motor map. These two algorithms require prior knowledge of the topological structure of the input data. During the learning process, it is necessary to calculate not only the distances between the input vector and the weight vectors but also the position distances from the winner neuron on the network. An alternative algorithm is the "neural-gas" algorithm[7]~[9]. This algorithm calculates only the distances between the input vector and the weight vectors in the learning process, thus improving the learning speed. However, there are no theoretical studies on whether the density of the weight vectors tends to approximate the density of the input vectors. When we apply the algorithm for a robot to learn the visuo-motor map, even if we set the input targets to be uniformly distributed in the work space, we may not expect the resultant map to be uniformly distributed. Therefore, if we utilize such map to generate the robot's movement, in some areas of the work space, large positioning errors due to the less of the neuron units may arise. Other studies on autonomous robot's learning problems can be found in [10] and [11].

In order for an autonomous robot to self-organize a high quality visuo-motor map, in this paper, we propose our original approach to control the learning process by taking into account the statistical property of the map (e.g. the spatial distribution of the units). Computer simulations show that our approach not only obtaines a fast learning process but also results in a high quality map with desired distribution of the units. Utilizing this map, we can expect the robot to realize more accurate movements in the work space than the above two algorithms.

In section 2, we discuss the extended self-organizing feature map algorithm and the neural-gas algorithm. Based on this, we proposed our approach to control the learning process in section 3. In section 4, we show the effectiveness of our approach through computer simulation. We then conclude our studies finally in section 5.

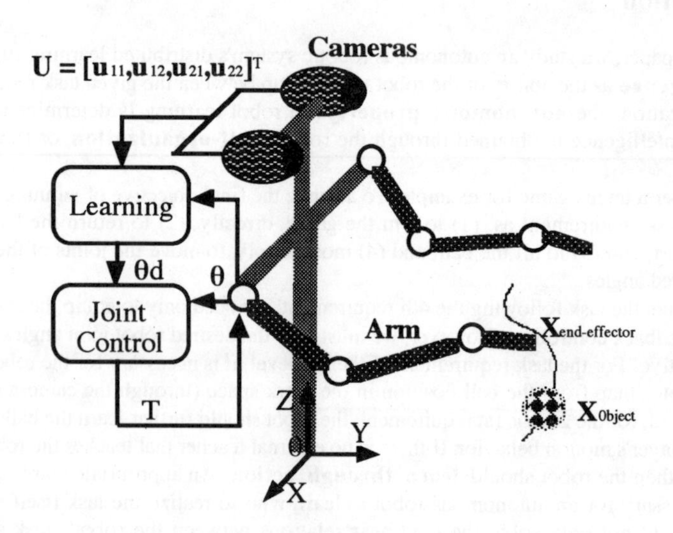

Fig.1 Autonomous robotic system

2 Self-Organization of Visuo-Motor Map

In this section, we focus our study on an autonomous robotic system's self-organization of its visuo-motor map, and discuss two distributed learning algorithms developed in [8], [9] and [10].

The over all robotic system is shown in Fig.1. Here, the target position X_t and the robot end-effector X_e in a 3-dimensional work space are observed by a pair of cameras. The two image coordinates on the cameras are combined as a 4×1 vector U and inputted to the neural network for learning the visuo-motor map. After the learning process, the robot realizes its movements based on the obtained map.

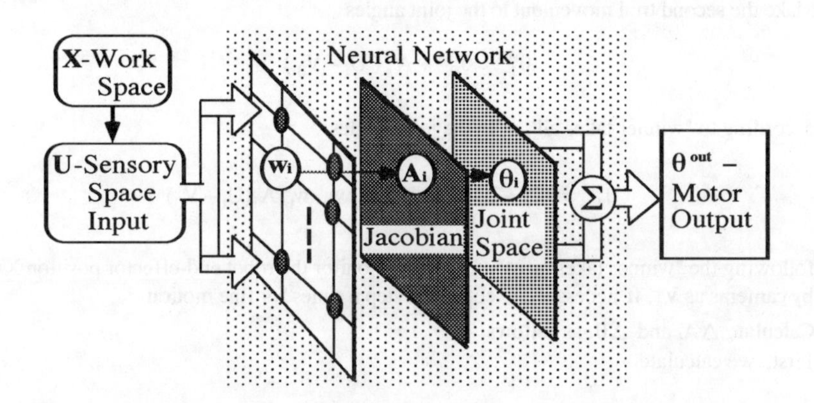

Fig.2 Neural network structure for learning the visuo-motor map

The main procedure in the learning process for the extended self-organizing feature map algorithm and the neural-gas algorithm are almost the same. We summarize them as follows[6],[7].

Step 1. Construct a neural network.

In the extended self-organizing feature map algorithm, the neuron unit **i** (for all i = 1, 2, ...N) is set at a node of a 3-dimensional lattice, as shown in Fig.2. And, in the "neural gas" algorithm, all neurons need not be fixed on any given structure.

Set randomly the 4×1 weight vector w_i, the corresponding 3×1 output joint angles' vector θ_i and the 3×4 Jacobian matrix A_i, with respect to the neuron **i**.

Step 2. Present a 3×1 target point X_t randomly in the robot's work space, and monitor it through a pair of cameras. Represent the two image coordinates as a 4×1 vector $U_t = [u_{11}, u_{12}, u_{21}, u_{22}]^T$ and input it to the neural network.

Step 3. Calculate the distance between the U_t and the weight vector w_i for all i = 1, 2, ...N and define a winner neuron unit **s** such that

$$\|U_t - w_s\| = \min\{\|U_t - w_i\|\} \tag{1}$$

The corresponding vector of the output joint angles and the Jacobian matrix are θ_s and A_s.

Step 4. Make the first trial movement to the following joint angles

$$\theta_0^{out} = \theta_s + A_s(U_t - w_s) \tag{2}$$

according to "winner takes all'.

or to the joint angles

$$\theta_0^{out} = L^{-1} \sum_{i=1}^{N} g^{mix}(U_t, w_s, w_i)[\theta_i - A_i(U_t - w_i)] \tag{3}$$

where

$$L = \sum_{i=1}^{N} g^{mix}(U_t, w_s, w_i) \tag{4}$$

for "winner takes most", and monitor the robot's end-effector position X_0 by cameras as V_0.

Step 5. Make the second trial movement to the joint angles

$$\theta_1^{out} = \theta_0^{out} + A_s (U_t - V_0) \tag{5}$$

according to "winner takes all', or to the joint angles

$$\theta_1^{out} = \theta_0^{out} + L^{-1} \sum_{i=1}^{N} g^{mix}(U_t, w_s, w_i) A_i(U_t - V_0) \tag{6}$$

following the "winner takes most". And then, monitor the robot end-effector position X_1 by cameras as V_1. If necessary, execute this step n times for fine motion.

Step 6. Calculate ΔA_i and $\Delta \theta_i$ as follows:
First, we calculate

$$\theta_i^* = \theta_0^{out} + A_i(V_0 - w_i) \tag{7}$$

and define

$$\Delta\theta_{01}^{out} = \theta_1^{out} - \theta_0^{out},$$

$$\Delta V_{01} = V_1 - V_0 \tag{8}$$

then we get,

$$\Delta\theta_i = \theta_i^* - \theta_i \tag{9}$$

$$\Delta A_i = ||\Delta V_{01}||^{-2}(\Delta\theta_{01}^{out} - A_i \Delta V_{01})\Delta V_{01}^T \tag{10}$$

ΔA_i decreases the cost function

$$E = \frac{1}{2}||\Delta\theta_{01}^{out} - A_i \Delta V_{01}||^2 \tag{11}$$

Step 7. Adjust the w_i, A_i and θ_i for all i = 1, 2, ..., N neuron units according to the following learning rules.

$$w_i \leftarrow w_i + \varepsilon \cdot g(U_t, w_s, w_i)(U_t - w_i) \tag{12}$$

$$\theta_i \leftarrow \theta_i + \varepsilon' \cdot g'(U_t, w_s, w_i)\Delta\theta_i \tag{13}$$

$$A_i \leftarrow A_i + \varepsilon' \cdot g'(U_t, w_s, w_i)\Delta A_i \tag{14}$$

and return to step 2.

In the above steps 4 to 7, the coefficients ε and ε' scale the overall correction size. The functions $g(.)$, $g'(.)$ and $g^{mix}(.)$ in eqs.(3), (12) and (13) determine the spatial variation region according to the relations between U_t, w_s, w_i and the learning step t. These functions greatly influence both the performance of the learning process (e.g. the convergence speed) and the learning result (e.g. the robustness of the output map etc.). The main difference between the extended self-organizing feature map algorithm and the "neural-gas" algorithm is due to the different selection of the functions $g(.)$, $g'(.)$ and $g^{mix}(.)$.

In the extended self-organizing feature map algorithm, the functions $g(.)$, $g'(.)$ and $g^{mix}(.)$ are chosen according to the distance from the winner unit's position **s** to the i-th unit position **r** on the 3D lattice as the following form

$$g(.) = \exp(-\|\mathbf{r}-\mathbf{s}\|^2/2\sigma^2(t)) \tag{15}$$

and in the "neural-gas" algorithm, these functions are chosen as the form

$$g(k) = \exp(-k/\lambda(t)) \tag{16}$$

respectively, where k is defined as the order of the distances from the target to the weight of each neuron unit. In detail, that is, for a given target U_t, we calculate the distance between all weight vectors w_i and U_t. We take $k = 1$ for the neuron unit that has the smallest distance, and $k = 2$ for the next smallest distance, and so on.

It is clear that, in the extended self-organizing feature map algorithm, we should calculate not only the distances from the target to the weight of each neuron unit in order to determine a winner, but also the distances from the winner to all other units' position $\|\mathbf{r}-\mathbf{s}\|$. This is not needed in the "neural-gas" algorithm. The order k in eq.(15) can be obtained when we determine a winner. On the other hand, in the "neural-gas" algorithm, the neuron units do not fix on any given structure, therefore, even if we present the input targets uniformly in the work space, we can not guarantee that the resultant map is uniformly distributed at the end of the learning process. This is greatly influenced by the parameters σ in eq.(15) and λ in eq.(16). We call these parameters the variance parameters and discuss their function in the next section.

From these two learning algorithms, it is understood that, during the learning process, for a given target in the work space, all neurons adjust their values simultaneously, in parallel. And, after the learning process, the robot moves its angle following the eq.(3) which is the weighted summation of all neuron units. Therefore, even if there are errors or destructions in some neurons, they do not influence the learning process and the map's output seriously. Distributed learning strategy improves not only the learning speed but also the system's robustness.

3 Control of the Learning Process

As mentioned in section 2, the variance parameters σ in eq.(15) and λ in eq.(16) determine the spatial variation region in each learning step. They play the most important role in controlling the learning processes. In this section, after analyzing the effect of these variance parameters, we propose our original approach to control the learning process in order to obtain the high quality visuo-motor map for the robot system.

In the extended self-organizing feature map algorithm, as seen in eq.(15), the parameter σ determines the region from the position of the winner on the lattice to those neuron units that should adjust their values a lot for a given target. In the "neural-gas" algorithm, the parameter λ determines the region of the order of neuron units that should adjust their values a lot for a given target. Large σ

and λ result in the adjustment of a lot of neurons and make the neurons very sensitive to every target input, and approach to the similar values. This reduces the effect of the distributive representation of the map. Conversely, if the σ and λ are taken to be very small, then, for a given input, only a few neurons adjust their values. A long learning time is then necessary to obtain a final map. For the two models mentioned above, σ and λ is varied with respect to the learning step t as

$$p(t) = p_i(p_f/p_i)^{t/t_{max}} \tag{17}$$

that uniformly decreases with respect to the learning step t. As pointed out in [7], initially, they are taken to be large (when t=0, $p(0)=p_i>1$) for rapid learning by many neuron units, and then gradually decreased (when $t \to t_{max}$, $p(t) \to p_f<1$) for the "fine-tuning". This approach did not consider the resultant distribution of the neurons' output in the work space when the learning process ended within a finite step t_{max}. Successful learning largely depends on the selection of p_i and p_f. As we will show in Fig. 6 of the next section, in some areas of the work space, the output vectors may be well mapped because of the existence of a lot of neurons in and/or near the areas, but in some other areas, the approximation using eq.(3) may fail with little neurons near the target. Therefore, when we use such map to generate a robot's movement following the eq.(3), there will be large position errors generated during motion through these areas.

Motivated by this, we present a new approach to control the above learning process by taking into account the statistical property of the learning result at every learning step.

We define the variance parameter here as

$$p(t) = C(t) \cdot f(v_d - v(t)), \text{ with } C(t) = a \cdot b^{t/c} \tag{18}$$

where a, b(<1) and c>0 are constants. v_d is the desired spatial distribution of the neurons' output vectors in the work space, and v is the spatial distribution at the learning step t. Based on this adjustment, the learning process not only depends on the time t but also on the error v_d-v. Therefore, the learning process will go on as long as $v \neq v_d$. As shown in the next simulation, we simply design

$$f(v_d - v(t)) = v_d - v(t) \tag{19}$$

4 Computer Simulations

Computer simulations have been done on the Unix-workstation to show the effectiveness of our approach. For simplicity, we considered a two-degree of freedom robot arm moving in a two-dimensional work space. The parameters of the above three algorithms are set as follows: N = 100 neural units have been used in our simulations; and, the maximum learning step is t_{max} = 4000. For the extended self-organizing feature map algorithm, we chose $\varepsilon_i = 1.0$, $\varepsilon_f = 0.01$, $\varepsilon_i' = \varepsilon_f' = 0.9$, $\sigma_i = \sigma_i' = 3$, $\sigma_i^{mix} = 1$, $\sigma_f = 0.05$, $\sigma_f' = 0.6$ and $\sigma_f^{mix} = 0.1$. For the "neural-gas" algorithm, we set $\varepsilon_i = 0.3$, $\varepsilon_f = 0.05$, $\varepsilon_i' = 0.9$, $\varepsilon_f' = 0.6$, $\lambda_i = 150$, $\lambda_i' = \lambda_i^{mix} = 50$, and $\lambda_f = \lambda_f' = \lambda_f^{mix} = 1$. And, in our approach, to make a clear comparison with the "neural gas" algorithm, we put all ε_i, ε_f, ε_i', ε_f', λ_i', λ_i^{mix}, λ_f', λ_i^{mix} as the same parameters as in the "neural-gas" algorithm. We only alternate the g(k) of the "neural-gas" algorithm using eq.(18), and set a = 1, b = 0.9 and c = 30. In order to calculate the spatial distribution of the neuron units at every learning step, we divide the work space into N (=100) meshes, and count those meshes in which there exists at least one neuron unit. For the example shown in Fig.3, the desired spatial distribution v_d = 100 and the spatial distribution v = 8.

Fig.4 shows the variations of p(t) with respect to the learning step t, where the extended self-organizing feature map algorithm and the "neural gas" algorithm use the eq.(17), and in our approach we use the eqs.(18) and (19). It is clear that the spatial variation region in our approach decreases faster than eq.(17). Since the learning process changes the neuron units' spatial distribution, and, by the eqs.(18), (19), the spatial variation controls the learning process, therefore, just from the results of the Fig.4, we can only say that our learning process is ended much faster, but we can not say that our learning result is better. However, from Fig.5 and 6, it is clear that the positioning error converges faster and the distribution is more uniform in our approach than using the eq.(17).

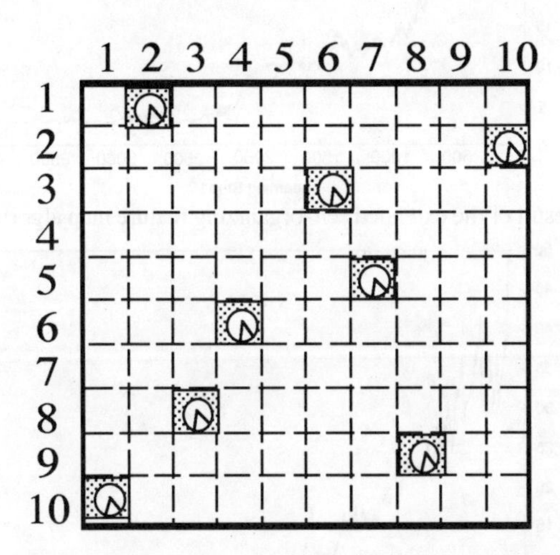

Fig.3 An example to calculate the spatial distribution of the neural units by counting those meshes in which there exists at least one neuron unit.

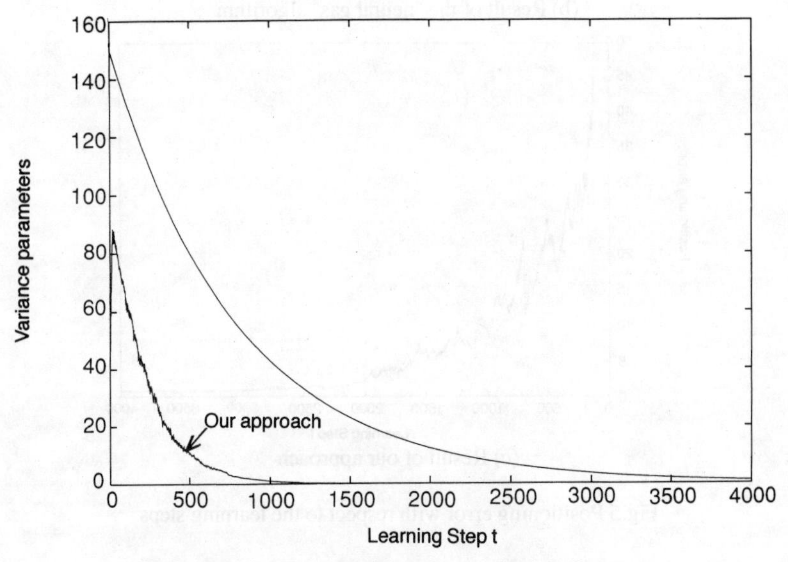

Fig.4 Variations of p(t) with respect to the learning step t

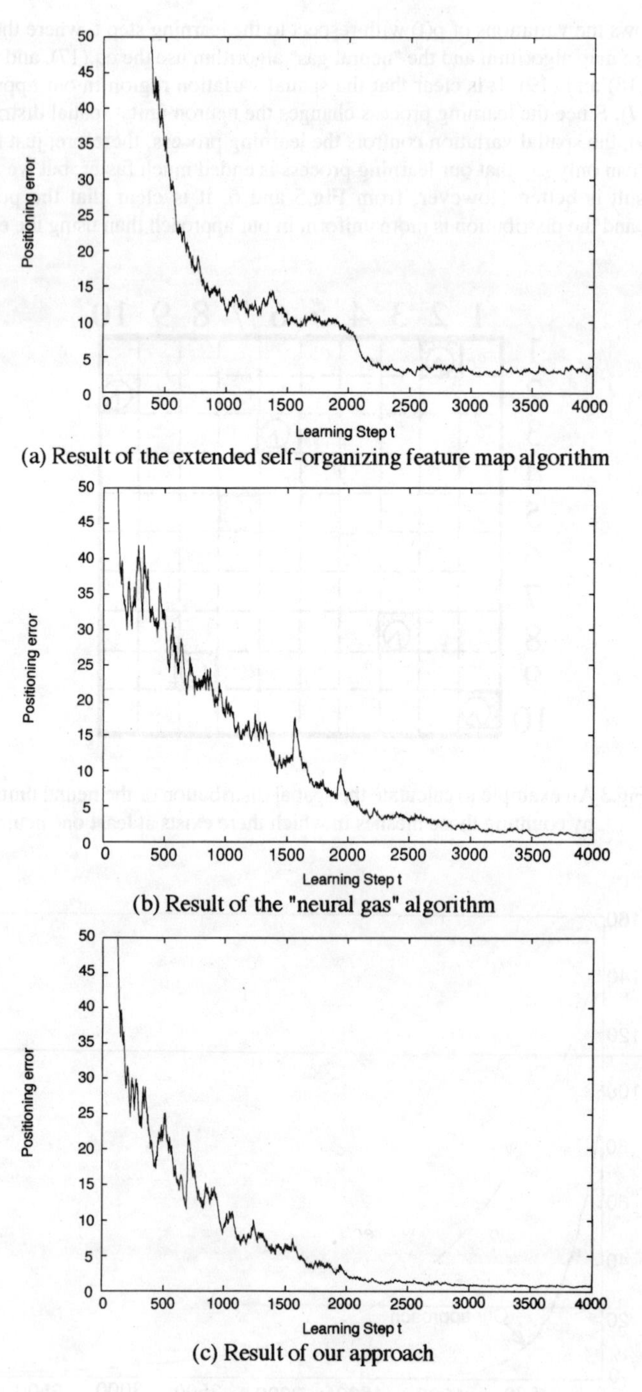

Fig.5 Positioning error with respect to the learning steps

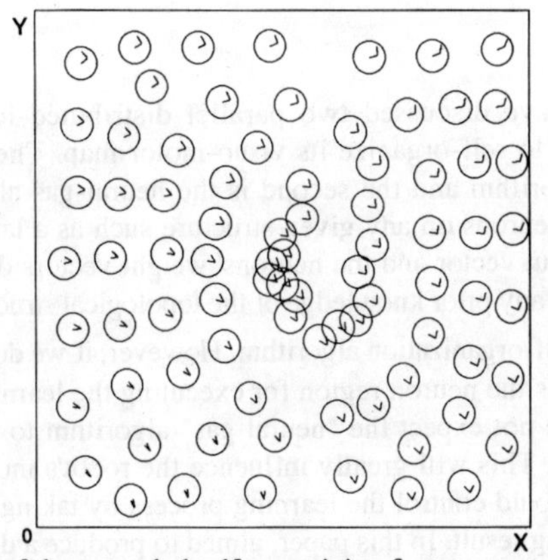

(a) Result of the extended self-organizing feature map algorithm

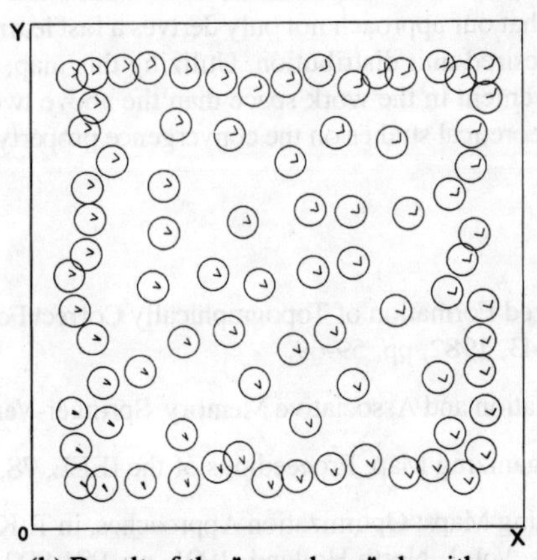

(b) Result of the "neural gas" algorithm

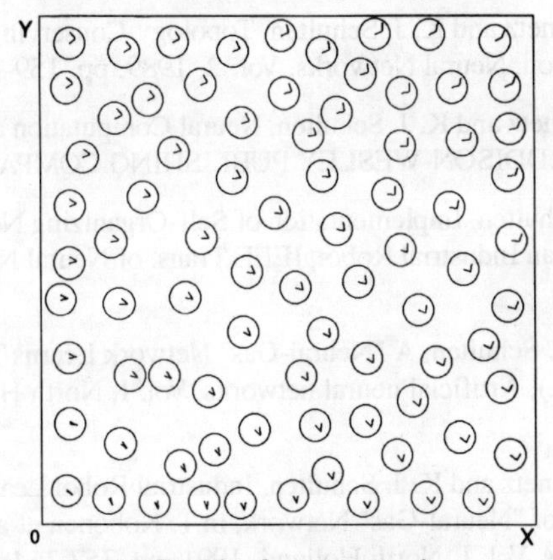

(c) Result of our approach

Fig.6 Spatial distribution of the neuron units in the work space after 4000 learning steps

5 Conclusion

In this paper, we have discussed two parallel distributed learning algorithms for an autonomous robotic system to self-organize its visuo-motor map. The first is the extended self-organizing feature map algorithm and the second is the neural-gas algorithm. The "neural gas" algorithm does not fix the neurons on any given structure such as a lattice. It only uses the error information between the input vector and the neurons' weight vectors during the learning process. Therefore, it does not require any prior knowledge of the topological structure of the input data which is required in the extended self-organization algorithm. However, if we do not carefully design $\lambda(t)$ -- the parameter that determines the neuron region for executing the learning process, at the end of a finite learning steps, we can not expect the "neural gas" algorithm to generate a map with some desired statistical properties. This will greatly influence the robot's movement when utilizing the resultant map. Hence, we should control the learning process by taking into account the statistical information about the learning result. In this paper, aimed to produce a desired visuo-motor map for the robot, we considered the spatial distribution of the neuron units to control the learning process. Computer simulations show that our approach not only derives a fast learning process but also results in a high quality map with desired unit distribution. Utilizing this map, we can expect the robot to achieve a more accurate movement in the work space than the above two algorithms. As the future research, we will make the theoretical studies on the convergence property of the proposed approach.

References

[1] T. Kohonen, Self-organized Formation of Topographically Correct Feature Maps, Biological Cybernetics, 43, 1982, pp. 59-69.

[2] T. Kohonen, Self-organization and Associative Memory, Springer-Verlag, Berlin, 1989.

[3] T. Kohonen, The Self-organizing Map, Proceedings of the IEEE, 78, 1990, pp. 1464-1480.

[4] T. Kohonen, Self-organizing Maps: Optimization Approaches, in T. Kohonen et al.(Eds.), Artificial neural networks, Vol. I, North Holland, 1991, pp. 981-990.

[5] H. J. Ritter, T. M. Martinetz and K. J. Schulten, Topology-Conserving Maps for Learning Visuo-Motor-Coordination, Neural Networks, Vol. 2, 1989, pp. 159-168.

[6] H. J. Ritter, T. M. Martinetz and K. J. Schulten, Neural Computation and Self-Organizing Maps: an introduction, ADDISON-WESLEY PUBLISHING COMPANY, 1992.

[7] J. A. Walter and K. J. Schulten, Implementation of Self-Organizing Neural Networks for Visuo-Motor Control of an Industrial Robot, IEEE Trans. of Neural Networks, Vol. 4, No. 1, 1993, pp.86-95

[8] T. M. Martinetz and K. J. Schulten, A "Neural-Gas" Network Learns Topologies, in T. Kohonen et al.(Eds.), Artificial neural networks, Vol. I, North Holland, 1991, pp. 397-402.

[9] J. A. Walter, T. M. Martinetz and K. J. Schulten, Industrial Robot Learns Visuo-motor Coordination by Means of "Neural-Gas" Network, in T. Kohonen et al.(Eds.), Artificial neural networks, Vol. I, North Holland, 1991, pp. 357-364.

[10] M. Kuperstein, Neural Model of Adaptive Hand-eye Coordination for Single Postures, Science, 239, 1988, pp. 1301-1311.

[11] P. Morasso and V. Sanguineti, Self-organizing body-schema for motor planning, J. of Motor Behavior, 1993.

Chapter 7
Multi-Robot Behavior

Behavior Control of Insects by Artificial Electrical Stimulation

YOSHIHIKO KUWANA, NAOMITSU WATANABE, ISAO SHIMOYAMA, and HIROFUMI MIURA

Dept. of Mechano-Informatics, The Univ. of Tokyo, Tokyo,113 Japan

Abstract

Recently microstructures are being built on silicon wafers, and they are often called micromachines or microrobots. To make an autonomous microrobot, there are several very difficult problems in supplying power, controlling their behavior, and making microactuators. We propose the use of living materials instead of silicon parts in this paper. A cockroach is used as an experimental material, and the muscles on its hind legs are stimulated by electrical impulses. To calculate the leg movement by electrical stimulation, we create the mathematical model of the muscles. This model provides the displacement of the joint angles of legs and the theoretical walking speed when the muscles are stimulated. A demonstration of a cockroach walking by computer produced electrical stimulation is shown.

Keywords: Cockroach, muscle, muscular modeling, electrical stimulation, walking

1 Introduction

Several microstructures employing an external skeleton and elastic joints similar to real insects[1] have been built on silicon wafers. These microstructures are often called micromachines or microrobots. However, to make an autonomous microrobot, power supply and behavior control remain very difficult problems to overcome. In this paper we propose the use of living materials instead of silicon parts as one possible solution.

It is interesting to consider bio-organisms as if built from mechanical elements. Muscles, one such bio-mechanism, contract when stimulated by the electrical activity of nerves, and it is possible to make muscles contract by using artificial electrical stimulation. If insect movement can be controlled in this way, they could be regarded as a new type of "robots," with an extremely wide range of possible applications. For example, insects have very sharp sensitivities to odors, sounds, and light. If they can be controlled as desired, their sensors could be used as noses, ears, and eyes. This is a new hybrid form of life combined living organisms with artificial machines.

The relation between stimulus and behavior has been thoroughly researched by neurobiologists. Various kinds of stimuli, electrical, visual, auditory, gustatory, and olfactory are used in this research. Most are concerned with the relation between behavior and nerves or between behavior and the type of stimulation. For example, Schmitz et al.[2] used an auditory stimulus for investigating the behavior of female crickets whose foreleg was amputated. Weber et al.[3] also used auditory and visual stimuli to test the female crickets' ability to track model calling songs with different syllable periods.

In this paper the use of insect muscles is discussed as a new method of actuation. A cockroach, scientific name *Periplaneta americana*, was used as living experimental material. The characteristics of cockroach's muscle were investigated, and then they were used to make a cockroach walk by means of an artificial electrical stimulus.

2 Muscles

2.1 Muscular Contraction Mechanism

Muscles contract as follows[4]: an action potential, brought via an axon, is transmitted to a muscle through an end plate and excites plasma membranes. Ca^{2+} ions isolate from the sarcorplasmic reticulum into muscle cells, and the muscle contracts. Finally the Ca^{2+} ions return to the sarcorplasmic recticulum and the muscle relaxes (Fig. 1). There are two kinds of muscles: extensor muscles and flexor muscles. Each contracts alternately[5] as an insect moves.

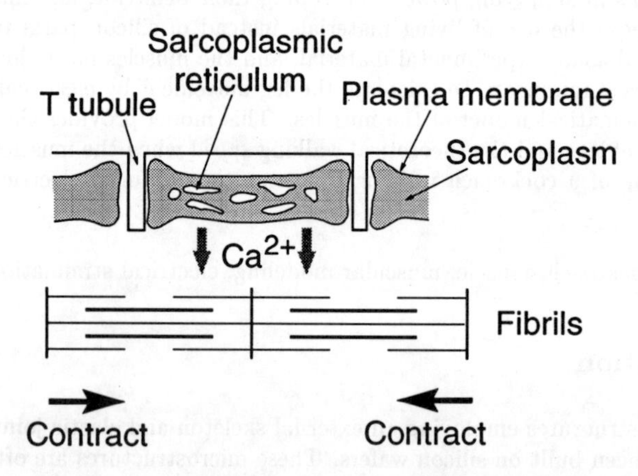

Fig. 1 Muscular Contraction Mechanism

Instead of a natural action potential from an axon, an artificial potential can be supplied through a thin wire inserted into a muscle as an electrode, and the muscle will contract. The potential placed into the muscle must be larger than that of a natural axon because the end of an axon is separated into several parts, each of which excites the muscle.

2.2 Muscle Structure of a Hind Leg

The structure of a cockroach's hind leg is shown in Fig. 2[6]. We stimulated muscles #171 and #181, both flexor muscles. The displacement of the joint angle changes in reaction to the stimulus. When the #171 muscles are stimulated, the joint angle θ between coxa and femur changes, and when the #181 muscles are stimulated, the joint angle ϕ between thorax and coxa changes as shown in Fig. 3.

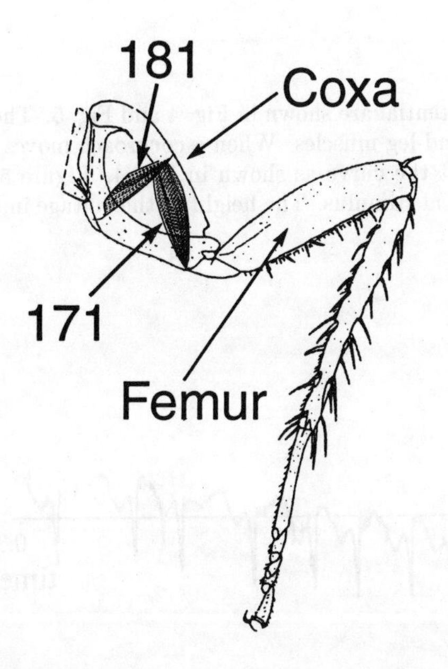

Fig. 2 Muscle Structure of a Cockroach's Hind Leg

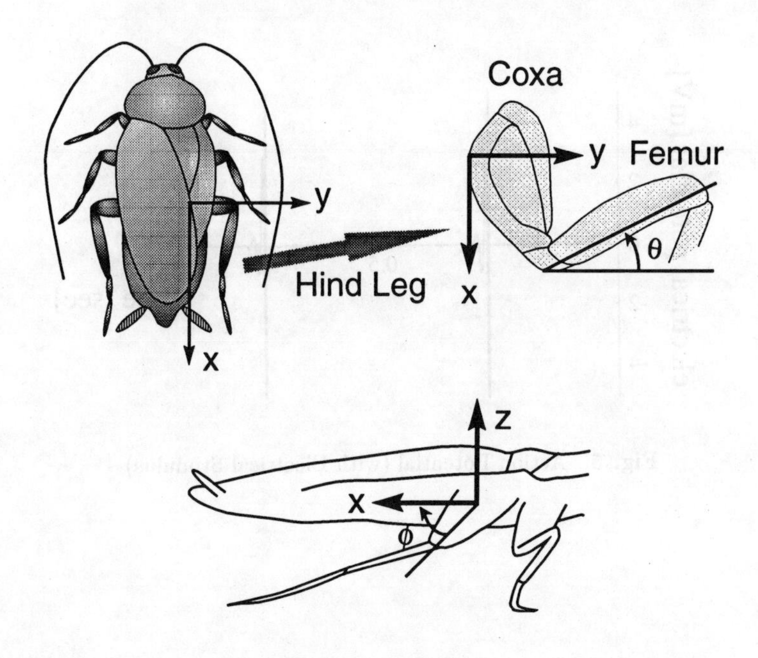

Fig. 3 θ and ϕ

2.3 Action Potential

The curves of the action potential are shown in Fig. 4 and Fig. 5. These curves show the potentials of a cockroach's hind leg muscles. When a cockroach moves its leg without any stimulus from outside, we got the curve as shown in Fig. 4. Figure 5 shows leg muscles excited by an artificial electrical stimulus. The height of the voltage impulse was 5[V], and the width was 1[msec].

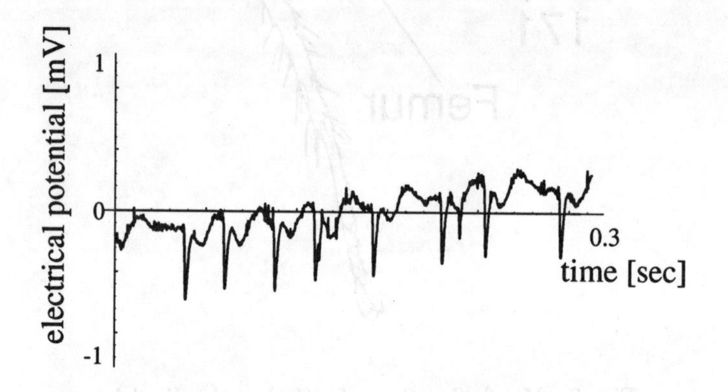

Fig. 4 Action Potential (without Stimulus)

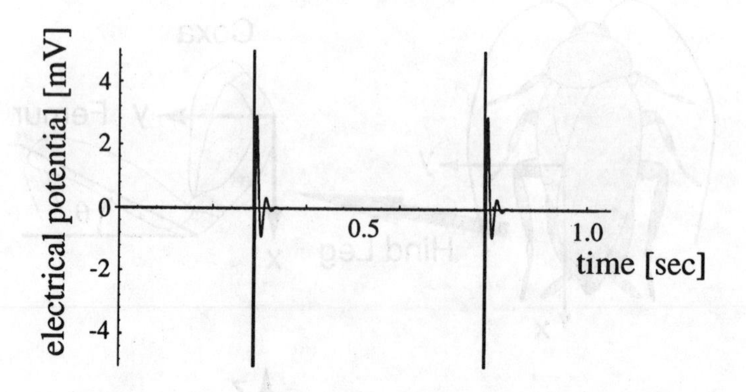

Fig. 5 Action Potential (with Electrical Stimulus)

3 Dynamic Model of Muscles

3.1 Mathematical Model

To calculate the motion of a cockroach's leg, the leg muscles are modeled with three elements: elastic, viscous, and contractile elements (Fig. 6).

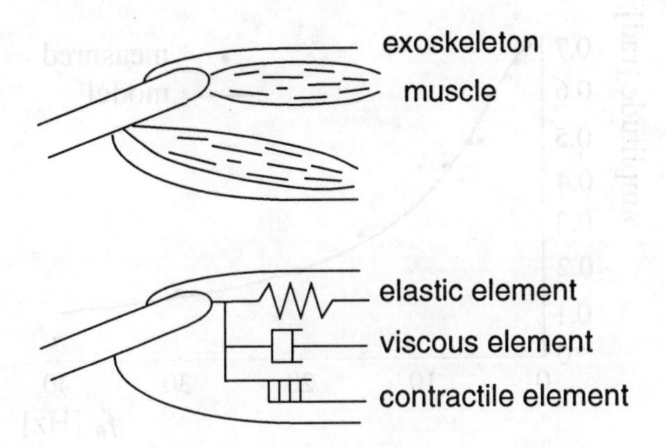

Fig. 6 Muscular Model

The following equations are mathematical model of a muscle:

$$I\ddot{\theta} + 2I\beta_\theta\dot{\theta} + I\omega_\theta^2\theta = IC_\theta F(t) \tag{1}$$

$$F(t) = \sum_{n=0}^{\infty} f(t - n/f_\theta) \tag{2}$$

$$f(t) = \begin{cases} 5 \;\; \text{[V]} & \text{if } 0 \le t \le 10^{-3} \text{ [sec]} \\ 0 \;\; \text{[V]} & \text{otherwise} \end{cases} \tag{3}$$

where I is the moment of inertia, θ is the joint angle between coxa and femur (in Fig. 3). When the #181 muscles (in Fig. 2) are stimulated, θ changes. $I\ddot{\theta}$ is the inertial force, $2I\beta_\theta\dot{\theta}$ is the attenuation force by viscosity, $I\omega_\theta^2\theta$ is the restitution force, $F(t)$ is the electrical stimulation function, C_θ is a proportionality constant, f_θ is the frequency of stimulation and $f(t)$ represents the one stimulation pulse.

As with θ, the joint angle ϕ changes when the #171 muscles (in Fig. 2) are stimulated, where ϕ is the joint angle between thorax and coxa (in Fig. 3). We can also formulate the leg movement as follows:

$$I\ddot{\phi} + 2I\beta_\phi\dot{\phi} + I\omega_\phi^2\phi = IC_\phi F(t) \tag{4}$$

where I is the moment of inertia, as in Eq. 1. $F(t)$ is the electrical stimulation function, the same as Eq. 2.

Table 1 Modeling Parameters (the Case of θ)

β_θ	18.0 [s^{-1}]
ω_θ	13.0 [s^{-1}]
C_θ	6.3×10^3 [$s^{-2}V^{-1}$]

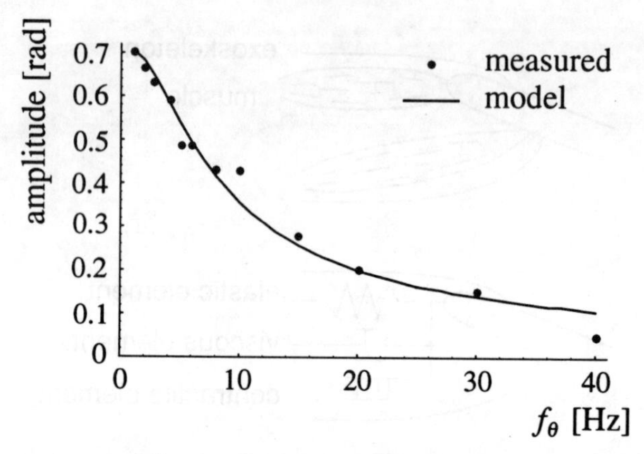

Fig. 7 Frequency Response of θ

Table 2 Modeling Parameters (the Case of ϕ)

β_ϕ	17.2 [s^{-1}]
ω_ϕ	16.4 [s^{-1}]
C_ϕ	0.89×10^3 [$s^{-2}V^{-1}$]

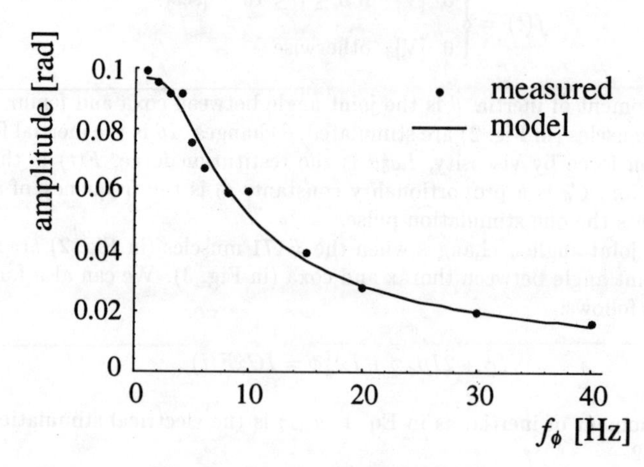

Fig. 8 Frequency Response of ϕ

3.2 Theoretical Leg Movement

The modeled frequency response of θ in Eq. 1 is shown in Fig. 7. The parameters β_θ, ω_θ, and C_θ of Eq. 1 were selected by fitting the curve to the measured data as shown in Table 1. The frequency response of ϕ in Eq. 4 can also be obtained, and is shown in Fig. 8. The parameters β_ϕ, ω_ϕ, and C_ϕ were selected to be consistent with the measured data in Table 2.

After one stimulation, at a frequency of 2Hz, the joint angles θ and ϕ change as shown in Fig. 9 and Fig. 10.

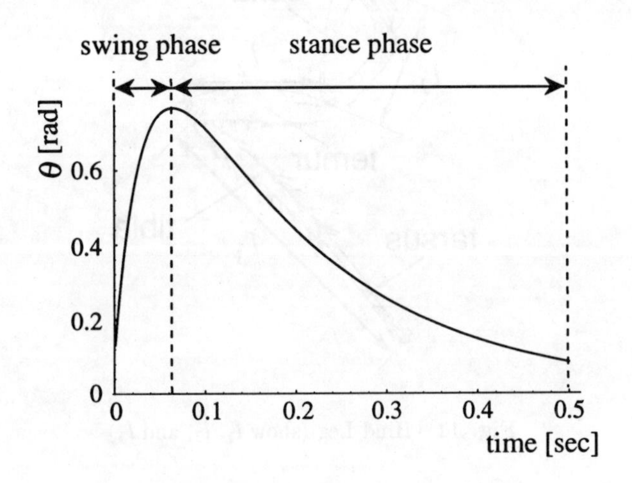

Fig. 9 Transition of θ after One Stimulation (f_θ=2Hz)

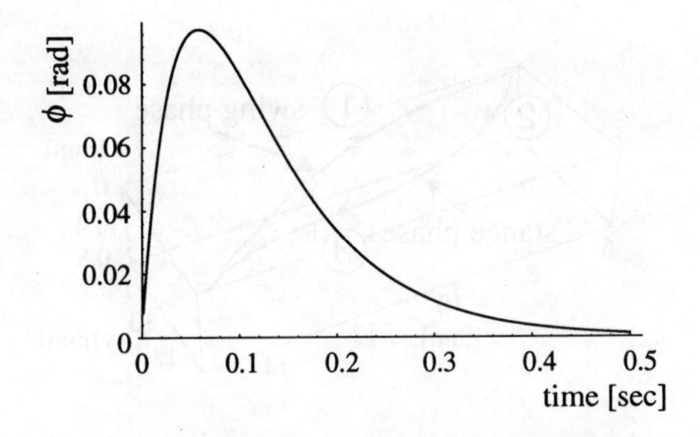

Fig. 10 Transition of ϕ after One Stimulation

When the x-y-z coordinates are given as shown in Fig. 3, the toe position (x, y, z) is calculated from Eq. 5, Eq. 6, and Eq. 7, where l_1 is the length of coxa, l_2 is the length of the femur, and l_3 is the length of tibia and tarsus as shown in Fig. 11.

$$x = (l_1 - l_2 \sin\theta + l_3 \cos\theta)\cos\phi \tag{5}$$
$$y = l_2 \cos\theta + l_3 \sin\theta \tag{6}$$
$$z = (l_1 - l_2 \sin\theta + l_3 \cos\theta)\sin\phi \tag{7}$$

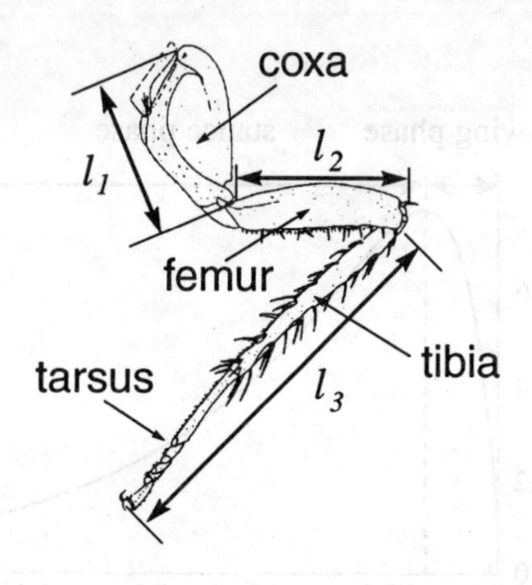

Fig. 11 Hind Leg (show l_1, l_2, and l_3)

Therefore, substituting the values from Fig. 9 and Fig. 10 into Eq. 5, Eq. 6 and Eq. 7, the toe motion by artificial electrical stimulation was obtained as shown in Fig. 12.

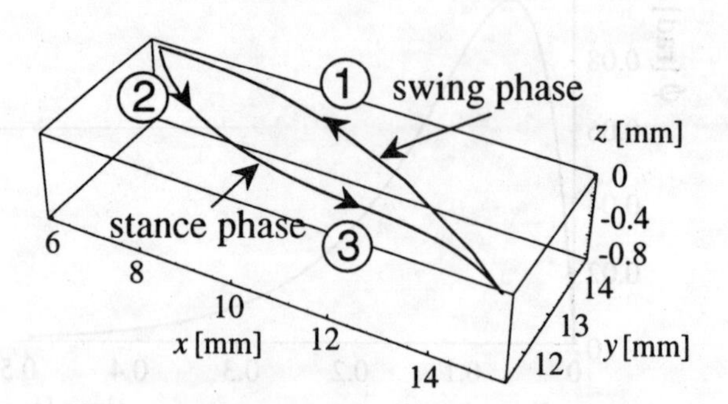

① The hind leg is brought up,
② Then placed on the ground,
③ And moves forward by kicking the ground.

Fig. 12 Trace of Hind Leg's Toe Position (f_θ =2Hz)

3.3 Walking Speed

The walking speed V is determined by stride, period, and duty factor as follows:

$$V = \frac{S}{T} \tag{8}$$

$$V = \frac{S_r}{\alpha T} \tag{9}$$

S	:	Absolute stride
S_r	:	Relative stride
T	:	Period
α	:	Duty factor

The absolute stride is the distance that a cockroach goes forward in one period while the relative stride is the distance that a leg moves in the swing phase. Since the absolute stride was difficult to calculate, Eq. 9 was used to determine the walking speed. Duty factor α is the ratio of the time in contact with the ground to the total time of one period, and is given as follows:

$$\alpha = \frac{t_{st}}{t_{sw} + t_{st}} = \frac{t_{st}}{T} \tag{10}$$

where t_{st} is the time of the stance phase, that is, the time in contact with the ground, and t_{sw} is the duration of the swing phase. The frequency response of the duty factor α is shown in Fig. 13.

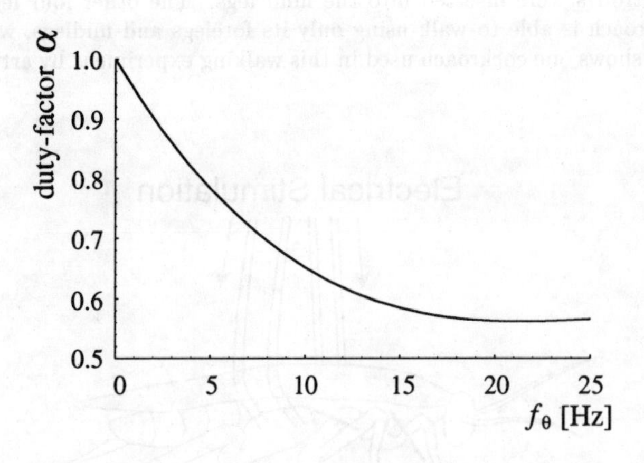

Fig. 13 Frequency Response of α

Several theoretical walking speeds are shown in Fig. 14. The theoretical maximum speed was calculated by measuring the stride ($S = 13.0$ mm) and assuming tripod walking with $\alpha = 0.5$. Insects can walk fastest with tripod walking. The walking period T is the reciprocal of the stimulus frequency. For the other two cases, Ⓘ and Ⓘ, S_r is found from the amplitude of θ, and α is given by Fig. 13. Ⓘ shows the results when only the coxal muscles are stimulated and Ⓘⓘ shows that both the coxal muscles and femoral muscles are stimulated.

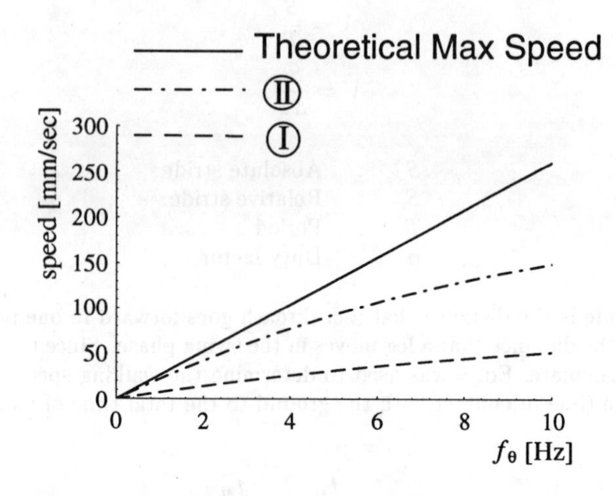

Fig. 14 Comparison of Walking Speeds

4 Walking Experiment

The walking speed of a cockroach using only the two hind legs were measured. As shown in Fig. 15, electrodes were inserted into the hind legs. The other four legs were cut off because a cockroach is able to walk using only its forelegs and midlegs, without its hind legs. Figure 16 shows one cockroach used in this walking experiment by artificial electrical stimulation.

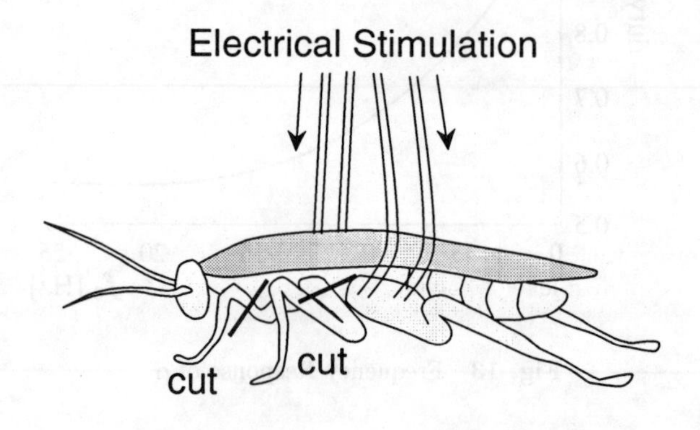

Fig. 15 Experimental Arrangement

Table 3 lists the walking speeds for several cases. The theoretical speeds, max, Ⓘ and

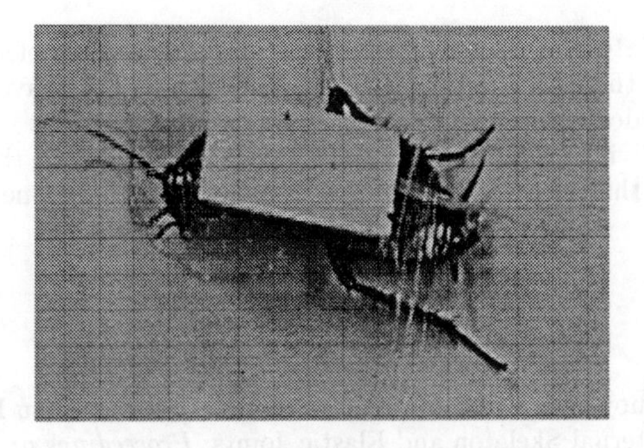

Fig. 16 Walking by Artificial Electrical Stimulation

Ⓘ, were found from Fig. 14. The experimental walking speed, by electrical stimulation, is much smaller than the theoretical speed because of the friction between the thorax and the floor and because only the two hind legs were stimulated. If more legs and muscles are stimulated, a cockroach will walk faster.

Table 3 Comparison of Walking Speeds

Frequency	Cockroach	Speed [mm/s]
	Theoretical Max Speed	52
2	Theoretical Ⓘ	42
[Hz]	Theoretical Ⓘ	14
	Experiment	5
	Theoretical Max Speed	103
4	Theoretical Ⓘ	77
[Hz]	Theoretical Ⓘ	26
	Experiment	7

5 Conclusion

This paper proposes the use of real insects as a new robotic system. Insects have feedback control systems for higher level behavior, such as obstacle avoidance. For lower level behavior like leg movement, however, they only have an ON - OFF type control system which uses stimulation by the nervous system.

It was possible to control the lower level behavior by artificial stimulation applied by a personal computer. By changing the period of stimulation, muscles were made to contract as desired. Cockroach legs have several kinds of muscles, and it is very difficult to control all of them. However, it is possible to control a few of them, thereby allowing insects to walk by artificial stimulation.

The muscles were modeled with three elements: elastic, viscous, and contractile. Using this model, we calculated the displacement of the joint angles after electrical stimulations, and determined the theoretical walking speed numerically. We have shown that insect muscles can be regarded as mechanical actuators.

When we think of living organisms as a set of mechanical elements, their abilities become very intriguing. In this paper, we proposed a method of using one of these elements, muscles.

References

[1] K. Suzuki, I. Shimoyama, H. Miura, and Y. Ezura, Creation of an Insect-based Microrobot with an External Skeleton and Elastic Joints, *Proceedings of 1992 IEEE MEMS*, 1992, pp. 190 – 195.

[2] B. Schmitz, H. U. Kleindienst, K. Schidberger, and F. Huber, Acoustic orientation in adult, female crickets(*Gryllus bimaculatus* de Geer) after unilateral foreleg amputation in the larva, *Journal of Comparative Physiology A*, Vol. 162, 1988, pp. 715 – 728.

[3] T. Weber, G. Atkins, J. F. Stout, and F. Huber, Female *Acheta domesticus* track acoustical and visual targets with different walking modes, *Physiological Entomology*, Vol. 12, 1987, pp. 141 – 147.

[4] R. F. Chapman, *The Insects, Structure and Function*, Harvard University Press, 1982.

[5] K. G. Pearson, Central Programming and Reflex Control of Walking in the Cockroach, *Journal of Experimental Biology*, Vol. 56, 1972, pp. 173 – 193.

[6] C. S. Carbonell, The Thoracic Muscles of the Cockroach *Periplaneta Americana, Smithsonian Miscellaneous Collections*, Vol. 107, No. 2, 1947, pp. 1 – 23.

Find Path Problem of Autonomous Robots by Vibrating Potential Method

KOJI YAMADA[1], HIROSHI YOKOI[2], and YUKINORI KAKAZU[1]

[1]Dept. of Precision Eng., Hokkaido Univ., Sapporo, 060 Japan
[2]National Inst. of Bioscience and Human-Tech., Tsukuba, 305 Japan

Abstract

We have been proposing vibrating potential method (VPM) as a new mathematical model for an intelligent robotic system that can solve various engineering problems. In the VPM, an information field as energy field is defined and multi-agents can observe and communicate each other through the field. The agents can move on the field by getting the energy from the field. Then, each robotic agent can execute position-control by local-area sensoring and path planning by detecting global information from the field. In this paper, a new robotic system is introduced as a mathematical model using the VPM. First, the concepts of the VPM are introduced. Next, a mathematical model of robotic system is defined by the VPM. Then find-path problem of autonomous robots on the system is discussed. Finally, computer simulations of the dynamic find-path process are shown.

Keywords: Vibrating Potential Method, multi-agents, path-planning

1. Introduction

In dynamic environment, flexibility and adaptability of information processing such as information acquisition, integration and decision are important for autonomous robotic system. Usually, a terrain map for a moving robot is constructed by distance information. The map such as a potential map is acquired by local search of terrain information. As perspective lacks, the path-planning couldn't avoid a local minima. On the other hand, global search can abstract the terrain information and make the abstracted map such as a graph map. However the graph map method simplifies the path-planning of robot as the graph search, it is impossible to avoid that a minute change of the solution makes a huge failure, because the local information is omitted. The question then arises about how to acquire and integrate the both local information and global information.

We, creatures can communicate with each other by using wave information such as vibration, sound and light. And a certain connection does not mediate between our communications. It is possible to interpret that our communication is made up of communicators that broadcast and receive information, and information field that is pool of information (Fig.1). From this interpretation, assuming that the information is described by wave functions, we propose a mathematical model of a new information processing system called vibrating potential method (VPM). In this method, the communicators and the information which the communicator deals with are defined as wave functions. The information field consists of a superposition of these wave functions. We use vibrating potential functions as the function, we call the information field a vibrating potential field (VPF). For an autonomous robotic system, we set an objective function to the communicator, we call this an agent. The agent can observe and select the information on the VPF. Then the information is processed and decides the agent's behavior as following its objective function. To behave the agent autonomously and distributively, we apply the dynamics of the life conservation to the objective function. In nature, creatures have energy interaction with each other and conserve each form in its environment. For example, creatures get heat capacity from food and move to warmer place. It is well known that these dynamics are based on the homeostasis of life mechanics. To construct a robotic system, we describe the energy interactions of robotic environment as the wave information. Our method has been applied to the other problems[1-3]. In this paper, we apply this method to solving the find-path problem of autonomous robots. We intend to solve this problem in dynamic environment. In our approach for this problem, given terrain information and robots are defined as the agents consist of wave functions we set. The robotic agent recognizes the environment by observing the wave from its surroundings on the VPF. Observing the propagating wave of the VPF, it is possible to expect that the robotic agent can recognize the both local-area information and global-area information. In our method, the only communication through the VPF is permitted. As the VPF is autonomously renewed, we can easily append other robots and obstacles to it. These features are effective to represent dynamic environment.

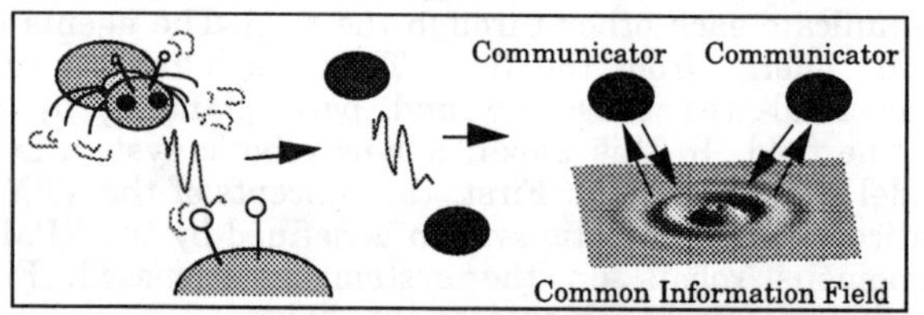

Figure 1: Communicators in the VPM

2. Find-Path Problem on VPF

To develop autonomous robotic system, finding-path problem is very important. This includes many problems such as dynamic recognition of environments, avoiding obstacles, path-planning and motion-planning. Terrain information which the robot is to recognize has represented as a potential map or a discrete graph map. These methods could not avoid the information omission and the local minima, we mentioned in the previous paragraph. And the both methods are not easy to represent dynamic environment in which there are many robots and moving obstacles[4,5]. Another approach we proposing is to map the terrain information to continuous space. We describe the terrain information as continuous wave information. Since our method deals with the wave energy, motion energy such as momentum of robots is dealt with as potential wave energy. That is, the both environmental representation and robot motion are described in one method. In our mathematical model, these wave information constructs a VPF. Our objective here is to find continuous motions and optimal paths for the autonomous agents from an initial configuration to a final target configuration. Consequently, the finding-path on VPF is to get wave information of goal from the field and to convert the energy to move.

3. Mathematical model of VPM

A fundamental mathematical model of problem-solving system using VPM is proposed here. In the VPM, the solution state of the problem is represented as the location of an agent (or agents). The agent is the smallest component of the system and defined by the continuous wave function. It can be characterized variously by the difference of its function such as moving robot, static obstacle and goal point. An agent is composed of a potential function h, and an output wave function w as below:

$$h_{ij}(\boldsymbol{r},\boldsymbol{n}) = \sum_{\mathrm{n}} \frac{(-1)^{\mathrm{n}} B_{\mathrm{n}}(\boldsymbol{r},\boldsymbol{n}) \cdot q_{\mathrm{n}}}{\alpha_{\mathrm{n}} + (\boldsymbol{r} - \boldsymbol{r}_{ij})^{2\mathrm{n}/k_{ij}(\boldsymbol{r},\boldsymbol{n})}}, \tag{1}$$

$$\mathrm{w}_{ij}(\boldsymbol{r},\boldsymbol{n}) = B_{\mathrm{n}}(\boldsymbol{r},\boldsymbol{n}) W^{|\boldsymbol{r}_{ij}-\boldsymbol{r}|} \exp\left(j\omega_0 |\boldsymbol{r}_{ij} - \boldsymbol{r}| - \eta t\right), \tag{2}$$

where, $\boldsymbol{r}$ is the position vector. $\boldsymbol{n}$ is the direction vector. $\boldsymbol{B}$ is the effective coefficient that determines an agent's form.
An example of potential function h is shown in Fig.2.

All agents are defined as constraints in the problem-space within VPF. The VPF is described by placing each agent's wave function on its own harmonic wave axes. To make the VPF as a common information field, all agents information is superposed on a VPF H as defined;

$$H(\boldsymbol{r},\boldsymbol{n}) = \sum_i \left\{ \sum_j h_{ij}(\boldsymbol{r},\boldsymbol{n}) \psi_{ij}(v) + \sum_k \mathrm{w}_{ik}(\boldsymbol{r},\boldsymbol{n}) \chi_{ik}(v) \right\}, \tag{3}$$

ψ, χ are the interaction axes and following equation guarantees their orthogonality.

$$\frac{\kappa(a)}{2} \frac{d^2 \zeta(v)}{dv^2} + E_v \zeta(v) = 0, \quad (\zeta(v) = \psi(v), \chi(v), \quad \zeta(\mu) = 0 \quad when |\mu| = a, b), \tag{4}$$

where, $\kappa(a)$ is the parameter which determines the spatial distance among unit coordinates axes. E is whole interaction energy. a, b is the initial coefficient.

k of equation (1) is the field recognition factor.

$$k_{ij}(\boldsymbol{r},\boldsymbol{n}) = k_0 + \frac{k_1}{2\pi} \left| \frac{d}{dt} \oint H(\boldsymbol{r},\boldsymbol{n}) \chi_{ij}(\phi) d\phi \right|. \tag{5}$$

The agent behaves satisfying Lagrange's equation. The motion momentum $\boldsymbol{p}$ and the angular momentum $\boldsymbol{q}$ are,

$$\dot{\boldsymbol{p}}(t) = \boldsymbol{M} \frac{\partial \oint H(\boldsymbol{r},\boldsymbol{n}) \psi(v) dv}{\partial \boldsymbol{r}}, \tag{6}$$

$$\dot{\boldsymbol{q}}(t) = \boldsymbol{I} \frac{\partial \oint H(\boldsymbol{r},\boldsymbol{n}) \psi(v) dv}{\partial \boldsymbol{n}}, \tag{7}$$

$\boldsymbol{M}$, $\boldsymbol{I}$ are inertia and inertial tensor of the agent.

To navigate the agents autonomously, we apply the objective function that executes the homeostasis of life mechanics. It is well known as the thermodynamic behavior of the life conservation that is minimizing the increase of entropy. The agent calculates its heat capacity both inside and outside of itself (Fig.3). The inflowing heat capacity E_{in} into agents, the internal heat capacity E and the outflowing heat capacity W are:

$$E_{in} = \oint_{r=z} \oint H(\boldsymbol{r},\boldsymbol{n}) \psi(v) dv dz, \tag{8}$$

$$\boldsymbol{E} = \sum_i \left(\frac{\boldsymbol{p}_i^2}{\boldsymbol{M}_i} + \frac{\boldsymbol{q}_i^2}{\boldsymbol{I}_i} \right), \tag{9}$$

$$W = \frac{\boldsymbol{P}^2}{\sum_i \boldsymbol{M}_i} + \frac{\boldsymbol{Q}^2}{\sum_i \boldsymbol{I}_i} \tag{10}$$

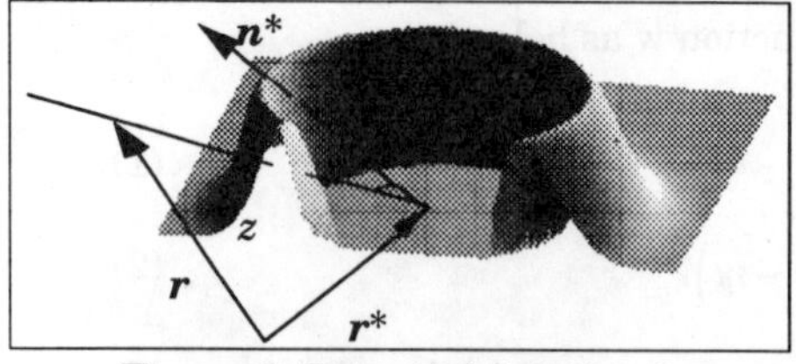

Figure 2: Potential function h

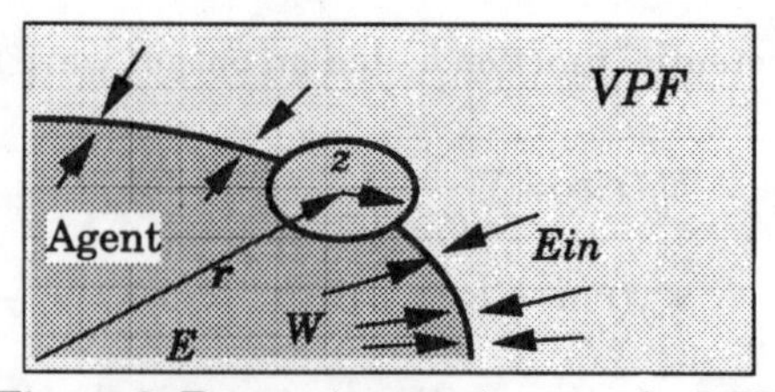

Figure 3: Energy interactions on the VPF

The entropy S is calculated as follows:

$$dS = \frac{d(E - W)}{E_{in}}. \tag{11}$$

The objective function of agents is:

$$\frac{\partial dS}{\partial \alpha} \leq 0. \tag{12}$$

$\boldsymbol{a}$ is the dominative coefficient for agents. We can navigate the agents by adjusting the coefficient.

In the VPF model, mutual interaction between the agents occurs only through the VPF, not through direct connections as in the connectionist model. The agent detects other agents wave information from its environmental surroundings. As a result, it is possible for an agent to communicate with unspecified agents. The VPM is able to represent these complex relations by way of the interactions between the agents and the VPF. Thus, each agent autonomously decides its own behavior according to the state of the environmental surroundings.

VPM's procedures are below:

Step 1: Each agent observes wave potential energy of its surroundings on the VPF.
Step 2: Each agent calculates its momentums and its potential energy.
Step 3: Superpose all agents energy information on the VPF.
Step 4: Go to step 1.

4. Experiments

Five computer experiments are demonstrated. In these experiments, the shape of an agent is a white rectangle. Dots represent the potential wave from the goal. The square patterned areas represent wall as obstacles. In this experiment, the objective function is set to move to the wave source.

First, we show the basic behavior of our model. An agent is set on the start point S bottom right of Fig.4, and a goal agent is point G. The goal agent doesn't have potential function, but it outputs only wave information to the VPF. The Information that which of the goal information should the agent chooses is provided for the agent.

Three cases of agent's size D are set. Fig.6 shows the agent approaching its goal and avoiding the obstacles. The case D=50, D=80, Fig.4(a)(c), the agent goes through the shortest path. In the case D=30 Fig.4(b), the agent can't find the shortest path. Because of the interference between the wave through the path P1 and P2 from G, the agent is navigated to the local narrow path. Fig.5 shows the inflow energy E_{in} of each agent. The nearer the agent approaches to the goal, the more E_{in} increases. Fig.6(d)-(f) shows the case that a different initial point is set to each agent. In each case, the agent can find a continuous motion and a sub-optimal path.

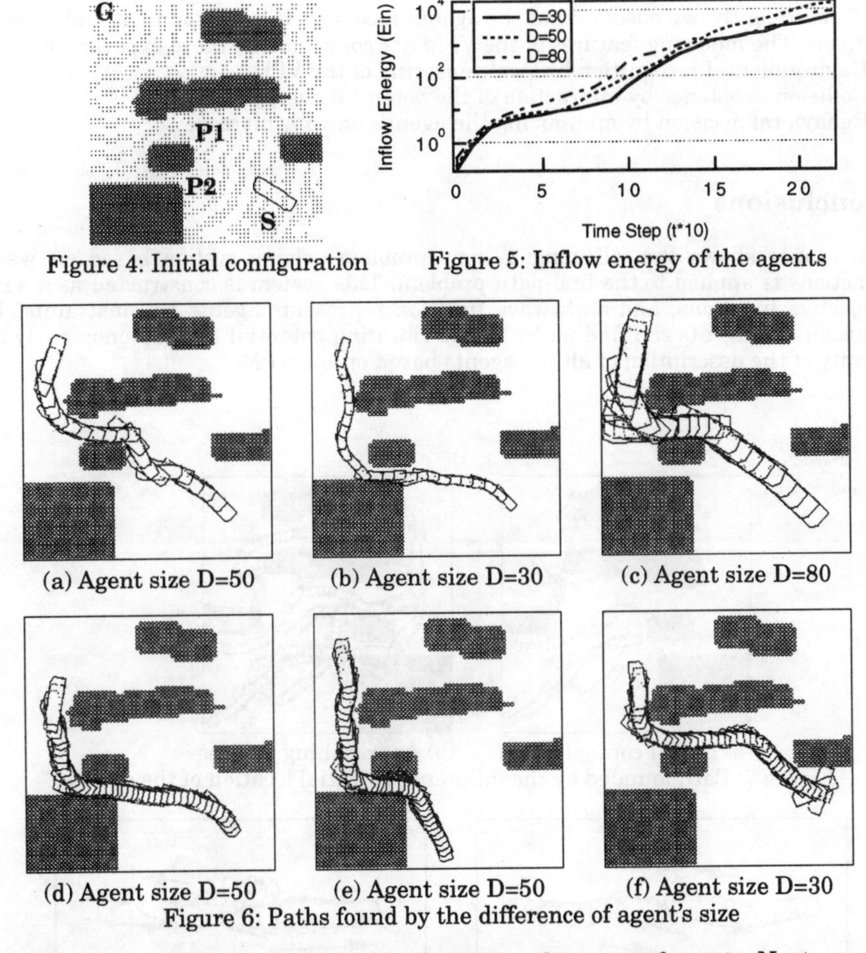

Figure 4: Initial configuration

Figure 5: Inflow energy of the agents

(a) Agent size D=50 (b) Agent size D=30 (c) Agent size D=80

(d) Agent size D=50 (e) Agent size D=50 (f) Agent size D=30

Figure 6: Paths found by the difference of agent's size

The same VPM can be applied to the behavior of a group of agents. Next, we show that there are eight robotic agents on the VPF. The linearity of the wave function makes the convolution of agent's wave functions possible. So we set D=30 and locate the agents at where Fig.7 (a) shows. The agents succeed approaching to the goal avoiding obstacles and agent each other. Thus, we can increase the number of agents or obstacles merely by adding their wave functions to the VPF.

Next, we set the two agents which have a different goal. Fig.8 shows that each agent can avoid and approach to each goal. The case eight agents have a different goal at the opposite corner is displayed in Fig.9. These show the adaptability of the VPM in dynamic environment.

Finally, we discuss about the properties of the VPM. In the VPM, robotic agent can continue moving and searching field as long as energy inflows. But the case that inflowing, outflowing and moving energy are canceled out each other, the agent fall into local minima. Fig.10 shows this case. Normally, the agent can arrive at the goal (a). But if the velocity coefficient is reduced, it comes into a deadlock (b). Conversely, we can say if deadlock occurs, it is possible to get out of the state by increasing the velocity rate.

In these computer simulations, we set the same parameters to the model in each case. Consequently, we observe that the agents recognize the environment and behave adaptively. The following features of the VPM are confirmed by these experiments:

- Environmental recognition by local sensoring of the VPF.
- Collision avoidance by interaction of the potential energy.
- Behavioral decision by minimizing the agent's entropy increase.

5. Conclusions

This paper shows the following: A new problem solving system based on wave interactions is applied to the find-path problem. This system is constructed as a VPF by objective functions, and each wave function represents agents as constraints. In conclusion, the agents can find paths on the vibrating potential field autonomously by the unity of the description of all the agents based on the VPM.

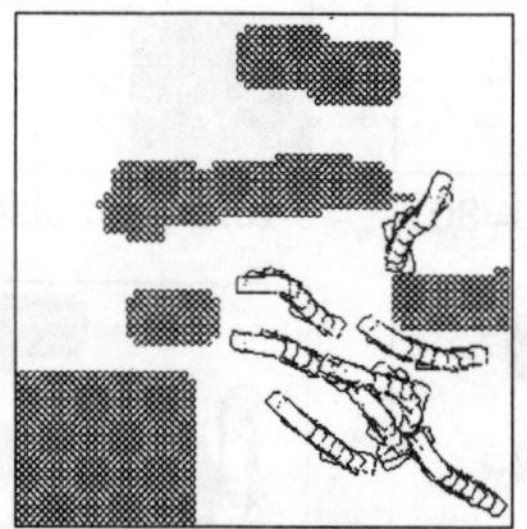

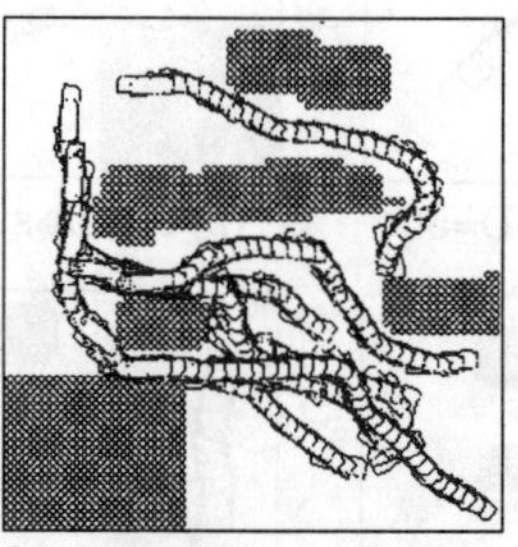

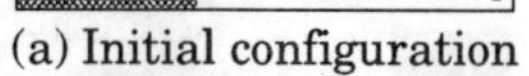

(a) Initial configuration (b) Approaching process

Figure 7: Path founded by the difference of initial location of the agents

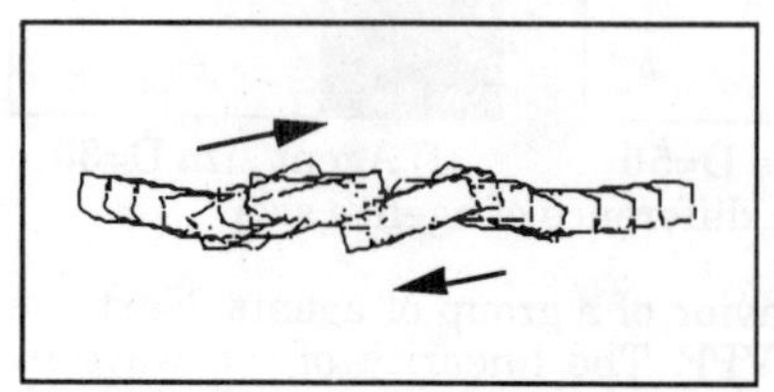

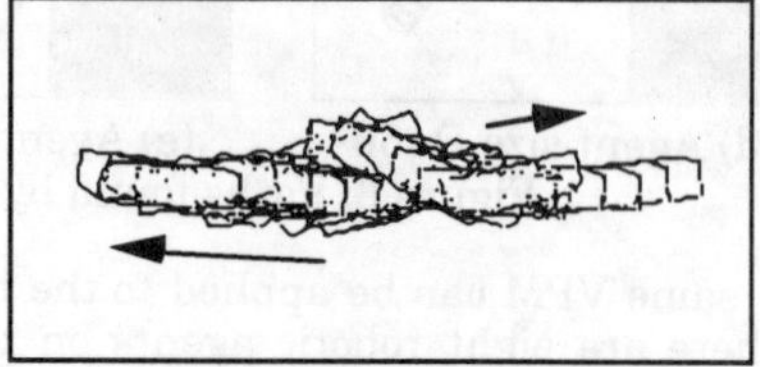

(a) Avoiding process (b) Approaching to each goal

Figure 8: Two agents which have a different goal

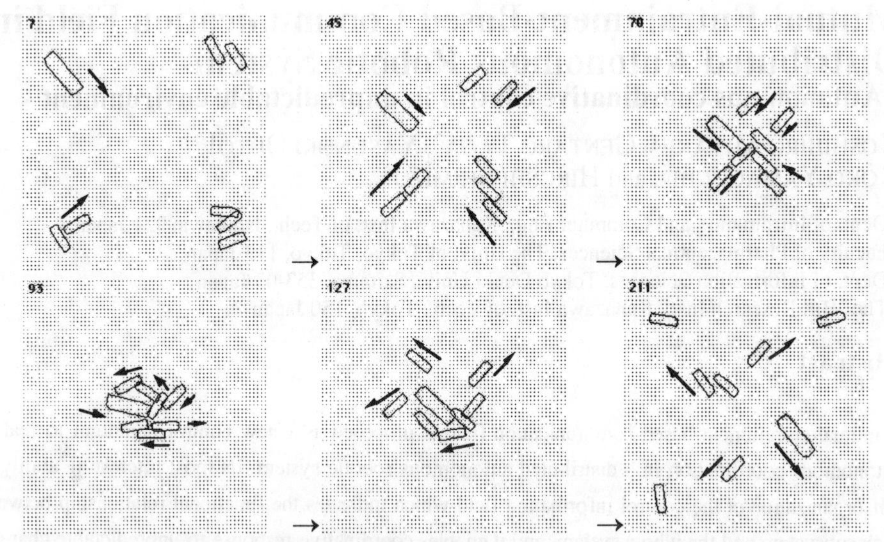

Figure 9: Eight agents which have a different goal

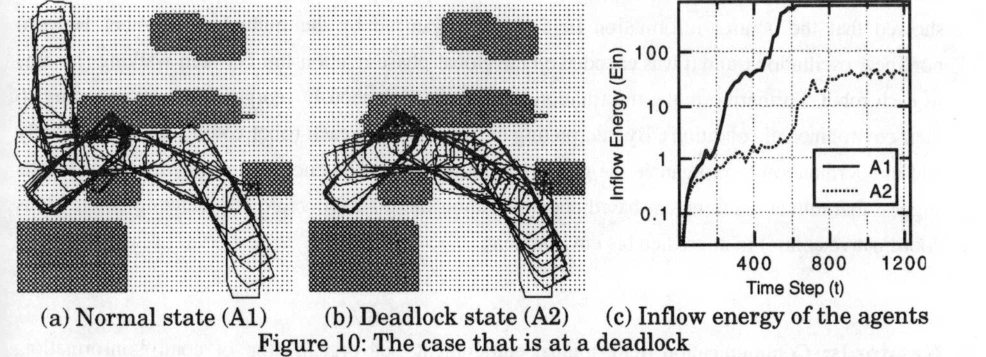

(a) Normal state (A1) (b) Deadlock state (A2) (c) Inflow energy of the agents

Figure 10: The case that is at a deadlock

REFERENCES

[1] H.Yokoi and Y.Kakazu, "An Approach to Traveling Salesman Problem by Bionic Model," ANNIE'91, pp.883-888, 1991

[2] Y. Kakazu and H. Yokoi, "The Behavior of Artificial Worms in a narrow path by the Bionic Model," ANNIE'92, pp.33-38, 1992.

[3] K. Yamada, H. Yokoi and Y. Kakazu, "Terrain Navigation by Vibrating Potential Method," Proc. of SMC'93,Vol.5, pp.218-222, 1993.

[4] Mitchell, J. S. B., "An Algorithmic Approach to Some" Problems in Terrain Navigation", Geometric Reasoning, The MIT Press, 1989.

[5] Schwartz, Jacob T. and Sharir, Mich "On the Piano Movers Problem: III. Coordinating the Motion of Several Independent Bodies: The Special Case of Circular Bodies Moving Amidst Polygonal Barriers," The International Journal of Robotics Research, Vol.2, No.3, Fall, pp.46-75, 1983.

Mutual-Entrainment-Based Communication Field in Distributed Autonomous Robotic System

- Autonomous coordinative control in unpredictable environment -

YOSHIHIRO MIYAKE[1], GENTARO TAGA[2], YASUNORI OHTO[1],
YOKO YAMAGUCHI[3], and HIROSHI SHIMIZU[4]

[1]Dept. of Infromation and Computer Eng., Kanazawa Inst. of Tech., Ishikawa, 921 Japan
[2]Faculty of Pharmaceutical Sciences, The Univ. of Tokyo, Tokyo, 113 Japan
[3]Dept. of Information Sciences, Tokyo Denki Univ., Saitama, 353-03 Japan
[4]The "Ba" Research Inst., Kanazawa Inst. of Tech., Tokyo, 150 Japan

Abstract

A mutual-entrainment-based communication field is proposed as a new control paradigm to realize autonomous coordination in a distributed autonomous robotic system. Its most interesting ability is self-organization of the control information field which indicates the functional relationship between each subsystem and the whole system, and it enables coordinative response to unpredictable changes of environment. By using such emergent information generation, we realized spontaneous and coordinative group formation in a multiple walking robot system as an example. As a result, we showed that the control information field is self-organized by the mutual entrainment between nonlinear oscillations, and it was encoded on the global phase relationship between walking rhythms in each robot. Furthermore, the information field was spontaneously reorganized corresponding to their environmental conditions. By interpreting this information, each robot could respond relevantly and in coordination to reorganize the group formation pattern as one whole system. These results suggest that mutual-entrainment-based communication field is a principle for realizing autonomous coordinative control in unpredictable environment.

Keywords: Communication field, mutual entrainment, self-organization of control information, autonomous coordinative control, distributed autonomous robotic system

1 Introduction

A mutual-entrainment-based communication field is proposed as a new control paradigm to realize autonomous coordination in a distributed autonomous robotic system. Its most interesting ability is self-organization of the control information field which indicates the functional relationship between each subsystem and the whole system, and it enables coordinative response to unpredictable changes of environment. This means that its essence is not in any definite and separated order, as is seen in a conventional control system but in the ability to self-organize a flexible and integrated order as one whole system.

Recently, spontaneous order generation in multi-agent systems [4] has been widely investigated, such as subsumption architecture [1], [2] and contract net protocol [17]. These studies are, however, based on message communication between subsystems similarly to the conventional control system. In other words, since system architecture is definitely and independently defined and the control algorithm is previously fixed, this kind of system cannot adapt to unpredictable changes of environment. To overcome this problem, the control system should not be completely fixed by the external designer. Through mutual communication between subsystems, the control information field which informs each subsystem of its functional relationship in the whole system should be self-organized depending on the environmental conditions.

A biological system is a good example of such an emergent control system. Thus, we have been studying the chemotaxis of *Physarum* plasmodium as an example and investigated the intracellular communication mechanism for autonomous coordinative migration. As shown in Fig.1, the organism migrates as a whole body toward the attractive stimulus (indicated by the arrow in the figure), and its shape is flexibly and coordinatively reorganized corresponding to such unpredictable change of the external environment. From our previous experiments with this organism [6]-[8], [11]-[15], we clarified the following three points. (i) The intracellular communication system is composed of coupled chemical oscillators. (ii) Local information from the environment is encoded on the period modulation in each oscillator. (iii) Control information for migration is self-organized as the phase gradient pattern by mutual entrainment between these oscillators. These experimental results strongly suggest that mutual entrainment between rhythms is essential for such emergent self-control.

Thus, we newly proposed mutual-entrainment-field-based communication as a candidate to enable such emergent control in an engineering system. Therefore, in the present paper, we focus our attention on the role of rhythmic interaction between distributed robots and attempt to realize such a communication field.

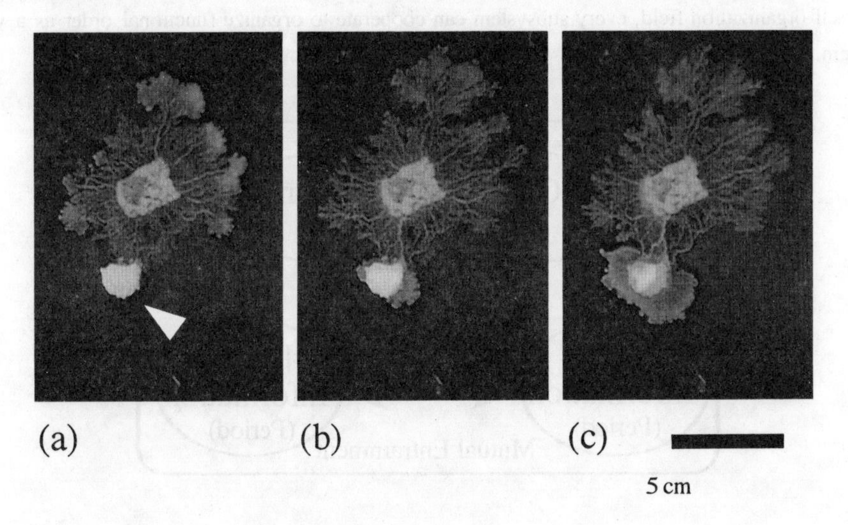

Figure 1: Taxis of *Physarum* plasmodium [10]

2 Communication Field

We construct the emergent communication field as a kind of self-organization system [3], [16]. The dynamics of this system can be described by using two kinds of information concerning local and global states, whose circulative interaction is indispensable for organizing global order as one whole system. The information encoded on this global ordered state is regarded as the control information, which is a kind of field information which indicates global relationships between local states in the whole system. We called this self-organization-based communication system the "communication field".

To realize this communication field, it should be composed of subsystems having nonlinear dynamics such as rhythmic property. In particular, based on our previous biological experiments, the system is constructed using mutual entrainment between nonlinear oscillations, as shown in Fig.2. In such a case, local information is encoded on the period of each oscillator, and the control information which indicates the functional relationship between each subsystem and the whole system is self-organized as the global phase relationship between local oscillators. Thus, depending on such field information for self-control, every subsystem can autonomously and coordinatively behave to organize the global function as one whole system. This kind of coupled oscillator model has been studied by our group [5], [10], [12] as a model of the *Physarum* plasmodium.

On the other hand, the conventional control system is designed based on the concept of message communication. Since it is achieved only by message transmission from a local sender to a local receiver and each communication subsystem is definitely and independently defined, such a system cannot include the self-organization process in itself. Thus, the message-communication-based system is not able to self-organize the information for self-control to achieve a global function, and it cannot respond to unpredictable changes of environment. Therefore, the autonomous coordinative control system should be designed from the viewpoint of the communication field. On the basis of this self-organization field, every subsystem can cooperate to organize functional order as a whole system.

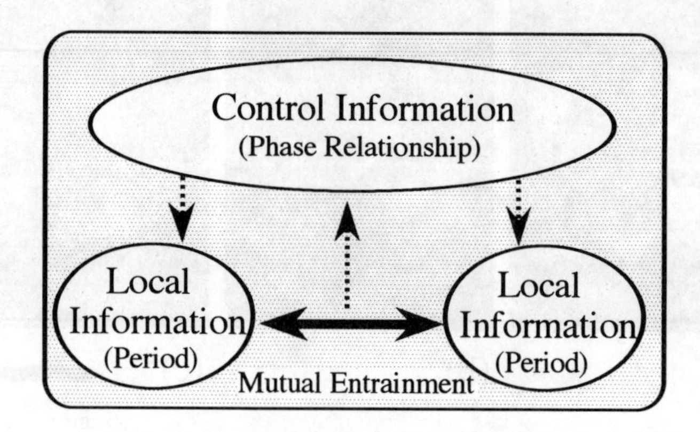

Figure 2: Communication field

3 System Configuration

3.1 Outline of the Model

In the present paper, coordinative group formation in a multiple walking robot system in a one-dimensional array is studied as an example of its engineering realization. The basic system structure consists of two hierarchical layers, as shown in Fig.3. One is the communication field which is composed of the rhythm generator system in each robot, and the other corresponds to nonrhythmic components. Thus, regardless of their physical differences, every robotic system can communicate by using rhythmic interaction. Especially, regarding the rhythm generator of stepping cycle as a kind of nonlinear oscillation, walking locomotion in every robot can interact with each other through such communication system.

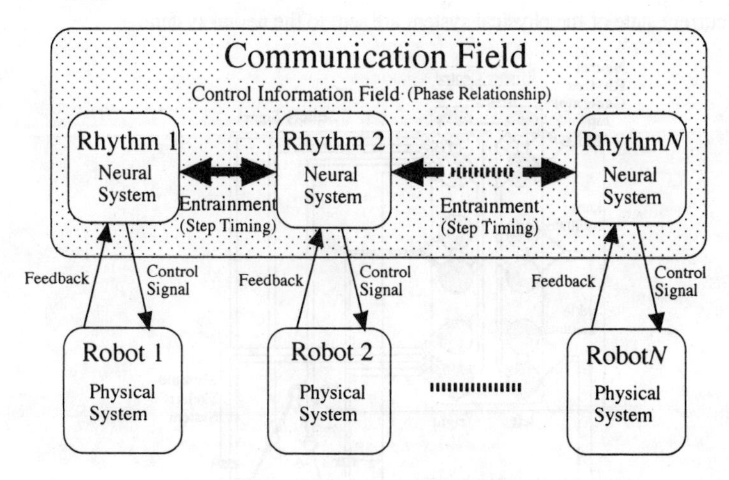

Figure 3: System configuration

3.2 Robotic System

As an elementary robot, any robot which shows nonlinear oscillation dynamics is thought to be applicable. Thus, we used a model of bipedal locomotion proposed by our group [18]-[20] as an example, and realized it by computer simulation. This model is composed of a neural rhythm generator and a physical system, as illustrated in Fig.4, and it generates locomotion as a completely autonomous oscillation through their mutual interaction.

The neural rhythm generator system is represented as

$$t_i \dot{u}_i = -u_i + \sum_{i,j=1}^{12} w_{ij} y_j - b v_i + u_{0i} + Feed_i(\mathbf{x}, \dot{\mathbf{x}}, \mathbf{Fg}(\mathbf{x}, \dot{\mathbf{x}})),$$

$$t'_i \dot{v}_i = -v_i + y_i,$$

$$y_i = f(u_i) \quad (f(u_i) = \max(0, u_i)) \quad (i = 1, \ldots, 12), \qquad (1)$$

where u_i is the inner state of the i-th neuron; y_i is the output of the i-th neuron; v_i is a variable representing the degree of adaptation or self-inhibition of the i-th neuron; u_{0i} is a signal from the higher center; w_{ij} is a connecting weight; τ_i and τ'_i are time constants of the inner state and the adaptation effect, respectively; and $Feed_i$ is a sensory signal.

The physical system moves according to its own dynamics and motor signals from the neural system. Its general form, derived by means of the Newton-Euler method, is written as

$$\ddot{\mathbf{x}} = P(\mathbf{x})[C(\mathbf{x})P(\mathbf{x})]^{-1}[\mathbf{D}(\mathbf{x},\dot{\mathbf{x}}) - C(\mathbf{x})\mathbf{Q}(\mathbf{x},\dot{\mathbf{x}},\mathbf{Tr}(\mathbf{y}),\mathbf{Fg}(\mathbf{x},\dot{\mathbf{x}}))] + \mathbf{Q}(\mathbf{x},\dot{\mathbf{x}},\mathbf{Tr}(\mathbf{y}),\mathbf{Fg}(\mathbf{x},\dot{\mathbf{x}})), \quad (2)$$

where $\mathbf{x}$ is a vector of inertial positions and angles of links; P and C are matrixes; $\mathbf{D}$ and $\mathbf{Q}$ are vectors; $\mathbf{Tr}$ is a vector of torques; $\mathbf{Fg}$ is a vector of forces on the ankle which depend on the state of terrain; and $\mathbf{y}$ is a vector of the output of the neural rhythm generator. The sensory signals which indicate the current state of the physical system are sent to the neural system.

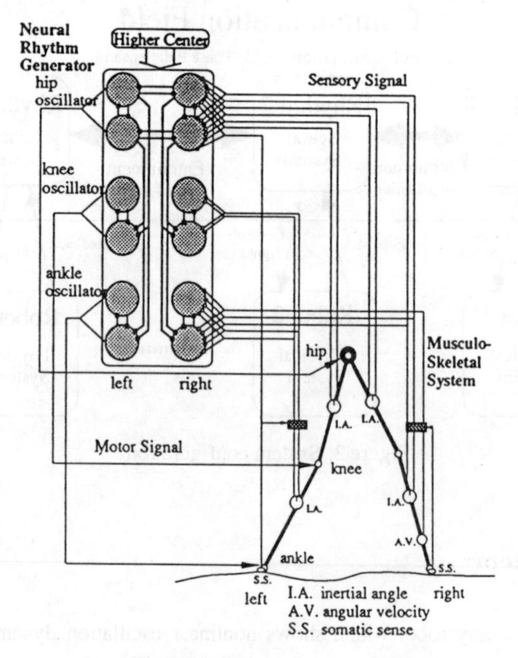

Figure 4: Robotic system [18]

3.3 Communication System

The communication field between these robots is composed of the neural rhythm generator in each robot and their mutual interactions. Everyone has experienced unconscious synchronization of stepping motion when walking with another person. Perhaps this phenomenon is a typical example of mutual entrainment in human locomotion through auditory interaction [9]. Thus, the interaction signal is assumed to be a periodic pulse, which can be imagined as a kind of stepping sound. Each pulse signal is generated at the time of contact of the ankle to the ground, and they are transmitted to the neural rhythm generator in neighboring robots [18]. The form of the interaction is represented as

$$pul_j = pu_{j-1} + pu_{j+1},$$
$$pu_j = \begin{cases} -A & for \quad z_{r,j} < z_g \quad and \quad T_{r,j} < B \\ A & for \quad z_{l,j} < z_g \quad and \quad T_{l,j} < B \\ 0 & otherwise \end{cases}, \tag{3}$$

where pul_j is the input signal to the neural system in the j-th robot. pu_j is the pulse signal which encodes the step timing of the j-th robot. A is pulse height and B is its duration. $z_{r,j}$, $z_{l,j}$ and z_g represent the height of the right ankle, left ankle and ground in the sagittal plane, respectively. $T_{r,j}$ and $T_{l,j}$ respectively stand for the time interval from the contact of the right ankle and left ankle in each walking cycle.

Spatiotemporal order self-organized in this communication field was analyzed as the period and phase gradient. The period was defined as the time interval between two successive steps of the same robot. The phase gradient was defined as the time difference of a corresponding two steps between the neighboring robots. These are represented as

$$pe_{j,i} = st_{j,i} - st_{j,i-1},$$
$$phg_{j,i} = st_{j+1,i} - st_{j,i}, \tag{4}$$

where $pe_{j,i}$ means the period, $phg_{j,i}$ stands for the phase gradient between neighboring robots, and $st_{j,i}$ is the step time of the j-th robot in the i-th walking cycle.

4 Communication-Field-Based Control

4.1 Self-Organization of Control Information Field

Time evolution of the walking pattern in the robotic system was calculated and is shown in Fig.5a. Under this condition, corresponding to our coupled oscillator model [5], [10], [12], the original period of the walking rhythm was fixed to be the same value in the whole system except for the top position (j=1). After the period decrease at the top position (right end in the array), their step timing gradually changes from the top to the rear of the system.

Temporal development of the period and the phase gradient in the communication field are shown in Figs.5b and 5c, respectively. After the period modulation (left end in the figure), the local response rapidly propagated to other regions, and finally the uniform distribution of the period and global phase gradient pattern were stably self-organized. Then, the phase gradient linearly decreased from the top to the rear of the system. Thus, the phase gradient pattern shows not only the global polarity but also the relative distance from the top position.

Furthermore, it was clarified that this phase gradient pattern is size invariant, independent of the total number of robots. This is because the variation range of the phase gradient does not change with system size, as shown in Fig.6. This means that the phase gradient value at each part represents not the absolute position but the relative one within the whole system. Thus, the space coordinate as a kind of control information is self-organized in this communication field. Therefore, by interpreting this control information, each robot can be informed of its relative position within the system, and group formation can be organized in a position-dependent manner.

In the present paper, since the phase gradient decreases linearly from the top to the rear, the control information field is interpreted by introducing some threshold values, as shown in Fig.7. Thus, some discrete regions of the phase gradient value can be defined according to the relative position within the system. Based on this information, the spatial distance between neighboring robots is modified and some clustered walking groups are organized. In the following cases, the number of groups was set at three.

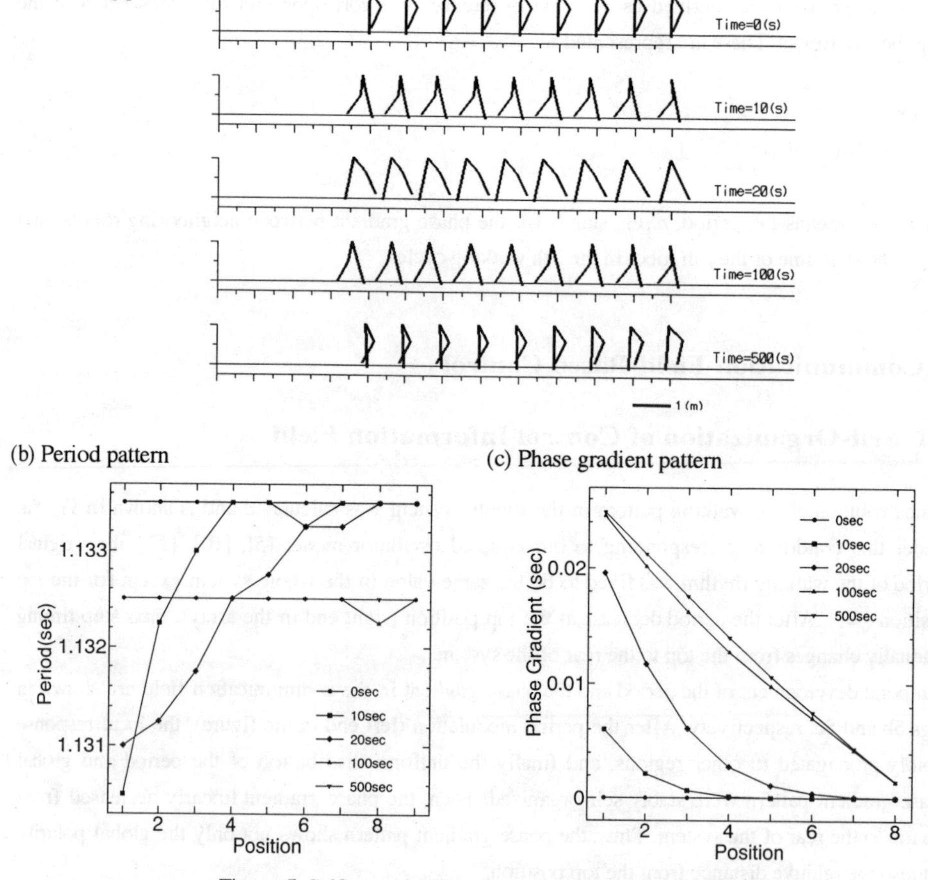

Figure 5: Self-organization of control information field

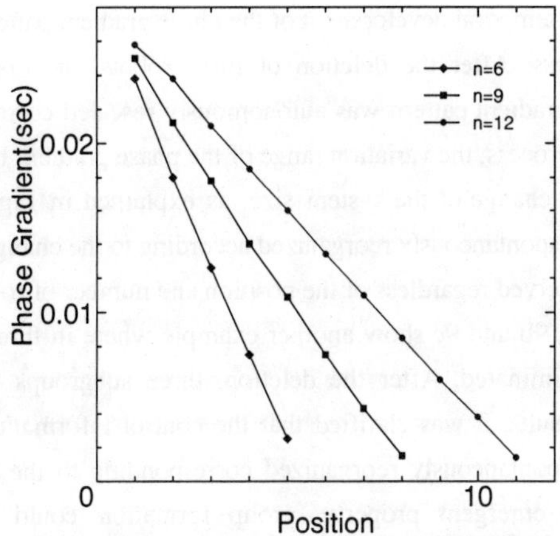

Figure 6: Relationship between phase gradient pattern and system size

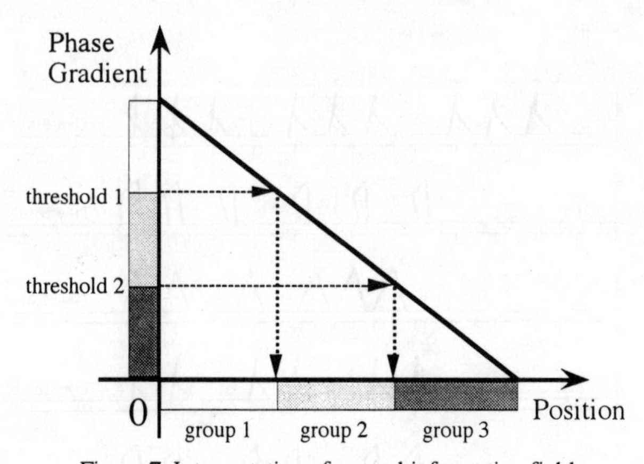

Figure 7: Interpretation of control information field

4.2 Autonomous Coordinative Response to Unpredictable Change

As an example of the autonomous coordinative response to unpredictable changes of environment, reorganization of group formation corresponding to a change in system size was studied. The system size is defined as the total number of robots, and we imagine the situation in which some robots are eliminated from the robotic system at a certain timing due to breakdown.

An example of the temporal development of a group formation pattern is shown in Fig.8a. In the top figure, three stable subgroups for 9 walking robots are observed. This is based on the interpretation of control information field, as explained in the above section. After the deletion of three robots at the 5th, 7th and 9th positions indicated by * in the figure, the group formation became disordered throughout the system. However, three subgroups for 6 robots were stably reorganized, as shown in the same figure. In spite of the unpredictable change of system size, the group formation pattern was spontaneously maintained as a whole system.

Figures 8b and 8c show the temporal development of the phase gradient pattern in the communication field under the same process. After the deletion of three robots, the disordered state appeared. However, the linear phase gradient pattern was autonomously rescaled corresponding to 6 robots as the total system. In such a process, the variation range of the phase gradient between the two terminal robots did not change with change of the system size, as explained in Fig.6. This means that the control information field is spontaneously reorganized according to the change of system size.

These phenomena were observed regardless of the position and number of robots eliminated from the robotic system. Figures 9a, 9b and 9c show another example where in three robots at the 3rd, 6th and 7th positions were eliminated. After the deletion, three subgroups for 6 robots were also reorganized. From these results, it was clarified that the control information field encoded on the phase gradient pattern is spontaneously reorganized corresponding to the change in system size. Therefore, by using this emergent property, group formation could be autonomously and coordinatively achieved even under unpredictable changes of environment.

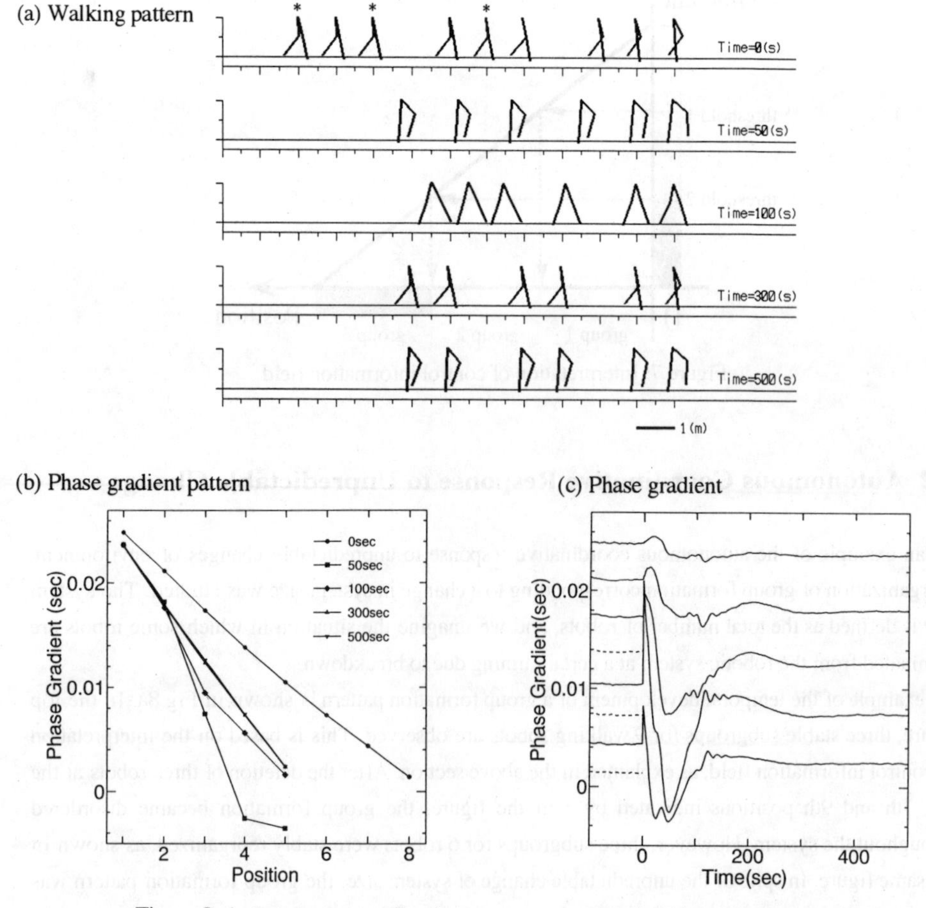

Figure 8: Autonomous coordinative response to unpredictable environment

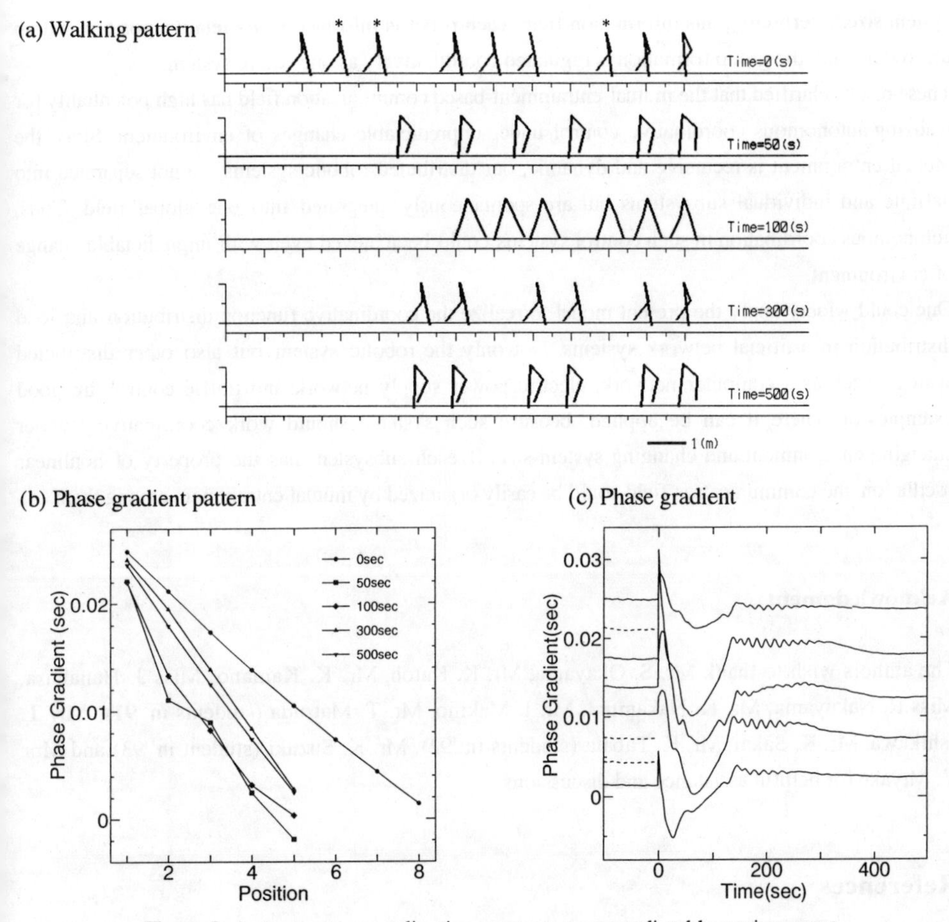

Figure 9: Autonomous coordinative response to unpredictable environment

5 Conclusion

In this paper, based on our previous researches of a biological communication system, a mutual-entrainment-based communication field was proposed as a new control paradigm to realize emergent control in a distributed autonomous robotic system. Its most interesting ability is self-organization of the control information field which indicates the functional relationship between each subsystem and the whole system, and realization of autonomous coordinative response to unpredictable changes of environment by using such emergent information.

Applying this approach to the present problem, we studied coordinative group formation in a multiple walking robot system. We clarified that the control information field is self-organized into the phase gradient pattern between distributed rhythms in the communication field. This phase relationship represents the relative positional relationship among individual robots in the whole system. Furthermore, this information field is spontaneously reorganized corresponding to the change of system size. Interpreting this information field, each robot is informed of its relative position within the system, and the group formation is regulated coordinatively as one whole system.

These results clarified that the mutual-entrainment-based communication field has high potentiality for realizing autonomous coordinative control under unpredictable changes of environment. Since the mutual entrainment is recursive and dynamic, our distributed robotic systems are not separated into definite and individual subsystems but are spontaneously integrated into one global field. Thus, autonomous coordination in such control systems could be achieved even with unpredictable change of environment.

One could widely apply the present model to realize the coordinative function distribution and load distribution in artificial network systems. Not only the robotic system but also other distributed systems such as a computer network, electric power supply network and traffic control are good examples of where it can be applied, because such systems should work coordinatively under changing environment and changing system size. If each subsystem has the property of nonlinear oscillation, the communication field could be easily organized by mutual entrainment among them.

Acknowledgment

The authors wish to thank Mr. S. Okayama, Mr. K. Katoh, Mr. K. Kamano, Miss J. Hanabusa, Miss R. Nakayama, Mr. K. Nakamura, Mr. I. Makino, Mr. T. Matsuda (students in '91), Mr. T. Ishikawa, Mr. K. Sakai, Mr. K. Tabata (students in '92), Mr. K. Suzuki (student in '93) and Mrs. Y. Miyake for helpful assistance and discussions.

References

[1] R. A. Brooks, "A robust layered control system for a mobile robot," *IEEE J. Robotics Automat.*, vol. RA-2, no. 1, pp. 14-23, Mar. 1986.

[2] R. A. Brooks, "New approach to robotics," *Science*, vol. 228, pp. 1227-1232, 1991.

[3] H. Haken, *Synergetics -- An introduction.* 3rd edn., Springer-Verlag, 1983.

[4] M. Minsky, *The Society of Mind*, Simon & Schuster, 1986.

[5] Y. Miyake, Y. Yamaguchi, M. Yano and H. Shimizu, "Environment-dependent self-organization of positional information in tactic response of *Physarum* plasmodium," (in preparation).

[6] Y. Miyake, S. Tabata, H. Murakami, M. Yano and H. Shimizu, "Environment-dependent positional information and information integration in chemotaxis of *Physarum* plasmodium. I. Self-organization of intracellular phase gradient pattern and coordinative migration," *J. Theor. Biol.* (submitted).

[7] Y. Miyake, H. Murakami, S. Tabata, M. Yano and H. Shimizu, "Environment-dependent positional information and information integration in chemotaxis of *Physarum* plasmodium. II. Artificial modulation of intracellular phase gradient pattern and response of migration," *J. Theor. Biol.* (submitted).

[8] Y. Miyake, H. Tada, M. Yano and H. Shimizu, "Relationship between intracellular period modulation and external environment change in *Physarum* plasmodium," *Cell Struct. Funct.* (in press).

[9] Y. Miyake and H. Shimizu, "Mutual entrainment based human-robot communication field," *Proc. of 3rd. IEEE Int. Workshop on Robot and Human Communication,* Nagoya, Japan, pp. 118-123, 1994.

[10] Y. Miyake, Y. Yamaguchi, M. Yano and H. Shimizu, "Environment-dependent self-organization of positional information in coupled nonlinear oscillator system --- A new principle of real-time coordinative control in biological distributed system ," *IEICE Trans. Fundamentals*, vol. E76-A, pp. 780-785, 1993.

[11] Y. Miyake, M. Yano, H. Tanaka and H. Shimizu, " Entrainment to external Ca^{2+} oscillation in ionophore-treated *Physarum* plasmodium," *Cell Struct. Funct.*, vol. 17, pp. 371-375, 1992.

[12] Y. Miyake, Y. Yamaguchi, M. Yano and H. Shimizu, "Environment-dependent positional information and biological autonomous control --- A mechanism of tactic pattern formation in *Physarum* plasmodium," *HOLONICS*, vol. 3, pp. 67-81, 1992.

[13] Y. Miyake, M. Yano and H. Shimizu, "Relationship between endoplasmic and ectoplasmic oscillations during chemotaxis of *Physarum polycephalum*," *Protoplasma*, vol. 162, pp. 175-181, 1991.

[14] K. Natsume, Y. Miyake, M. Yano and H. Shimizu, "Information propagation by spatio-temporal pattern change of Ca^{2+} concentration throughout *Physarum polycephalum* with repulsive stimulation," *Cell Struct. Funct.*, vol. 18, pp.111-115, 1993.

[15] K. Natsume, Y. Miyake, M. Yano and H. Shimizu, " Development of spatio-temporal pattern of Ca^{2+} on the chemotactic behavior of *Physarum* plasmodium," *Protoplasma*, vol. 166, pp. 55-60, 1992.

[16] G. Nicolis and I. Prigogine, *Self-organization in nonequilibrium systems*, John Wiley & Sons, 1977.

[17] R. G. Smith and R. Davis, "Frameworks for cooperation in distributed problem solving," *IEEE Trans. Syst., Man, Cybern.*, vol. SMC-11, no. 1, pp. 61-70, 1981.

[18] G. Taga, Y. Miyake, Y. Yamaguchi and H. Shimizu, "Generation and coordination of bipedal locomotion through global entrainment," in *Proc. Int. Symp. Autonomous Decentralized Systems*, Kawasaki, Japan, IEEE Computer Society Press, pp.199-205, 1993.

[19] G. Taga, Y. Miyake, Y. Yamaguchi and H. Shimizu, "Generation of bipedal locomotion through action-perception cycle of entrainment in unpredictable environment," in *Proc. Int. Workshop Mechatronical Computer System for Perception and Action*, Halmstad Univ. Sweden, pp. 383-389, 1993.

[20] G. Taga, Y. Yamaguchi and H. Shimizu, "Self-organization control of bipedal locomotion by neural oscillators in unpredictable environment," *Biol. Cybern.*, vol. 65, pp. 147-159, 1991.

Cooperative Behavior of Parent-Children Type Mobile Robots

TAKANORI SHIBATA, KAZUYA OHKAWA, and KAZUO TANIE

Robotics Dept., Mecanical Eng. Laboratory, MITI, Tsukuba, 305 Japan

Abstract

This paper proposes a parent-children type robot system that moves in unstructured environments. The parent robot works as a leader of the system. The children robots work as sensors to sense their environments while touching them. The parent collects the sensory information of its environment and generates a map. In order to express the map effectively, this paper applies a structured neural network to memory. The neural network learns the sensory information incrementally. While using the network, the parent robot determines their behavior. On the other hand, the children are disposable. When some of the children malfunction because of their dangerous environment, the remaining children compensate for them and continue to work. Simulations are performed to show the effectiveness of the proposed system.

Key words: Behavior-based robot, multiple robots, incremental learning, mobile robot

1 Introduction

When robots perform complicated tasks in various environments, they need a sensory system, reasoning mechanism, and a controller that is adaptable to the environment. Intelligent control of robots is possible by providing them with such functions. To give these functions systematically to robots, hierarchical control architecture is useful [1].

It is important for robots to obtain information on their environment by means of their sensory system and decide their own behavior when they move to uncertain environments [2]. We have already proposed the hierarchical intelligent control system in which each robot has been intellectualized [1]. This system consists of a learning level, skill level, and adaptation level. In addition to this, we have suggested the way to hierarchize and decentralize the multiple autonomous robots system for efficient structure of the system. Especially in the previous report, the effectiveness of the following way to decide the behavior of the system was shown [3]. It was the way in which a high level robot constructed the environment model as a result of integrating much information given by the cooperative work of scattered low level robots.

The ability to arrange and learn the given information is important for robots to accomplish intellectual work efficiently. When the robots move in an uncertain environment, human operator cannot give them an environment model in advance; therefore, the robots must have functions so that it can construct and innovate the model while moving.

The previous paper has proposed the way to use the neural network for the learning function of the robots. This paper suggests the way to let robots construct an environment model while moving autonomously by using the structured neural network that makes their additional learning possible [4-8]. Though this way can be applied to a single autonomous robot, this paper applies it to the group of parent-children type robot system because this system is hard to break down. Moreover, the parent robot makes the environment model by the network while compressing sensory information efficiently. Simulations are performed to show that the robots can behave intelligently while learning efficiently in an uncertain environment.

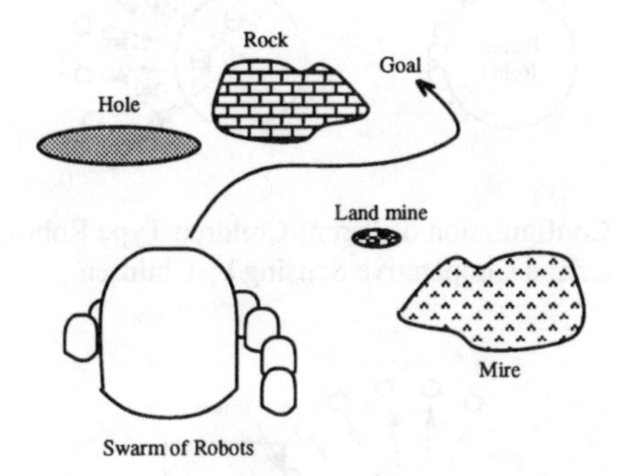

Figure 1: Idea of Parent-Children Type Robot System Behaving in Uncertain Environments with Dangerous Obstacles

2 The Structure and Function of Parent-Children Type Robot System

The previous paper proposed the parent-children type robot system whose structure was hierarchical and hard to break down [3]. This system mixed the decentralization and centralization of robots in order to optimize itself. As a leader, the parent robot needs to plan and execute the behavior of robots for completion of the mission. The parent must manage the other robots as well as decide it's own work for optimization of the whole system behavior. As the characteristic of this system, it is assumed that the mission can be achieved only if the parent remained.

On the other hand, the autonomy of each child robot is weak because the parent decides its behavior. Its function is kept on a low level in order to decrease the burden of the parent in planning. These children are in charge of the touch sensing and they are disposable. That is to say, even if some of the children malfunction by contacting dangerous obstacles in an uncertain environment, the other robots compensate for the broken robots and continue to work. Moreover, the price of the child robot is cheap.

The parent robot constructs an environment model on the basis of the information on environment given by the multiple children that work as cooperative sensing elements, and then, plans a path according to the model.

At that time, the parent needs ability to renew the environmental model while moving. It must remember and learn the given information efficiently. For this purpose, the parent robot uses a structured neural network for incremental learning for a memory of the environmental model. Detailed account of the method is given in the next chapter.

3 Acquisition and Memorization of Information on an Environment by Cooperative Sensing and Incremental Learning

This chapter concentrates on the algorithm of the parent robot to decide behavior of the group of robots. The parent acquires information of the environment from the children, and makes the above-mentioned decision according to the model of its environment that it constructs by the incremental learning.

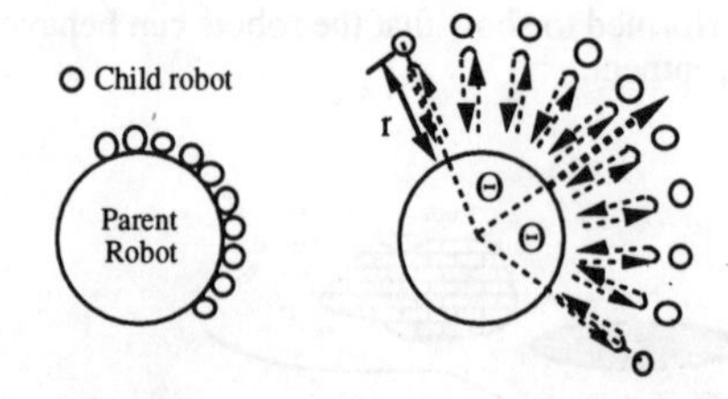

Figure 2: Configuration of Parent-Children Type Robot System and Cooperative Sensing by Children

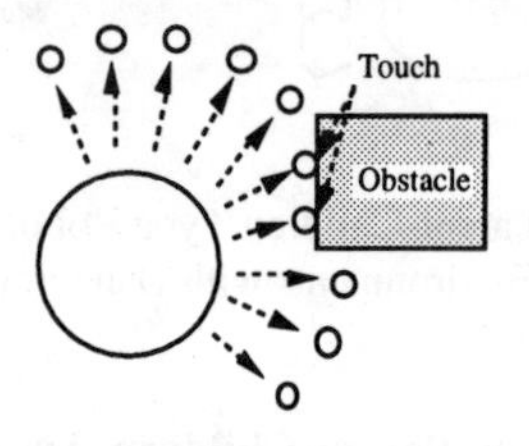

Figure 3: Touch Sensing an Obstacle by Children

3.1 Characteristics of the Children Robots

The children move on the orders of the parent. As the Fig. 2 shows, the children have mobility to move forward in a straight line to sense their environment and move back when they cover the aimed distance. When they touch the obstacle, they no longer advance (as shown in Fig. 3), and recognize the distance between the obstacles and the parent according to that of their own movements. They alternately return to the parent and report the distance by correspondence.

3.2 Characteristics of the Parent Robot

The parent robot has two functions in a general way: one is management of the children to sense the unknown environment, the other is to decide a path of all robots in a group based on information given by the children. This paper deals mainly with the latter function. The parent decides a spread space of the children and a range of their sensing according to the number of children and gives instructions to them. Simultaneously, the parent can communicate with the children through wire or wireless message, but this paper assumes that they communicate through wireless message only.

Much research has dealt with path planning problems. Since this paper deals with unknown environments of robots, the human operator cannot give models of the environment in advance. Therefore, the parent robot collects information on its environment by using the children, and then, memorizes the information by a neural network incrementally. This neural network is used to determine a path based on the potential method as a sensor-based behavior, as follows: the operator only gives the destination to the parent robot, the parent robot generates a potential field that expresses a workspace of the system. This field is used as a map of the environment. While collecting sensory information on obstacles and dangerous areas from the children, the parent adds them to the potential field using Gaussian basis functions of the neural network. This map is the acquired model of the environment.

3.3 Neural Network for Incremental Learning

Fukuda, Shiotani et. al, 1992 and 1993, have proposed a neural network for incremental learning [7, 8]. As for the structure of the neural network, a radial basis function is implemented into each neuron in the hidden layer. While adding a unit with the radial basis function to the hidden layer and modifying the position and form of the radial basis functions, the neural network memorizes new patterns one by one incrementally without forgetting previous memorized patterns. This neural network was applied to pattern recognition problem and showed its effectiveness.

In this paper, a Gaussian basis function is implemented to each neuron in the hidden layer of the neural network as shown in Fig. 4. This neural network expresses a model of the environment of the robot system. One Gaussian basis function means an obstacle that is sensed by a child robot. When the obstacle is large, it is possible for multiple robots to sense it at the same sensing period, even though the sensed positions are different. In this case, the number of units in the hidden layer becomes numerous This is because multiple units are added to the hidden layer for the one obstacle. While repeating this kind of process, the structure of the neural network should be excessive. To avoid this, suppression of the information on the environment is performed as generalization by integrating the Gaussian basis functions which neighbors each other as shown in Fig. 5. To express the boundary of the obstacles or dangerous area clearly, the Gaussian basis function has a dead band.

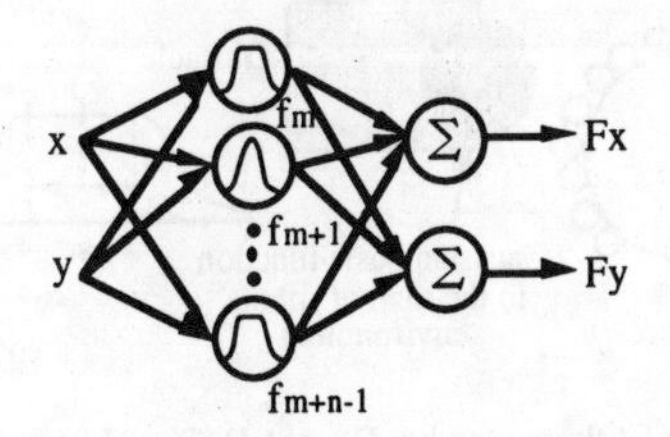

Figure 4: Structure of Neural Network for Incremental Learning

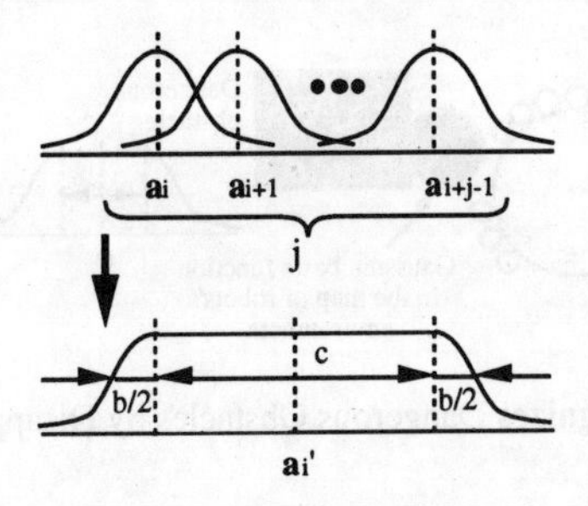

Figure 5: Integration of Neighbor Potentials of Obstacles using Gaussian Basis Function with a Dead Band

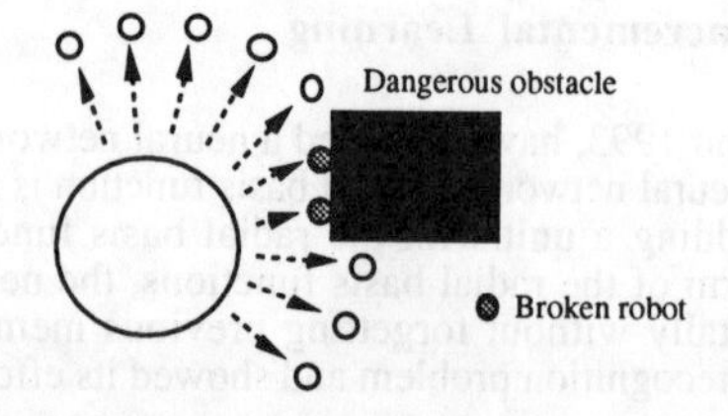

Figure 6: Touching a dangerous obstacle

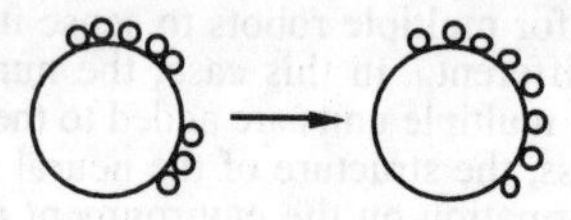

Figure 7: System self-restoration function

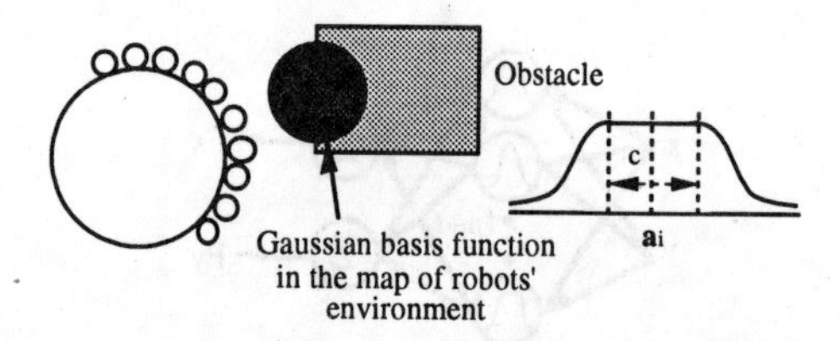

Figure 8: Recognition of Obstacles by Parent Robot through Sensory Information Given from the Children -Gaussian Basis Functions used to Express Obstacles in the Environment Map

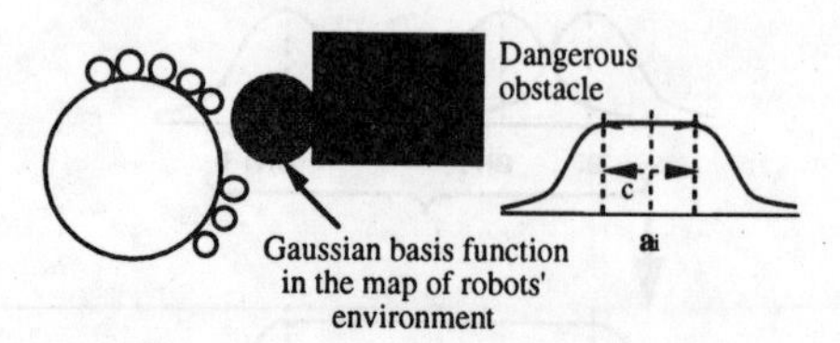

Figure 9: Parent Robot Recognizes Dangerous Obstacles by Disappearance of Some Children

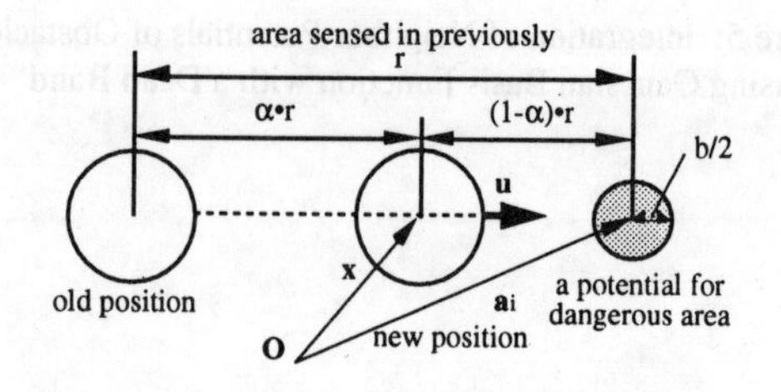

Figure 10: Expression of Dangerous Obstacles

3.4 Algorithms of Management, Learning, Modeling, and Motion Planning for Parent Robot

3.4.1 Management of the Children by the Parent

The parent robot indicates the course and the distance of movement to the children. Fig. 2 shows the angular width of range in sensing is "$\pm\Theta$" toward the direction of the movement. If the number of robot is "n," the space between the children can be calculated as follows:

$$\theta = \frac{2\Theta}{n-1} \tag{1}$$

The distance of the straight movement of the children is $r = r_0$ as an initial value. When some of the children are broken by contact with dangerous obstacles as indicated in Fig. 6, this system compensates by such self-restoration function as to change the spread space between children as shown in Fig. 7 . However, there is a fear of making a mistake in sensing because the distance between the robots becomes too wide at the tip of its course which is similar to a shape of a fan;wide at the end In order to make up for this and keep the reliability of sensing, the parent shortens the distance of the children's movement according to the decrease of the children. Thus, if the number of broken robots is assumed as "p," the spread space of the remaining robots in sensing is changed as follows:

$$\theta = \frac{2\Theta}{n-1-p} \tag{2}$$

Added to this, the distance "r" is altered as follows:

$$r = (r_0 - r_b) \bullet \frac{n-p}{n} + r_b \tag{3}$$

Provided that "r_b" is a bias value.

3.4.2 Navigation by the Parent Robot

Various studies have been performed on the path planning of the mobile robot [9-17]. For example, the way to decide the optimum path by structurization of space and using a graph has been proposed. In this case, however, since the movement is in an unknown environment, it requires a way to decide the behavior based on sensory information. For this purpose, this paper applies the potential method to decide behavior of the robots as a method to avoid obstacles [11]. In this method, the human operator gives a goal in advance. The parent robot makes a potential field that expresses its environment. Information on obstacles and danger sent by children is expressed on assumption by the Gaussian basis functions in the potential field. The parent robot uses this potential field as a map of its environment for path planning. To avoid danger, the distance of movement of all robots in a group is shortened by safety rate.

The following explains the algorithms for navigation of the swarm of robots. First, a potential function U between the position of the parent robot and the goal is defined.

$$U(\mathbf{x}) = \frac{1}{2} \bullet k_g \bullet (\mathbf{g} - \mathbf{x})^2 \tag{4}$$

where k_g is a coefficient for gradation, g is the position vector of the goal, and x is the position vector of the parent robot. Then, the force f_g to let the robot approach the goal is defined as follows:

$$\begin{aligned}\mathbf{f}_g &= -grad(\mathrm{U}(\mathbf{x})) \\ &= \mathrm{k}_g \bullet (\mathbf{g} - \mathbf{x})\end{aligned} \tag{5}$$

When some children inform the parent of obstacles, the parent robot creates the Gaussian basis function to express the obstacles in the environment map as shown in Fig. 8. Then, the parent robot calculates a resultant force with a power to obtain the goal. To calculate a collision-free path reactively, the following function and eq. (5) is used:

$$\mathbf{f}_i = \mathrm{k}_o \bullet \frac{\mathbf{x} - \mathbf{a}_i}{\|\mathbf{x} - \mathbf{a}_i\|} \bullet \exp\left(-\frac{(\mathbf{x} - \mathbf{a}_i)^2}{\mathrm{b}^2}\right) \tag{6}$$

where f_i is a repulsive force from an obstacle i, a_i is a position of the obstacle and the center of the Gaussian basis function, b is a deviation of the Gaussian basis function, and k_o is a coefficient of the repulsive force. To calculate this repulsive force and memorize the information on the obstacle in a model of the environment, a structured neural network is used as shown in Fig. 4. The Gaussian basis function is implemented in a unit at the hidden layer of the neural network.

There are two ways to determine the center of the Gaussian function in a map of work space. One is that the child robot touches an obstacle and returns to the parent robot to inform the existence of the obstacle. In this case, the center position of the function is determined as follows:

$$\mathbf{a}_i = \left(r_i + \frac{b}{2}\right) \bullet \mathbf{u} + \mathbf{x} \tag{7}$$

and

$$\|\mathbf{u}\| = 1 \tag{8}$$

The other is that the child robot touches a dangerous obstacle and breaks down. In this case, the child robot cannot return to the parent robot so the parent robot recognizes existence of the danger, but cannot recognize how far the dangerous area exists. Then, the parent robot puts the Gaussian function in the direction that the broken child robot moved as shown in Fig. 9. The center position of the Gaussian basis function is as follows (Fig. 10):

$$\mathbf{a}_i = \mathbf{r} \bullet (1 - \alpha) \bullet \mathbf{u} + \mathbf{x} \tag{9}$$

where α is a safety rate related to the distance of path that has already been sensed.

When some robots touch obstacles in the same sensing period, the same number of units may be generated in the neural network. However, this may let the neural network be very large in size. To avoid this, when j neighbor robots sense an obstacle at the same period and the positions of the sensed obstacles are in the distance b, the obstacles are considered as one large obstacle. Therefore, the new center position of the obstacle, a_i', is calculated as shown in Fig. 5 and is as follows:

$$\mathbf{a}_i' = \frac{\mathbf{a}_i + \mathbf{a}_{i+1} + \ldots \mathbf{a}_{i+j-1}}{j}\ , \tag{10}$$

and the dead band of the Gaussian basis function, c, is as follows:

$$c = b \times (j-1) \tag{11}$$

While using this generalized information, one unit is added to the hidden layer of the neural network. The repulsive force by the generalized Gaussian basis function is defined as follows:

If $\|\mathbf{x} - \mathbf{a}_i\| \le \frac{c}{2}$, then

$$\mathbf{f}_i = \mathrm{k_o} \bullet \frac{\mathbf{x} - \mathbf{a}_i}{\|\mathbf{x} - \mathbf{a}_i\|}, \tag{12}$$

else if $\|\mathbf{x} - \mathbf{a}_i\| > \frac{c}{2}$, then

$$\mathbf{f}_i = \mathrm{k_o} \bullet \frac{\mathbf{x} - \mathbf{a}_i}{\|\mathbf{x} - \mathbf{a}_i\|} \bullet \exp\left(-\frac{\left\{ \|\mathbf{x} - \mathbf{a}_i\| - \frac{\mathrm{c}}{2} \right\}^2}{\mathrm{b}^2} \right) \tag{13}$$

The output units of the neural network calculate the repulsive force by the sensed obstacles and at a dangerous area.

The parent robot determines the path while mixing the gravity force to the goal and the repulsive force is as follows:

$$\mathbf{F} = \mathbf{f}_g - \sum_{i=1}^{m} \mathbf{f}_i \tag{14}$$

where $\boldsymbol{F}$ is a force and direction of the robots to move, m is the number of the Gaussian functions that express obstacles.

A path for one sensing is divided into some steps. Each step for moving is determined as follows:

$$\mathbf{v}_i = \frac{\alpha \bullet r}{j} \bullet \frac{\mathbf{F}}{\|\mathbf{F}\|} \bullet \frac{\|\mathbf{F}\|}{\|\mathbf{f}_g\|} \tag{15}$$

$$= \frac{\alpha \bullet r}{j} \bullet \frac{\mathbf{F}}{\|\mathbf{f}_g\|} \tag{16}$$

where v_i is a vector of one step for moving and j is the number of steps per one sensing. While summing the steps, the path is expressed as follows:

$$\mathbf{v} = \sum_{i=1}^{j} \mathbf{v}_i \tag{17}$$

4 Simulation

Simulations are performed to show that the parent-children type robot system behaves in an unstructured dangerous environment while learning its sensory information

4.1 Conditions of Simulation

In the environment of the robots, there are two kinds of obstacles. One is the object that obstructs the movement of robots such as a rock and a wall. This object does not damage the robots even if they touch it. The other, such as a hole, a mire, land mine, has the danger to cause the robots to malfunction. These robots do not know their environment in advance but only the parent robot knows its initial position and the goal given by the human operator. As explained in previous chapters, the parent robot collects sensory information of the environment from cooperative sensing from the children and produces a map of the environment while using a neural network with the Gaussian basis functions. Then, the parent robot plans a path by the potential method using the map. For simplicity, it is assumed that the robots move on a plane and the children investigates the same length from the parent. The obstacles are expressed by circles as shown in Fig. 11. In simulations, there is a single parent robot and 7 children initially ($n = 7$). The angular width of range in sensing Θ, is equal to $\pi/3$.

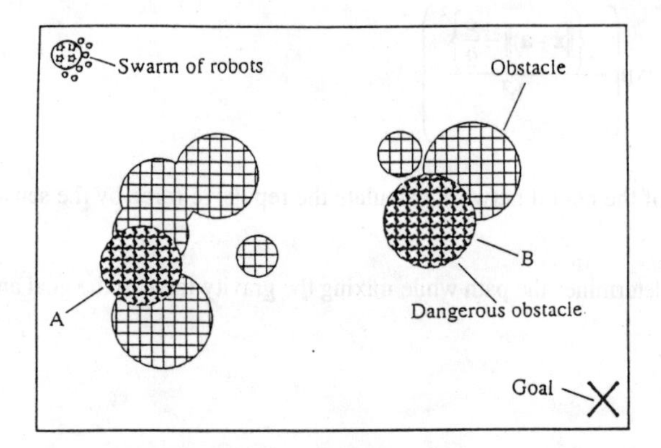

Figure 11 (a): Parent-Children Type Robot System and Their Environment including Dangerous Areas

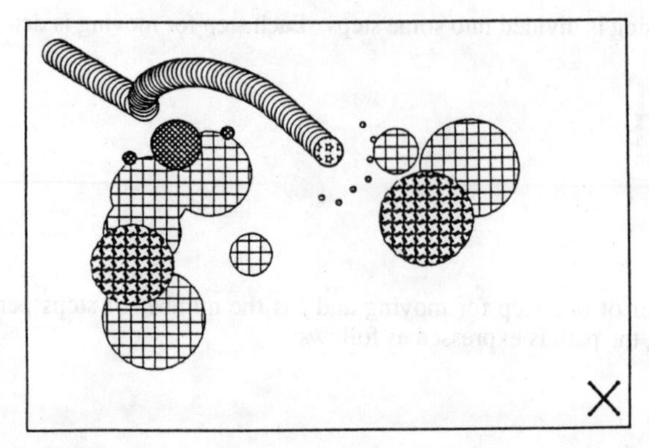

Figure 11 (b): Children Sense their Environment while Touching Obstacles

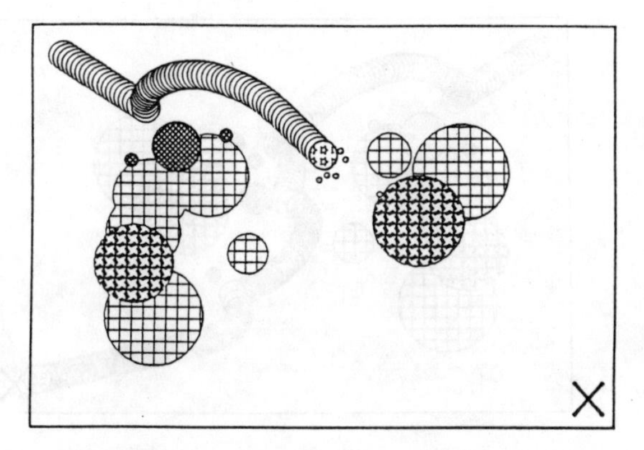

Figure 11 (c): A Broken Child Robot Caused by Touching a Dangerous Obstacle

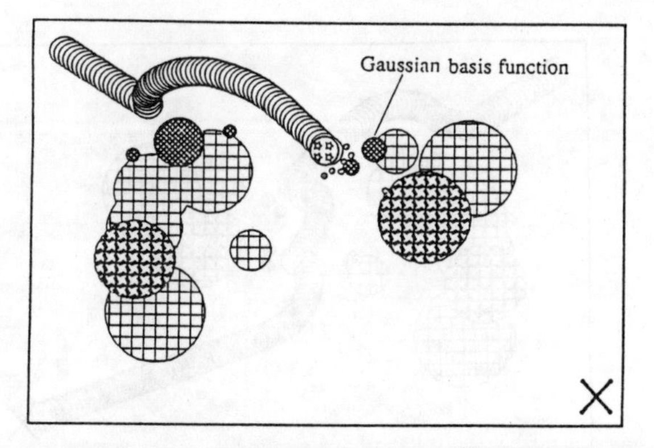

Figure 11 (d): Recognizing the existence of an Obstacle and Danger Followed by Added Gaussian Basis Functions to the Neural Network.

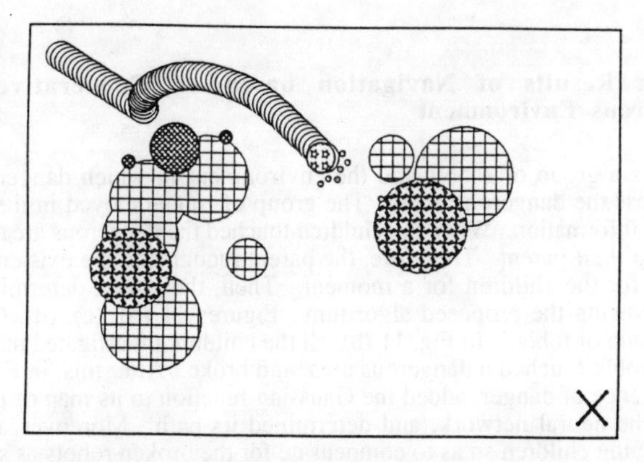

Figure 11 (e): Reorganizing the Children by the Parent Robot to Compensate for the Broken Robot and Continues its Task

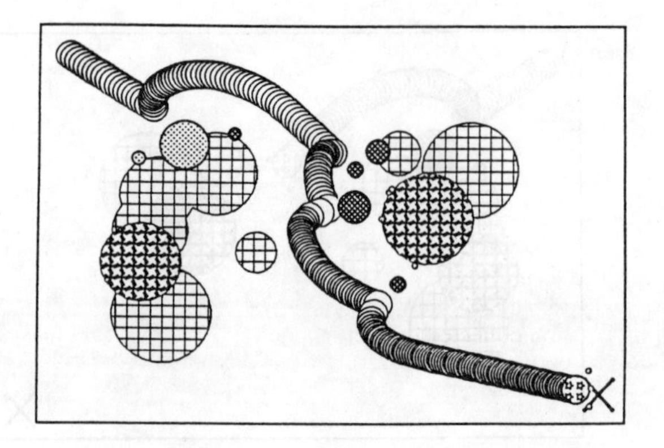

Figure 11 (f): Goal Reached by the Parent Robot and Environment Expressed by the Neural Network with Seven Hidden Units

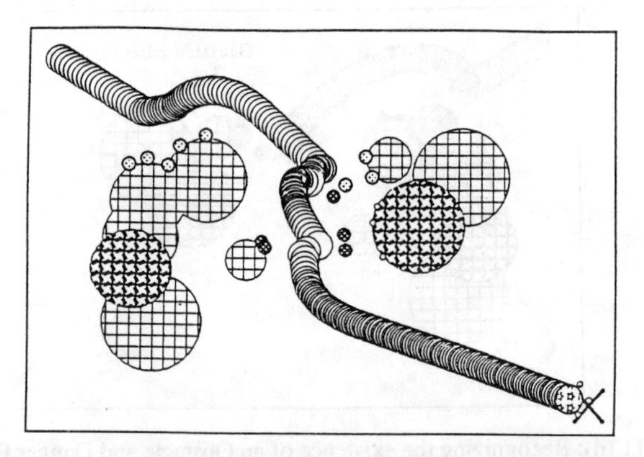

Figure 12: Unused Compression of the Sensory Information by the Parent Robot Resulted in 13 Hidden Units

4.2 Simulation Results of Navigation based on Cooperative Sensing in an Uncertain Dangerous Environment

Figure 11 (a) shows a group of robots and the environment in which dangerous obstacles exist. Deep patterns express the dangerous areas. The group of robots moved in the plane based on the cooperative sensory information. When the children touched the dangerous area, they malfunctioned and did not return to their parent. Therefore, the parent recognized the existence of the dangerous areas after waiting for the children for a moment. Then, the parent determined a path to avoid dangerous areas by using the proposed algorithm. Figures 11 (b), (c), (d), (e) and (f) show the movement of the group of robots. In Fig. 11 (b), all the children investigated their environments. In Fig. 11 (c), a child robot touched a dangerous area, and broke. After this, in Fig. 11 (d), the parent recognized the existence of danger, added the Gaussian function to its map of the environment that was expressed by the neural network, and determined its path. Moreover, the parent robot re-organized the remaining children so as to compensate for the broken robots as shown in Fig. 11 (e). That is, the self-restoration function changed the spread space between the children. Then, they moved while avoiding dangerous areas. The group of the robots repeated such process. As a result, the parent robot arrived at the goal as shown in Fig. 11 (f). The obtained neural network consisted of

7 units at hidden layer. Figure 12 shows a behavior of the robot system that did not use the compressed neural network. In this case, the obtained neural network consisted of 13 units at hidden layer. Therefore, the proposed method is effective with respect to the size of the network to express the sensed environment.

5 Conclusion

This paper proposed a way to apply the neural network for incremental learning to store the environmental information which children bring one by one. The parent robot manages a swarm of robots, obtains the information collected by their cooperative sensing, and makes the neural network memorize information to be used for path planning. The parent obtains the map of surrounding environment when it is moving in an unknown environment. In addition, it also shows that the parent can store the environmental map in the compact network. Thus, the parent-children type robot system arrived at the destination efficiently while adapting to its unstructured environment.

References

[1] T. Shibata, T. Fukuda, Hierarchical Intelligent Control of Robotic Motion, IEEE Trans. on Neural Networks (to appear) (1992)

[2] R. C. Luo, M. G. Kay, Multisensor Integration and Fusion in Intelligent System, IEEE Trans. on Systems, Man, and Cybernetics, Vol. 19, No. 5, pp. 901-931 (1989)

[3] T. Shibata, K. Ohkawa, K. Tanie, Sensor-Based Behavior using A Neural Network for Incremental Learning in Family Mobile Robot System, Proc. of ICNN, (1994) (to appear)

[4] J. Moody and C. J. Darken, Fast Learning in Networks of Locally-Tuned Processing Units, Neural Computation, 1, pp. 281-294 (1990)

[5] D. F. Specht, Probabilistic Neural Networks, Neural Networks, vol. 3, pp. 109-118 (1990)

[6] T. Poggio and F. Girosi, Regularization Algorithm for Learning that are Equivalent to Multilayer Networks, Science, 247, pp. 978-982 (1990)

[7] T. Fukuda, S. Shiotani, et al., A New Neuron Model for Additional Learning, Proc. of IJCNN'92-Baltimore, Vol. 1, pp. 938-943 (1992)

[8] S. Shiotani, T. Fukuda, T. Shibata, An Architecture of Neural Network for Incremental Learning, Neurocomputing, (1993) (to appear)

[9] N. J. Nilsson, A Mobile Automation, an Application of Artificial Intelligence Techniques, Proc. of IJCAI, pp. 509 (1969)

[10] T. Lozano-Perez and M. A. Wesley, An Algorithm for Planning Collision Free Paths among Obstacles, Communication ACM, 22, pp. 560 (1979)

[11] O. Khatib, Real Time Obstacle Avoidance for Manipulators and Mobile Robots, Proc. of the IEEE Int'l Conf. on Robotics and Automation, pp. 500-505 (1985)

[12] R. A. Brooks, A Robust Layered Control System for a Mobile Robot, IEEE J. of Robotics and Automation, RA-2, pp. 14-23 (1986)

[13] V. J. Lumelsky, A. A. Stepanov, Dynamic Path Planning for a Mobile Automation with Limited Information on the Environment, IEEE Trans. on Automatic Control, Vol. AC-31, No. 11, pp. 1058-1063 (1986)

[14] T. L. Anderson, M. Donath, Animal Behavior as a Paradigm for Developing Robot Autonomy, Designing Autonomous Agents, pp. 145-168 (1990)

[15] M. K. Habib and H. Asama, Efficient Method to Generate Collision Free Paths for Autonomous Mobile Robot Based on New Free Space Structuring Approach, Proc. of IEEE/RSJ IROS '91, Vol. 2, pp. 563-567 (1991)

[16] R. C. Arkin, Behavior-Based Robot Navigation for Extended Domains, Adaptive Behavior, Vol. 1, No. 2, pp. 201-225 (1992)

[17] T. Shibata and T. Fukuda, Coordinative Behavior by Genetic Algorithm and Fuzzy in Evolutionary Multi-Agent System, Proc. of IEEE Int'l Conf. on R&A, Vol. 1, pp. 760-765 (1993)

Driving and Confinement of A Group in A Small Space

KENJI KUROSU[1], TADAYOSHI FURUYA[2], MITSURU SOEDA[2], JIFENG SUN[3] and AKIRA IMAISHI[3]

[1]Dept. of Industry Eng. and Management, Kinki Univ., Iizuka, 820 Japan
[2]Kitakyushu College of Tech., Kitakyushu, 803 Japan
[3]Faculty of Information Eng., Kyushu Inst. of Tech., Iizuka, 820 Japan

Abstract

We propose a strategy for driving a group to a designated direction and confining the members in a small space. There is one object called a leader, who guides such a group towards a destination. The leader detects the member which is furthest from the designated space, approach it behind and repulses it towards the destination. On the other hand, there are some forces, e.g. propulsive forces and attractive/repulsive forces, among the members of the group. Simulations are done based on both usual type controllers and fuzzy controllers, which shows that the fuzzy controller works better than the ordinary one because of its flexibility of the parameter tuning.

Keywords: Group behavior, leader, confinement, destination, distance-depended force, angle-depended force, fuzzy control

1 Introduction

It is always important to guide and confine a group to a destination, such as leading lots of automobiles out from a parking area, guiding people to a stadium, and driving flock of sheep out of a sheepfold. A group in such a guidance would take a complicated movement, because of various influences existed inside and outside. Interactions between the guide and the group or among the members of group can be described as some kinds of general forces, e.g. physical forces or messages.

The early work in the group behaviors can be divided into two kinds; one is to simplify the group behaviors into very few dynamic equations with some characteristic parameters of the group such as shapes, center movements, speeds, etc. Another is trying to obtain the movements of every member in a group by assuming dynamic difference of differential equations of individual members.

Kurosu and Furuya [1] [2] discussed the guided movement of a group, and developed a fuzzy control strategy, in which dynamic equations of individual members are given together with their leader's action patterns. Hirai [3] analyzed the group behaviors in panic, and proposed a method of controlling a group based on the global information. Mitsumiya [4] [5] discussed the action modeling of a shoal of fishes.

Considering the case that a leader drives and confines a group in a constrained space, this paper proposes some fundamental action models for a leader and individual members of the group. The action models of the leader are presented by leader's motive forces, including (1) distance-depended force, (2) angle-depended force, and (3) patrol-moving force.

For the analysis of member's dynamic characteristics, the members of the group are considered as physical objects which has masses and viscous ressitances. They interact each other and react to the leader's repulsive forces, which decrease exponentially as the distances between the leader and the members become large.

With some simulation results, we make a comparison between the fuzzy control and the ordinary control. In the case of the fuzzy control, fuzzy rules are introduced to determine the forces related to the relative distances or angles. It is proved that the fuzzy control is better both in the execution time and the success rate.

2 Dynamics of Individual Members in a Group

When a leader drives the members of a group toward a destination, the motion equation of the ith member is given as follows:

$$\begin{cases} m_i\ddot{x}_i + v_i\dot{x}_i = F_i \\ F_i = F_{1i} + F_{2i} + F_{3i} \end{cases}, \tag{1}$$

where m_i, x_i, and v_i represent the mass, the position vector, and the viscous resistance of the ith member, respectively. F_i is the applied force of the ith member. As described in Eq(1), F_i consists of three components, i.e. F_{1i}, F_{2i}, and F_{3i}. F_{1i} is the propulsive force which makes the members stick together and move forward. F_{1i} can be written as:

$$F_{1i} = F_{ai} + F_{bi} + F_{ci} + F_{di} , \tag{2}$$

where F_{ai}, F_{bi}, F_{ci}, and F_{di} are the forward-moving force, the attractive force from other members, the force making the members move in the same direction, and the force avoiding collison, refer to [1] [2] in detail.

F_{2i} in Eq(1) is a repulsive force from the environment, including F_{wi} from the wall and F_{mi} from the leader (CCO).

$$F_{2i} = F_{wi} + F_{mi}, \tag{3}$$

where:

$$F_{wi} = \begin{cases} 0 & for\ d < d_i \\ \{w_0 w_{wi}\dfrac{(d-d_i)}{d} + w_1\}e_{wi} & for\ d \geq d_i, v_{wi} > 0 \\ w_1 e_{wi} & for\ d \geq d_i, v_{wi} \leq 0 , \end{cases} \tag{4}$$

where v_{wi} is the vertical component of the ith member's velocity, d_i is the distance from the ith member to the wall, e_{wi} is a unity vector the direction of which is vertical to the wall, and d, w_o, and w_1 are positve constants.

F_{mi} in Eq((3) is represented as:

$$F_{mi} = -c_i(\frac{x_c - x_i}{|x_c - x_i|}) , \tag{5}$$

where:

$$c_i = c * exp(-|x_c - x_i|^2) \tag{6}$$

and c is a positive constant.

F_{3i} in Eq(1) is the disturbance as follows:

$$F_{31} = q * e_{ri}, \tag{7}$$

where q is a positive constant and e_{ri} is a random vector.

3 Control Strategy for Driving and Confining a Group

Assumed there is a CCO driving a group towards a exit, as shown in Figure 1. The action space of CCO and the group is limited by a wall. CCO moves with some motive forces itself and drives the group according to the some control rules. As one member had go out of the exit, it can not go back again.

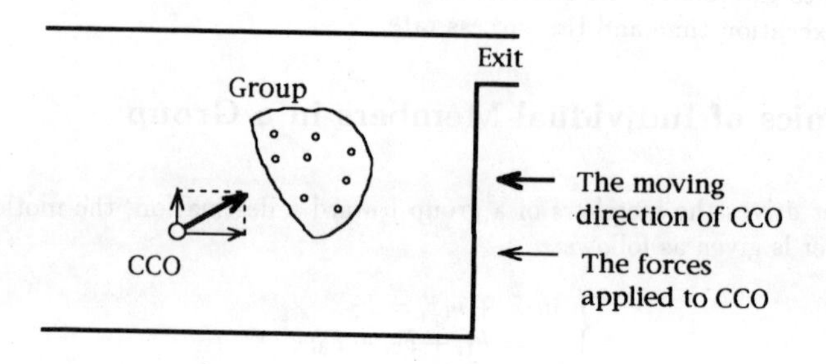

Figure 1: Driving a group toward a exit

3.1 Control Rule

In order to drive a group toward a exit, CCO must locate in the back of the group. We set the initial state of CCO to be such a state. On the other hand, the operation of driving and confining must be started from the member which is furthest from the exit. Before discussing the driving control strategy, we give two fundamental rules as follows:

Rule 1: Related to the exit, CCO moves to the back of the group, including the initial state.

Rule 2: CCO detects the furthest member (MLO) from the exit, and drives it toward the exit.

3.2 Action patterns of CCO

We use some action patterns to represent the dynamics of CCO. These patterns conclude (1) moving to the back of MLO (Pattern 1), (2) jumping to the back of MLO instantly (Pattern 2), (3) motivated by the distance-depended force and the angle-depended force (Pattern 3), (4) The member departing away the exit become MLO (Pattern 4), and (5) turning around the member when the member is behind CCO (Pattern 5).

3.2.1 Pattern 1

As shown in Figure 2, for approaching to MLO, CCO in Pattern 1 has the propulsive forces f_x and f_y in x and y direction. f_x and f_y are given as:

$$\begin{cases} f_x = k_x * (x_m - x_c) \\ f_y = k_y * (y_m - y_c) \ , \end{cases} \tag{8}$$

where x_m and y_m are the coordinates of MLO , x_c and y_c are the coordinates of CCO, and k_x and k_y are constants.

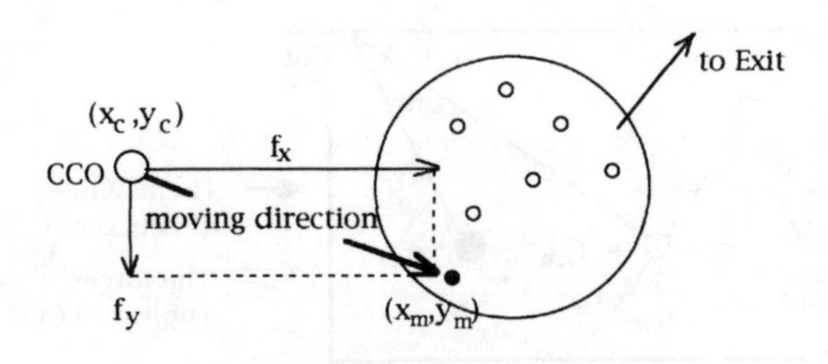

Figure 2: Pattern 1 of CCO

3.2.2 Pattern 2

CCO in Pattern 1 can drive single members to the exit, but doesn't work well in the case of driving a multi-member group, because MLO changes time by time. In order to solve this problem, we define Pattern 2 , in which CCO instantly moves to the back of new MLO with a designated distance, as shown in Figure 3. Here, the moving time from the state of Figure 3(a) to the state of Figure 3(b) is neglected. Therefore, CCO in Pattern 2 is not limited by Rule 1 and Rule 2 above.

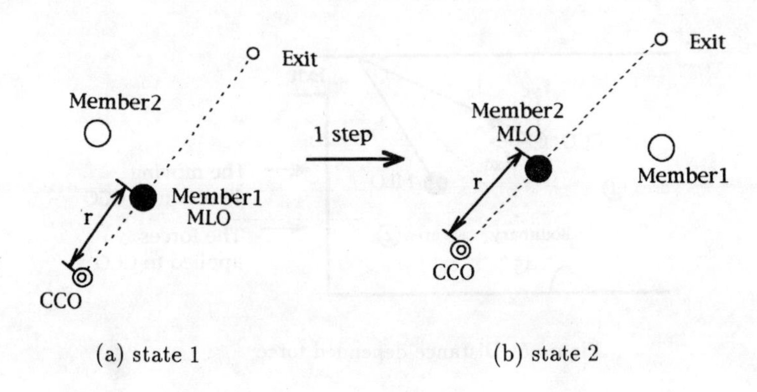

Figure 3: Pattern 2 of CCO

3.2.3 Pattern 3

The instant move in Pattern 2 is desirable, but unrealistic. We can make a trade-off between Pattern 1 and Pattern 2, and develop Pattern 3, in which CCO moves from the back of an old MLO to the back of a new MLO with help of an angle-depended force and a distance-depended force.

As shown in Figure 4, the angle-depended force f_m is used to make CCO, MLO, and the exit locate in the same line. x and y components of f_m are given as:

$$f_{xm} = f_{ym} = k_\theta * (\theta_m - \theta_c), \tag{9}$$

where θ_c is the angle of $\angle$ CCO-exit-wall, θ_m is the angle of $\angle$ MLO-exit-wall, and k_θ is a constant.

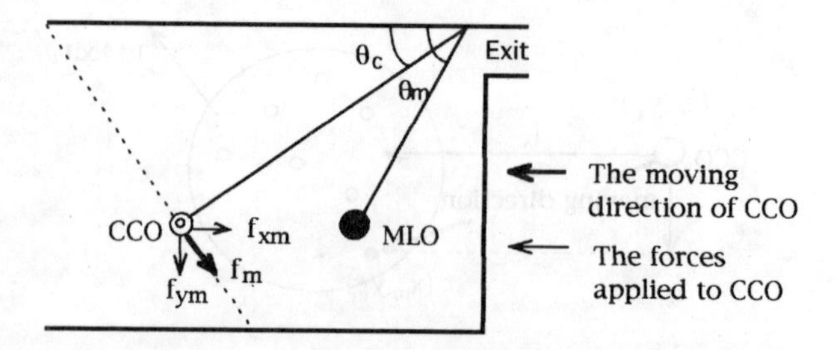

Figure 4: Angle-depended force

Although f_m can make CCO, MLO, and the exit locate in the same line, it can not let CCO drive MLO toward the exit. Therefore, we must design a distance-depended force f_c for driving MLO. The distance-depended force make the difference between the CCO-Exit distance and the MLO-Exit distance to be a designated value l_{cm}, as shown in Figure 5. x and y components of the distance-depended force f_c are given as follows:

$$(f_{xc} \;\; f_{yc}) = \begin{cases} (k_1 * (l_c - l_m) \;\; 0) & for\ l_c - l_m > l_{cm}\ and\ CCO \in area1 \\ (0 \;\; -k_1 * (l_c - l_m)) & for\ l_c - l_m > l_{cm}\ and\ CCO \in area2 \\ (0 \;\; 0) & for\ l_c - l_m \leq l_{cm}\ . \end{cases} \tag{10}$$

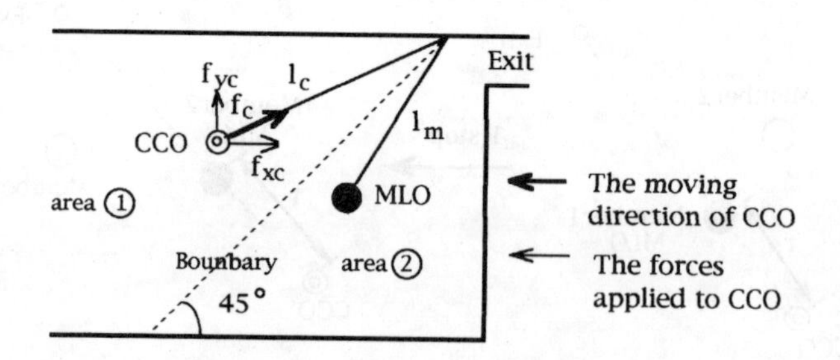

Figure 5: Distance-depended force

Hence, CCO in Pattern 3 has the propulsive force as follows:

$$f = f_m + f_c\ . \tag{11}$$

3.2.4 Pattern 4

If a member departs away from the exit, CCO in Pattern 4 makes this member as a new MLO, instead of the member which is furthest from the exit. As shown in Figure 6, it can be said that $Member_i$ moves toward the exit if it move to $PartI$; otherwise it departs from the exit.

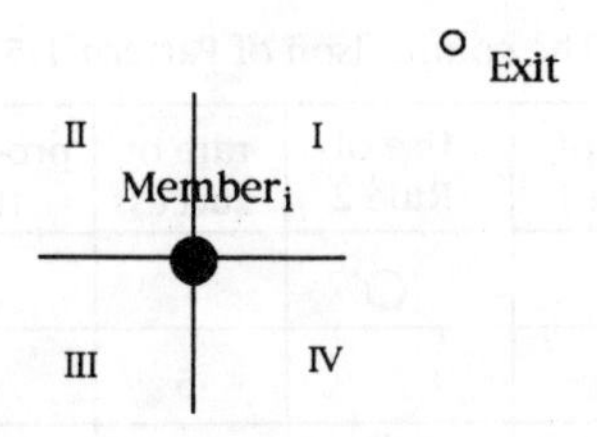

Figure 6: Moving direction of $Member_i$

3.2.5 Pattern 5

If CCO in Pattern 1 or Pattern 2 or Pattern 3 locates in the front of MLO, it would fail to drive MLO toward the exit. In Pattern 5, the propulsive force in Pattern 4 is improved by adding the force which make CCO turn back and locate in the back of MLO, as shown in Figure 7. Hence, the propulsive force in Pattern 5 is written as:

$$f = f_m + f_c - f_e \ , \tag{12}$$

where f_m and f_c are the same as those in Pattern 4. As shown in Figure 7, according to whether CCO is in area 1 or area 2, f_e is determined as follows:

$$(f_{xe} \;\; f_{ye}) = \begin{cases} (0.0 \;\; k_e) & for \; l_m > l_c \; and \; CCO \in area1 \\ (0.0 \;\; -k_1 \times (l_c - l_m)) & for \; l_m > l_c \; and \; CCO \in area2 \\ (0.0 \;\; 0.0) & for \; l_m \leq l_c \; . \end{cases} \tag{13}$$

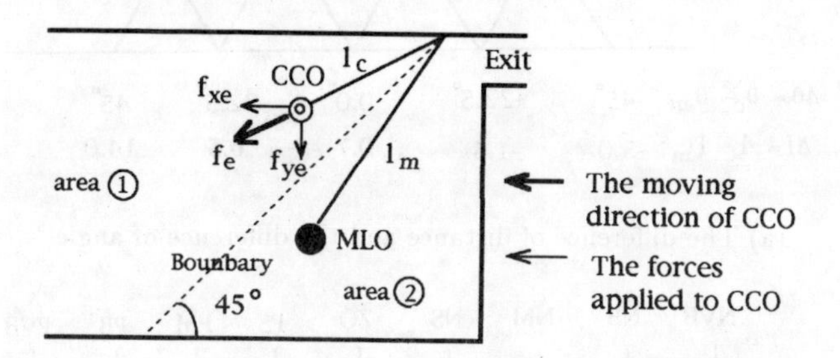

Figure 7: Pattern 5 of CCO

Surpose the member in a group has the motion equation given in Eq(1), and CCO acts in Patterns 1-5. Figures 9-10 are the simulation results when CCO in different patterns drives the group which has different number of members. The results show that ,in the case of CCO in Pattern 3 or Pattern 5, if f_c is big, the execution time is short and the success rate is low; if f_c is small, the execution time is long and the succes rate is high. In Figures 9-10, the thin curve is the trajectory of CCO and the thick curves represent the trajectory of the Members. Table 1 shows a comparison among Patterns 1-5 based on the simulation.

Table.1 The comparison of Pattern 1-5

	Use of Rule 1	Use of Rule 2	rate of success	processing time	capacity
Pattern 1	O	O	1	4	1
Pattern 2	O	–	5	5	5
Pattern 3	O	O	3	2	2
Pattern 4	–	O	2	3	3
Pattern 5	O	–	4	1	4

4 Fuzzy Control for Driving a Group

As described in section 3, when CCO drive a group, it can move in various patterns. In order to evaluate the results of driving, we consider several criteria, such as (1) all members reach the exit, (2) The execution time, and (3) the minimun energy for driving. Such criteria rank in the priority of (1), (2), and (3).

For improving the criteria above, we introduce a fuzzy controller to CCO in Pattern 5. Membership functions about inputs and outputs of the controller are given in Figure 8(a)(b). θ_c, θ_m, l_m, l_c, f_x, and f_y are the same as those in Figure 4 and Figure 5.

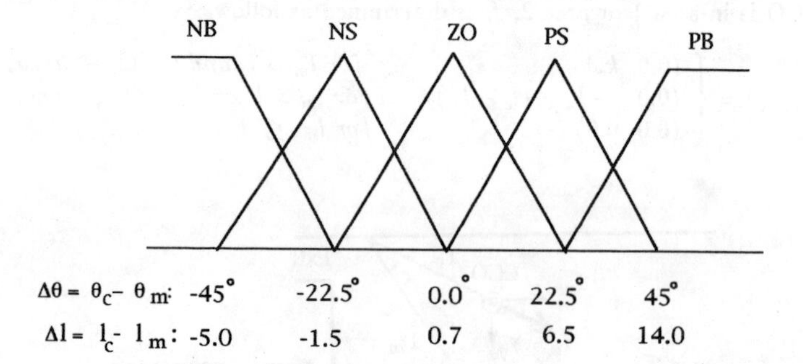

(a) The difference of distance and the difference of angle

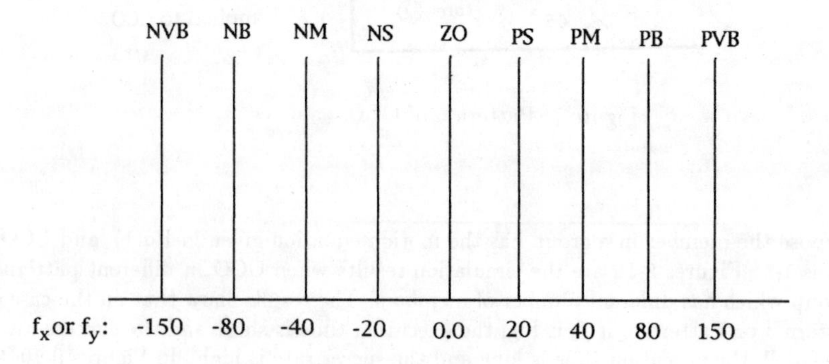

(b) The output of forces

Figure 8: Membership functions

Fuzzy rules when CCO is in the area 1 is shown in Table 2.

Table 2 Fuzzy rules

	Δl \ $\Delta\theta$	NB	NS	ZO	PS	PB
f_x	NB	NVB	NB	NVB	ZO	PM
	NS	NVB	NM	NVB	NS	PS
	ZO	NB	NM	ZO	NS	PVB
	PS	NB	NM	PS	PS	PVB
	PB	NVB	NS	PM	PB	PVB
f_y	NB	NVB	ZO	ZO	PVB	PVB
	NS	NB	ZO	NS	PB	PVB
	ZO	NB	NM	ZO	PM	PVB
	PS	NVB	NB	ZO	PM	PVB
	PB	NVB	NB	NS	PM	PB

We developed the fuzzy rules when CCO is in area 2 in the same way.

Figure 11 shows the simulation results as a fuzzy-controlled CCO drives a two-memeber group. Figure 12 shows the results of such a CCO driving a six-member group. The thick curve in Figures 11-12 is the trajectory of CCO and the thin ones are those of the Members. All of these examples success.

5 Conclusion

We have proposed a control strategy for driving and confining a group toward a destination. The leader (CCO), which approaches and drives the group, can take five kinds of action patterns. Each member in the group is driven by repulsive force from both CCO and the wall, and the inter-member forces. CCO in simple action patterns, e.g. Pattern 1 or Pattern 2, moves fast but has a low success rate. CCO in complex pattern, e.g. Pattern 5, move slowly but has a high success rate. The execution time of driving and confining is proportional to the number of individual members in the group. In a fuzzy control strategy of driving and confining, the execution time decreases when the value kinds of the input increase in some extent. However, the necessary time does not decrease when the value kinds of the input reaches a theshold. Simulation results show that the execution time of the fuzzy-controlled driving is less than that of the ordinary control in the same conditions. It can leads to some conclusions that a fuzzy-controlled CCO has a higher capacity than that with the ordinary control.

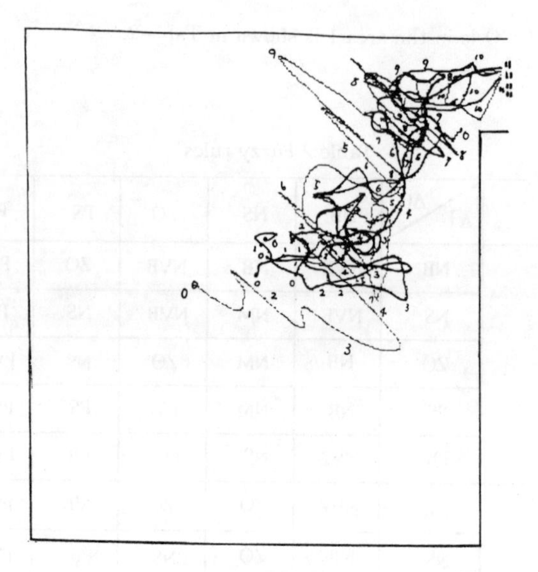

Figure 9: Example of the ordinary control CCO in Pattern 4 (five-member group)

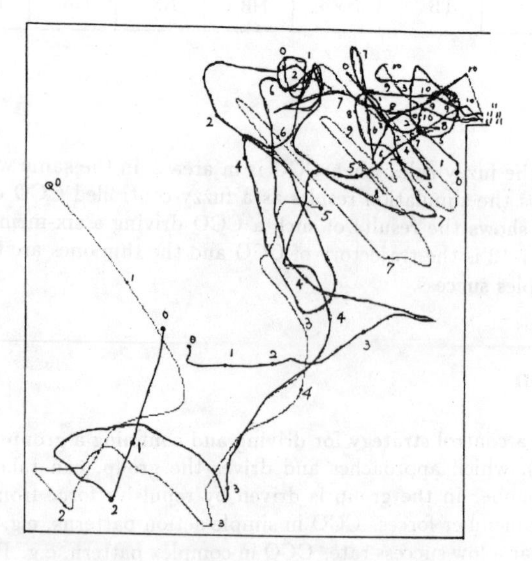

Figure 10: Example of the ordinary control CCO in Pattern 5 (five-member group)

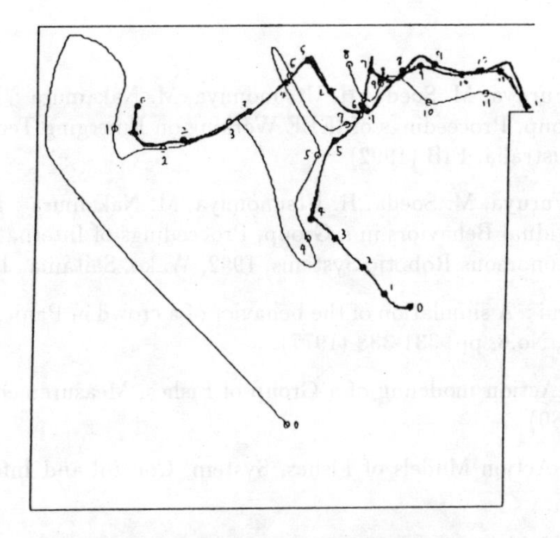

Figure 11: Example of the fuzzy control CCO in Pattern 5 (two-member group)

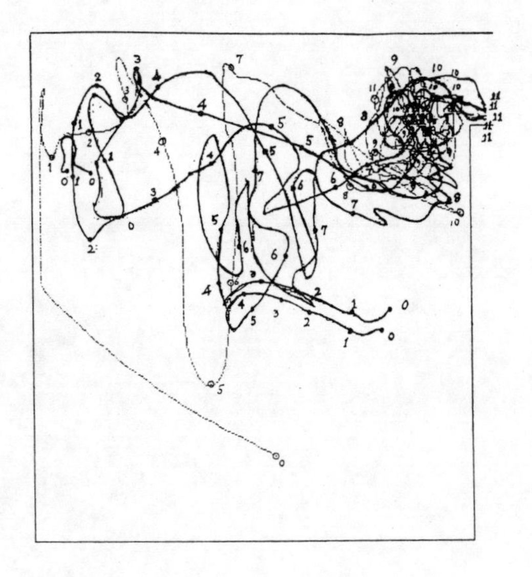

Figure 12: Example of the fuzzy control CCO in Pattern 5 (six-member group)

References

[1] K. Kurosu, T. Furuya, M. Soeda, H. Utsunomiya, M. Nakamura : Dynamics and Fuzzy Control of a Group, Proceedings of IEEE Working on Emerging Technology and Factory Automation, Australia, FIB (1992)

[2] K. Kurosu, T. Furuya, M. Soeda, H. Utsunomiya, M. Nakamura : Dynamics and Fuzzy Control of Individual Behaviors in a Group, Proceedings of International Symposium on Distributed Autonomous Robotic Systems, 1992, Wako, Saitama, Japan, pp. 11-18

[3] K. Hirai, K. Tarui : A simulation of the behavior of a crowd in Panic, System and Control (Japan), Vol.21, No.6, pp. 331-338 (1977).

[4] S. Mitsumiya : Action modeling of a Group of Fishes, Measurement and Control, 19-7, pp. 704-707 (1980)

[5] S. Mitsumiya : Action Models of Fishes, System, Control and Information, 37-12, pp. 696-703 (1993)

Chapter 8
Coordinated Control

Cooperative System between a Position-controlled Robot and a Crane with Three Wires

HISASHI OSUMI[1], TAMIO ARAI[1], NAOMI YOSHIDA[1], YUSI SHEN[1],
HAJIME ASAMA[2], HAYATO KAETSU[2], and ISAO ENDO[2]

[1]Dept. of Precision Machinery Eng., The Univ. of Tokyo, Tokyo, 113 Japan
[2]The Inst. of Physical and Chemical Research (RIKEN), Wako, 351-01 Japan

Abstract

This paper proposes a strategy for handling a heavy object by cooperation of a PID position controlled robot and a new type crane with three wires. Installation of free joints or flexibility is essential to avoid excessive inner forces caused by the positioning error of the robot and the crane. First, the necessary mechanical conditions to achieve the cooperation among position-controlled robots are shown. Second, a free joint mechanism designed for the cooperation of the robot and the crane is introduced. Since the cooperative system with the free joint mechanism has kinematic redundancy, it can be used for the optimization of some criteria determined from the characteristics of tasks. This system is controlled by master/slave type controllers. After the robot grasps a suspended object, the robot is position-controlled as a master. The crane is velocity-controlled as a slave to eliminate the displacements of each free joint generated by the motion of the robot. Eventually, the transfer of the suspended object to the target position is achieved. The results of experiments using a prototype crane and a popular industrial robot verified the effectiveness of the free joint mechanism and the cooperative control algorithm. The strategy proposed here is very effective and practical for cooperation of not only conventional industrial robots but also distributed autonomous mobile robots with non-holonomic constraints.

Keywords: Cooperative control, free joint mechanism, position control, crane with multiple wires

1. Introduction

Cooperation of multiple robots can improve the performances of robot systems, such as dexterity, pay load capacity, gasping function and so on. There are two types of cooperation, which are the cooperation without interactions of forces between robots and the cooperation with them. In the former type cooperation, task planning is the most important technical issues, but the same position controller as that of single robot can be used and it can be realized easily. This type has been practically used for welding today[1]. On the other hand, under the interactions of forces, control of inner forces between robots or assurance of the stability of the controllers becomes one of the most critical items. For these problems, many algorithms have been proposed until now.

Uchiyama[2] designed a hybrid control system for dual manipulators. Hu[3] proposed a coordinated control algorithm for redundant manipulator. The stability of the system was ensured by linearizing the error system. Xi[4] used hybrid position/force controllers for each robot, and proposed "the event-based planning" for the coordination. Almost all the studies have dealt with dynamics of the control system and their algorithms have been designed based on torque controllers. Their theoretical performances are good enough for cooperation, but very few control algorithms have been applied to actual robot systems, because almost all the robots used today are controlled by PID position controllers and torque controlled robots are not so popular in practice. Kosuge[5] used a virtual internal model for cooperative control of multiple manipulators. In this system, the desired positions of manipulators with PID position controller are modified according to the feedback signals and virtual internal models to compensate the positioning errors of manipulators. Cooperative algorithms for decentralized autonomous robotic systems have also been proposed in recent years[6],[7]. In the case of the cooperation between mobile robots, non-holonomic constraints caused by wheel mechanisms become one of the most serious problems. Hashimoto[6] installed an additional actuator on each mobile robot and designed a cooperative controller. In such decentralized systems, cooperation will be accomplished by means of communication between robots, whose transmission ratio is generally not enough for servo level cooperation. So, cooperative algorithms based on torque controllers also may not be applied.

For the practical use of cooperation of robots or realizing cooperative tasks in decentralized robotic systems, methods applicable to position controlled systems are essential. Installation of flexibility or free joints between two robots is one of the most simple and effective solutions to avoid excessive inner forces for PID position-controlled robots or autonomous mobile robots. However, the installation of flexibility decreases the positioning accuracy due to the vibrations or deflections of mechanisms.

There are some studies dealing with the cooperative control between manipulators with flexibility. Ahmad[8] designed a vibration control system for cooperative system. Arimoto[9] designed a cooperative control system for multi-fingered hand. Rubber at the finger tip was modeled as a stiffness matrix and the asymptotical stability of the cooperative system with the proposed algorithm was verified. Guo[10] proposed a decomposition of finger tip forces of a multi-fingered robot hand into a dynamic equilibrating force and an internal force and determined the former force considering the deflections of the fingers. Authors[11] designed a cooperative control system for multiple manipulators with flexibility at their tips. Authors[12] also developed a cooperative system between a crane and an industrial robot with a PID controller. A flexible arm was installed at the tip of the robot. The robot was position-controlled as a master and the crane was velocity-controlled as a slave by the feedback signal of the deflection of the arm.

Most of them have focused on the control of dynamics or vibration of their systems, however, the vibrations or deflections caused by installed flexible mechanisms do not appear in case that the flexible mechanisms are designed and installed in an adequate manner[13],[14],[15]. Moreover, the stability problem will not be so critical for slow motion control in practical use. In these cooperative systems with well-designed flexibility, both robots can be controlled by their own servo controllers according to the sequence level command for cooperation, and servo level control algorithms are not so important.

Thus, this paper proposes a cooperative control algorithm for the manipulation of a rigid object by two position-controlled robots. First, the method to install the flexibility for cooperation will be shown. Using the proposed method, a cooperative system, between a robot and a crane with three wires, is designed and developed for automating heavy parts assembly tasks.

2. Installation of flexibility to cooperative system

If the springs installed between two manipulators do not satisfy adequate conditions, excessive inner forces generate because the geometrical constraint caused by a mutual positioning error is imposed on two rigid manipulators. On the other hand, if there are too many springs installed, vibrations occur by disturbance.

<u>The condition to avoid both the excessive inner forces and vibrations is that there are linearly independent six springs between two manipulators.</u> The proof is shown as follows. Figure 1 shows the model of a cooperative system with springs between two manipulators. $\mathbf{s}_i$ denotes the unit vector in the direction of i-th spring's eigen vector and a_i denotes the spring deflection in the direction of $\mathbf{s}_i$. When there are linearly independent six springs, even if the mutual positioning error $\mathbf{e}$ appears, it can be realized by six deflections calculated as Eq.(1).

$$\mathbf{d} = S^{-1}\mathbf{e} \tag{1}$$

Where $\mathbf{d} = [\, d_1\, d_2\, d_3\, d_4\, d_5\, d_6 \,]^T$, $S = [\, \mathbf{s}_1\, \mathbf{s}_2\, \mathbf{s}_3\, \mathbf{s}_4\, \mathbf{s}_5\, \mathbf{s}_6 \,]$

Thus, excessive inner forces do not occur. Additionally, the deflections a_i are determined by only geometrical constraint of $\mathbf{e}$ and there is not any other combination of deflections realizing $\mathbf{e}$. So, the position and orientation of the object are determined independent of force disturbance, and consequently vibrations do not occur.

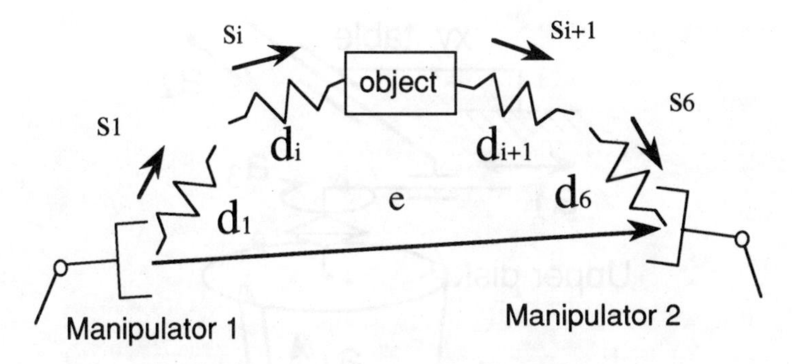

Fig.1 Cooperative system with six springs installed adequately

Free joint mechanisms, instead of flexible mechanisms, can also be used for the cooperation by modeling the displacement of each joint as a deflection of the corresponding spring. In this case, inner force between two manipulators is always zero, thus, pushing or deforming handling objects is impossible.

3. Cooperative system between a robot and a crane

The robot illustrated in Fig.2 has seven degrees of freedom including three wires[16],[17]. Each arrow expresses a moving direction of each actuator. The upper part of the robot consists of a x-y table and a rotational mechanism about z axis. Three wire feed mechanisms are placed on the upper disk and each wire length can be controlled respectively. The actuator 7 is installed in order to increase motion controllability in the torsional direction. Using seven actuators, the robot can achieve not only the desired position but also the orientation of the suspended object by controlling these seven actuators respectively. Moreover, the robot can reduce the residual vibrations in all directions in 3D space.

In order to use the proposed method for the cooperation between a crane and a robot, the suitable crane model for the method will be derived. Since the motion of a suspended object can be modeled as that of a spring mass system as shown in Fig.3, the crane can be regarded as a rigid manipulator with springs at the tip, which is the equilibrium point of the crane suspension system. The model of the springs depends on the mass and the shape of the object, the number of wires and the manner of suspension. For example, in the case of a conventional crane with single wire, the spring mass model consists of five springs[14].

In the case of the crane illustrated in Fig.2, there are three vibration modes shown in Fig.4. From the figure, it can be understood that this crane system can be modeled as a seven D.O.F. rigid manipulators with three springs at the tip. The directions of the springs with respect to the object coordinate system Σ_o are as Eq.(2).

$$
\begin{aligned}
\mathbf{s}_{c1} &= [\,0\ -\alpha\ 0\ 1\ 0\ 0\,]^T \\
\mathbf{s}_{c2} &= [\,\alpha\ 0\ 0\ 0\ 1\ 0\,]^T \\
\mathbf{s}_{c3} &= [\,0\ 0\ 0\ 0\ 0\ 1\,]^T
\end{aligned}
\tag{2}
$$

Where α is a variable determined through kinematic parameters of the crane. So, three additional springs or free joints in the directions independent of $\mathbf{s}_{c1}$-$\mathbf{s}_{c3}$ are needed for the cooperative robot.

Figure 5 illustrates the designed mechanism with three free joints. One is a prismatic joint in the z direction and the others are rotational joints about the x and the y axes.

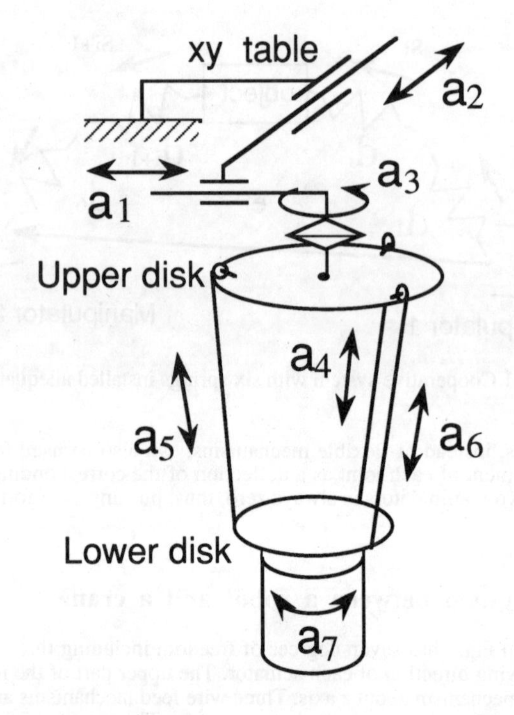

Fig.2 Crane with three wires

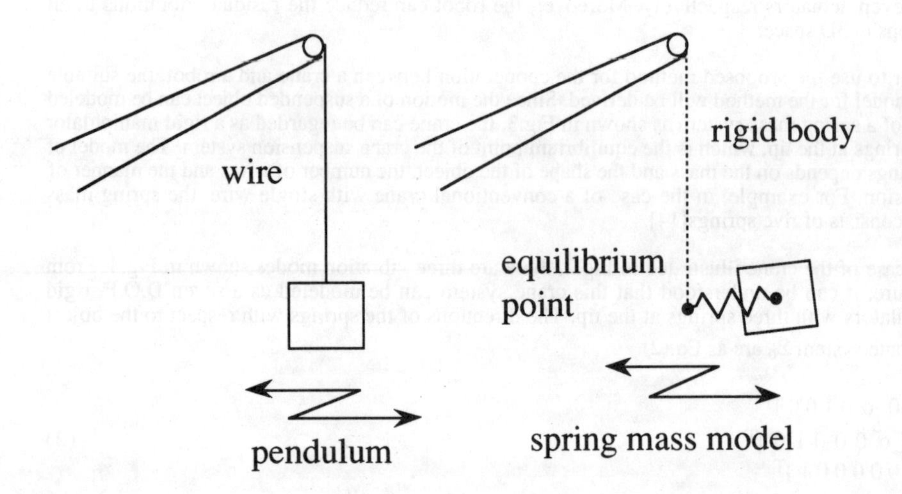

Fig.3 Spring model of a suspension system

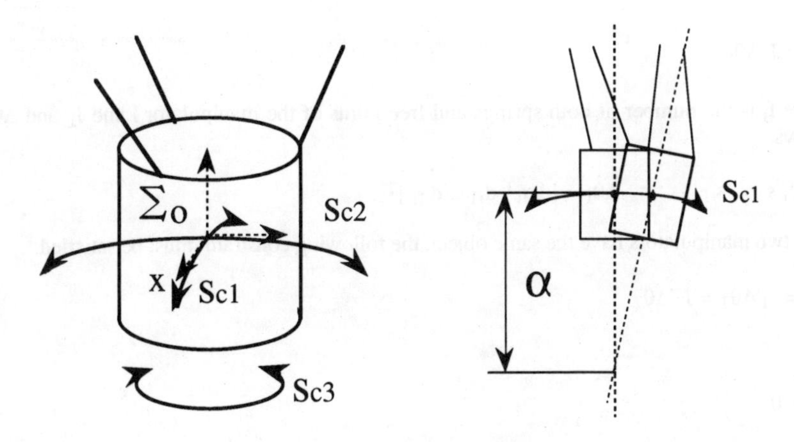

Fig.4 Vibration of a suspended object

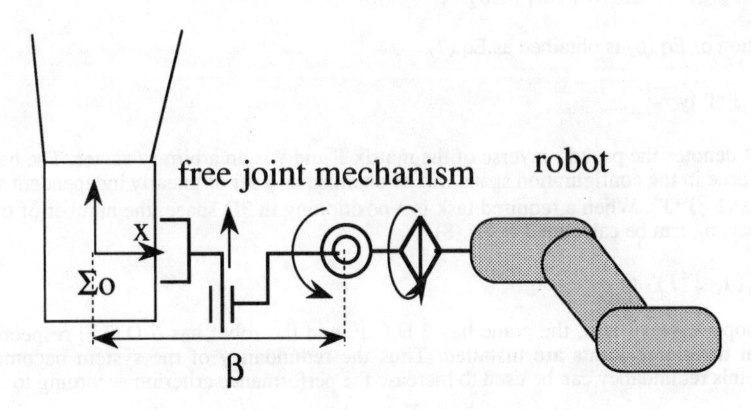

Fig.5 Free joint mechanism for the cooperation

The three direction vectors of installed free joints, $\mathbf{s}_{r1}$-$\mathbf{s}_{r3}$ are expressed as Eq.(3).

$$\mathbf{s}_{r1} = [\,0\;0\;0\;1\;0\;0\,]^T$$
$$\mathbf{s}_{r2} = [\,0\;0\;-\beta\;0\;1\;0\,]^T \qquad (3)$$
$$\mathbf{s}_{r3} = [\,0\;0\;1\;0\;0\;0\,]^T$$

Where β is the distance between the origin of the object coordinate and the free joint corresponding to $\mathbf{s}_{r2}$. Since these six vectors are linearly independent, it can be verified that the mechanism satisfies the necessary condition.

4. Kinematic analysis of the cooperative system

The way of deriving the degrees of freedom of a general closed loop structure including springs installed in the above-mentioned manner, is shown. Let J_i be the Jacobean matrix of manipulator i and $\mathbf{q}_i$ be its joint angle vector. By considering the deflections of springs as virtual joint variables, the velocity of the object, $\Delta\mathbf{x}_m$ is expressed by $\Delta\mathbf{q}_i$ and the eigen vectors $\mathbf{s}_{ik}$(k=1..f_i) as Eq.(4).

$$\Delta \mathbf{x}_m = J_i'\Delta\theta_i \tag{4}$$

Where f_i is the number of both springs and free joints of the manipulator i and J_i' and $\Delta\theta_i$ are as follows.

$$J_i' = [\, J_i \; \mathbf{s}_{i1} \,..\, \mathbf{s}_{if_i} \,], \qquad \Delta\theta_i = [\Delta\mathbf{q}_i^T \; d_{i1} \,..\, d_{if_i} \,]^T$$

Since two manipulators have the same object, the following constraint must be satisfied.

$$\Delta \mathbf{x}_m = J_1'\Delta\theta_1 = J_2'\Delta\theta_2 \tag{5}$$

So,

$$J'\Delta\theta = 0 \tag{6}$$

Where

$$J' = [\, J_1' \; -J_2' \,], \qquad \Delta\theta = [\, \Delta\theta_1^T \; \Delta\theta_2^T \,]^T$$

The solution of Eq.(6) is obtained as Eq.(7).

$$\Delta\theta = (\, I - J'^{+}J' \,)\mathbf{y} \tag{7}$$

Where J'^{+} denotes the pseudo-inverse of the matrix J' and $\mathbf{y}$ is an arbitrary vector. The basis of the moving space in the configuration space can be obtained as a set of linearly independent vectors of the matrix $(I - J'^{+}J')$. When a required task is a positioning in 3D space, the number of the system redundancy, n_r, can be calculated as Eq.(8).

$$n_r = \mathrm{rank}\,(\, I - J'^{+}J') - 6 \tag{8}$$

In our cooperative system, the crane has 7 D.O.F. and the robot has 6 D.O.F. respectively and additional three free joints are installed. Thus the redundancy of the system becomes 4. For example, this redundancy can be used to increase the performance criterion according to a required task.

5. Statics analysis of the cooperative system

Figure 5 shows the kinematic model of the robot and the installed hand mechanism. Since there are three free joints between the base of the robot and the object, the forces applied to the object by the robot are restricted in three dimensional sub-space. The basis of the sub-space can be derived as follows. Let the joint angle vector of the robot be $\mathbf{q}_r \in R^6$ and that of free joint mechanism be $\mathbf{d}_r \in R^3$. Moreover, the vector $\theta_r \in R^9$ consisting of $\mathbf{q}_r$ and $\mathbf{d}_r$ is defined as Eq.(9).

$$\theta_r = [\, \mathbf{q}_r^T \; \mathbf{d}_r^T \,]^T \tag{9}$$

The relationship of infinitesimal displacements between the position and orientation vector of the object, $\mathbf{x}_o \in R^6$, and θ_r, can be expressed as Eq.(10).

$$\Delta\mathbf{x}_o = J\Delta\theta_r = [\, J_r \; J_d \,][\, \Delta\mathbf{q}_r^T \; \Delta\mathbf{d}_r^T \,]^T \tag{10}$$

J is the Jacobean matrix between the motion of the object and that of the joint angle vectors and J_r, J_d are matrices corresponding to $\Delta\mathbf{q}_r$, $\Delta\mathbf{d}_r$ respectively. Then, the relationship between the force generated at the tip of the hand, $\mathbf{f}$, and the torque vectors of the robot, τ_r, is expressed as Eq.(11).

$$[\ J_r\ J_d\ ]^T \mathbf{f} = [\ \tau_r^T\ \mathbf{0}^T\ ]^T \tag{11}$$

From Eq.(11), it is understood that the force vectors generated by the robot actuators is constrained by Eq.(12).

$$J_d^T \mathbf{f} = \mathbf{0} \tag{12}$$

So, **f** generated by the robot must be expressed as Eq.(13).

$$\mathbf{f} = (\ I_6 - J_d^{+T} J_d^T)\mathbf{y} = J_d' \mathbf{y} \tag{13}$$

Where $I_6 \in R^{6x6}$ means the identity matrix and **y** is an arbitrary vector. J_d' is as follows.

$$J_d' = (\ I_6 - J_d^{+T} J_d^T)$$

From Eq.(13), it can be understood that after the mechanism with three free joints are installed at the tip of the robot, the robot can only generate the forces in the sub-space spanned by the column vectors of the matrix J_d'. Since the number of the free joints is three, the rank of J_d' becomes three. The basis of the sub-space can be obtained as the three linearly independent column vectors of J_d'.

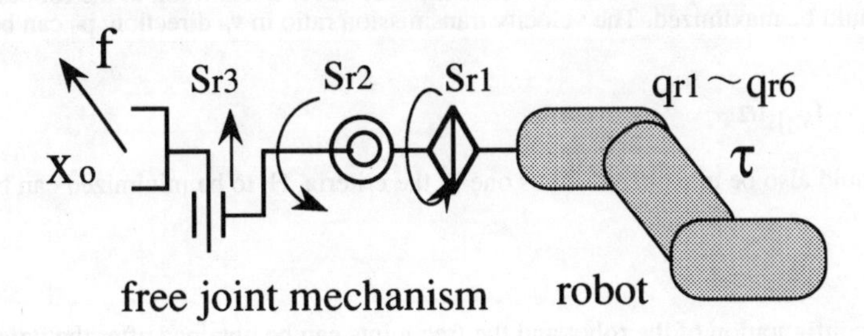

Fig.6 Kinematic model of the robot and the hand mechanism with three free joints

The relationship between the force vector and the torque vector of the robot is obtained as Eq.(14).

$$\tau_r = J_r^T J_d' \mathbf{y} = J_{rd}^T \mathbf{y} \tag{14}$$

Where J_{rd} is as follows.

$$J_{rd}^T = J_r^T J_d'$$

The singular decomposition of J_{rd} can be obtained as Eq.(15).

$$J_{rd} = U \Sigma_{rd} V^T \tag{15}$$

Where U and V are 6x6 orthogonal matrices and Σ_{rd} is expressed as Eq.(16).

$$\Sigma_{rd} = \mathrm{diag}(\ \varepsilon_1\ \varepsilon_2\ \varepsilon_3\ 0\ 0\ 0\ ) \tag{16}$$

Where ε_1-ε_3 are the singular values of J_{rd}^T. Defining the three dimensional ellipsoid formed by all the force vectors generated by the robot actuators as "active force ellipsoid", the axes of the active force ellipsoid are expressed as $\varepsilon_i^{-1}\mathbf{u}_i(i=1,2,3)$, where $\mathbf{u}_i$ is the i-th column vector of the matrix U.

6. Use of redundancy

The redundancy of the robot and the free joints will be used to optimize a criterion specified by a desired task here. When the robot guides the object to the target position, we assume that the force generated at the tip of the robot in the motion direction of the robot should be small and also that it should be easy for the robot to move to the target position. Under these specifications, the criterion of the task can be expressed as follows. Let the motion vector of the robot be $\mathbf{v}_d$. Then, the velocity transmission ratio in $\mathbf{v}_d$ direction, p_v, can be expressed as Eq.(17).

$$p_f = [\mathbf{v}_d^T(J_{rd}J_{rd}^T)\mathbf{v}_d]^{-1/2} \tag{17}$$

In usual cases, if the rank of J_{rd} is not full, p_f become infinite. This is because the mechanism can generate infinite forces in its non-movable directions. However, in case that free joints are included in joint variables, the mechanism can never generate forces in such directions. Thus, the equation (17) can not be used in this case, and other modified critera should be needed. Here, the criterion expressed as Eq.(17) is modified as Eq.(18).

$$p_f = [\mathbf{v}_d^T(J_dJ_d^T)\mathbf{v}_d]^{1/2} \tag{18}$$

p_f expresses the norm of the projection of $\mathbf{v}_d$ onto the sub-space spanned by the column vector of J_d. Since the forces at the tip of the hand mechanism are never transmitted to the robot in this sub-space, p_f should be maximized. The velocity transmission ratio in $\mathbf{v}_d$ direction, p_v can be shown as Eq.(19).

$$p_v = [\mathbf{v}_d^T(J_rJ_r^T)^{-1}\mathbf{v}_d]^{-1/2} \tag{19}$$

Where p_v should also be maximized. Thus one of the criteria, H, to be minimized can be obtained as Eq.(20).

$$H = (p_f p_v)^{-1} \tag{20}$$

The optimal configuration of the robot and the free joints can be obtained after the iteration of the calculation of Eq.(21).

$$\Delta\theta_r = J^+\mathbf{v}_d + k(I_9 - J^+J)\partial H/\partial\theta_r \tag{21}$$

Where k is a negative constant.

7. Control algorithm for cooperation

Figure 7 shows the proposed cooperative system between a robot and a crane. In this system, the crane compensates the gravity of suspended objects and the robot compensates the positioning accuracy instead of human workers. Since the system has the same payload capacity as the crane and the same positioning accuracy as the robot, it will be very effective for positioning heavy objects.

In this system, the robot is position-controlled as a master, and the crane is velocity-controlled as a slave. The redundancy of the system will not be used here. When the robot moves, displacements of three springs of the crane and the three free joints of the robot are generated shown in Fig.8. The crane is controlled so as to make all the displacements zero.

The detail of the control algorithm is as follows.

1) Calculate equilibrium point $\mathbf{x}_c$ from the crane joint angles.

2) Measure d_{r1}-d_{r3} by potentio-meters and d_{c1}-d_{c3} by gyro scope and two axis inclinometers. Calculate $\Delta \mathbf{x}_c$ from d_{r1}-d_{r3} and d_{c1}-d_{c3}
3) Calculate Jacobean matrix of the crane, J_c, numerically.
4) Calculate the movement of the crane actuators $\Delta \mathbf{q}_c$ according to the following equations.

$$\Delta \mathbf{q}_c = J_c^{-1} \Delta \mathbf{x}_c \ , \ \mathbf{q}_c = \mathbf{q}_c + \Delta \mathbf{q}_c$$

5) Return to 1).

As a result, the object is positioned at the point where the robot stops.

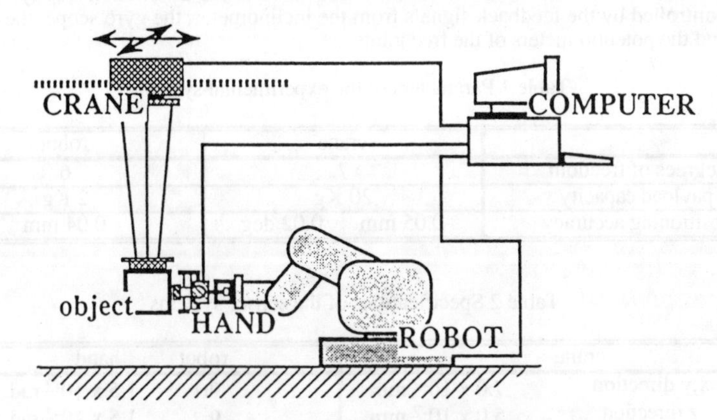

Fig.7 Cooperation system between a robot and a crane

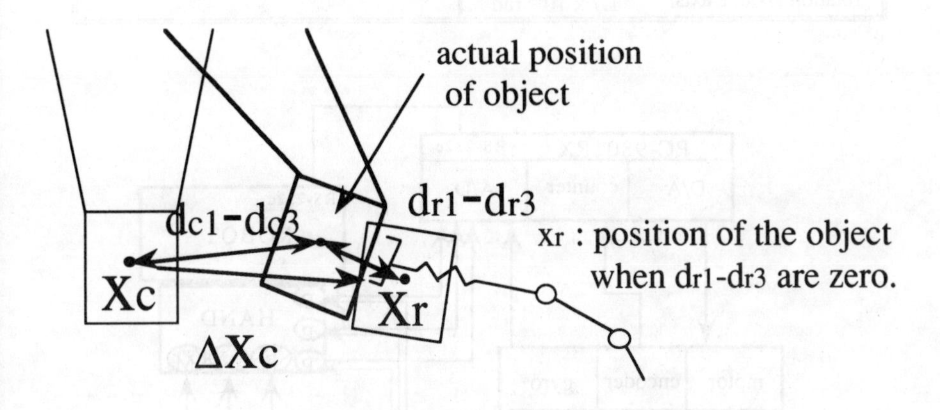

Fig.8 Displacements generated by the motion of the robot

8.Experiment of cooperation

Figure 9 shows the configuration of the system. As a controller, PC-9801RX(10 MHz, processor 8086, 8087) is used. The robot is MOVE MASTER RV-E2 (Mitsubishi Electronic Corporation) which has 6 D.O.F. and is controlled by the position commands from the computer. The working space of the crane is 1.0 m high, 0.9 m wide and 0.5 m deep, and the crane is velocity-controlled by the input voltages from the D/A converters of the computer. The weight of the suspended object is 15 Kg. The parameters of the crane and those of the robot are shown in Table.1 and the specifications of the sensor systems are shown in Table.2. Figure 10 illustrates the mechanism of the free joint hands. Potentio-meters are installed in each free joint and the displacements of the angles can be measured. Electro-magnetic breaks are also installed in the two rotational joints and they are locked during the free motion of the robot.

Assuming that the positioning of a heavy object is a desired task, the task is executed through three phases as follows.

phase 1: The object is roughly positioned by the crane. Since the crane has three wires, all the residual vibration modes can be reduced by only the crane.
phase 2: The robot catch the object. During the free motion of the robot, all the break are locked.
phase 3: After all the breaks are unlocked, the robot move to the destination slowly. The crane is feedback controlled by the feedback signals from the inclinometer, the gyro scope, the encoders of the crane and the potentio meters of the free joints.

Table 1 Parameters of the experimental system

	crane	robot
degrees of freedom	7	6
payload capacity	20 Kg	2 Kg
positioning accuracy	0.05 mm 0.02 deg	0.04 mm

Table 2 Specifications of the sensor systems

crane		robot hand	
x,y direction	7.0×10^{-2} mm	d_{r1}	1.5×10^{-2} rad
z direction	5.0×10^{-2} mm	d_{r2}	1.5×10^{-2} rad
rotation about x,y axes	5.0×10^{-5} rad	d_{r3}	5.0×10^{-1} mm
rotation about z axis	1.7×10^{-3} rad		

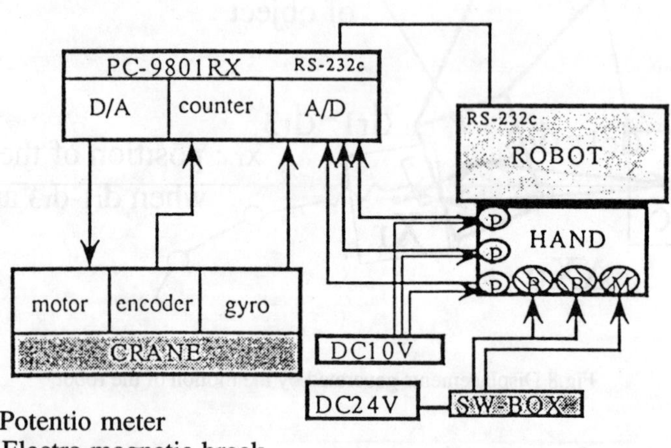

Fig.9 Configuration of the experimental system

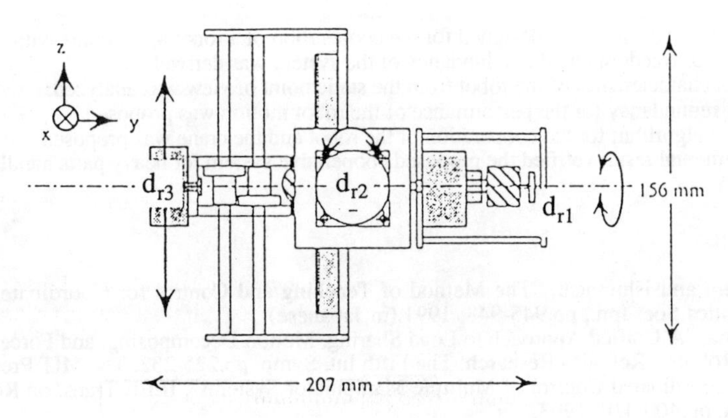

Fig.10 Free joint mechanism

Figure 11 shows the result of the experiment of cooperation in the phase 3. The robot moved $\Delta \mathbf{x}_r$ for the first 1.7 seconds, while the crane was feedback controlled from the beginning to the end. Since the maximum speed of the crane was limited for safety, the displacements became large during the motion of the robot. But after the robot stopped, all the displacements became zero and vibration did not appear. From the results, the proposed method was verified.

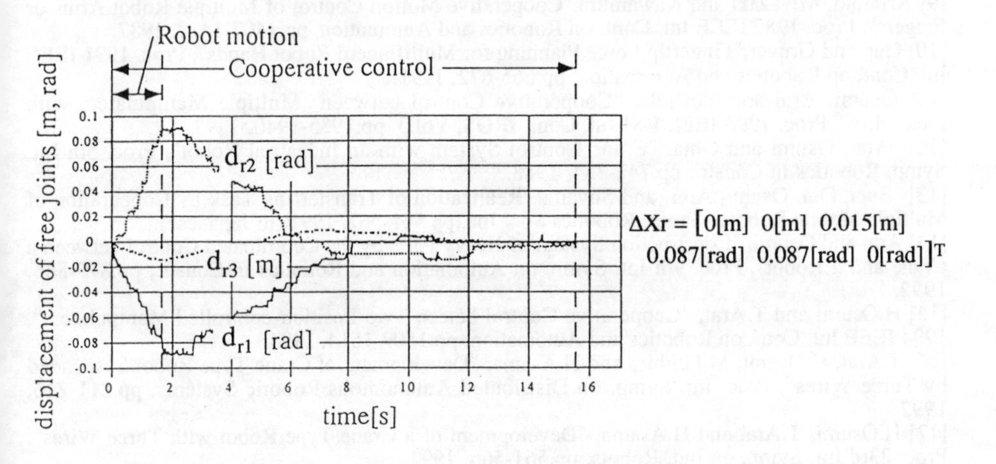

Fig.11 Experimental result

9. Conclusions

A simple but very practical method using flexible mechanisms were proposed for cooperation of position-controlled manipulators.

- The kinematic condition to be required for the cooperation between two position-controlled manipulators was shown.

- Free joints mechanisms were designed for the cooperation of a robot and a crane with three wires.
- The degrees of freedom and the redundancy of the system was derived.
- The control characteristics of the robot from the static point of view was analyzed.
- The use of redundancy for the performance of the robot motion was proposed.
- Cooperative algorithm for the cooperation of the robot and the crane was proposed.
- The experimental result verified the proposed cooperative system for heavy parts handling system.

References

[1] Kasagami and Ishimatsu, "The Method of Teaching and Control for Coordinated Motion", Prepr. Robotics Soc. Jpn., pp.945-948, 1991.(in Japanese)
[2] Uchiyama, "A Unified Approach to Load Sharing, Motion Decomposing, and Force Sensing of Dual Arm Robots", Robotics Research: The Fifth Int. Symp. pp.225-232, The MIT Press, 1990.
[3] P.Hsu,"Coordinated Control of Multiple Manipulator Systems", IEEE Trans. on Robotics and Automation, pp.400-410, 1993.
[4] N.Xi, T.J.Tarn and A.K.Bejczy, "Event-Based Planning and Control for Multi-Robot Coordination", Proc. IEEE Int. Conf. on Robotics and Automation, pp.251-258, 1993.
[5] Kosuge, Ishawa, Furuta and Sakai,"Control of Single-Master Multi-Slave Manipulator System Using VIM", Proc. 1990 IEEE Int. Conf. on Robotics and Automation, pp.1172-1177, 1990.
[6] Hashimoto, Oba and Eguchi,"Dynamic Control Approach for Motion Coordination of Multiple Wheeled Mobile Robots Transporting a Single Object", Proc. 1993 IEEE/RSJ Int. Conf. on IROS, vol.3, pp.1944-1951, 1993.
[7] D.J.Stilwell and J.S.Bay, "Optimal Control for Cooperating Mobile Robots Bearing a Common Load", Proc. 1994 Int. Conf. on Robotics and Automation, pp.58-63, 1994.
[8] Ahmad and Guo, "Dynamic Coordination of Dual-Arm Robotic Systems with Joint Flexibility",Proc.1988 IEEE Int. Conf. on Robotics and Automation, pp.332-337, 1987.
[9] Arimoto, Miyazaki and Kawamura,"Cooperative Motion Control of Multiple Robot Arms or Fingers", Proc. 1987 IEEE Int. Conf. on Robotics and Automation, pp.1407-1412, 1987.
[10] Guo and Gruver,"Fingertip Force Planning for Multifingerd Robot Hands", Proc. 1991 IEEE Int. Conf. on Robotics and Automation, pp.665-672, 1991.
[11] Osumi, Arai and Yoshida, "Cooperative Control between Multiple Manipulators with Flexibility", Proc. 1993 IEEE/RSJ Int. Conf. IROS, Vol.3, pp.1935-1940, 1993.
[12] Arai, Osumi and Ohta, "Crane Control System with an Industrial Robot", Proc. 5th Int. Symp. Robotics in Constr., pp.747-754, 1988.
[13] Buei, Ota, Osumi, Arai and Suyama,"Realization of Transferring Task by Cooperation of Multiple Mobile Robots", Prepr. Robotics Soc. Jpa.,pp.561-562, 1991.(in Japanese)
[14] Arai and Osumi, "Construction System of Heavy Parts by the Coordinated Control between a Crane and a Robot", Proc. 9th Int. Symp. on Automation and Robotics in Constr., pp.879-886, 1992.
[15] H.Osumi and T.Arai, "Cooperative Control beteen Two Position-controlled Manipulators", 1994 IEEE Int. Conf. on Robotics and Automation, pp.1509-1514, 1994.
[16] T.Arai, H.Osumi, M.Fujihira and H.Asama, "Development of Crane Type Robot Suspended by Three Wires", Proc. Int. Symp. on Distributed Autonomous Robotic Systems, pp.211-216, 1992.
[17] H.Osumi, T.Arai and H.Asama, "Development of a Crane Type Robot with Three Wires", Proc. 23rd Int. Symp. on Ind. Robots, pp.561-566, 1992.

Manipulability Indices in Multi-wire Driven Mechanisms

YUSI SHEN, HISASHI OSUMI, and TAMIO ARAI

Dept. of Precision Machinery Eng., The Univ. of Tokyo, Tokyo, 113 Japan

Abstract

To manipulate a heavy object, one effective method is to use a multi-wire suspended system containing several cranes that can manipulate the object cooperatively. In this paper, the manipulability for multi-wire driven mechanisms is discussed. For kinematics of a multi-wire driven mechanism, we need not only the geometrical constraints of wire lengths but also the force constraints, such as the wire tension which should be always greater than or equal to zero. First, the sufficient and necessary mechanical conditions corresponding to the force constraints are shown. Under these conditions, we can obtain the "manipulability ellipsoid" of a multi-wire driven mechanism using almost the same methods as for a popular multi-link manipulator. Second, a more practical evaluation index "set of manipulating forces" is introduced and the way of calculation of this index is explained. We also derive both manipulability indices as for multi-wire suspended mechanisms in the gravity field by modeling the gravitational force as a special wire tension. Numerical examples are given to explain the proposed indices and to show their validity. The method is expected to be effective in designing and task planning of multi-wire driven systems including multi-wire suspended mechanisms.

Key words: manipulability, set of manipulating forces, multi-wire driven mechanism, wire tension, multi-wire suspended mechanism

1 Introduction

Generally ordinary manipulators have small load capacity in comparison with their own weight. For this reason, they are not suitable for manipulating heavy objects. Comparatively, wire suspended mechanism such as a crane can be used to lift heavy objects. However, one conventional crane mechanism consists of only one wire, which makes impossible any control of the orientation of the object. To overcome this problem, human help is necessary to perform this control, which appears to be a dangerous work.

One effective method to automate this kind of dangerous work is to use several cranes that can manipulate the object cooperatively. In this way, by using only some wires, we can control not only the position of the object but also its orientation.

We can see the research [1] in which a dynamic model has been developed for the closed chain mechanism consisting of two rotary cranes holding a rigid object in a two-dimensional workspace, however, the control of rotation of the object has not been discussed. Also we can see some research in which researchers do not use multi-crane system, and propose idea to control heavy object by special crane system consisting of several wires. The authors [2][3] have recently developed a *7* degrees of freedom (DOF) crane type manipulator with *3* wires. This manipulator can control the position and orientation of a suspended object statically. However, this structure can not be used for controlling dynamics because of its complexity of dynamics. The control of position and orientation of an object is done in [4] for *3* DOF in a plane and in [5] for *6* DOF in general space. Although all of these researches meet no problems mathematically, in practice, it is necessary to find a general measurement for multi-wire driven systems which may evaluate the manipulability. There are few researches about this measurement problem. In [6] we can find a dexterity measure consisting of a "condition number" and a "force mapping" of the mechanism. Although the calculating method is given for that problem, we need to have a more precise geometric image representing the manipulability of the system so that we can use it effectively for designing multi-wire driven mechanisms.

On the other hand, manipulation of a heavy object should be considered including gravity because gravity plays an important role to control the motion of object for multi-wire suspended system, unlike a ordinary manipulator. We can find few works [7] that regard the gravity as an actuator for manipulating an object.

In this paper, we propose manipulability indices for manipulating a rigid object with multi-wires. At first we show up the sufficient and necessary mechanical conditions for manipulating an object corresponding to the force constraints. Using these conditions, we can find that the manipulability ellipsoid of multi-wire driven mechanism is almost the same as the manipulability ellipsoid of a manipulator as in chapter 2. Then we propose a more practical evaluation index called "set of manipulating forces" which is derived from constrains of wire tensions that produce driving forces for the object and we show the way of calculation in chapter 3. In chapter 4 we derive manipulability indices of multi-wire suspended mechanisms in the gravity field by modeling the gravitational force as a special wire tension.

2 Manipulability Ellipsoid of Multi-wire Driven System

When an object in n DOF space is manipulated by m wires, we can describe the relationship between multi-wire tension vector $\mathbf{t} = [t_1\ t_2\ \dots\ t_m]^T$ and the general force (force and moment) vector $\mathbf{f} \in R^{n\times 1}$ acting on mass center of the object as following:

$$\mathbf{f} = \mathbf{A}\mathbf{t} \tag{1}$$

where

$$\mathrm{A} = [\mathbf{a}_1\ \mathbf{a}_2\ \dots\ \mathbf{a}_m]$$

$$\mathbf{a}_i = \begin{bmatrix} \mathbf{c}_i \\ \mathbf{p}_i \times \mathbf{c}_i \end{bmatrix}$$

$\mathbf{c}_i$: unit vector of ith wire tension
$\mathbf{p}_i$: position of ith connection point, see Fig. 1.

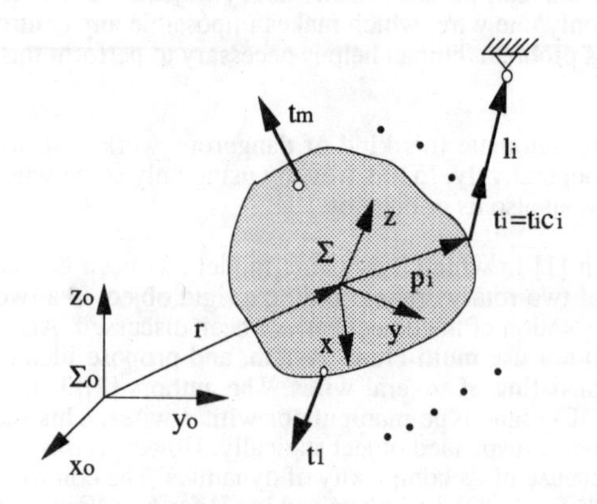

Fig. 1: A multi-wire driven system consisting of m wires

From instantaneous mechanics the matrix $A = J^T$ is well known, where J is the Jacobian. It is necessity for multi-wire driven system to keep all tensions always ≥ 0. We can describe the condition as theorem 1.

Theorem 1:

A rigid object with n degrees of freedom can be manipulated by at least $n+1$ wires, if and only if the following conditions are satisfied.

1. $$\mathrm{rank}(A) = n \tag{2}$$

2. $$\sum_{i=1}^{n+1} \alpha_i \mathbf{a}_i = \mathbf{0}, \qquad \alpha_i > 0 \text{ for all of } i \tag{3}$$

The proof of theorem 1 is known by both using the pseudo-inverse of Jacobian matrix and analyzing the vector in the null space of it [4] [6] [8], or by using only vector analysis [5] [9].

From theorem 1 we can know clearly that the directions of $\mathbf{a}_i$, which depended on the directions of each wire, carry out a very important role in the system. We can do manipulation only when there are at least $n+1$ wires and these $\mathbf{a}_i$ completely satisfying theorem 1. For simplicity we let $m = n+1$ in chapter 2 and chapter 3.

According to $A = J^T$ the relationship between velocity of wire lengths and the object is:

$$\dot{\mathbf{l}} = A^T \dot{\mathbf{x}}_0 \tag{4}$$

where $\mathbf{l} = [l_1\ l_2\ ...\ l_m]^T$, l_i is length of ith wire and $\mathbf{x}_0$ is the position and orientation vector of the object. Although there are m wires, each wire can not be controlled independently. Theorem 2 shows the condition that needs to be satisfied for the velocities of wire lengths.

Theorem 2:

If theorem 1 is satisfied by a multi-wire driven system, than the sufficient and necessary condition for velocities of wire lengths is:

$$\sum_{j=1}^{m} \alpha_j \dot{l}_j = 0 \qquad \alpha_j > 0 \text{ for all of } j \tag{5}$$

Proof:

According to linear algebra theory, sufficient and necessary condition for a solution of equation (4) about $d\mathbf{x}_0/dt$ is:

$$\mathrm{rank}(A^T) = \mathrm{rank}(A^T \quad d\mathbf{l}/dt) \tag{6}$$

since $\mathrm{rank}(A^T) = \mathrm{rank}(A)$, we just need to prove:

$$\mathrm{rank}(A) = \mathrm{rank}\begin{bmatrix} A \\ \dot{\mathbf{l}}^T \end{bmatrix}$$

Necessity:

Using the basic transformation which preserves the rank of the matrix we have:

$$\begin{bmatrix} A \\ \dot{l}^T \end{bmatrix} \rightarrow \begin{bmatrix} \alpha_1 \mathbf{a}_1 & \alpha_2 \mathbf{a}_2 & \cdots & \alpha_m \mathbf{a}_m \\ \alpha_1 \dot{l}_1 & \alpha_2 \dot{l}_2 & \cdots & \alpha_m \dot{l}_m \end{bmatrix} \rightarrow \left[\begin{matrix} \alpha_1 \mathbf{a}_1 & \alpha_2 \mathbf{a}_2 & \cdots & \alpha_n \mathbf{a}_n \\ \alpha_1 \dot{l}_1 & \alpha_2 \dot{l}_2 & \cdots & \alpha_n \dot{l}_n \end{matrix} \middle| \mathbf{b} \right] = [B | \mathbf{b}]$$

where

$$\mathbf{b} = \begin{bmatrix} \mathbf{0} \\ \sum_{j=1}^{m} \alpha_j \dot{l}_j \end{bmatrix}$$

Equation (6) has the same meaning with:

$$\text{rank}(A) = \text{rank}([B \mid \mathbf{b}]) \tag{7}$$

Let $A_p = [\mathbf{a}_1\ \mathbf{a}_2\ \ldots\ \mathbf{a}_n]$, then by using the relation of

$$\text{rank}(A_p) = n = \text{rank}(A) \tag{8}$$

we can find that:

$$\sum_{j=1}^{m} \alpha_j \dot{l}_j = 0 \tag{9}$$

must be satisfied.

Sufficiency:

If equation (5) is satisfied, with rank(A_p) = n we are sure that:

$$\text{rank}([B \mid \mathbf{b}]) = \text{rank}(B) = n \tag{10}$$

is satisfied too. This implies that:

$$\text{rank}(A) = \text{rank}\begin{bmatrix} A \\ \dot{l}^T \end{bmatrix} \tag{11}$$

□

It means that wire lengths can variate in a *n* DOF sub-space called "velocity sub-space" constrained by equation (5), not yet in all *m* DOF space. In this sub-space wire lengths can move like pistons, and the manipulability ellipsoid of a multi-wire driven system can be treated using the same procedure as for a rigid manipulator. Let the velocities of wire lengths be "control variables", we have the equation of manipulability ellipsoid:

$$\dot{\mathbf{x}}_0^T J^T J \dot{\mathbf{x}}_0 = \dot{\mathbf{x}}_0^T A A^T \dot{\mathbf{x}}_0 \leq 1 \tag{12a}$$

and the manipulability measure as:

$$w = [\det(AA^T)]^{1/2} \tag{12b}$$

For explaining the proposed methods and showing their validity, we take a wire driven mechanism made by three wires manipulating a particle as an example in Fig. 2. The coordinates of three wire support points are respectively (0, 1), (0, -1) and ($\sqrt{3}$, 0). The kinematic equations are:

$$l_i = \| \mathbf{x}_0 - \mathbf{x}_i \|, \quad (i = 1, 2, 3) \tag{13}$$

where $\mathbf{x}_i$ is coordinate of the ith wire support point. By doing partial differentiation with $\mathbf{x}_0$ and composing the Jacobean matrix, the manipulability ellipsoids on the x axis can be obtained in Fig. 2.

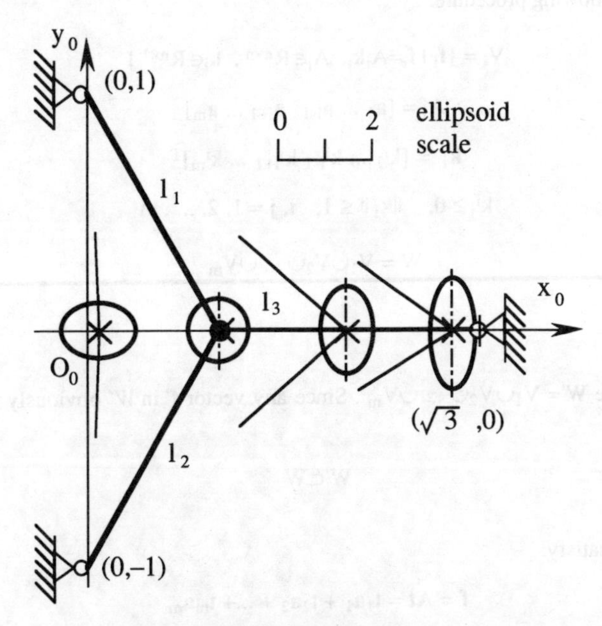

Fig. 2: Example of Manipulability Ellipsoid

When considering any wire driven systems, regardless of that we have to assure that wire tensions are always ≥ 0, as a result of the analysis of velocity of the object, there are some situations which differ from our feelings. In this system we can see that the manipulability near $(\sqrt{3}, 0)$ becomes better along y axis, we do not need to take into account the value of the wire tension when analyzing the velocity of the system only in the velocity sub-space. we can also find that the manipulability becomes better not only at the point near $(\sqrt{3}, 0)$, but also near the sides of equilateral triangle constructed by *3* wire support points along the outward direction. For an example, the point near origin has good manipulability along *-x* direction when analyzing the velocity of the system only in the velocity sub-space, However, in order to drive the object along that direction, it will be necessary to generate extremely strong tensions at *2* wires connected to the nearest two supporting points. We can hardly say that equation (12) is a preferable evaluation index. In next chapter we propose a more practical evaluation index under constrains of tensions.

3 Set of Manipulating Forces

Let us consider that when tension vector $\mathbf{t} = [t_1\ t_2\ \dots\ t_m]^T$ is constrained by $\|\mathbf{t}\| \leq 1$, what is the reasonable $\mathbf{f} \in R^{n\times 1}$ and what kind of shape its set should be. We define evaluation index called "set of manipulating forces" and written by:

$$W = \{\ \mathbf{f} \mid \mathbf{f} = A\mathbf{t},\quad \|\mathbf{t}\| \leq 1\ \ \&\ \ t_i \geq 0\ (\ i = 1, 2, \dots, m)\ \} \tag{14}$$

We can calculate the magnitude of the $\mathbf{f}$ and the shape of the W by using theorem 3.

Theorem 3:

Within *m* wires driven system, the set of manipulating forces written by *W* in equation (14) can be calculated by the following procedure.

1. $$V_i = \{\mathbf{f}_i \mid \mathbf{f}_i = A_i\mathbf{k}_i,\ A_i \in R^{n\times n},\ \mathbf{k}_i \in R^{n\times 1}\} \tag{15}$$

where
$$A_i = [\mathbf{a}_1 \ldots \mathbf{a}_{i-1}\ \ \mathbf{a}_{i+1} \ldots \mathbf{a}_m] \tag{16}$$

$$\mathbf{k}_i = [k^i_1 \ldots k^i_{i-1}\ k^i_{i+1} \ldots k^i_m]^T \tag{17}$$

$$k^i_j \geq 0, \quad \|\mathbf{k}_i\| \leq 1, \quad i, j = 1, 2, \ldots, m$$

2. $$W = V_1 \cup V_2 \cup \ldots \cup V_m \tag{18}$$

Proof : in *2* steps

1)

Let union set *W'* be $W' = V_1 \cup V_2 \cup \ldots \cup V_m$. Since any vector **f** in *W'* obviously satisfies $\mathbf{f} \in V_i$, we have:

$$W' \subset W \tag{19}$$

2)

Any **f** in *W* should satisfy:

$$\mathbf{f} = A\mathbf{t} = t_1\mathbf{a}_1 + t_2\mathbf{a}_2 + \ldots + t_m\mathbf{a}_m \tag{20}$$

and from theorem 1:

$$\mathbf{0} = q_1\mathbf{a}_1 + q_2\mathbf{a}_2 + \ldots + q_m\mathbf{a}_m \tag{21}$$

where $t_i > 0$ and $q_i > 0$ for $i = 1, 2, \ldots, m$.

If $t_i = 0$, it is obvious that:

$$\mathbf{f} \in V_i \subset W' \tag{22}$$

Thus, we consider the case of $t_i > 0$ in following.

By linear algebra theory we know that the solution of equation (21) can be written in form of:

$$\mathbf{y} = h\mathbf{q}, \qquad h \in R \tag{23}$$

where *h* is any real number and $\mathbf{q} = [q_1\ q_2 \ldots q_m]^T$. From equation (20) and (23) we have:

$$\mathbf{f} = A(\mathbf{t} - \mathbf{y}) = A(\mathbf{t} - h\mathbf{q}) \tag{24}$$

If we choose:

$$h = \min_{1 \leq j \leq m} (t_j / q_j) = t_i / q_i > 0 \tag{25}$$

we rewrite the term $(\mathbf{t} - h\mathbf{q})$ in equation (24) as:

$$(\mathbf{t} - h\mathbf{q}) = \mathbf{s} = [s_1\ s_2 \ldots s_m]^T \tag{26}$$

We have:

$$s_i = 0,$$

$$s_j = t_j - t_i\,(q_j / q_i) \geq 0,\ j = 1, 2, ..., m, \ \&\ j \neq i \tag{27}$$

Let :

$$\mathbf{s} = [k^i{}_1 \ ...\ k^i{}_{i-1}\ \ 0\ \ k^i{}_{i+1}\ ...\ k^i{}_m]^T \tag{28}$$

Then, equation (24) becomes:

$$\mathbf{f} = A(\mathbf{t} - \mathbf{y}) = A\mathbf{s} = A_i\mathbf{k}_i \in V_i \subset W' \tag{29}$$

equation (29) means:

$$W \subset W' \tag{30}$$

By equation (19) and (30) we can prove that:

$$W = W' \tag{31}$$

□

Then, we consider how to calculate volume of set of manipulating forces. Each of the partial ellipsoids is defined as:

$$V_i = \{\mathbf{f}_i \mid \mathbf{f}_i = A_i\mathbf{k}_i,\ \|\mathbf{k}_i\| \leq 1,\ k^i{}_j \geq 0,\ (\ j = 1, ..., m,\ \&\ j \neq i\ )\ \} \tag{32}$$

and the volume is calculated by the following integration:

$$\int_{V_i} d\mathbf{f}_i = \int_{V_{ti}} \left|\det(A_i)\right| dt_i = \frac{\pi^{\frac{n}{2}}}{2^n \Gamma(\frac{n}{2}+1)} \left|\det(A_i)\right| = c_n \left|\det(A_i)\right| = c_n v_i \tag{33}$$

where $V_{ti} = \{\ \mathbf{k}_i \mid \|\mathbf{k}_i\| \leq 1,\ \ k^i{}_j \geq 0,\ \ (\ j = 1, 2, ..., m,\ \ \&\ \ j \neq i\ )\ \}$ is a partial sphere in the wire tension space and c_n is a parameter only dependent on number of DOF.

In equation (33), we calculate the volume of a partial sphere in *m* DOF wire tension space by using the known formula and transform this convex volume to *n* DOF workspace by matrix A, as shown in Fig. 3 which is an example of *n = 2, m = 3*.

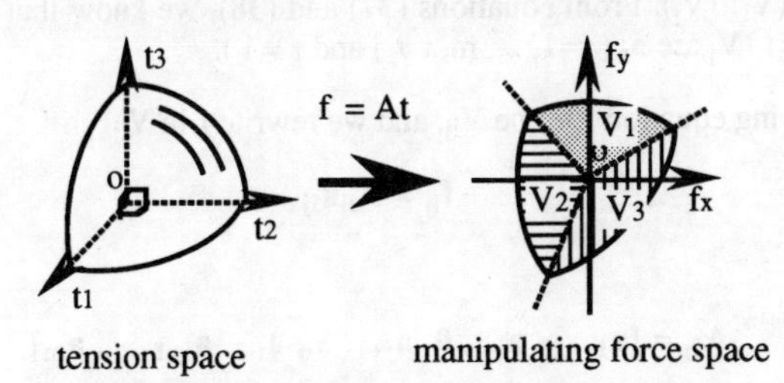

Fig. 3. Transformation of Convex Volume by Matrix A

Because c_n is constant in a determined working space, we neglect c_n in theorem 4, in which we calculate the whole volume of set of manipulating forces.

Theorem 4:

The volume of the set of manipulating forces *W* can be calculated by summing all *m* volumes of partial ellipsoids as:

$$v = \sum_{i=1}^{m} v_i \tag{34}$$

Proof:

If we can prove that the volume of intersection $(V_i \cap V_j)$ of arbitrary two partial sets V_i and V_j is zero, then theorem 4 is proved.

Let $\mathbf{f} \in (V_i \cap V_j)$, $(1 \le i, j \le m, i \ne j)$. Because **f** is a vector in V_i, it satisfies

$$\mathbf{f} = A_i \mathbf{k}_i = A[k^i_1, ..., k^i_{i-1}, 0, k^i_{i+1}, ..., k^i_m]^T. \tag{35}$$

On the other hand, from theorem 1, we have

$$\mathbf{a}_j = -\sum_{r=1 (r \ne j)}^{m} \alpha_r \mathbf{a}_r \qquad (\alpha_r > 0). \tag{36}$$

Again we write equation (35) with the $\mathbf{a}_j$ described in equation (36), the following relationship can be obtained.

$$\mathbf{f} = (k^i_1 - k^i_j\alpha_1)\mathbf{a}_1 + ... + (k^i_{j-1} - k^i_j\alpha_{j-1})\mathbf{a}_{j-1} + (k^i_{j+1} - k^i_j\alpha_{j+1})\mathbf{a}_{j+1} + ... + \\ + (-k^i_j\alpha_i)\mathbf{a}_i + ... + (k^i_m - k^i_j\alpha_m)\mathbf{a}_m \tag{37}$$

Also the **f** in equation (37) is a vector in V_j. That means the coefficients of all vectors $\mathbf{a}_r$ (r=1, ..., m, $r \ne j$) must be ≥ 0, so the coefficient of $\mathbf{a}_i$ should be ≥ 0. From $\alpha_i > 0$ and $k^i_j \ge 0$, we know that $k^i_j = 0$ must be true.

In the case of

$$\mathbf{f} = k^i_1\mathbf{a}_1 + ... + k^i_{j-1}\mathbf{a}_{j-1} + k^i_{j+1}\mathbf{a}_{j+1} + ... + \\ + k^i_{i-1}\mathbf{a}_{i-1} + k^i_{i+1}\mathbf{a}_{i+1} + ... + k^i_m\mathbf{a}_m, \tag{38}$$

it is obvious that $\mathbf{f} \in (V_i \cap V_j)$. From equations (37) and (38), we know that the base vectors of the intersection space $V_i \cap V_j$ are $\mathbf{a}_r$ (r=1, ..., m,. $r \ne j$ and $r \ne i$).

Let the set of **f** satisfying equation (38) be V_{ij}, and we rewrite **f** in V_{ij}

$$\mathbf{f}_{ij} = A_{ij}\mathbf{k}_{ij} \tag{39}$$

where

$$A_{ij} = [\mathbf{a}_1, ..., \mathbf{a}_{j-1}, \mathbf{0}, \mathbf{a}_{j+1}, ..., \mathbf{a}_{i-1}, \mathbf{a}_{i+1}, ..., \mathbf{a}_m] \tag{40}$$

$$\mathbf{k}_{ij} = [k^i_1, ..., k^i_{j-1}, 0, k^i_{j+1}, ..., k^i_{i-1}, k^i_{i+1}, ..., k^i_m]^T \tag{41}$$

Because the volume of V_{ij} is

$$v_{ij} \propto |\det(A_{ij})|, \tag{42}$$

and $\det(A_{ij}) = 0$, the volume v_{ij} is equal to zero. That is to say the volume of intersection $V_i \cap V_j$ is zero.
□

We defined "v" in equation (34) as "manipulating force measure". It corresponds to the concept of "manipulating force ellipsoid" in multi-link manipulator[10]. We can use the measure to express the capacity of manipulating forces of a multi-wire driven mechanism. In the case of manipulator, the volume of force manipulability ellipsoid becomes zero at a singular configuration. Differently, the manipulating force measure of multi-wire driven system is not equal to zero even at a singular configuration, when the manipulating force can not be generated in certain directions. We can say that it is a large difference between a multi-wire driven system and a manipulator. In set of manipulating forces W, the contribution of each partial ellipsoid V_i can be understood easily. Using the smallest v_i as a measure, we can evaluate the distance from the boundary of the working space. Using the ratio of the largest v_i and the smallest one, v_{max}/v_{min}, we can evaluate the isotropy.

Fig. 4 is the example of the set of manipulating forces in the same mechanism as in Fig. 2. It is obvious that the set of manipulating forces in the plane consists of *3* pieces of partial ellipsoids, and each of them is generated by *2* wire tensions. From the shape of these sets of manipulating forces we can also know that it is much more difficult to produce a manipulating force at the point near $(\sqrt{3}, 0)$ in the outward direction from the equilateral triangle determined by the *3* wire support points. In contrast to the ellipsoids in Fig. 2, the result in Fig. 4 shows the manipulating capacity as a more intuitive measure.

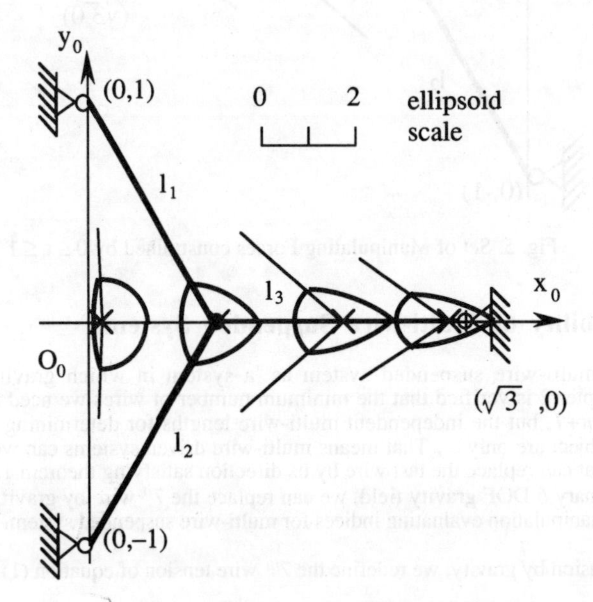

Fig. 4. Example of Set of Manipulating Forces

The meaning of theorem 3 is: Starting with m vectors $\mathbf{a}_i$ in a n DOF work space, we construct convex areas by grouping any n vectors. As a result, we obtain the set of manipulating forces by calculating the union set of these m convex areas. Using same method, we can find easily another manipulating measure constrained by:

$$0 \le t_i \le 1, \qquad i = 1, 2, ..., m \tag{43}$$

as equation (44):

$$\int_{V_i} df_i = \int_{V_{ti}} |\det(A_i)|\, dt_i = |\det(A_i)| = v_i \tag{44a}$$

$$v = \sum_{i=1}^{n+1} v_i \tag{44b}$$

Fig. 5 shows the example of this set of manipulating forces in the same mechanism as in Fig. 2. The shape of the set consists of *3* parallelograms.

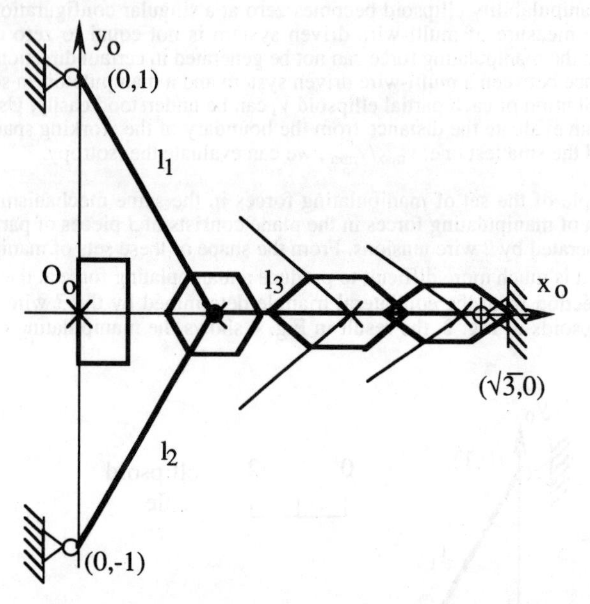

Fig. 5. Set of Manipulating Forces constrained by $0 \le t_i \le 1$

4 Manipulability of Multi-wire Suspended Systems

We can define a multi-wire suspended system as "a system in which gravity influences its kinematics". In chapter 2 is verified that the minimum number of wires we need to manipulate an object in *n* DOF is *n+1*, but the independent multi-wire lengths for determining the position and orientation of the object are only *n* . That means multi-wire driven systems can work if there are *n* wires and a force that can replace the last wire by its direction satisfying theorem 1. Considering the problem in the ordinary *6* DOF gravity field, we can replace the *7th* wire by gravity. In this chapter we will derive the manipulation evaluating indices for multi-wire suspended system.

Replacing a wire tension by gravity, we redefine the *7th* wire tension of equation (1) as:

$$t_7 = m_0 g \tag{45}$$

$$\mathbf{a}_7 = [0\ \ 0\ \ -1\ \ 0\ \ 0\ \ 0]^T \tag{46}$$

where m_0 is the mass of the object, g is magnitude of gravity acceleration.

First we consider the manipulability ellipsoid of a multi-wire suspended mechanism. Because control variables are only *6* wire lengths, we can calculate the manipulability ellipsoid by just constraining these *6* velocities in an unit sphere: $\| d\mathbf{l}/dt \| \le 1$, where $\mathbf{l} = [l_1\ l_2\ ...\ l_6]^T$. Than the calculating procedure is the same as in chapter 2.

We replace wire l_2 in Fig. 2 with gravity, and make a multi-wire suspended mechanism as in Fig. 6. The coordinates of two wire support points are respectively (0, 1) and ($\sqrt{3}$, 0). The kinematic equations are:

$$l_i = \| \mathbf{x}_0 - \mathbf{x}_i \|, \quad (\ i = 1,\ 3\ ) \tag{47}$$

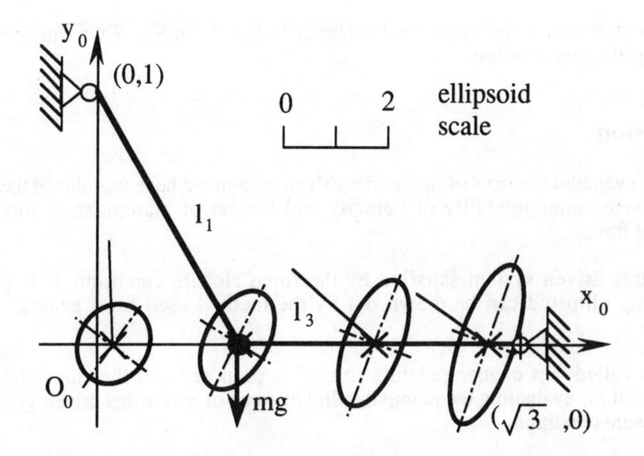

Fig. 6. Example of Manipulability Ellipsoid in Wire Suspended System

We can draw the manipulability ellipsoids as in Fig. 6. Comparing with Fig. 2, we can see that l_1 and l_3 affect the velocity of the object more than before.

Second we consider a set of manipulating forces of the multi-wire suspended mechanism. With the redefinition (45) (46) we can rewrite the equation (1) as:

$$\mathbf{f} = A_7\mathbf{t}' + m_0 g \mathbf{a}_7 \tag{48}$$

where $\mathbf{t}' = [t_1\ t_2\ ...\ t_6]^T$ and A_7 is defined by equation (16). As a first step, we can calculate equation (49) the set of manipulating forces without considering the second term of equation (48), because it is a constant vector:

$$V_7 = \{\ \mathbf{f}' \mid \mathbf{f}' = A_7\mathbf{t}'\ ,\ \|\ \mathbf{t}'\ \| \leq 1\ \} \tag{49}$$

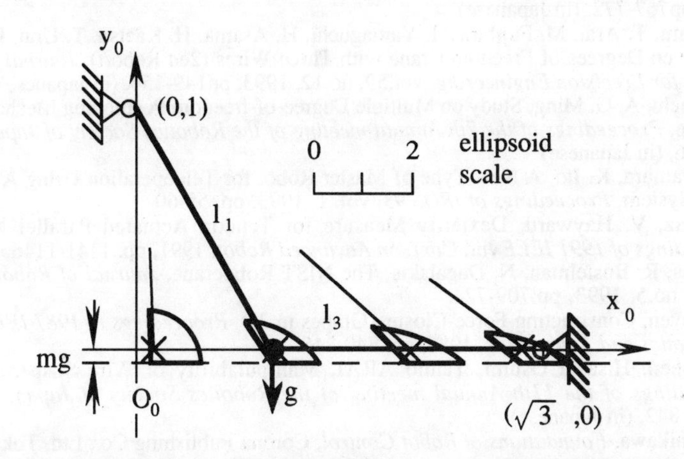

Fig. 7. Example of Set of Manipulating Forces in Wire Suspended System

As a second step, we add the constant vector to this V_7 and we obtain the set of manipulating forces just by translating the picture of V_7 in magnitude of $m_0 g$ and in direction of $\mathbf{a}_7$. If the set of manipulating forces can not include the point where we calculate this set, it means that the wire tensions constrained by $\|\ \mathbf{t}'\ \| \leq 1$ can not suspend the object.

We use again as an example the same mechanism as in Fig. 6. In Fig. 7 we can check whether the wires can support the object or not.

5 Conclusion

As manipulating evaluation indices of multi-wire driven system we have introduced the manipulability ellipsoid to describe manipulability of velocity and the set of manipulating forces to describe manipulability of force.

- For *n+1* wires driven system satisfied by the force closure condition, it is proved that the manipulability ellipsoid can be drawn out by the method used for a general rigid link type manipulator.

- The concept called "set of manipulating forces" is proposed and the calculating procedure is given. It is used for evaluating the manipulability of force of *n+1* wires driven system satisfied by the force closure condition.

- We show that it is valid to evaluate a multi-wire suspended system with the general method for multi-wire driven system by replacing a tension by gravity.

A multi-wire driven system is differing from a rigid link mechanism because the direction of the force is determined by the path of the wire and its point of attachment. For this reason, there is no sense to use a manipulability measure derived only from geometric relationships. The proposed concept of "set of manipulating forces" is an effective concept for designing the structure of multi-wire suspended mechanisms in which gravity can be used as an active manipulating force.

References

[1] R. Souissi, A. J. Koivo, Modeling and Control of Two Co-operating Planar Cranes, *Proceedings of 1993 IEEE Int. Conf. on Robotics and Automation*, vol.3, 1993, pp. 957-962,

[2] H. Osumi, T. Arai, H. Asama, Development of a Seven Degrees of Freedom Crane with Three Wires (1st Report), *Journal of the Japan Society for Precision Engineering*, vol.59, no.5, 1993, pp767-772. (in Japanese)

[3] H. Osumi, T. Arai, M. Fujihira, H. Yamaguchi, H. Asama, H. Kaetsu, T. Urai, Development of a Seven Degrees of Freedom Crane with Three Wires (2ed Report), *Journal of the Japan Society for Precision Engineering*, vol.59, no.12, 1993, pp149-154. (in Japanese)

[4] T. Higuchi, A. G. Ming, Study on Multiple Degree-of-freedom Positioning Mechanism Driven by Wire, *Proceedings of the 7th Annual meeting of the Robotics Society of Japan*, 1989, pp. 603-606. (in Japanese)

[5] S. Kawamura, K. Ito, A New Type of Master Robot for Teleoperation Using A Radial Wire Drive System, *Proceedings of IROS'93*, vol. 1, 1993, pp. 55-60.

[6] R. Kurtz, V. Hayward, Dexterity Measure for Tendon Actuated Parallel Mechanisms, *Proceedings of 1991 IEEE Int. Conf. on Advanced Robot*, 1991, pp. 1141-1146.

[7] J. Albus, R. Bostelman, N. Dagalakis, The NIST Robocrane, *Journal of Robotics Systems*, vol.10, no.5, 1993, pp.709-724.

[8] V. Nguyen, Constructing Force-Closure Grapes in 3D, *Proceedings of 1987 IEEE Int. Conf. on Robotics and Automation*, 1987, pp. 240-245.

[9] Yusi Shen, Hisashi Osumi, Tamio ARAI, Manipulability of Wire Suspension System, *Proceedings of the 11th Annual meeting of the Robotics Society of Japan*, vol.3, 1993, pp.839-842. (in Japanese)

[10] T. Yoshikawa, *Foundations of Robot Control*, Corona Publishing Co., Ltd, Tokyo, 1988. (in Japanese)

Cooperating Multiple Behavior-Based Robots for Object Manipulation
- System and Cooperation Strategy -

ZHI-DONG WANG, EIJI NAKANO, and TAKUJI MATSUKAWA

Dept. of Mechanical Intelligence, Tohoku Univ., Sendai, 980-77 Japan

Abstract

A multiple vehicle robot system for cooperative object manipulation is described. This system consists of a host and several distributed behavior-based vehicle robot agents. A strategy on cooperatively organizing the behavior-based robots which only have limited ability in manipulation is discussed. An extended subsumption architecture, which has some layers for manipulation and cooperation, is built for each vehicle robot. A simulator and some simulation results are shown also.

Keywords: Multi-Agent Robot System, Cooperative Manipulation, Manipulation by Vehicles, Behavior-Based Robots System

1 Introduction

In recent years, there have been many studies focusing on cooperation of distributed autonomous multi-robot systems and robot swarm intelligence [1][2][5][6][10][15][18][19]. Most of them are concentrating on methods in which the cooperation strategy is based on distributed problem solving theory and distributed decision theory. The characteristic of these studies is that each robot gets some information about other robots by communication, then uses the information for making a distributed decision, for example, the studies on dynamic collision avoidance among mobile robots, and on respective tasks accomplishment via multiple robots. Compared with the above, studies on cooperative tasks which have dynamic constraints, such as object manipulation by multiple robots, are few.

On the existence of dynamic constraints, the cooperation among robots should have more dynamic factors than the cooperation for collision avoidance. When a pure distributed autonomous robots system is constructed for object manipulation, some problems make the physical system difficult to realize. The increase in communication quantity required to guarantee the quality of cooperation, and achieving harmony between moving motion and manipulating motion in a robot are amongst the problems experienced.

In this paper the scheme of a multiple robots system, consisting of a host and distributed vehicle robots, for performing a cooperative task (as illustrated in Fig.1), will be discussed.

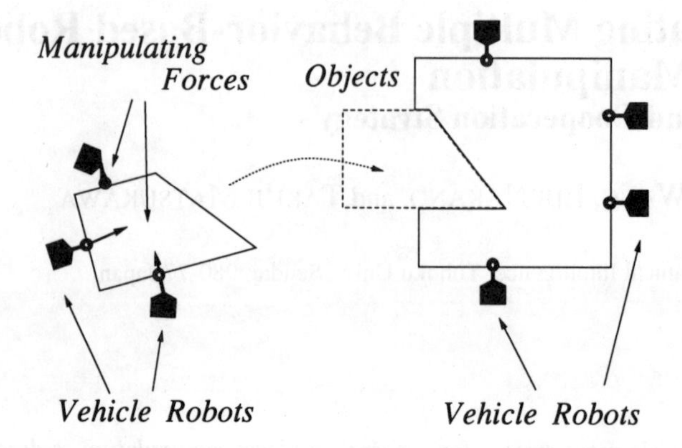

Figure 1: Cooperative Task of Objects Manipulating and Assembling by Multiple Robots

A strategy on cooperatively organizing these behavior-based robots for manipulation will also be discussed. To control this vehicle robot an extended subsumption architecture, in which communicated information is also acting as input (just like the signals from sensors do), will be described. A simulator and some simulation results will be shown.

2 Behavior-Based Multiple Robot System with Host(BeRoSH)

For a multiple robot system, there are two extremes (Fig.2), a centralized robots system and distributed robots system. In a centralized robots system, each robot is only a collection of sensors, actuators and some local feedback loops. Almost all tasks are processed in a host. The communication between the host and robots only involves sending data from sensors to the host and receiving detailed commands from the host. Conversely, in a distributed robots system, each robot plans and solves a problem(task) "independently" while communicating with the other robots. Contents of communication is information which has been processed in each robot.

A multiple robot system which aims at cooperative work, always has some tasks which are common to the whole system rather than to the individual robot, for example, a task for planning of manipulation, or a task for global cooperation, etc. These tasks are fit to be processed in a host rather than in each robot. However, in a centralized robots system, the defects, such as the limitation of processing ability, reliability on fault-tolerant, etc, might become more conspicuous as the system becoming larger, because the processing of all of tasks is performed by the host. Moreover, since each robot is distributed physically, processing of some tasks are suited to be done separately rather than being concentrated at the host. Then, it can be said that centralized processing is not quite suitable to cooperative work via multiple robots.

The trend towards studies in this area seems to indicate that a distributed autonomous robots system is superior to a centralized robots system from the view points of flexibility, robustness and fault-tolerant ability. In a pure distributed system, the processing of a

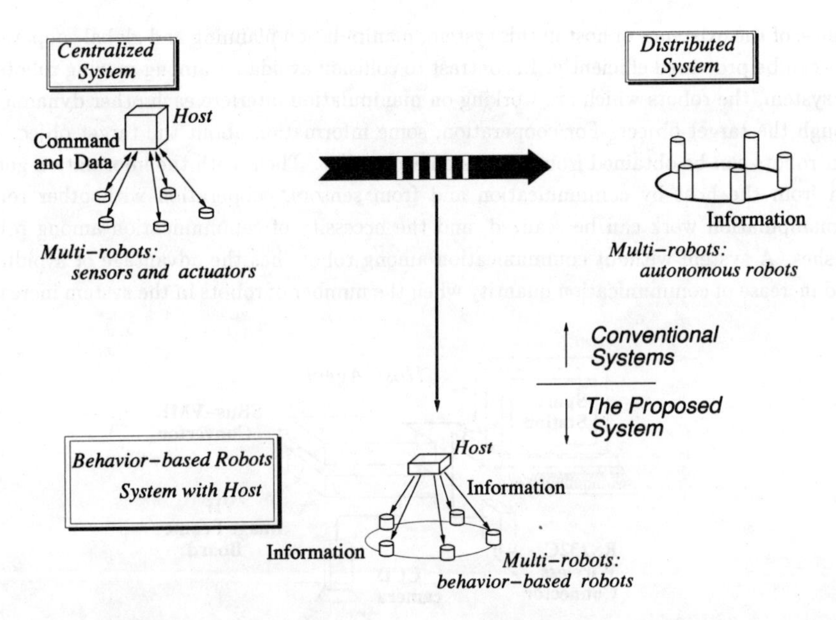

Figure 2: Centralized System, Distributed System and the Behavior-Based Robots System with Host (BeRoSH) which we proposed

cooperative manipulation task which is common to the whole system, as aforesaid, is also done separately. In this situation, cooperation among each distributed agent to do this task become necessary. The cost of this kind of cooperation is the requirement for excessive robot intelligence, excessive communication and tautological processing. These are unnecessary and unnatural for constructing a multiple robot system and can be thought of as a type of loss to the whole system.

In this study, rather than aiming at constructing a pure distributed robots system, we propose a multiple robots system which has a host as a leader and has a certain extent of distributed processing ability. This distributed processing ability is realized by behavior-based architecture in each robot. The system is believed to be suitable for cooperative work and has the following characteristics:

- The system is constructed of a host and multiple robot agents.
- The object of cooperation is to complete a manipulating task with multiple robots.
- Each robot has limited intelligence and limited ability for manipulation, which are realized by an extended subsumption architecture.
- The host also has a visual sensor, and can generate global planning and information for leading the robots.
- Communication occurs only from the host to robots. There is not any communication among robot agents.

Because of the existence of host in this system, manipulation planning and global cooperative tasks can be processed efficiently. In contrast to collision avoidance among moving robots, in this system, the robots which are working on manipulation interfere each other dynamically through the target object. For cooperation, some information about the target object and other robots can be obtained from sensors on each robot. Then, with the information gotten both from the host by communication and from sensors, cooperation with other robots for manipulation work can be realized, and the necessity of communication among robots vanishes. A system without communication among robots has the advantage of avoiding a rapid increase of communication quantity when the number of robots in the system increases.

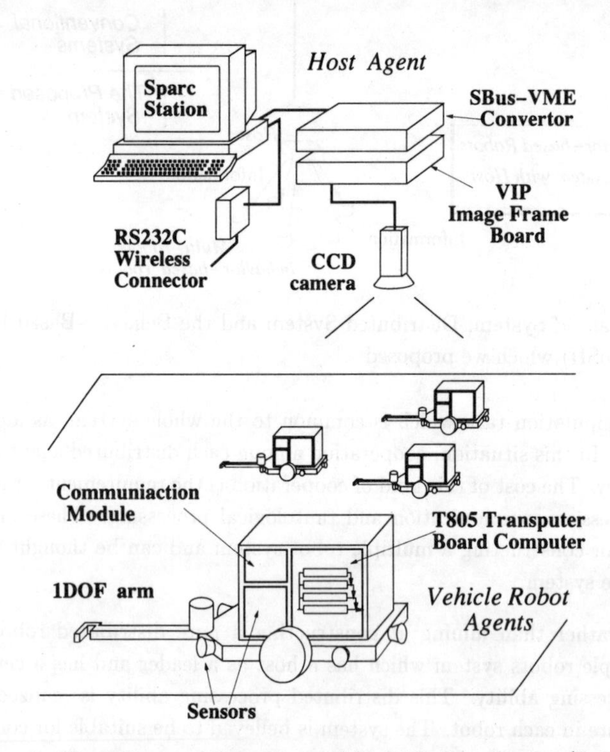

Figure 3: The Scheme of Experimental Multi-Robots System(BeRoSH)

In this paper, the system we proposed is named BeRoSH (*Be*havior-based Multiple *Ro*bot *S*ystem with *H*ost for Object Manipulation). In this system, the host has the ability to observe workspace, target object and all robots, and the ability to decide the goal and subgoal for each robot or for the robot group. But the host does not do detailed path planning for each robot. Each robot is of a behavior-based vehicle type and has a 1-DOF arm for manipulation. That is, each robot has limited ability in manipulation and is locally autonomous. A real experimental robots system will be built soon as Fig.3 and some simulation work has been done towards proving our concepts of cooperation and system construction.

We want to emphasis that our system is not just a simply mixing of two types of systems, centralized system and distributed system, in order to obtain an average of the advantages of each system. We also want to emphasis that this method is not a temporary measure resulting from the current difficulty in constructing a real distributed autonomous robots system for object manipulation. The host, leader of the system, is not only for compensating the incompleteness of intelligence (caused by each robot only having limited intelligence), but also for organizing robots cooperatively. A host makes system cooperation easily realizable and allows it advance smoothly, because it is very easy for a host to observe the overall situation of a system, to make decisions for the overall system, and to lead robot agents. An illustration is that in a lot of cases, even amongst human beings (which are believed to be the most intelligent life on earth), better quality of cooperation appears in a group with a leader or a supervisor when a difficult task is being performed.

3 What is Cooperation ?

In a BeRoSH,

- How to design the ability of robots
- How to design the task that the host must do
- How to allocate tasks between the host and robots
- What is the policy for organizing robots

are very important and interesting problems. These problems are related to a basic concept "What is the cooperation required here?"

In our system, each robot only has limited ability. This means that a whole task can not be completed by one robot alone because each robot is just able to perform a limited part of the action required to complete a task. For example, a robot in BeRoSH can only generate a 1-DOF contacting force to target the object along the direction of the 1-DOF arm.

It is known that, in the manipulation of an object, there is a manipulating space, having the same structure as the acceleration space of the target object, showing the state of accelerations which are generated. Here, we define the *ability direction* of a robot in this space, to show both the directions in which the robot can act and the manipulating ability which the robot has in these directions.

Then, the meaning of cooperation is that: cooperation is achieved if every agent in a system has the nature that it is active in its *ability direction* and is passive in its non-*ability direction*, and agents are organized to work together. The passive nature means that, as mush as possible, a robot agent doesn't give negative influence to the object's motion in its non-*ability direction*.

The most basic and essential point of the cooperation between multiple robots is to allow them to perform a task together without conflict. In this cooperation, the passive nature on non-*ability directions* of each robot prevents conflicts from occurring, and the correct organization of each robot's *ability direction* enables all robot to work on performing the cooperative task, manipulation. The host plays the role of ensuring that the *ability direction* of all robots covers the whole manipulation space. The concept of Form Closure[7], which is usually used in multi-finger grasping control, can be used to do this planning by the

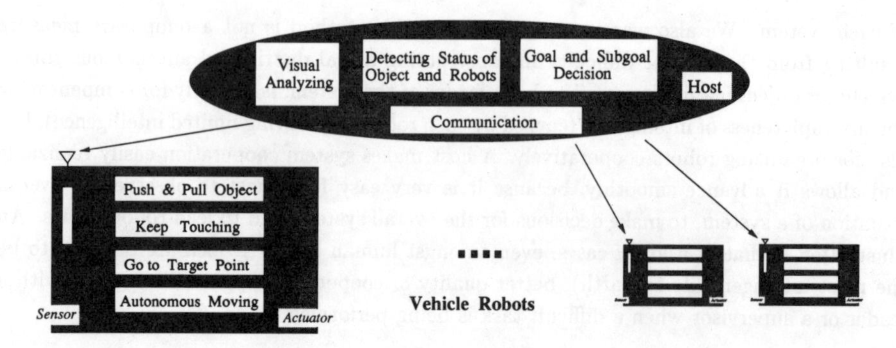

Figure 4: The System Consists of a Host and Behavior-Based Vehicle Robots

host. The robot's nature, of being active in its *ability direction* and passive in its non-*ability directions*, is realized by behavioral attributes fixed on each robot which will be described in detail in the next section. Each robot does not have any other ability related to planning or analyzing for manipulation.

The cooperation in BeRoSH is different to that in studies on distributed problem solving where, in general, each robot gets the same answer separately according to some problem decision functions, while information is exchanged among robots. The cooperation is also different to that in studies about multiple manipulator control such as force-position control[13], hybrid control[16] and impedance control, because these studies focus on designing a control scheme to realize controllability and stability in a system. It is also different from the cooperation in studies of the grasping control by multiple arms or fingers[12][17], in which the main point is solving the inverse dynamic problem of target object and robots with some optimizing indexes(e.g. internal force).

Not being related with any analysis and plan based on models (such as force distribution to robots or path planning to the object and robots) makes this cooperative manipulation flexible to changes in mass or inertia of the object, change in the number of robots performing the manipulation, and unknown disturbances in manipulation of the object. Compared to the studies mentioned above, we feel that the cooperation we propose here is more natural to a multiple robot system.

4 Behavior-Based Vehicle Robot and Communication

In this study, an extended subsumption architecture (shown in Fig.4) is introduced to distributed robot agents as the basic control model for the following reasons.

- The moving motion and the manipulation motion, which are really different from each other in their nature, can be integrated easily.
- Robots can be built incrementally because they can function with only the first layer.
- By changing the number of layers of the robot's subsumption architecture, the degree of distribution of a whole system can be easily changed. This is useful for studying various constructions and forms of a multiple robots system.

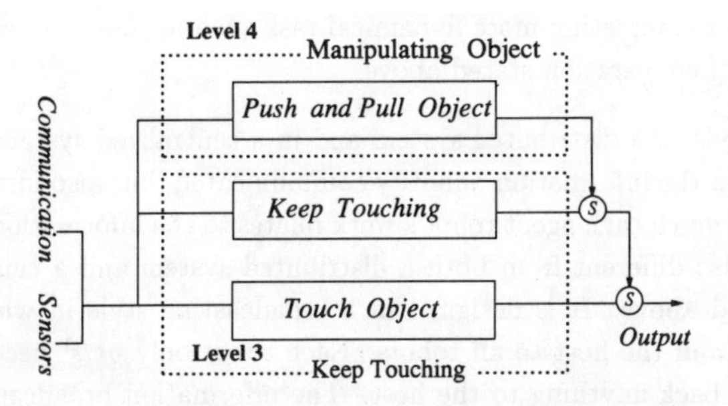

Figure 5: Basic Behavioral Layers for Manipulation: *Touch Object* (connecting a robot's arm with the correct place on the target object). *Keep Touching* (maintaining contact with the object and keeping the contacting force as small as possible). *Push and Pull* (pushing or pulling at the touching point to the target position).

In the original subsumption architecture proposed by Brooks[3][4], input signals are only from sensors, and the robot's behavior is only activated by those signals which are stimulations coming from the environment. In this study, an extended subsumption architecture in that its input is not only the signals from sensors but also the information communicated from the host(Fig.4) is built for each vehicle robot. Each robot agent has two characteristics, reacting to the environment and obeying directions from the host. Some dynamic action characteristics of each robot can emerge in cooperative manipulation while they are following the host's lead. One of the key points in this study is to discuss how to organize behavior-based robots to let the cooperative nature of a system emerge. Our concept is different from those researches which adopt ways to combine behavior-based systems with traditional AI systems, such as connecting a symbolic system as a virtual layer to a subsumption architecture[9].

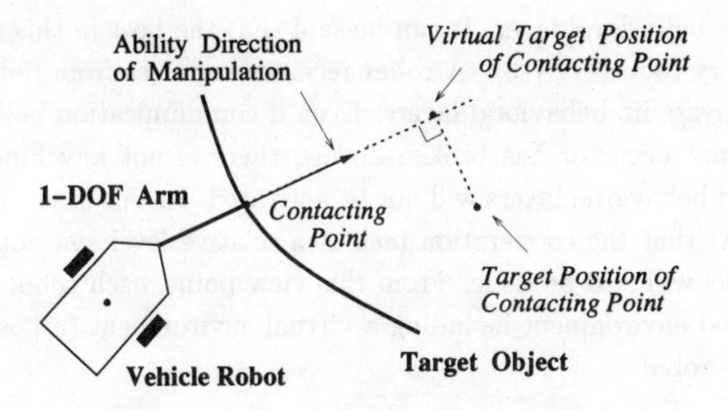

Figure 6: *Ability Direction* of a Vehicle Robot in Pushing and Pulling Target Object

In the subsumption architecture, basic behavioral layers for manipulation are built as shown by Fig.5. To realize the cooperation described in the last section, *Push and Pull* behavioral layer is designed so that its output is only acting in the *Ability Direction* of a robot (the direction of robot's arm(Fig.6)). *Keep Touching* behavioral layers is working in all the other "directions" except the *Ability Direction.* We are also designing some high-level cooperation behavioral layers for completing more dynamical task such as object assembly, which follow the same concept of cooperation stated above.

Communication style in a distributed system and in a centralized system is quite different. It not only differs in the information quantity communicated, but also in the nature of information and in how much each agent robot's work relates to the information. Communication in our system is also different from both a distributed system and a centralized system in both aspects stated above. It is designed in a broadcasting style in which information is only broadcasted from the host to all robots. Each robot only *gets*[1] necessary information but does not send back anything to the host. The information broadcasted from the host includes goals, subgoals, status of robots and target object seen from a global view point, and so on. It is more accurate to refer to information as some status and some guidance to the robots, not as detailed commands from a host. It is not necessary to send all of this information to each robot at every process sampling time.

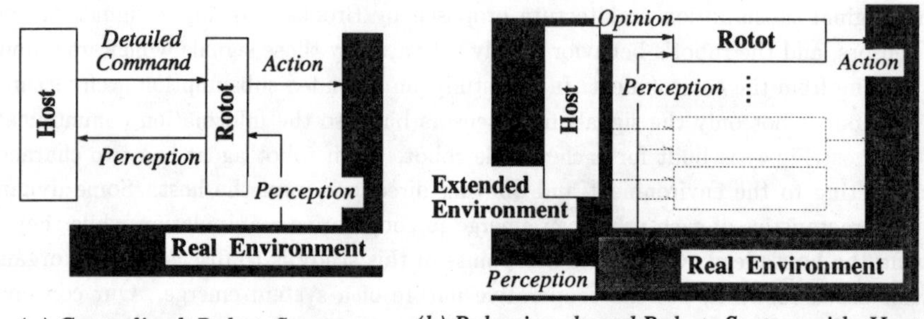

(a) Centralized Robot System *(b) Behavior–based Robots System with Host*

Figure 7: Communication Styles in Centralized System and in BeRoSH

As mentioned above, in BeRoSH, each robot is behavior-based and communication is only an input signal to a behavioral layer. It can be said that the host in this system is a virtual environment to every robot(Fig.7-b). A robot receives impulses from both virtual and real environment to activate its behavioral layers. Even if communication between the host and some robots does not occur(or has broken down), there is not any hindrance to a robot except some related behavioral layers will not be activated. In this case, the influence on the whole system is just that the cooperation task at a relative level can not be accomplished, but lower level tasks will still be done. From this view point, each robot is an autonomous robot in an extended environment including a virtual environment (a host). So we call it a locally autonomous robot.

5 Simulator and Simulation Results

Some simulation work has been done towards proving our concepts of cooperation and system construction. A 2-DOF plane manipulation is assumed, and polygons are used as the target objects to be manipulated.

[1]We prefer to say that information activates behavioral layers that a robot has. Here the information communicated from the host is an input of the extended subsumption architecture, similar to sensors' signal.

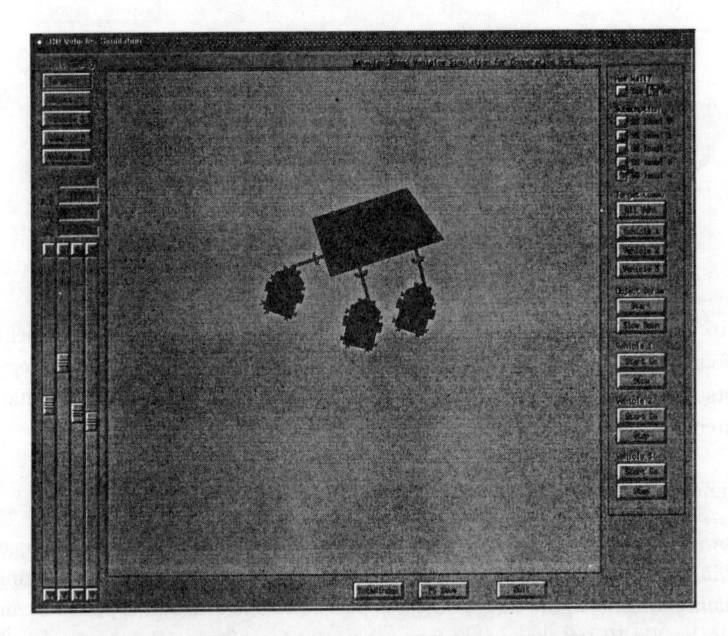

Figure 8: Simulator of BeRoSH for Cooperative Manipulation

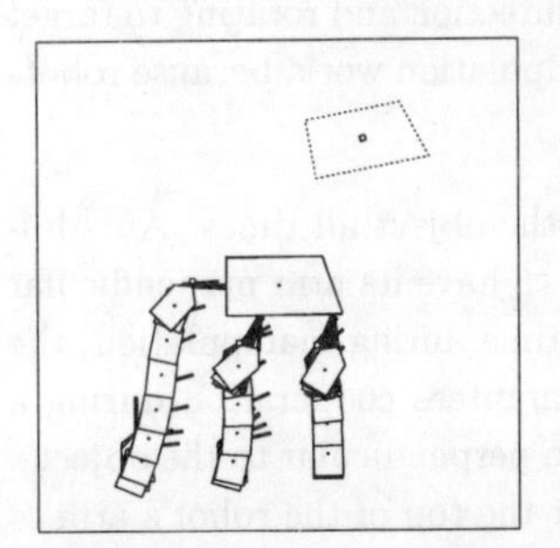
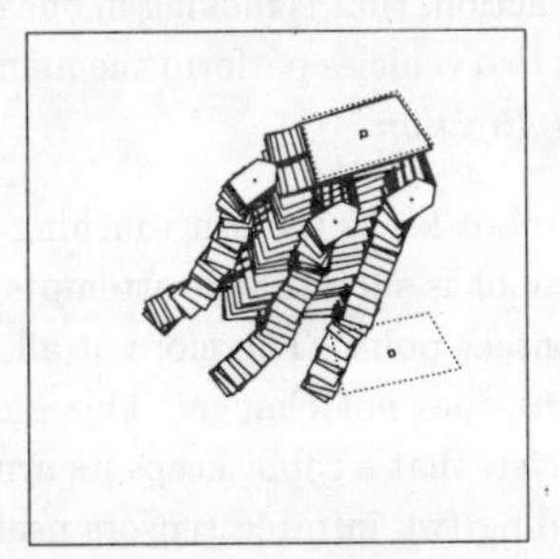
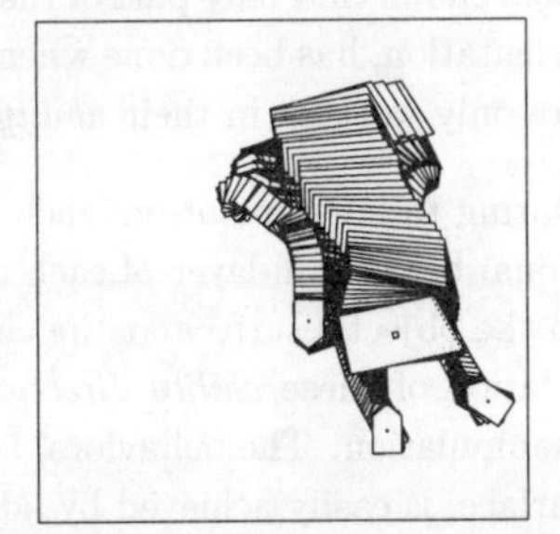

Figure 9: Simulation Result: three robots manipulate a target object cooperatively. Robots proceed to sub-goals and touch the target object at first. Then, by organizing the three robots' *ability directions*, the target object is pushed to one goal and is pulled to another goal. (The small squares are subgoals to each robot. The objects, drawn with broken lines, are goals of the target object given to all robots)

Our simulation was done on a SUN SparcStation. A Simulator (Fig.8) was developed on InterViews3.1. In the simulator, dynamics of the target object, kinematical constraints of vehicles, and friction with the ground is simulated. In order to simulate errors that exist in the world of a real robot, random errors with normal distribution are added to all signals from sensors used for simulating a vehicle robot.

In the simulation robots run in a 2m×2m space. The mass of the target object is 5[Kg]. The inertia moment to the mass center of the object is 0.108[Kgm^2]. The mass of each vehicle is

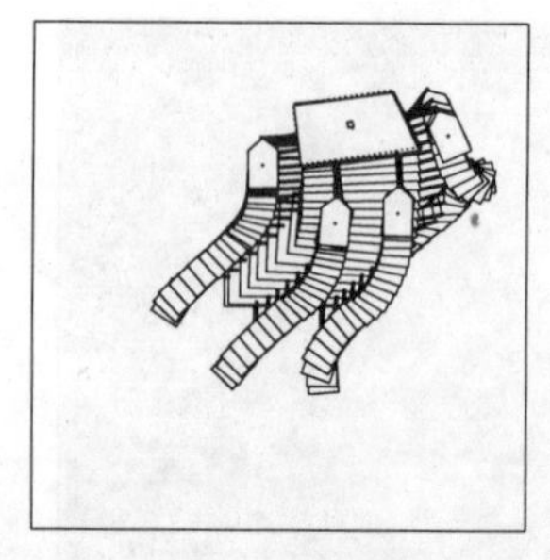

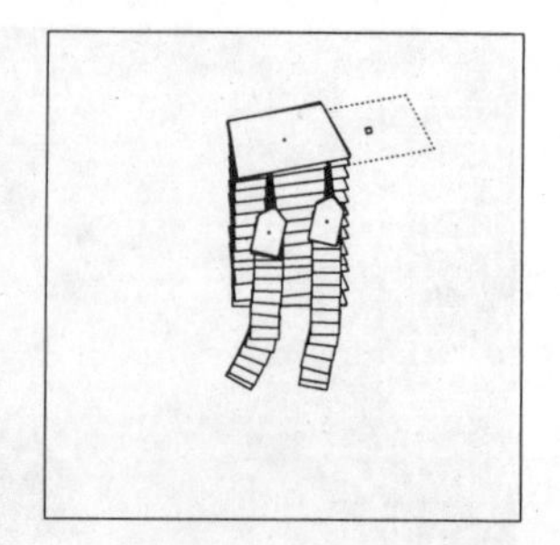

Figure 10: Simulation Result: *Left*: without any change to each robot's behavioral attribute, four robots manipulate a target object using the same cooperation strategy. *Right:* two robots manipulate a target object. Only part of the manipulation action, that is moving in one direction and rotating to target orientation, has been done.

1.5[Kg], and the inertia moment to a vehicle's mass center is 0.0065[Kgm^2].

Fig.9 shows three vehicle robots proceeding to each sub-goal, touching the object, keeping the touching state and manipulating the target object cooperatively. By organizing the three robots' *ability directions*, the target object is pushed to one goal and is pulled to another goal cooperatively. Fig.10-*left* shows that four robots are transferring the target object using the same cooperation strategy without any change to each robot's behavioral attribute. Fig.10-*right* shows that only part of the action, that is moving in one direction and rotating to target orientation, has been done when two vehicles perform the manipulation work, because robots are only working in their *ability direction*.

During the manipulation, each robot keeps its arm touching the object all times. An additional behavioral layer of each robot is such that it attempts to have its arm perpendicular to the object's surface at the contact point. Therefore at all time during manipulation, the relation of these *ability directions* does not change. This guarantees cooperation during a manipulation. The behavioral layer, that a robot keeps its arm perpendicular to the object's surface, is easily achieved by adding two infrared sensors near the top of the robot's arm.

6 Discussion and Conclusions

A distributed multiple autonomous robot system is different from non-multiple robot systems not only in the number of robots but also in the characteristics of the target task. This has resulted in active studies on new problems which were not previous met by robot researchers, such as parallel processing among multiple robots, communication among robots, fault tolerance in a robot swarm, and heterogeneous robot systems. On the other hand, there are some other topics in distributed multiple autonomous robot system, such as object manipulation, which a lot of researchers have discussed, but only in the context of conventional robotic systems. Of course, some conventional methods can be used to build a multiple robot system for object manipulation, however, they may not be the most suitable ones. Some rethinking and new strategies are necessary according to the environment where robots are working, and the tasks that robots are accomplishing. Our research is aimed in this direction. In the studied approach, instead of decomposing the manipulation task into many steps, as is done in multi-finger robot grasping[2], an alternative strategy in which task accomplishment can be decomposed into two steps,"designing robot's behavioral attributes including some dynamical cooperating ability" and "organizing all robots abilities", is proposed. As discussed in this paper, this strategy is suitable for organizing large scale robot systems because it emphasizes robot behavior designing, and it uses a simple but effective cooperation strategy for accomplishing a manipulation task.

A multiple robot system(BeRoSH) has been proposed for cooperative object manipulation.

1. *System*: The system is a hybrid robot system which has a host as the leader and many robots with limited ability to perform a task and cooperation. It is different from either a centralized robots system or a distributed robots system.
2. *Cooperation*: The meaning of cooperation is that: cooperation is achieved if every agent in a system has the nature that it is active in its *ability direction* and is passive in its non-*ability direction*, and agents are organized to work together. The cooperation we discussed is more natural to a multiple robot system.
3. *Locally Autonomous Robot*: A subsumption architecture which has both behavioral attributes of autonomous moving motion and those of cooperative manipulation is built for each vehicle robot. It is different from Brooks's original subsumption architecture because the subsumption architecture's input is not only from sensors but also from communication with a host. Some dynamic action characteristics of each robot can emerge in a cooperative manipulation while they are following the host's lead.
4. *Communication*: The communication in our system is designed in a broadcasting style in which information is only broadcasted from the host to all robots. The information to each robot is guidance and status seen from a global view, rather than detailed commands from the host. To each robot, the host is a virtual environment. Each robot is an autonomous robot in an extended environment including the virtual environment.
5. We agree with the opinion that intelligence emerges from parts having no intelligence[11]. Intelligence can emerge not only from the interaction of some behavioral characteristics, but also from the cooperation of multiple behavior-based robots, as shown by the results of the cooperative work by BeRoSH.

Finally a simulator and simulation results are shown to support the view point we proposed.

Acknowledgments

Some of the concepts presented in this paper were fostered through frequent discussions with several members of Advanced Robotics Laboratory at Tohoku University. The authors would also like to thank P.Velletri and X.Chen for their assistance in preparing this paper.

---

[2]Some of the steps, which are according to some physical parameters and models, are: planning the object path and acceleration, distributing forces for generating the path and acceleration, motion planning for each finger and constructing a controller for each finger.

Reference

[1] T.Arai, H.Ogata, T.Suzuki, Collision Avoidance among Multiple Robots Using Virtual Impedance, *Proc. 1989 IEEE Int. Workshop on Intelligent Robots and Systems(IROS'89)*, 1989, pp.478-485.

[2] H.Asama, A.Matsumoto, T.Ishida,Design of an Autonomous and Distributed Robot System, *Proc. 1989 IEEE Int. Workshop on Intelligent Robots and Systems(IROS'89)*, 1989, pp.283-290.

[3] R.A.Brooks, A Robust Layered Control System For A Mobile Robot, *IEEE J. Robotics and Automation*, Vol.RA-2, No.1, 1986, pp.14-23.

[4] R.A.Brooks, Intelligence Without Representation, *Artificial Intelligence*, 47, 1991, pp.139-159.

[5] T.Fukuda, M.Buss and Y.Kawauchi, Communication System of Cellular Robot: CEBOT, *Proc. IEEE Industrial Electronics Conf.*, 1989, pp.634.

[6] O.Khatib, Real-Time Obstacle Avoidance for Manipulators and Mobile Robots, *Int. J. Robotics Research*, Vol 5-1, 1986, pp.90-98.

[7] K.Lakshminarayana, Mechanics of Form Closure, *ASME Paper*, 1978, 78-DET-32.

[8] P.Maes, R.A.Brooks, Leaning to Coordinate Behaviors, *Proc. of AAAI*, 1990, pp.796-802.

[9] C.Malcolm, T.Smithers, Symbol grounding via a hybrid architecture in an autonomous assembly system, *Designing Autonomous Agents*, MIT Press, 1990, pp.123-144.

[10] M.Mataric, Minimizing Complexity in Controlling a Mobile Robot Population, *Proc. 1992 IEEE Int. Conf. on Robotics and Automation*, 1992, pp.830-835.

[11] M.Minsky, *The Society of the Mind*, Simon and Schuster, New York, 1986.

[12] Y.Nakamura, K.Nagai and T.Yoshikawa, Dynamics and Stability in Coordination of Multiple Robotic Mechanisms, *Int, J.Robotics Research*,Vol 8-2, 1989, pp.44-61.

[13] E.Nakano, S.Ozaki, T.Ishida and I.Kato, Cooperational Control of the Anthropomorphous Manipulator 'MELARM', *Proc. of 4th Int. Symp. on Industrial Robots*, 1974, pp.251-260.

[14] G.Ogasawara, T.Omata and T.Sato, Multiple Movers Using Decision-theoretic Control, *Japan/USA Symp. on Flexible Automation*, Vol 1, 1992, pp.623-630.

[15] L.E.Parker, Designing Control Laws for Cooperative Agent Teams, *Proc. of ICRA*, *Proc. 1993 IEEE Int. Conf. on Robotics and Automation*, 1993, pp.582-587.

[16] M.H.Raibert and J.J. Craig, Hybrid Position/Force Control of Manipulators, Trans. ASME J. Dynamic Systems, Measurement, and Control, Vol 102, 1981, pp.126-133.

[17] M.Uchiyama, A Unified Approach to Load Sharing Motion Decomposing and Force Sensing of Dual Arm Robots, *Robotics Research : The Fifth Int. Symp.(MIT Press)*, 1990, pp.225-232.

[18] J.Wang, G.Beni, Distributed Computing Problems in Cellular Robotic Systems, *Proc. 1990 IEEE Int. Workshop on Intelligent Robots and Systems(IROS'90)*, 1990, pp.57-61.

[19] S.Yuta and S.Premvit, Consideration on Cooperation of Multiple Autonomous Mobile Robot - Introduction to Modest Cooperation -, *Proc. 1991 IEEE Int. Conf. on Advanced Robotics (ICAR'91)*, 1991, pp.545-550.

Dynamically Reconfigurable Robotic System
- Assembly of New Type Cells as a Dual-Peg-in-Hole Problem -

GUOQING XUE[1], TOSHIO FUKUDA[1], FUMIHITO ARAI[1], HAJIME ASAMA[2], HAYATO KAETSU[2], and ISAO ENDO[2]

[1]Dept. of Mechano-Informatics and Systems, Nagoya Univ., Nagoya, 464-01 Japan
[2]The Inst. of Physical and Chemical Research (RIKEN), Wako, 351-01 Japan

Abstract:

In this paper, we first introduce a new type cell(module) of Self Organizing Manipulator that was developed by authors, then present a method for assembling this type of cell using manipulators. Since the new type of cells has two guides for connecting, it can also be considered as a dual-peg-in-hole issue. To obtain the effective assembly sequence, we observe the sequence of work and record force/torque data, psychological data of the people, while human was performing the same work. By analyzing the result of the experiment, we obtained a reasonable sequence for manipulator to perform such a work. We apply this result into the assembly of cells by using cooperation of camera and force/torque sensor. The analysis of force and geometric condition in the Dual-Peg-in-Hole issue, and results of experiments are also presented.

1. Introduction

Recently, reconfigurable modular robot that can flexibly adapt to variety of task is required due to the needs of variety of products. We have proposed CEBOT (**CE**llular ro**BOT**ic Systems), and studied a kind of modular robot-Self Organizing Manipulator since 1990[1]. In Fig.1 we show the concept of Self Organizing Manipulator. This system are able to build a manipulator automatically by composing several kinds of cells according to a given task. By now, we have developed 3 types of prototype cells. The power line and signal bus of the first prototype cells are made outside cells, and the second prototype of cells used a connector in an irregular shape. Both two types of cells were very difficult to assemble. A new prototype of cells with two-guide is developed to overcome these problems. In Fig.2 we show the photo of these new type of cells. For the detail about the hardware of this type of cells, see the paper[2].

Since the new type of cells has two pillered guides, the assembly of this type of cells can be modelled as a Dual-peg-in-hole problem. In this paper, first, we extract the sequence of assembly by observing the human work, then analyze the geometric and force conditions in performing such a work. The experiments of assembly that use force sensor and camera are shown at the end of this paper.

2. Dual-peg-in-hole problem

The problem of Peg-in-Hole has been studied by many researchers for a long time. Whitney proposed the concept of RCC, analyzed geometry and force condition of a 2-dimensional problem in detail[5]. Research about contact state, assembly sequence, skill in peg-in-hole problem is also presented[3][4][6].

Here, first we analyze the clearance of Dual-peg-in-hole problem, then analyze the restriction condition of Dual-peg-in-hole, and show the relation with Single-peg-in-hole. The sequence of

Fig.1 Concept of Self-Organizing Manipulator

Fig.2 New Type Cells

assembly is obtained by observation of human work. The 3-dimensional model of pillar is used in analysis of geometric and force condition. The model of Dual-peg-in-hole used in this paper is shown in Fig.3.

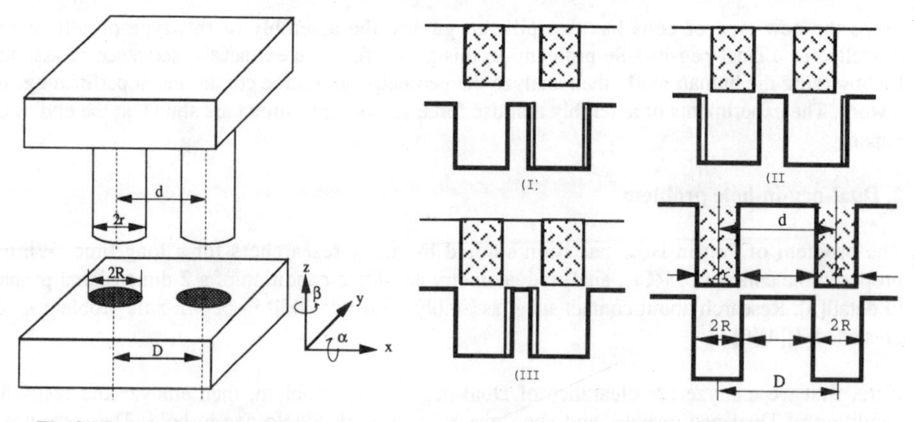

Fig.3 Model of Dual-peg-in-hole

Fig.4 Clearance of Dual-peg-in-hole

2.1 Clearance of Dual-peg-in-hole

If the peg is in the shape of pillar, clearance of single-peg-in-hole is similar in any direction. However, in Dual-peg-in-hole problem, clearance is difference in direction. As shown in (I) of Fig.4, two pegs are different in clearance, the general clearance is similar to the little one. In the case of (II), the inside clearance is little than outside, the general clearance is similar to inside. Similarly, in the case of (III), the general clearance is similar to outside.
Consequently, the translational clearance can be calculated as (1)

$$\begin{aligned} C_1 &= D + 2R - d - 2r \geq 0 \\ C_2 &= d - 2r - D + 2R \geq 0 \\ C_3 &= 2R - 2r \geq 0 \\ C &= Min(C_1, C_2, C_3) \end{aligned} \qquad (1)$$

Where, C1, C2, C3, is parameter determined by (I),(II),(III), C is general translational clearance. Similarly, The rotational clearance can be calculated as (2),(3) according to the Fig.5 and Fig.6.

$$\begin{aligned} e_x^2 + e_y^2 &\leq \varepsilon^2 \\ (D - dCos\beta + e_x)^2 + (dSin\beta + e_y)^2 &\leq \varepsilon^2 \end{aligned} \qquad (2)$$

$$\alpha \leq Sec^{-1}(\frac{R}{r}) \qquad (3)$$

Where, ε is translational clearance, ex, ey, is the distance between the center of hole in the direction of x, y respectively. β is the angel between the center line of peg and center line of hole. D, d is the distance of peg and hole respectively. α is the angle between center face of hole and center face of peg. 2R and 2r is the diameter of peg and hole respectively.

2.2 The Relationship between Dual-peg-in-hole and Single-peg-in-hole

The center column of Fig.7 shows the contact state of Dual-peg-in-hole. The left column shows the contact state assuming that area between the holes is treated as a imaginational peg. The right column shows the contact state assuming that two pegs together with the area between two pegs aretreated as a imageinational peg. From this fig, it is not difficult to understand that the restriction of center column is the sum of left and right column. If we take restriction set of left imaginational peg as R1, restriction set of right imaginational peg as R2, the restriction set of Dual-peg can be shown as (4)

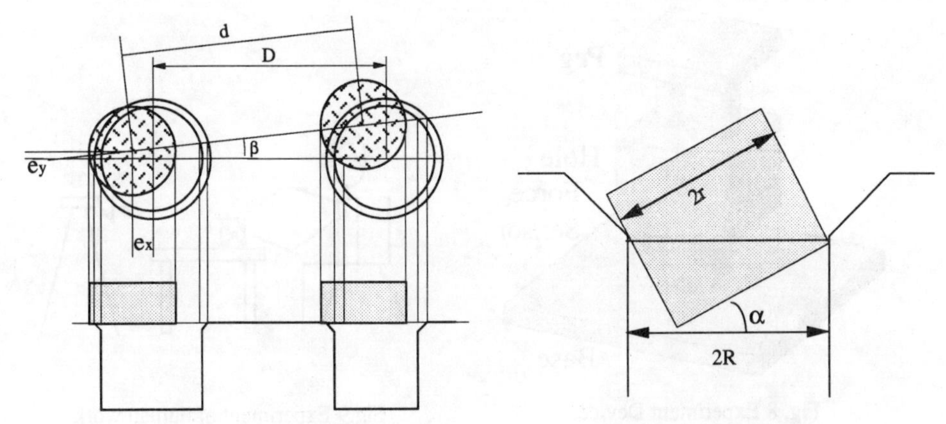

Fig.5 Clearance in the direction of β

Fig.6 Clearance in the direction of α

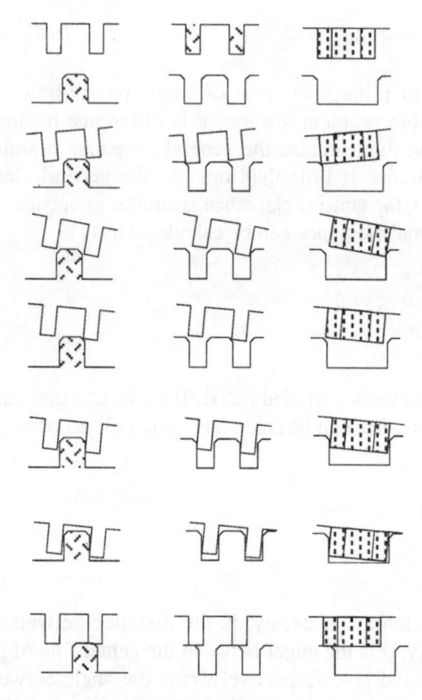

Fig.7 Relationship between Dual-peg-in-hole and Single-peg-in-hole

$$\Re = \Re_1 \cup \Re_2 \qquad (4)$$

Thus, the Dual-peg-in-hole problem can be decomposed to two Single-peg-in-hole problems.

3. Observation of Human work

The experimental device for observation of human work is shown in Fig.8. A 6-Axis Force/Torque sensor is located on the base and hole and pegs are located on the Force/Torque sensor. The scene of experiment is shown in Fig.9.

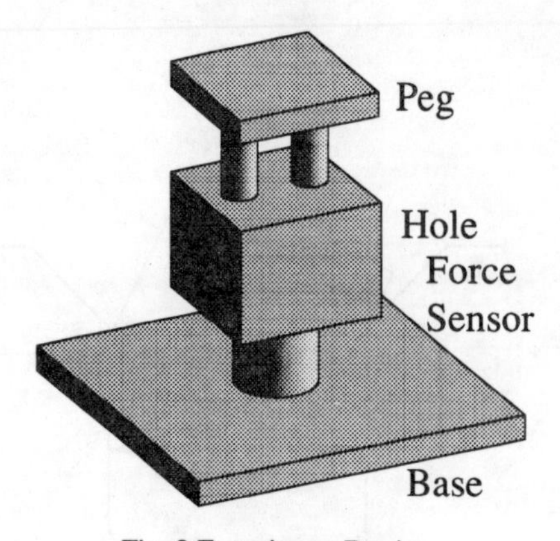

Fig. 8 Experiment Device

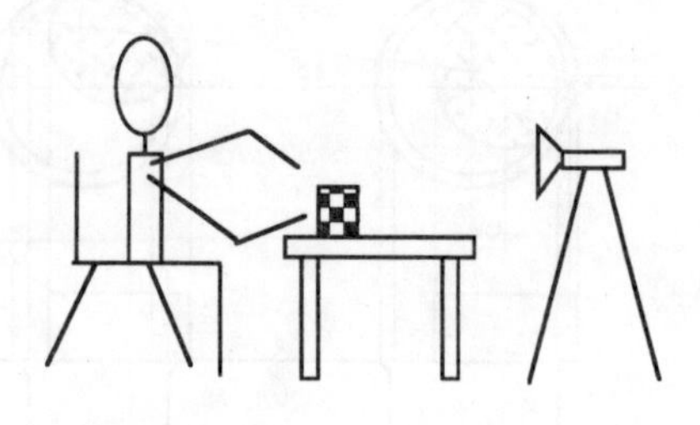

Fig.9 Experiment of human work

Experiment:
1. Recording the insertion sequence of all the people using video camera -> Information about work sequence and insertion strategy.
2. Recording the Force/Torque data using experimental device -> Information of force/torque
3. Recording one's impression by taking a questionnaire survey to every testee --> Psychological information
Data processing(16 testee):
Video information processing: Inspect the work sequence of every testee
Force/Torque data processing: Plot all the 6 axis force/torque data into a graph(Sampling time is about 8ms)
An example is shown in Fig.10
Together the above result with psychological information, we get the result shown in Table 1.

Here, an interesting feature must be noticed --> why "diagonally search"?
Two reasons are considered.
1. Use diagonally search for the purpose of eye that can see object easily.
2. The restriction in the direction of horizontal is lager, and easy to be detected.
As a result, the extracted insertion sequence of Dual-peg is shown in Fig.11

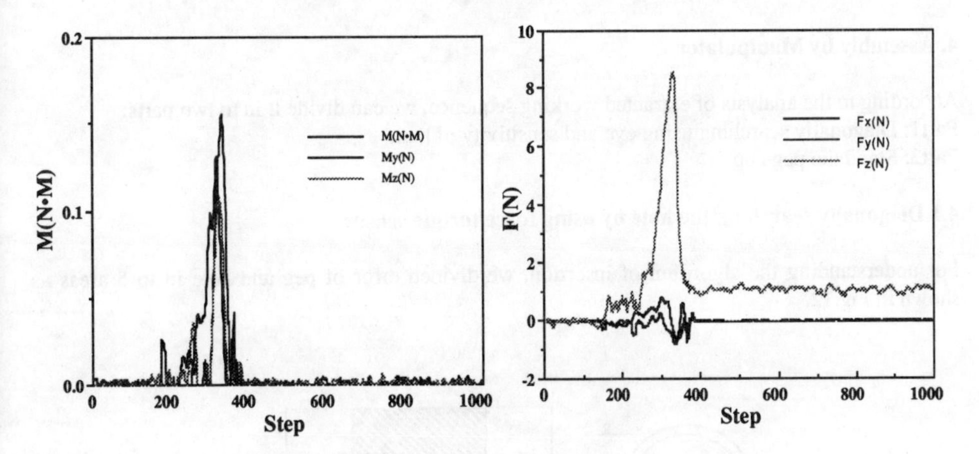

Fig.10 An Example of Force/Torque Data of human work

Table 1. Insertion strategy and characteristic of testee

	Method	Characteristic	Common
Hand First	Roughly approach guided by eye, precisely move using hand	long time, little moment and force	• High speed ,soft insertion can be achieved by using both hand and eye • Almost every testee diagonally search the hole at same angle, and then stand peg up
Eye First	Roughly approach and precisely move guided by eye	Short time, large moment, force	

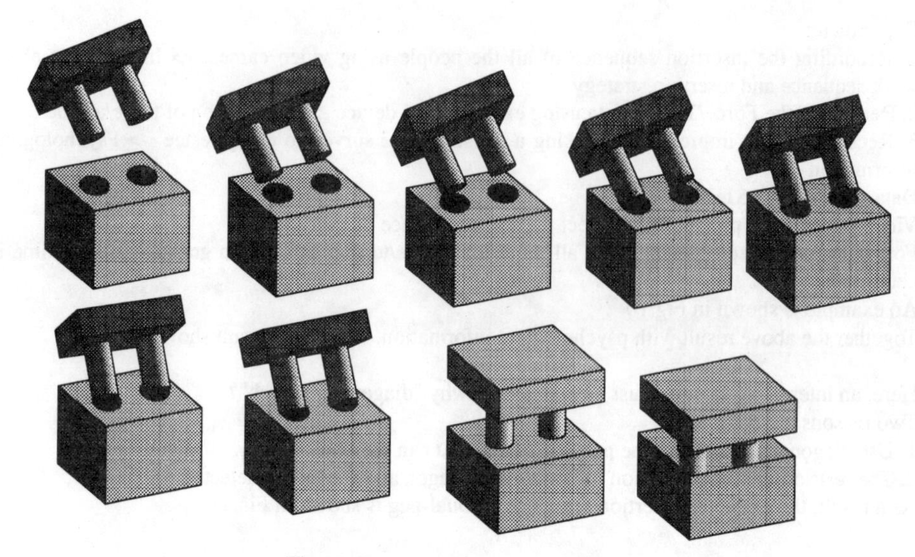

Fig.11 Extracted Sequence of testee

4. Assembly by Manipulator

According to the analysis of extracted working sequence, we can divide it in to two parts:
Part1: Diagonally searching using eye and sensitivity of hand
Part2: Stand the pegs up

4.1 Diagonally searching the hole by using force/torque sensor.

For understanding the algorithm of insertion, we divided error of peg and hole in to 5 areas as shown in Fig.12.

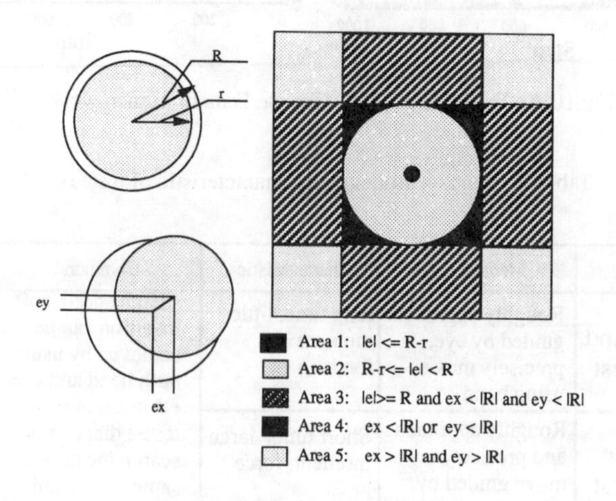

Fig.12 Relation of Peg and Hole

Insertion sequence:

Step1: Insert the peg in the direction of Z until Fz > Fzd. If error is in the area of Area1, the insertion will be succeeded.

Step2: If Fz>Fzd, keeping the position of top of peg unchanged, rotate the peg in the direction of y. Insert the peg until Fz = Fzd. Keep the Fz unchanged, move the peg in the direction of x. If error is in the area of Area2, the horizontal restriction force will be detected. Use the restriction force to guide the motion of peg to the center of hole.

Step3: In step2, if the horizontal restriction did not be detected, slide the peg in the direction of x. If error is in the area of Area3, the horizontal restriction force will be detected. Use the restriction force to guide the motion of peg to the center of hole.

Step4: If the peg's sliden distance in the direction x exceeded (+/- r) and the horizontal restriction force still not be detected, slide the peg in the direction of y. If error is in the area of Area4, the horizontal restriction force will be detected. Use the restriction force to guide the motion of peg to the center of hole.

Step5: If the horizontal restriction for is not detected while slide in the direction of x and y, the error is in the Area5.

In the case of Step 5, camera is needed for measuring the position of hole.

4.2 Measuring the position of peg and hole by image processing

In the case of 3-dimensional measurement, the accuracy of image processing is depended very much on the viewpoint of camera[7]. Here, to measure the position of peg and hole, we move the cell and camera to make the camera perpendicular to the cell.

Image processing:

Extract the contour of peg: noise processing, differential, threshold, line thinning, labeling, center of weight, line approximation

Extract the contour of hole: noise processing, differential, threshold, line thinning, labeling, center of weight, circle approximation

The result of image processing is shown in Fig.13, Fig.14.

4.3 Force Analysis while standing up

Most research treats the pillar peg-in-hole problem as a 2-dimentional problem. Actually, in single-peg-in-hole problem, it is no problem because of the symmetry. But in Dual-peg-in-hole problem, it is necessary to treat it as a 3-dimensional problem. As shown in Fig.15, in this case, peg and hole is in the 3-point contact. According to the geometric condition, the 2-dimensional corresponding

Fig.13 Image processing result(peg)

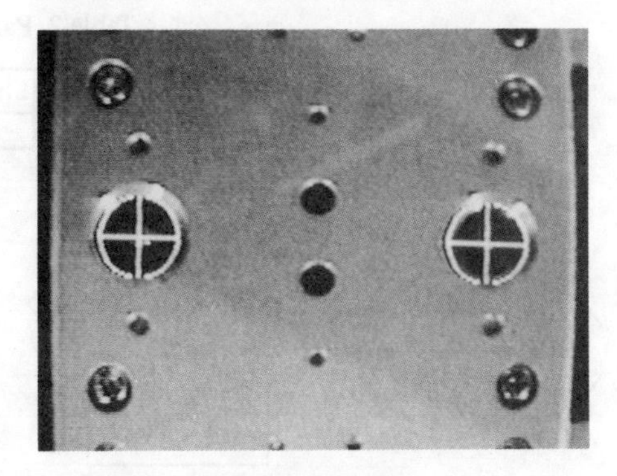

Fig.14 Image processing result(hole)

diameter H shown in Fig.16 can be calculated as (5)

$$H = R + \frac{R - 2rCos\beta + RCos^2\beta}{Sin^2\beta} \quad \text{.....(5)}$$

To avoid the drastic change of force, it is necessary to keep the contact state unchanged while standing up the peg. Thus the force condition(6) must be met

$$\begin{cases} f2 + \dfrac{f1}{\mu} + F \bullet Sin(\beta + \theta) + p \bullet Cos\beta = 0 \\ f1 + \dfrac{f2}{\mu} + F \bullet Cos(\beta + \theta) + p \bullet Sin\beta = 0 \\ p \bullet PO + F \bullet DO + f2 \bullet H \bullet Cos\beta + f1 \bullet H \bullet Sin\beta = 0 \end{cases} \quad \text{.....(6)}$$

Where μ is the coefficient of friction, f1, f2 is frictional force, F is the external force, p is the gravity of peg. PO, DO is the working distance of external force and gravity respectively. The geometric condition is shown as (7)

$$\begin{aligned} & x_0 = H \bullet Sin^2\beta, \quad y_0 = H \bullet Sin\beta \bullet Cos\beta \\ & x_d = L2f \bullet Sin\beta + (H \bullet Cos\beta - r + L1f)Cos\beta \\ & y_d = L2f \bullet Cos\beta - (H \bullet Cos\beta - r + L1f)Sin\beta \\ & x_p = L2p \bullet Sin\beta + (H \bullet Cos\beta - L1p)Cos\beta \\ & y_p = L2p \bullet Cos\beta - (H \bullet Cos\beta - L1p)Sin\beta \\ & PO = x_p - x_0 \\ & DO = (x_0 - x_d)Sin\theta - (y_0 - y_d)Cos\theta \end{aligned} \quad \text{.....(7)}$$

The result is shown in (8)

$$F = \frac{p\left(PO\mu^2 - PO - H\mu^2 + H\mu Sin(2\beta)\right)}{DO - DO\mu^2 - H\mu Cos(\theta) + H\mu^2 Sin(2\beta + \theta)} \quad \text{.....(8)}$$

We inspect the relation between F, β, θ using formula (8). Parameter is shown in table 2. The result is shown in Fig.17 and Fig.18. Fig.17 illustrate the force by altitude, Fig.18 illustrate the

Table 2. Parameters

μ	p(g)	R(mm)	r(mm)	L1p(mm)	L1f(mm)	L2p(mm)	L2f(mm)
0.1	200	10.0	9.95	4	200	5	200

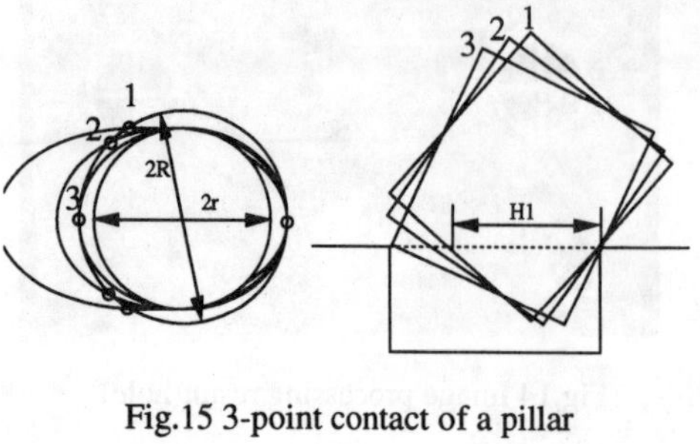

Fig.15 3-point contact of a pillar

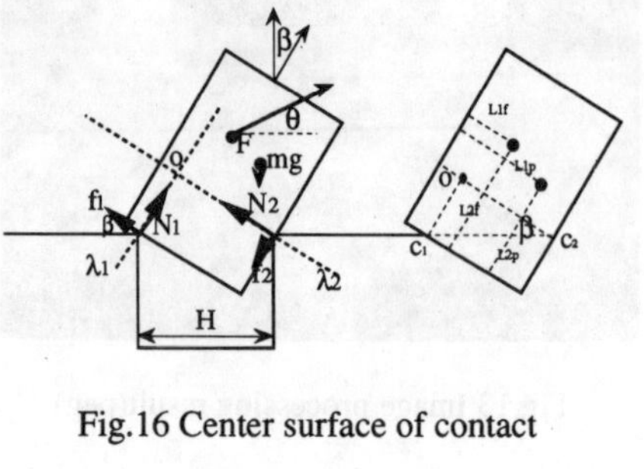

Fig.16 Center surface of contact

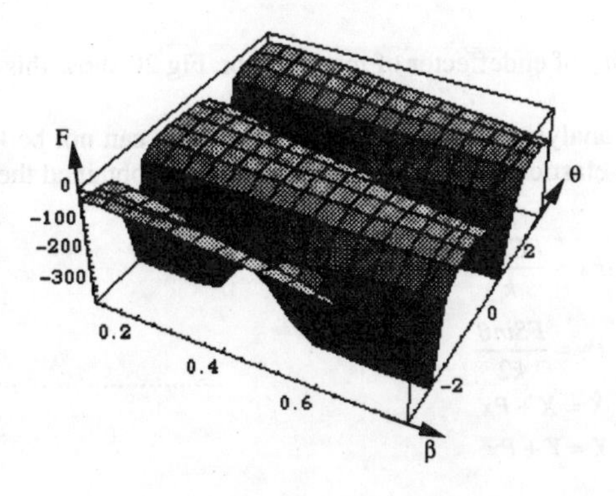

Fig.17 Relation Among F,β,θ

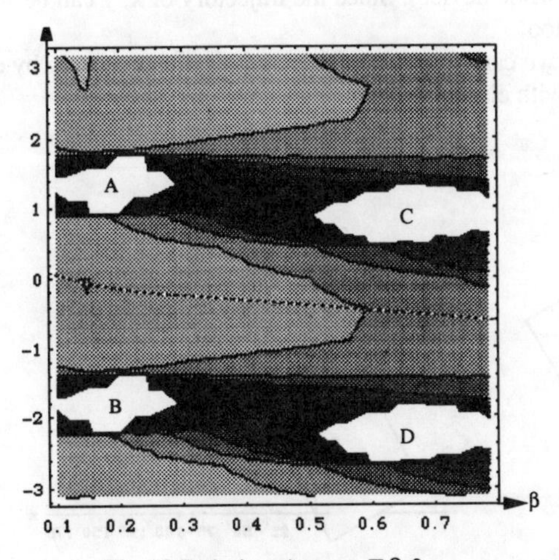

Fig.18 Relation Among F,β,θ

force by depth. The A, B, C, D in Fig.18 is the area that infinite force is needed. The dotted line shows the best direction of force.

4.4 Trajectory of Manipulator

For the conventinal industrial robot that only with positional controller, to realize the geometric condition, the trajectory of manipulator is needed.
From the Fig.19, the geometric condition can be obtained as (9)

$$
\begin{aligned}
&C_2 t = r - L \\
&Mt = L2f - H \bullet Sin\beta \\
&MC_2 = \sqrt{C_2 t^2 + Mt^2} \\
&\gamma = ArcTan(\frac{Mt}{C_2 t}) \\
&x = -MC_2 \bullet Cos(\gamma + \beta) \\
&y = MC_2 \bullet Sin(\gamma + \beta)
\end{aligned}
\qquad \text{.........(9)}
$$

Where x, y is the trajectory of endeffector of manipulator. Fig 20 show this trajectory.

From the result of force analysis we see that the contact state can not be kept only by controlling position. Here we add an elastic device as shown in Fig.21, we obtained the relation as (10)

$$\begin{aligned} Px &= \frac{FCos\theta}{k_1} \\ Py &= \frac{FSin\theta}{k2} \\ X &= X + Px \\ Y &= Y + Py \end{aligned} \qquad \text{.........(10)}$$

Where, Px, Py is the displacement of elastic device in the direction of x, y. X, Y is the position of manipulator (include elastic device). Since the trajectory of x, y can be solved from(9), we can get the trajectory of X, Y too.

By using this method, we can control the force and it's direction only by controlling the position of manipulator installed with elastic device.

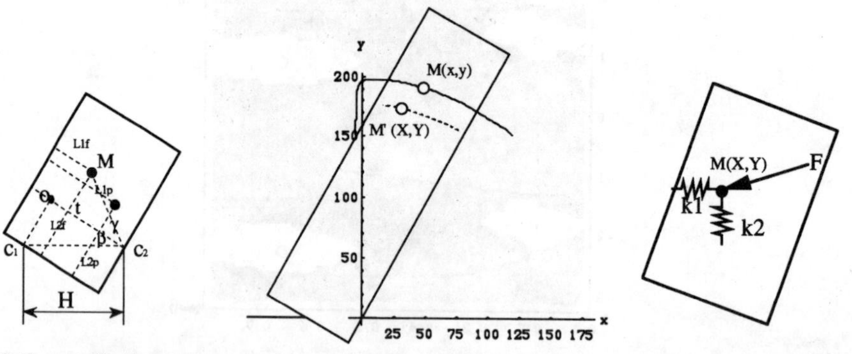

Fig.19 Geometric Condition Fig.20 Trajectry of Manipulator Fig.21 Elastic mechanism

5. Experiment

5.1 System Construction

The construction of experimental system is shown in Fig.22. two 5-D.O.F manipulator (MoveMaster) is used. One is for assembly and the other is for moving camera. Manipulator is linked to PC by RS232C, and PC are linked to workstation by Ether-net.

5.2 Experimental Result and Discussion

The experiment can be divided into 3 parts.(1) Position measuring by image processing. (2) Diagonally searching using force/torque sensor. (3) Stand the peg up while keeping the contact state.

(1) The result of image processing is shown in Fig.13, Fig.14. Though the error of image processing is 0.1~0.2mm, the error of motion of cell and especially the influence of motion of camera, the total error may up to 2mm.

(2) Initial error is set as 2mm in both directions. Experiment is carried accordance to the sequence

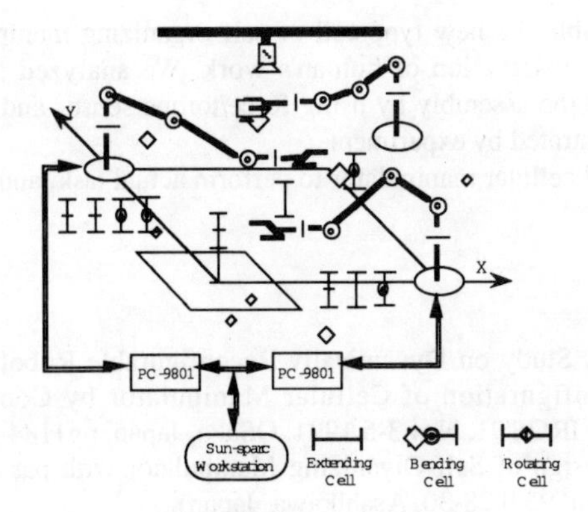

Fig.22 System Hardware

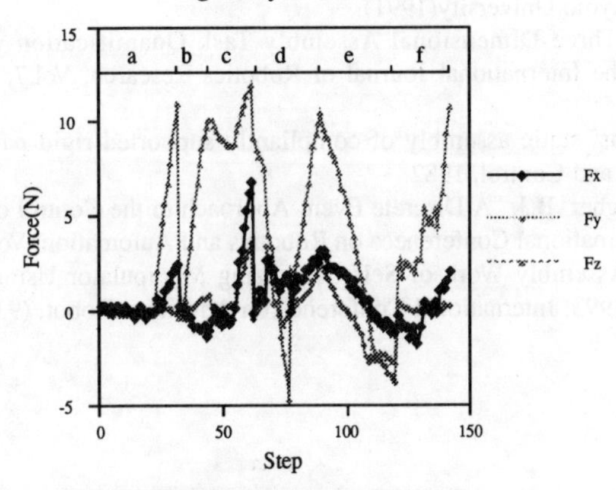

Fig.23 Experimental Result

of 4.1. The force data of whole process is shown in Fig.23.
a. vertical insertion.
b. diagonally insertion.
c. diagonally searching in the direction of x.
d. diagonally searching in the direction of y.
e.Stand the peg up while keeping the contact state.
f. vertical insertion.

Discussion:
Most research of peg-in-hole is assumed that initial error is little enough to use compliance. Our methods do not need this condition since we used both camera and force/torque sensor.
Comparing this result with the human result which is shown in Fig.12, we can see that there is significant difference between two figures. The reason may be.
1. The difference of sensitivity between human hand and force/torque sensor
2. The difference of sampling time
3. The difference of softness between human and manipulator joint.

6. Conclusion

In this research, to assemble the new type cell of self organizing manipulator, we extracted the work sequence from the observation of human's work. We analyzed the force and geometric condition, and carried out the assembly by using force/torque sensor and camera. The effience of proposed algorithm is illustrated by experiment

We will use the assembled cellular manipulator to perform actual task, and inspect the efficiency of system in the future.

Reference

[1] Fukuda.T., Xue.G., A Study on Dynamically Reconfigurable Robotic Systems(Assembling, Disassemblin and Reconfiguration of Cellular Manipulator by Cooperation of two Robot Manipualtors), IEEE/RSJ, IROS'91, Nov.3-5.1991, OSaka, Japan. pp1184-1189

[2] Kotosaka, Asama,"Design of Self-Organizing Manipulator with passive conect mechanism", JSME 3rd FAN symposium('93.9.28-30, Asahikawa, Japan).

[3] Hirai.S., "Analysis and Planning of Manipulation Using the Theory of Polyhedral Convex Cones", Ph.D Thesis, Kyoto University(1991)

[4] Sturges.R.H., "A Three-Dimensional Assembly Task Quantification with Application to Machine Dexterity", The International Journal of Robotics Research, Vol.7, No.4, August 1988, pp34-78

[5] Whitney.D.E, "Quasi-static assembly of compliantly supported rigid parts.", J. of Dynamic Systems, Measurement and Control, 1982

[6] Asada.H., McCarragher .B.J, "A Discrete Event Approach to the Control of Robotic Assembly Tasks", 1993 IEEE International Conferenece on Robotics and Automation, Vol.1, pp331-336.

[7]Xue.G, Fukuda.T, Assembly Work of Self-Organizing Manipulator Using Active Visual and Force Sensing", Proc. 1993. International Conference on Advanced Robot, (93'ICAR), pp197-202, (1993).